THE MARS PLASMA ENVIRONMENT

Springer

THE MARS PLASMA ENVIRONMENT

Edited by

C. T. RUSSELL
University of California,
Los Angeles, CA, USA

Reprinted from *Space Science Reviews*, Volume 126, Nos. 1–4, 2006

Springer

A.C.I.P. Catalogue record for this book is available from the Library of Congress

ISBN: 978-0-387-70941-3

Published by Springer
P.O. Box 990, 3300 AZ Dordrecht, The Netherlands

Sold and distributed in North, Central and South America
by Springer,
101 Philip Drive, Norwell, MA 02061, U.S.A.

In all other countries, sold and distributed
by Springer,
P.O. Box 322, 3300 AH Dordrecht, The Netherlands

www.springer.com

Printed on acid-free paper

TABLE OF CONTENTS

FOREWORD

Mars sits very exposed to the solar wind. Ironically Mars possesses the strongest remanent magnetization of any body thus far visited in the solar system, yet the scale size of this magnetization is so small that it provides an insignificant shield against the solar wind. Compared to Venus that is eight times as massive, Mars has but a weak hold on its atmosphere. Mars has been the subject of intense study over the last four decades and we have learned much about its surface and lower atmosphere but studies of the solar wind interaction with its upper atmosphere and ionosphere have been much more rare. Mars 3 and 5 provided the first significant data on the induced magnetosphere, deflection of the solar wind and erosion of the atmosphere. PHOBOS-2 extended these measurements with a magnetometer and a plasma package, ASPERA (Automatic Space Plasma Experiment with a Rotating Analyzer). It increased our understanding of the interactions, but lasted far too short a time. Mars Global Surveyor carried a magnetometer and an electron reflectometer and discovered the martian magnetic anomalies but added only slightly to our understanding of the interplay between the solar wind and the atmosphere.

When the European Space Agency embarked on its Mars exploration strategy, it chose to include a comprehensive plasma package, on its pilot mission, Mars Express. In retrospect it should have complemented this package with a magnetometer but it did not. Nevertheless despite this handicap, the Mars Express mission has contributed greatly to the understanding of the Mars plasma environment, with its analyzer of space plasmas and energetic atoms (ASPERA-3).

In early 2006 (Feb 27–March 1) a workshop was convened on "The Solar Wind Interaction and Atmosphere Evolution of Mars" in Kiruna, Sweden by S. Barabash and H. Gunell. On the basis of the presentations at the conference we solicited papers for a special volume. These papers were not restricted to papers from the conference, nor were the papers by authors who attended the conference restricted to the material they presented. The result is a very comprehensive look at the Mars-solar wind interaction and the evolution of its atmosphere.

Herein we document this advance in understanding of the martian plasma environment with a series of articles from theoreticians and modelers, data interpreters and experimentalists. The volume begins with the treatment by B.E.Wood of the ancient sun and solar wind because of the importance of knowing how conditions have evolved over the history of the planet. This review is followed by two hybrid modeling papers by S. H. Brecht and S. A. Ledvina and by E. Kallio and coworkers describing how the solar wind interacts with the presently observed martian atmosphere. This is followed in turn by the discussion of the results of MHD modeling

Space Science Reviews (2006) 126: 1–2
DOI: 10.1007/s11214-006-9125-7

by M. W. Liemohn and colleagues, and a paper by D. A. Brain on the MGS measurements of the interaction. These papers set the stage for the main event, the new results from ASPERA-3.

The ASPERA-3 papers are led by the star herself, a description of the instrument by S. Barabash and the ASPERA team. This is followed by a discussion of how the moments of the plasma distribution are calculated, accompanied by a display of those results by M. Fränz. This is followed by a discussion of the plasma morphology at Mars by E. Dubinin *et al.*, and a paper by M. Yamauchi *et al.*, on how the properties of the ion distributions can be used to infer the magnetic field direction. Then begins a series of papers on energetic neutral atoms. A. Galli begins with a paper on energetic hydrogen and oxygen atoms on the night side. A. Grigoriev *et al.*, discusses a subsolar ENA jet. Y. Futaana reports on the Martian response to an interplanetary shock, including the production of ENAs. Next the volume includes four articles on phenomena at lower altitudes. R. Lundin and colleagues discuss auroral acceleration above the magnetic anomalies; H. Nilsson looks at the influence of magnetic anomalies on ion distributions; E. Nielsen and coworkers report on observations by the radar/ionosonde on the top side ionosphere and R. Frahm and colleagues report on photoelectron peaks from the Mars atmosphere. Finally the topic switches to X-rays with a review by K. Dennerl; to the effects of asymmetries in the exosphere on X-rays by M. Holmström; to the exosphere itself with a paper by A. Galli on the results of ASPERA-3's neutral particle detector; and a paper on ENA effects on the martian (and Venusian) exosphere by H. Lichtenegger. This volume documents an impressive leap forward in our comprehension of this complex environment.

The editor wishes to thank first of all the authors themselves who assembled these papers and responded well to the comments of the referees. He also is grateful to the many referees who volunteered to assist in the undertaking by spending their time improving the contents of this volume. These referees include C. Bertucci, D. A. Brain, T. E. Cravens, R. Gladstone, C. Mazelle, D. L. Mitchell, P. C. Brandt, D. G. Mitchell, E. Kallio, D. Hinson, K. Macgregor, J. Linsky, S. Brecht, S. Ledvina, B. Jakosky, Y. Yung, A. Nagy, W. Kazsprazak, H. Wei, G. Delory, R. Strangeway, H. Lammer, J. Leisner, E. Moebius, U. Motschmann, R. Modolo, D. Crider, E. Dubinin, D. Young, R. Goldstein, J. Luhmann, J-A. Sauvaud, V. A. Krasnopolsky, E. Sittler, S. Vennerstrom. The editor also wishes to thank the staff at Springer including Silvia Iviglia, Randy Cruz and Fiona Routley, as well as Marjorie Sowmendran at the University of California, Los Angeles, who handled all the communication with the authors, reviewers and the publisher.

November 13, 2006

C. T. Russell
E-mail: ctrussell@igpp.ucla.edu

THE SOLAR WIND AND THE SUN IN THE PAST

BRIAN E. WOOD
JILA, University of Colorado, Boulder, CO 80309-0440
(E-mail: woodb@origins.colorado.edu)

(Received 18 February 2006; Accepted in final form 14 July 2006)

Abstract. Exposure to the solar wind can have significant long term consequences for planetary atmospheres, especially for planets such as Mars that are not protected by global magnetospheres. Estimating the effects of solar wind exposure requires knowledge of the history of the solar wind. Much of what we know about the Sun's past behavior is based on inferences from observations of young solar-like stars. Stellar analogs of the weak solar wind cannot be detected directly, but the interaction regions between these winds and the interstellar medium have been detected and used to estimate wind properties. I here review these observations, with emphasis on what they suggest about the history of the solar wind.

Keywords: solar wind, stellar winds, ultraviolet spectroscopy

1. Introduction

On long timescales the solar wind can alter the character of planetary atmospheres in our solar system. Mars is potentially the most dramatic example of this, since there is substantial evidence that Mars lost most of its atmosphere in the distant past (Carr, 1996; Jakosky and Phillips, 2001), and erosion by the solar wind is a leading candidate for the cause of this loss (Luhmann *et al.*, 1992; Perez de Tejada, 1992; Jakosky *et al.*, 1994; Kass and Yung, 1995; Lundin, 2001; Lammer *et al.*, 2003). The lack of a global magnetic field makes the Martian atmosphere more vulnerable to solar wind sputtering processes than Earth's atmosphere, which is largely shielded from the solar wind by a protective magnetosphere. Mars apparently once had a global magnetic field, but it disappeared $\sim$3.9 Gyr ago (Acuña *et al.*, 1999). The thicker Martian atmosphere dissipated not long after (e.g., Jakosky and Phillips, 2001), consistent with the solar wind being the culprit.

In order to theoretically investigate the plausibility of solar wind erosion removing the greater part of the Martian atmosphere, it is necessary to know what the solar wind was like in the distant past when this is believed to have occurred. After all, there is no reason to believe that the young solar wind was identical to the Sun's current wind.

Space Science Reviews (2006) 126: 3–14
DOI: 10.1007/s11214-006-9006-0

2. The Solar Wind and Corona

The solar wind arises within the Sun's hot corona ($T \approx 2 \times 10^6$ K). The heating processes that yield these remarkably high atmospheric temperatures (considering that the Sun's surface temperature is "only" 5800 K) are still not well understood (e.g., Walsh and Ireland, 2003), but there is no doubt that magnetic fields are responsible, meaning that the solar corona is one of many atmospheric phenomena (e.g., sunspots, flares, prominences, etc.) that are controlled by magnetic fields generated in the solar interior. The dynamo mechanism that generates the magnetic field is not fully understood (Ossendrijver, 2003; Charbonneau, 2005), but its origin is widely believed to be near the boundary between the Sun's radiative interior and convective outer regions, roughly 70% of the distance from Sun center to the surface. After their initial generation, the magnetic fields are strengthened by shearing processes induced by the Sun's differential rotation.

Regardless of exactly how the magnetic energy is generated and then converted to thermal energy, wind acceleration models that assume simple thermal expansion from the resulting hot corona reproduce the observed properties of the solar wind surprisingly well (Parker, 1958), although additional acceleration from coronal MHD waves is sometimes invoked to explain the high speed streams that are often observed, especially at high ecliptic latitudes where they are ubiquitous in solar minimum conditions (MacGregor and Charbonneau, 1994; Suzuki, 2004). Coronae are copious producers of X-ray emission, so stellar coronae have been detected and studied by X-ray observations from past satellites such as *Einstein* and *ROSAT*, and currently operating satellites *Chandra* and *XMM-Newton*. These X-ray studies have shown that stellar coronae are a universal property of solar-like stars, but their properties are highly variable (Schmitt and Liefke, 2004). For example, X-ray surface fluxes from solar-like stars have been observed to cover a range from 10^3 – 10^7 ergs cm^{-2} s^{-1}, with the relatively inactive Sun having a rather low value of $10^{4.5}$ ergs cm^{-2} s^{-1}.

Coronal properties are correlated with stellar age and rotation rate (Skumanich, 1972; Pallavicini *et al.*, 1981; Walter, 1982, 1983; Soderblom *et al.*, 1993; Ayres, 1997; Güdel *et al.*, 1997). Qualitatively, these correlations are well understood. Stars are initially formed by the gravitational collapse of interstellar clouds. Conservation of angular momentum during this collapse typically results in very rapid rotation for newly born stars. Rapid rotation enhances the magnetic dynamo and young stars therefore have very active coronae that are bright X-ray sources. In a process called "magnetic braking" the magnetic field of a rotating star drags against the wind flowing from the star. In time this slows the rotation rate, which weakens the magnetic dynamo and lowers the coronal X-ray flux. The bottom line is that young stars are rapid rotators that are coronally very active, while mature stars like the Sun are relatively slow rotators whose coronae are comparatively inactive.

Based on these stellar observations, the solar corona would have been very different ~3.5 Gyr ago when most of the Martian atmosphere is believed to have disappeared. Therefore, there is every reason to believe that the solar wind would have also been quite different as well. However, it is not clear at all whether the more active corona of the young Sun would have produced a stronger or weaker wind. One might naively expect that a more active corona should yield a stronger coronal wind, but this is not necessarily the case. The X-ray fluxes that are commonly used as the measure of coronal activity are associated with closed magnetic field regions, while the solar wind will flow from open field regions. If a more active Sun results in more of the solar surface being covered by closed field regions that crowd out open fields, then the result might actually be a *weaker* solar wind. The current Sun itself provides evidence for such an effect. During the course of the Sun's 11-year activity cycle, it has been found that the solar wind pressure and mass loss rate are slightly lower during the maximum of the cycle when coronal activity is highest, at least in the ecliptic plane (Lazarus and McNutt, 1990). Thus, determining the coronal properties of the young Sun via observations of young stellar coronae is not enough to estimate the properties of the young solar wind.

3. Detecting Solar-like Stellar Winds

3.1. Attempts at Direct Detection

The only way to determine what the solar wind was like in the distant past is to detect and study winds of young, solar-like stars. Unfortunately, detecting analogs for the solar wind around other stars is very difficult. Other types of stellar winds are very easy to detect. The massive, radiation-pressure driven winds of hot stars and the cool, massive winds of red giants and supergiants both produce P Cygni emission line profiles in spectra of these stars, which allows the measurement of wind properties with reasonable precision (Harper *et al.*, 1995; Mullan *et al.*, 1998; Kudritzki and Puls, 2000). However, these are not solar-like stars and these winds are not analogous to the much weaker solar wind, which provides no such spectral diagnostics.

Astronomers have searched for radio emission from nearby stars that could presumably be from a wind, since ionized winds like the Sun should be sources of free-free emission at some level. However, with current radio arrays a solar-like wind around even a very nearby star will only be detected if it has a mass loss rate orders of magnitude higher than the current solar wind, so there have been no clear detections (Brown *et al.*, 1990; Lim *et al.*, 1996; Gaidos *et al.*, 2000). Claims of very high mass loss rates for a few very active stars based on radio detections have been met with skepticism, as the detected emission is more likely to be from the stellar corona rather than from a wind (Mullan *et al.*, 1992; Lim and White, 1996; van den Oord and Doyle, 1997).

Another novel technique that has been used to search for winds is via X-ray emission. Charge exchange between an outflowing ionized wind and inflowing interstellar neutral atoms should yield X-ray emission in the same way that charge exchange with the solar wind leads to X-rays from comets and planets (Lisse *et al.*, 2001; Cravens, 2002; Dennerl, 2002; Gunell *et al.*, 2004). However, though potentially more sensitive than the radio technique, initial attempts to detect circumstellar wind-induced X-ray emission around nearby stars have not been successful (Wargelin and Drake, 2002).

3.2. Stellar Astrospheres

The only clear detections of coronal stellar winds like that of the Sun are not of the winds themselves, but rather detections of the interaction regions between the winds and the interstellar medium (ISM), which are called "astrospheres," analogous to the "heliosphere" that surrounds the Sun. Models of the solar wind/ISM interaction began soon after the discovery of the solar wind (Parker, 1961). Recent reviews of heliospheric modeling efforts include Holzer (1989), Baranov (1990), and Zank (1999). The large scale structure of the heliosphere is defined by three boundaries. Moving outwards from the Sun they are the termination shock (TS), where the solar wind is slowed to subsonic speeds, the heliopause (HP), where the solar wind and ISM plasma are deflected away from each other, and finally the bow shock (BS), where the interstellar wind flow is decelerated to subsonic velocities. The location of the first of these boundaries was recently established when *Voyager 1* crossed the TS at a distance of 94 AU from the Sun (Stone *et al.*, 2005). The upwind (relative to the ISM flow) directions to the HP and BS are not known observationally, but models place them at distances of about $\sim$140 AU and $\sim$240 AU, respectively.

The ISM immediately surrounding the Sun is only partially ionized. In the wind/ISM collision, the neutral atoms in the ISM do not interact as strongly as the ions, but they still take part through charge exchange. Modeling neutrals in the heliosphere is not easy because the charge exchange sends them entirely out of thermal and ionization equilibrium. Nevertheless, many modern heliospheric modeling codes have become sufficiently sophisticated to properly model the neutrals (Baranov and Malama, 1993, 1995; Zank *et al.*, 1996). These models predict that the heliosphere will be permeated by a population of hot hydrogen atoms (H I), especially between the HP and BS where the interstellar H I is decelerated, compressed, and heated. This region in the outermost heliosphere has been called the "hydrogen wall."

This hot H I, particularly in the hydrogen wall, produces a detectable absorption signature in UV spectra of the H I Lyman-α lines of nearby stars from the *Hubble Space Telescope* (HST). However, the lines of sight to these nearby stars not only pass through our heliosphere, but they also pass through the astrospheres of the observed stars. Thus, it is also possible to detect astrospheric Lyman-α absorption, and thereby indirectly detect solar-like stellar winds.

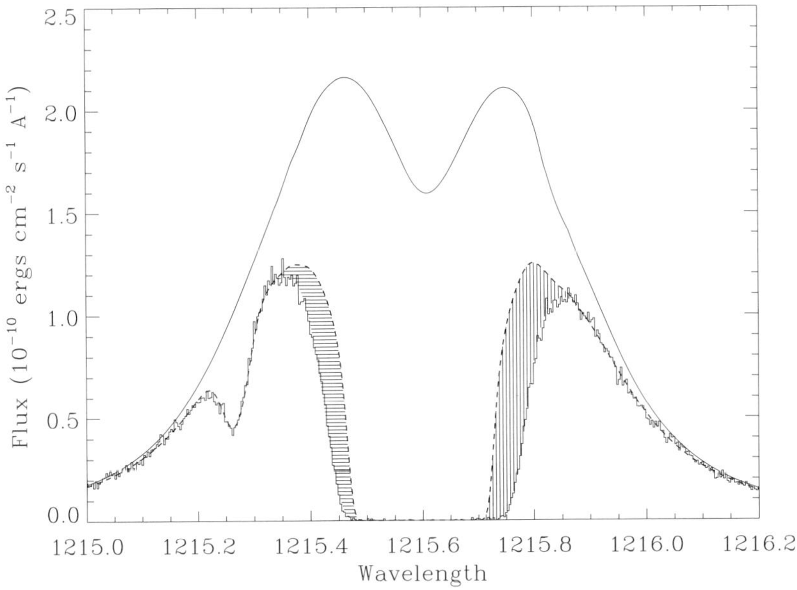

Figure 1. HST Lyman-α spectrum of α Cen B, showing broad H I absorption at 1215.6 Å and D I absorption at 1215.25 Å. The upper solid line is the assumed stellar emission profile and the dashed line is the ISM absorption alone. The excess absorption is due to heliospheric H I (vertical lines) and astrospheric H I (horizontal lines). From Linsky and Wood (1996).

Figure 1 shows the HST Lyman-α spectrum of the very nearby star α Cen B (Linsky and Wood, 1996). The upper solid line is an estimate of the intrinsic Lyman-α emission line profile from the star. Intervening H I gas between HST and the star absorbs much of this Lyman-α emission, resulting in the very broad absorption line centered at about 1215.61 Å in the figure. Much narrower and weaker absorption is also seen from neutral deuterium (D I) at 1215.27 Å. Most of the intervening H I and D I between us and the star is interstellar, but the ISM cannot account for all of the H I absorption. When the H I absorption line is forced to have a temperature consistent with the temperature suggested by the width of the D I Lyman-α absorption, the ISM H I absorption ends up too narrow to fit the data. Thus, Figure 1 indicates that there is excess H I absorption on both sides of the line that cannot be interstellar. The excess absorption on the blue (i.e., short-wavelength) side of the absorption line is the astrospheric Lyman-α absorption signature, and the excess absorption on the red (i.e., long-wavelength) side of the line is from our own heliosphere. The primary reason that heliospheric and astrospheric absorption are shifted away from the ISM absorption, but in opposite directions, is that ISM neutrals are decelerated and deflected as they cross the BS. From within the heliosphere we see the resulting heliospheric absorption as being redshifted, while from our position outside the astrospheres we see the resulting astrospheric absorption as being blueshifted.

Many HST Lyman-α observations of solar-like stars have been analyzed to identify those with detectable heliospheric and/or astrospheric absorption (Linsky and Wood, 1996; Wood *et al.*, 1996, 2005b; Dring *et al.*, 1997; Wood and Linsky, 1998; Izmodenov *et al.*, 1999). Even though all observed lines of sight will pass through the heliosphere and the astrosphere of the observed star, the absorption signatures of these structures are not always detectable. The most common reason for a non-detection is a high ISM H I column density, which leads to broad ISM Lyman-α absorption that obscures the heliospheric/astrospheric absorption. Another factor is the orientation of the line of sight with respect to the upwind direction of the ISM flow. Heliospheric absorption is found observationally to be significantly easier to detect in upwind directions, consistent with model predictions (Wood *et al.*, 2005b). A final major factor that applies solely to astrospheric detectability is the nature of the ISM surrounding the star. Although neutrals are present in the ISM around the Sun, many regions in the "Local Bubble" where the Sun is located will be fully ionized, meaning that an astrosphere in such a location will contain no neutral H to produce Lyman-α absorption.

4. Wind Measurements from Astrospheric Absorption

The current tally of observed lines of sight with detections of heliospheric and astrospheric absorption is 8 and 13, respectively (Wood *et al.*, 2005b). The heliospheric detections are useful for testing heliospheric models and constraining local ISM properties, but we are here more interested in the astrospheric detections, since they represent indirect detections of the coronal winds of these stars. A stronger stellar wind will result in a larger astrosphere and higher hydrogen wall column densities. Thus, a stronger stellar wind will yield more astrospheric Lyman-α absorption, indicating how stellar mass loss rates can be estimated for the stars with detected absorption.

Extracting a stellar mass loss rate from the Lyman-α data requires the assistance of hydrodynamic models of the astrosphere, using the same codes used to model our heliosphere. Models are computed assuming different stellar wind densities, corresponding to different mass loss rates, and the Lyman-α absorption predicted by these models is compared with the data to see which best matches the observed astrospheric absorption. Figure 2 shows the astrospheric absorption predicted by four models of the α Cen astrosphere, assuming four different stellar mass loss rates. The model with twice the solar mass loss rate (i.e., $\dot{M} = 2.0\,\dot{M}_{\odot}$) fits α Cen's blue-side excess absorption best (Wood *et al.*, 2001). Mass loss rate estimates have been made in this way for all of the astrospheric detections (Wood *et al.*, 2002, 2005a). Uncertainties in these mass loss rates are probably of order a factor of two, mostly due to uncertainties in the stellar wind speeds.

The mass loss measurements assume that all coronal winds have velocities similar to the $V_w \sim 400$ km s^{-1} speed of the slow solar wind. Astrospheric size

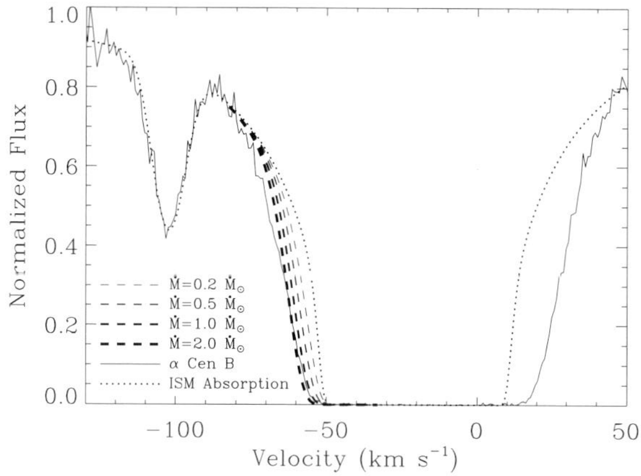

Figure 2. The α Cen B spectrum (thin solid line) and inferred ISM absorption (dotted line) from Figure 1. The dashed lines show the blue-side excess Lyman-α absorption predicted by 4 models of the α Cen astrosphere, assuming 4 different mass loss rates. The 2.0 $\dot{M}_{\odot}$ model fits the α Cen spectrum well. From Wood *et al.* (2001).

and the amount of astrospheric absorption should scale roughly as the square root of the wind ram pressure, $P_w \propto \dot{M} V_w$. If ram pressure was the quantity we were after, to first order the inferred pressure would be independent of the assumed wind velocity. But for many astrophysical purposes the mass loss rate is of more interest. That will be the quantity focused on here, but as a consequence it must be noted that mass loss rates derived from astrospheric absorption will vary inversely with the assumed stellar wind speed (Wood and Linsky, 1998). A justification for the assumption of constant wind speed among solar-like stars is that all cool main sequence stars have roughly the same surface escape speed, and stellar winds of *all* types are generally found to be within a factor of 2 of the surface escape speed.

With these caveats aside, Figure 3 shows mass loss rates (per unit surface area) plotted versus coronal X-ray surface flux (Wood *et al.*, 2005a). For the low-activity main sequence stars, mass loss increases with activity in a manner consistent with the $\dot{M} \propto F_X^{1.34\pm0.18}$ power law relation shown in the figure. The three evolved giant or subgiant stars in the figure do not have mass loss rates consistent with those of the main sequence stars.

It was noted in Section 2 that during the solar activity cycle, the solar wind actually *weakens* slightly at solar maximum, seemingly in contradiction with the apparent mass-loss/activity correlation seen in Figure 3. However, the solar wind is more associated with the large scale dipole component of the magnetic field instead of the small scale active regions responsible for most of the Sun's X-ray emission. The dipole field actually weakens at solar maximum along with the wind. However, the interior magnetic dynamo is ultimately responsible for both the small scale and large scale fields. Thus, it is not unreasonable to expect *both* field components to increase with increasing dynamo activity, at least when averaged over any activity

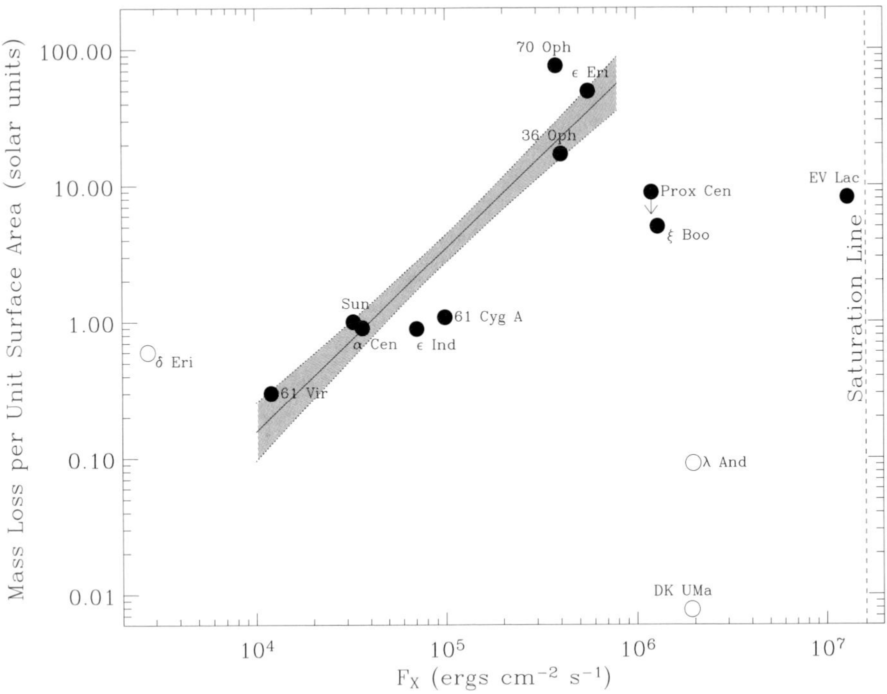

Figure 3. Mass loss rates per unit surface area plotted versus stellar X-ray surface fluxes. Filled symbols are for main sequence stars, while open symbols are for evolved stars. A power law has been fitted to the main sequence stars with $F_X < 8 \times 10^5$ ergs cm^{-2} s^{-1}. From Wood *et al.* (2005a).

cycles (Schrijver *et al.*, 2003). In this sense, the correlation in Figure 3 is not surprising, and not in contradiction with the solar cycle behavior.

The mass-loss/activity relation in Figure 3 does not seem to extend to high activity levels. This may indicate a fundamental change in magnetic field topology as stellar activity is increased. Such a change is also suggested by observational evidence that very active stars usually have stable, long-lived polar starspots (Strassmeier, 2002), in contrast to the solar example where sunspots are only observed at low latitudes. Perhaps the polar spots are indicative of a particularly strong dipolar magnetic field that envelopes the entire star and inhibits stellar wind flow, thereby explaining why the very active stars in Figure 3 apparently have surprisingly weak winds. The most solar-like of the stars in the high activity regime in Figure 3 is ξ Boo, which is actually a binary (G8 V + K4 V). It is therefore worth noting that high latitude starspots have been detected for ξ Boo A (Toner and Gray, 1988). Furthermore, Petit *et al.* (2005) have detected magnetic field structures that are significantly different from solar, including a 40 G global dipole field and a 120 G toroidal field component.

Recalling that young stars are more active than old stars (see Section 2), the correlation between mass loss and activity indicated in Figure 3 implies an anticorrelation

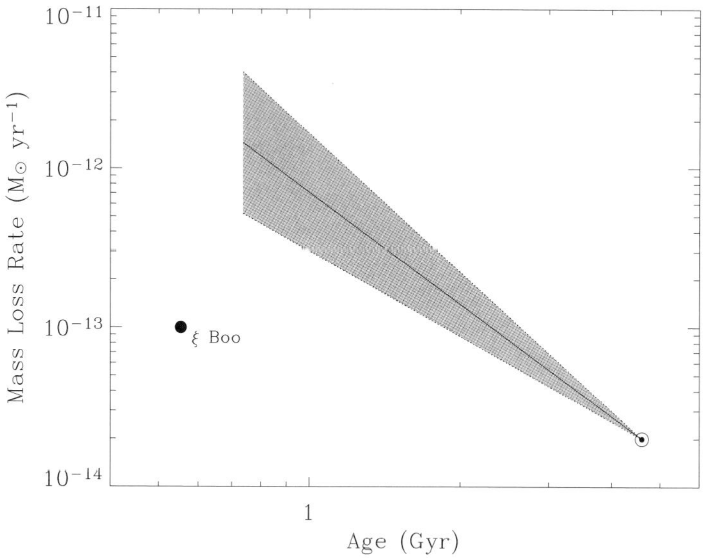

Figure 4. The mass loss history of the Sun inferred from the power law relation in Figure 3. The truncation of the relation in Figure 3 means that the mass-loss/age relation is truncated as well. The low mass loss measurement for ξ Boo suggests that the wind weakens at $t \approx 0.7$ Gyr as one goes back in time. From Wood *et al.* (2005a).

of mass loss with age. Ayres (1997) finds the following relation between stellar X-ray flux and age: $F_X \propto t^{-1.74\pm0.34}$. Combining this with the power law relation from Figure 3 yields the following relation between X-ray flux and age: $\dot{M} \propto t^{-2.33\pm0.55}$ (Wood *et al.*, 2005a). Figure 4 shows what this relation suggests for the history of the solar wind. The truncation of the power law relation in Figure 3 leads to the mass-loss/age relation in Figure 4 being truncated as well at about $t = 0.7$ Gyr. The plotted location of ξ Boo in Figure 4 indicates what the solar wind may have been like at times earlier than $t = 0.7$ Gyr.

The magnetic braking mechanism that slows stellar rotation with time relies on the drag of the magnetic field against the stellar wind, so the time dependence of the wind suggested by Figure 4 has important ramifications for the angular momentum evolution of solar-like stars. These implications are explored in more detail by Wood *et al.* (2002). One interesting result is that theoretical descriptions of magnetic braking are consistent with $\dot{M} \propto t^{-2.33\pm0.55}$ only if disk-averaged stellar magnetic fields decrease at least inversely with age.

5. Planetary Implications

The stellar wind measurements suggest that the mass loss rate of the solar wind was generally higher in the distant past. Analyses of lunar surface soils have also suggested a stronger solar wind in the past, though quantifying the effect is difficult from such data (e.g., Geiss, 1974). This potentially makes the solar wind an even more important factor in the evolution of planetary atmospheres in our solar system. For example, when Mars lost its global magnetic field $\sim$3.9 Gyr ago (i.e., t $=$ 0.7 Gyr in Figure 4), the Martian atmosphere would have been exposed to a solar mass loss rate about 80 times higher than today, which makes the solar wind a more likely cause of the disappearance of much of the Martian atmosphere. It is interesting that the time when Mars lost its global magnetosphere is also near the time when the astrospheric data suggest that the solar wind strengthened abruptly (at $t \approx$ 0.7 Gyr), entering the low activity regime where the power law mass-loss/age relation applies. The poor Martian atmosphere was therefore doubly unlucky. First its protective magnetosphere disappears and at about the same time the eroding solar wind strengthens significantly.

Other atmospheres within the solar system besides Mars may have also been significantly affected by the solar wind, especially Venus and Titan (Chassefière, 1997; Lammer *et al.*, 2000). Wind erosion is also potentially important for many of the extrasolar planets that have been detected around other stars, especially since most of the planets that have been discovered so far are very close to their stars (Grießmeier *et al.*, 2004).

In addition to its eroding effects, a strong young solar wind has also been proposed as a solution for the so-called "Faint Young Sun Paradox." The paradox arises from solar evolution models that suggest the Sun would have been as much as 30% fainter in the distant past, making it difficult to explain why the climates of Earth and Mars were not correspondingly colder (Sagan and Mullen, 1972). It has been pointed out that if the young solar wind were sufficiently strong to significantly lower the mass of the Sun, then the young Sun would have been more massive in the past and therefore more luminous, thereby eliminating the "faint young Sun" (Guzik *et al.*, 1983; Sackmann and Boothroyd, 2003). Unfortunately, though the astrospheric measurements do suggest higher mass loss rates for the young solar wind (see Figure 4), the wind is still not strong enough to have lowered the Sun's mass significantly (Wood *et al.*, 2002). Thus, this solution to the faint young Sun problem is not supported by the data. It has also been proposed that a strong solar wind could lead to a warmer Earth by attenuating cosmic ray inflow into the Earth's upper atmosphere (Shaviv, 2003). However, this proposal relies on controversial claims of a correlation between cosmic ray flux and global temperature.

Before concluding, it should be said that inferences about the history of the solar wind from stellar astrospheric observations would certainly benefit from more data. The mass-loss/activity and mass-loss/age relations presented here are based on only a handful of astrospheric detections. More detections would be especially desirable at high activity levels to better determine what the solar wind was like when the Sun was very young and active. However, the Space Telescope Imaging Spectrograph (STIS) instrument on HST, which is the source of most of the Lyman-α spectra used to detect the astrospheres, failed in 2004 August. This means that additional observations will not be possible in the near future, unless STIS is repaired.

Acknowledgements

Support for this work was provided by NASA through grant NNG05GD69G to the University of Colorado.

References

Acuña, M. H., *et al.*: 1999, *Science* **284**, 790.
Ayres, T. R.: 1997, *J. Geophys. Res.* **102**, 1641.
Baranov, V. B.: 1990, *Space Sci. Rev.* **52**, 89.
Baranov, V. B., and Malama, Y. G.: 1993, *J. Geophys. Res.* **98**, 15157.
Baranov, V. B., and Malama, Y. G.: 1995, *J. Geophys. Res.* **100**, 14755.
Brown, A., Vealé, A., Judge, P., Bookbinder, J. A., and Hubeny, I.: 1990, *Astrophys. J.* **361**, 220.
Carr, M. H.: 1996, Water on Mars (New York: Oxford Univ. Press).
Charbonneau, P.: 2005, Living Rev. Solar Phys. 2, 2, URL: http://www.livingreviews.org/lrsp-2005-2.
Chassefière, E.: 1997, *Icarus* **126**, 229.
Cravens, T. C.: 2002, *Science* **296**, 1042.
Dennerl, K.: 2002, *Astron. Astrophys.* **394**, 1119.
Dring, A. R., *et al.*: 1997, *Astrophys. J.* **488**, 760.
Gaidos, E. J., Güdel, M., and Blake, G. A.: 2000, *Geophys. Res. Lett.* **27**, 501.
Geiss, J.: 1974, in Conference on Lunar Interactions: Interactions of the Interplanetary Plasma with the Modern and Ancient Moon, Criswell, D. R., and Freeman, J. W. (eds.) (Houston: Lunar Science Institute), 110.
Grießmeier, J. -M., *et al.*: 2004, *Astron. Astrophys.* **425**, 753.
Güdel, M., Guinan, E.F., and Skinner, S. L.: 1997, *Astrophys. J.* **483**, 947.
Gunell, H., Holmström, M., Kallio, E., Janhunen, P., and Dennerl, K.: 2004, *Geophys. Res. Lett.* **31**, L22801.
Guzik, J. A., Willson, L. A., and Brunish, W. M.: 1987, *Astrophys. J.* **319**, 957.
Harper, G. M., Wood, B. E., Linsky, J. L., Bennett, P. D., Ayres, T. R., and Brown, A.: 2004, *Astrophys. J.* **452**, 407.
Holzer, T. E.: 1989, *Ann. Rev. Astron. Astrophys.* **27**, 199.
Izmodenov, V. V., Lallement, R., and Malama, Y. G.: 1999, *Astron. Astrophys.* **342**, L13.
Jakosky, B. M., Pepin, R. O., Johnson, R. E., and Fox, J. L.: 1994, *Icarus* **111**, 271.
Jakosky, B. M., and Phillips, R. J.: 2001, *Nature* **412**, 237.
Kass, D. M., and Yung, Y. L.: 1995, *Science* **268**, 697.
Kudritzki, R. -P., and Puls, J.: 2000, *Ann. Rev. Astron. Astrophys.* **38**, 613.
Lammer, H., *et al.*: 2003, *Icarus* **165**, 9.
Lammer, H., Stumptner, W., Molina-Cuberos, G. J., Bauer, S. J., and Owen, T.: 2000, *Planet. Space Sci.* **48**, 529.
Lazarus, A. J., and McNutt, R. L., Jr.: 1990, in Physics of the Outer Heliosphere, ed. Grzedzielski, S., and Page, D. E., (New York: Pergamon), 229.
Lim, J., and White, S. M.: 1996, *Astrophys. J.* **462**, L91.
Lim, J., White, S. M., and Slee, O. B.: 1996, *Astrophys. J.* **460**, 976.
Linsky, J. L., and Wood, B. E.: 1996, *Astrophys. J.* **463**, 254.
Lisse, C. M., *et al.*: 2001, *Science* **292**, 1343.
Luhmann, J.G., Johnson, R. E., and Zhang, M. H. G.: 1992, *Geophys. Res. Lett.* **19**, 2151.
Lundin, R.: 2001, *Science* **291**, 1909.
MacGregor, K. B., and Charbonneau, P.: 1994, *Astrophys. J.* **430**, 387.
Mullan, D. J., Carpenter, K. G., and Robinson, R. D.: 1998, *Astrophys. J.* **495**, 927

Mullan, D. J., Doyle, J. G., Redman, R. O., and Mathioudakis, M.: 1992, *Astrophys. J.* **397**, 225.
Ossendrijver, M.: 2003, *Astron. Astrophys. Rev.* **11**, 287.
Pallavicini, R., Golub, L., Rosner, R., Vaiana, G. S., Ayres, T., and Linsky, J. L.: 1981, *Astrophys. J.* **248**, 279.
Parker, E. N.: 1958, *Astrophys. J.* **128**, 664.
Parker, E. N.: 1961, *Astrophys. J.* **134**, 20.
Perez de Tejada, H.: 1992, *J. Geophys. Res.* **97**, 3159.
Petit, P., *et al.*: 2005, *Mon. Not. R. Astron. Soc.* **361**, 837.
Sackmann, I. -J., and Boothroyd, A. I.: 2003, *Astrophys. J.* **583**, 1024.
Sagan, C., and Mullen, G.: 1972, *Science* **177**, 52.
Schmitt, J. H. M. M., and Liefke, C.: 2004, *Astron. Astrophys.* **417**, 651.
Schrijver, C. J., DeRosa, M. L., and Title, A. M.: 2003, *Astrophys. J.* **590**, 493.
Shaviv, N. J. 2003, *J. Geophys. Res.* **108**, 1437.
Skumanich, A.: 1972, *Astrophys. J.* **171**, 565.
Soderblom, D. R., Stauffer, J. R., MacGregor, K. B., and Jones, B. F.: 1993, *Astrphys. J.* **409**, 624.
Stone, E .C., Cummings, A. C., McDonald, F. B., Heikkila, B. C., Lal, N., and Webber, W. R.: 2005, *Science* **309**, 2017.
Strassmeier, K. G.: 2002, *Astronomische Nachrichten* **323**, 309.
Suzuki, T. K.: 2004, *Mon. Not. R. Astron. Soc.* **349**, 1227.
Toner, C. G., and Gray, D. F.: 1988, *Astrophys. J.* **334**, 1008.
van den Oord, G. H. J., and Doyle, J. G.: 1997, *Astron. Astrophys.* **319**, 578.
Walter, F. M.: 1982, *Astrophys. J.* **253**, 745.
Walter, F. M.: 1983, *Astrophys. J.* **274**, 794.
Walsh, R. W., and Ireland, J.: 2003, *Astron. Astrophys. Rev.* **12**, 1.
Wargelin, B. J., and Drake, J. J.: 2002, *Astrophys. J.* **578**, 503.
Wood, B. E., Alexander, W. R., and Linsky, J. L.: 1996, *Astrophys. J.* **470**, 1157.
Wood, B. E., and Linsky, J. L.: 1998, *Astrophys. J.* **492**, 788.
Wood, B. E., Linsky, J. L., Müller, H. -R., and Zank, G. P.: 2001, *Astrophys. J.* **547**, L49.
Wood, B. E., Müller, H. -R., Zank, G. P., and Linsky, J. L.: 2002, *Astrophys. J.* **574**, 412.
Wood, B .E., Müller, H. -R., Zank, G. P., Linsky, J. L., and Redfield, S.: 2005a, *Astrophys. J.* **628**, L143.
Wood, B. E., Redfield, S., Linsky, J. L., Müller, H.-R., and Zank, G. P.: 2005b, *Astrophys. J. Supp.* **159**, 118.
Zank, G. P.: 1999, *Space Sci. Rev.* **89**, 413.
Zank, G. P., Pauls, H. L., Williams, L. L., and Hall, D .T.: 1996, *J. Geophys. Res.* **101**, 21639.

THE SOLAR WIND INTERACTION WITH THE MARTIAN IONOSPHERE/ATMOSPHERE

STEPHEN H. BRECHT[1,*] and STEPHEN A. LEDVINA[2]
[1]*Bay Area Research Corporation, P.O. Box 366, Orinda, CA 94563, USA*
[2]*Space Sciences Lab., Univ. of California, Berkeley, CA, USA*
*(*Author for correspondence: E-mail: sbrecht@pacbell.net)*

(Received 27 April 2006; Accepted in final form 18 October 2006)

Abstract. The interaction of the solar wind with the Martian exosphere and ionosphere leads to significant loss of atmosphere from the planet. Spacecraft data confirm that this is the case. However, the issue is how much is actually lost. Given that spacecraft coverage is sparse, simulation is one of the few ways for these estimates to be made. In this paper the evolution of our attempts to place bounds on this loss rate will be addressed. Using a hybrid particle code the loss rate with respect to solar EUV flux is addressed as well as a variety of numerical and chemical issues. The progress made has been of an evolutionary nature, with one approach tried and tested followed by another as the simulations are improved and better estimates are produced. The results to be reported suggest that the ion loss rates are high enough to explain the loss of water from Mars during earlier solar epochs.

Keywords: hybrid simulations, Martian ionosphere loss, solar wind interaction with Mars

1. Introduction

The solar wind interaction with a planet unprotected by an extensive intrinsic field has been the subject to considerable study. The research has come in two forms; missions to the planets Mars and Venus, and numerical simulations. The number of papers from just the Pioneer Venus Mission numbers in the hundreds. The number of papers from missions to Mars is rapidly approaching a similar magnitude if not larger. One of the features seen in the data from these missions to Mars and Venus is that the solar wind interacts directly with the atmosphere/exosphere/ionosphere of these planets. This interaction is found to result in the loss of ions from these two planets. The current understanding of the solar wind interaction with Mars has been reviewed by Mazelle *et al.* (2004) and Nagy *et al.* (2004). Numerical simulations have been undertaken to make an estimate of these loses.

A significant focus has been on the planet Mars as the issue of water loss is intertwined with the issue of possible life on that planet. The current questions revolve around issues of the amount of water and where it seems to have gone. There are a variety of theories on this topic. One of the more interesting and complete papers concerning this topic is by Lammer *et al.* (2003). In this paper the authors discuss the various loss mechanisms for water loss and the issue of water being tied up in the soil. One of the major loss mechanisms proposed is that of pick up of oxygen ions via the solar wind interaction with the Martian ionosphere/exosphere. Fox (1997) produced an estimate of what it would take to achieve the 2:1 ratio of hydrogen to oxygen for water loss. In her estimate the rate needed to be roughly 1×10^{26} oxygen atoms per second or roughly 1.2×10^{8} cm^{-2} s^{-1}. Data from the solar maximum period that Phobos -2 was orbiting Mars indicates a loss rate of 3×10^{25} ions s^{-1}.

To estimate the global loss of oxygen from Mars scientists have employed two distinct approaches to make these estimates. One is the MHD formalism (cf. Liu *et al.*, 2001 Ma *et al.*, 2002, 2004) and the second is the kinetic formalism, specifically the hybrid particle code (Kallio and Janhunen, 2002; Modolo *et al.*, 2005). Most if not all of the results to date have produced O^+ and O_2^+ loss rates that are consistent or less than those measured by the ASPERA instrument on Phobos-2. These results will be compared to the results reported in this paper during the discussion section.

The research to be reported in this paper has made use of hybrid particle simulations using the HALFSHEL code. The HALFSHEL code has been used in simulations of unmagnetized bodies for many years (cf. Brecht, 1990; Brecht and Ferrante, 1991; Brecht *et al.*, 1993; Brecht, 1995a, b; Brecht *et al.*, 2000; Ledvina *et al.*, 2004). However, for the current simulations a variety of modifications to the code were required. While the algorithm used in the simulations is a well tested and highly accurate one, the additional physics and chemistry models necessary to make estimates of the global loss of oxygen atoms and molecules from Mars are complex and require simplifications that may or may not be justified. The models included a planetary atmosphere/ionosphere/exosphere set of chemistry equations and neutral profiles. Collision models and various loading strategies were also required. A variety of methods for including these new models into the code have been tried, with the effort extending over approximately 3 years for testing, improvement, and further testing. In this paper the evolution of the predictions of the lose rates from Mars will be presented as new and more complex modeling strategies have been undertaken. In addition, simulations of the lose rates under differing EUV flux conditions will be presented. However, one of the main goals is to determine how sensitive the answers are to different aspects of the solar wind interaction physics so that future research will be better guided.

In the next section of the paper, a brief description of the simulation model used in this work will be presented. Following this section loading strategies will be discussed. Next a discussion of numerical issues addressed during the course of this effort will be presented. Finally, results as they have evolved will be presented, followed by some conclusions from this work, and plans for further improvement of these simulations.

2. Simulation Model

The hybrid model assumes charge neutrality and uses the Darwin approximation (Birdsall and Langdon, 1985; see Appendix I) to remove the displacement current from Maxwell's equations, thus removing light waves from the simulation. The simulations contain multiple ion species, typically protons and ionic states of atomic and molecular oxygen. A full suite of electromagnetic waves exist in these simulations up to and including the Whistler waves. With these approximations the model equations become: Ampere's law

$$\nabla \times \mathbf{B} = (4\pi/c)\,\mathbf{J} \tag{1}$$

Faraday's law

$$c\nabla \times \mathbf{E} = -\partial \mathbf{B}/\partial t \tag{2}$$

the ion particle equations of motion

$$m_i d\mathbf{v}_i/dt = q_i \mathbf{E} + q_i \mathbf{v}_i \times \mathbf{B}/c - q_i \eta \mathbf{J} \tag{3}$$

$$d\mathbf{x}_i/dt = \mathbf{v}_i \tag{4}$$

the inertialess electron momentum equation

$$0 = -e\,n_e\,\mathbf{E} + \mathbf{J}_e \times \mathbf{B}/c - \nabla p_e + e\,n_e\,\eta \mathbf{J} \tag{5}$$

and the requirement of quasi-neutrality

$$n_e = n_i. \tag{6}$$

The electron temperature is found by solving the following equation

$$\partial T_e/\partial t = -\mathbf{u}_e \bullet \nabla T_e - 3/2\, T_e \nabla.\mathbf{u}_e + 2/3\, \eta \mathbf{J}^2/n_e \tag{7}$$

where η is the plasma resistivity.

Use of Equation (5) allows direct calculation of the electric field via the electron momentum equations. This equation also points out the need for three different current systems. The total current as calculated from Equation (1), the ion current obtain from the particles directly, and the electron currents, obtained by subtracting the ion currents from the total currents.

Before addressing the models and results it is useful to consider the simulations being performed. The simulations are in Cartesian coordinates. The solar wind velocity is taken to be 425 km s^{-1}, the solar wind density is 2 protons/cm^3. The IMF is advected in with a Parker spiral of 56 degrees in the ecliptic plane and a total field strength of 3 nT. The nominal solar wind UV photo ionization frequencies are taken to be solar maximum values of 2.73×10^{-7} s^{-1} for O^+ and 7.3×10^{-7} s $^{-1}$ for $CO_2{}^+$ (Liu *et al.*, 2001; Ma *et al.*, 2002). The O^+ EUV flux is slightly less than Modolo *et al.* (2005) rate for solar maximum of 3.1×10^{-7} s^{-1}. The electron temperature necessary for chemistry and electron impact ionization is provided by

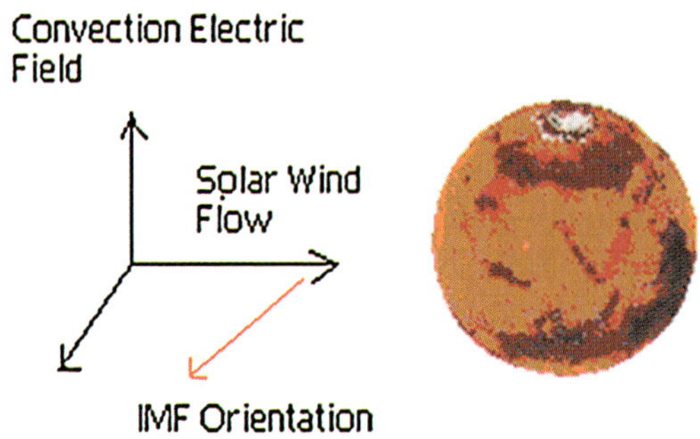

Figure 1. Schematic of the computational orientation of the simulations with the Parker Spiral magnetic field.

Equation (7) above 500 km. Below 500 km the electron temperature is fixed with an altitude profile from Shinagawa and Cravens (1989). The spatial resolution for field solves is $\Delta x = \Delta y = 248$ km/cell and $\Delta z = 300$ km/cell in a Cartesian coordinate system. This leads to a total grid extent of 7.51 R_m in the solar wind flow direction and 8.84 R_m in the transverse directions. The planet is essentially centered in the simulation.

The algorithm is a predictor-corrector as developed by Harned (1982) and discussed by Brecht and Thomas (1988). This particular 3-D code conserves energy in simulations of over 20 000 steps (112 ω_{ci}) with an error of less than 1% using 4 particles per cell. This is to be contrasted with the algorithm by Matthews (1994) where energy conservation is 4% with 32 particles per cell at 100 ω_{ci}. At 300 ω_{ci} loss of conservation using the Mathews method goes from 14% with 32 particles per cell to 47% with 16 particles per cell to 180% with 8 particles per cell. It is worth noting that 500s represents 145 ω_{ci} for protons in a 3 nT magnetic field and well over 300 ω_{ci} in and near the magnetic pile up boundary(cf: Vignes *et al.*, 2000; Trotignon *et al.*, 2006).

The interpolation of density to the particles and to the grid is done using a logarithmic interpolation. This is much more accurate than standard linear interpolation for simulations with exponential density gradients. On the upstream side of the simulations inflow boundary conditions bring in IMF and solar wind plasma. On all other edges of the simulation box the conditions are outflow with $\nabla \cdot \mathbf{B} = 0$ maintained. The planet boundary is at a radius of 3395 km and has electromagnetic field and plasma absorbing boundary conditions. There are no assumptions of plasma outflow and the ionosphere is loaded at essentially zero velocity.

The flow is coming from the left and the convection electric field is pointing north (upward). Figure 1 illustrates the orientation of the simulation and the direction of the relevant fields with respect to the planet in this simulation.

3. The Model for the Atmosphere/Ionosphere/Exosphere

3.1. Initial Simulations

The first sets of simulations were performed in 2003 (Brecht and Ledvina, 2003) using profiles of the ionospheric constituents O^+ and O_2^+. The profiles were set up so that after a certain number of time steps more particles of each species were added to the simulation thus gradually building up a full profile. The strategy used for these simulations was to continually build up the Shinagawa and Cravens (1989) profile over a period of 200 seconds (10 000 steps).

The results of these simulations strongly suggested that in order to achieve a decent estimate of the ion loss rate from Mars much more detailed simulations were required. Several major issues were evident from these simulations. The O_2^+ loss rate was found to be much higher than the O^+. This was surprising but consistent with some MHD simulations performed by Liu *et al.* (2001). Given this result the pickup altitude for these ions was investigated and found to be round 250–300 km; consistent with the recent ASPERA-3 observations reported by Lundin *et al.* (2004). This was also consistent with the findings of Liu *et al.* (2001). The hybrid simulations found loss rates of 1×10^{26} ions s^{-1} for O^+ and 5×10^{27} ions s^{-1} for O_2^+. Liu *et al.* (2001) reported from MHD simulations 2.61×10^{25} ions s^{-1} loss rate for O_2^+ in the tail and 0.45×10^{25} ions s^{-1} loss rate in the tail for O^+ for a total rate of 3.06×10^{25} ions s^{-1}. Liu *et al.* (2001) stated that the rates measured by Lundin (1989) with the ASPERA instrument about Phobos-2 agree with their results.

The issues seen in the early simulations could not be ignored. It was clear that the simulations needed to run for a longer period of time. Further, the rates were clearly still changing, suggesting that either we were artificially eroding the loaded profiles or that a steady state had not been reached. Both of these features argued for longer running times. However, the simulations were out to 340 seconds (17000 time steps) and the ionospheric profiles were moving outward, suggesting that we were over driving the system. The density contours of the lower altitude regions suggested that our goal of reaching a fully loaded ionosphere were reached in 200 seconds. In summary these results motivated considerable improvement and change to the simulation approach to be taken.

3.2. New Simulation Model

It was decided that a new and much more complex approach needed to taken in the simulations. The new approach required that a real set of chemistry equations be used to drive the loading of new particles into the simulation, while eroding the density represented by the existing particles via recombination. For a variety of reasons it was decided that the set of chemistry equations used by Liu *et al.* (2001) and Ma *et al.* (2002, 2004) would be appropriate. It would mean that when

TABLE I
Chemical reactions for Simulations.

Reactions	Rates
$CO_2 + h\nu \rightarrow CO_2^+ + e$	$k = 7.3 \times 10^{-7}\ s^{-1}$
$O + h\nu \rightarrow O^+ + e$	$k = 2.73 \times 10^{-7}\ s^{-1}$
$O + e \rightarrow O^+$	k, Cravens *et al.* (1987)
$CO_2{}^+ + O \rightarrow O_2{}^+ + CO$	$k = 1.64 \times 10^{-10}\ cm^3\ s^{-1}$
$CO_2{}^+ + O \rightarrow O^+ + CO_2$	$k = 9.6 \times 10^{-11}\ cm^3\ s^{-1}$
$O^+ + CO_2 \rightarrow O_2{}^+ + CO$	$k = 1.1 \times 10^{-9}\ cm^3\ s^{-1}$
$O_2{}^+ + e \rightarrow O + O$	$k = 7.38 \times 10^{-8}\ cm^3\ s^{-1}$
$O^+ + e \rightarrow O$	$k = 3.7 \times 10^{-12}\ (250/T_e\ (K))^{0.7}\ cm^3\ s^{-1}$

comparing to the MHD runs being performed at the University of Michigan, we at least would have the same chemistry, Table I. However, one additional equation was added to the chemistry set and is the last equation seen on Table I. This is recombination of the oxygen ions (Schunk and Nagy, 2000). In addition the low pickup altitude strongly suggested that we needed to include ion-neutral drag. It had been hoped and supposed that most of the pickup would be at higher altitudes, where ion-neutral drag would be minimal. The results of the initial sets of simulations proved that this was not the case. Further, results from the ASPERA instrument on Mars Express, also indicated pickup at these altitudes as well (cf. Lundin *et al.*, 2004).

Table I contains the set of chemistry equations and rates placed into the production routines of the code. These new production routines were tested extensively with regard to timing of their execution, and the resolution that was needed for a load. The particle loading is performed in a random fashion and loading as fine as 1.5 degrees of resolution. It was found that performing a spherical load with an angular resolution of 2.5 degrees, and a radial resolution of approximately 20 km, provided adequate coverage, no obvious ripples, and kept the ultimate particle total under about 8 million particles per ionospheric species: O^+ and O_2^+. In order to complete the chemistry neutral profiles needed to be assumed. It was again decided to follow Liu *et al.* (2001) and Ma *et al.* (2002) and include the same profiles they had used:

$$(CO_2) = 1 \times 10^{10} e^{-(z-140)/15.8}$$
$$(O) = 3 \times 10^{8} e^{-(z-140)/43.5}$$

The chemistry used was tested using a one-dimensional photochemical simulation. The simulation was performed from the surface of the planet up to an altitude of 500 km with 2 km cell resolution. The same electron temperature profile (Shinagawa and Cravens, 1989) was used as in the hybrid simulations. It was found that the ionospheric peak of the O_2^+ reached equilibrium around 320 seconds, regardless of the time step chosen. Photo-chemical equilibrium was not reached for the O^+ at higher altitudes, implying that ion dynamics are important for this species. The peak density values did show some sensitivity to the chosen time step. Using a time step of 1.42 seconds produced an O_2^+ peak density of 1.3×10^5 cm^{-3} at an altitude of 150 km. The O^+ reached a value of 3000 cm^{-3} after 400 sec. at an altitude of 300 km. This is in good agreement with the values reported by Ma *et al.* (2004). The peak density values decreased slightly for smaller time steps. The hybrid simulations evaluate the chemistry every 1.4 seconds. This set of chemistry equations lead to peak levels of ionization consistent with models and data.

The next part of the program was to place the ion-neutral drag into the simulations. It was decided that using the electric field to apply the effects of the drag would make the most sense, since the electrons are a fluid, and it was much more computationally efficient to apply a grid based collisionally modified electric field to the ion motion equations. The thermalization of the ions via the drag was not of interest to the problem at hand. One could have simply put the drag into the ion momentum term without using Monte Carlo approach as was done by Bößwetter *et al.* (2004). However, if the ion neutral drag was important, then the electron collisions with ions and neutrals must be considered as well. Therefore, it was felt a more accurate description of what was occurring at these lower altitudes was best represented by including the changes in the conductivity of the plasma via collisions. For this reason it was decided to use the generalized electric field which included the Hall and Pedersen conductivities rather than a fluid like collision term in the ion momentum. The generalized electric field (Mitchner and Kruger, 1973) is:

$$\mathbf{E} = \frac{\mathbf{J} + \beta_e \mathbf{J} \times \mathbf{B} + s\mathbf{B} \times (\mathbf{J} \times \mathbf{B})}{\sigma} - \frac{1}{en_e}\nabla p_e$$

where the $\boldsymbol{\sigma}$ is the conductivity, $\mathbf{J}$ is the current density, β_e is the electron Hall parameter, $(\omega_{\text{en}}/\nu_{\text{eH}})$ with $\omega_{\text{en}} = \text{eB/m}_\text{e}$, the electron cyclotron frequency, and ν_{eH} is the electron collision frequency (electron-neutral and electron-ion). $\mathbf{B}$ is the magnetic field and the ion slip factor, s, is $((\rho_n/\rho)^2\beta_e\beta_I)$ where ρ_n is the neutral density, ρ is the total density (ion + neutral), and β_I is the ion Hall parameter $(\omega_{\text{in}}/\nu_{\text{in}})$. In the limit that the collision frequencies go to zero, this equation returns to the normal electric field equation found in a hybrid particle code, where the electron current is needed, as well as the electron pressure gradient (Brecht and Thomas, 1988). The electron-ion collision frequencies were taken from Mitchner and Kruger (1973). The electron-oxygen (neutral) and electron-CO_2 (neutral) collision frequencies were taken from Strangeway (1996). The ion-neutral collision frequencies were taken from the NRL Plasma Formulary (1990). The goal of this research was to determine the sensitivity of the ion loss rates at Mars to various numerical assumptions. It was not to try and match the boundary features seen in the data. However, it is worth showing some of the general features seen in the simulations. Figure 2 and

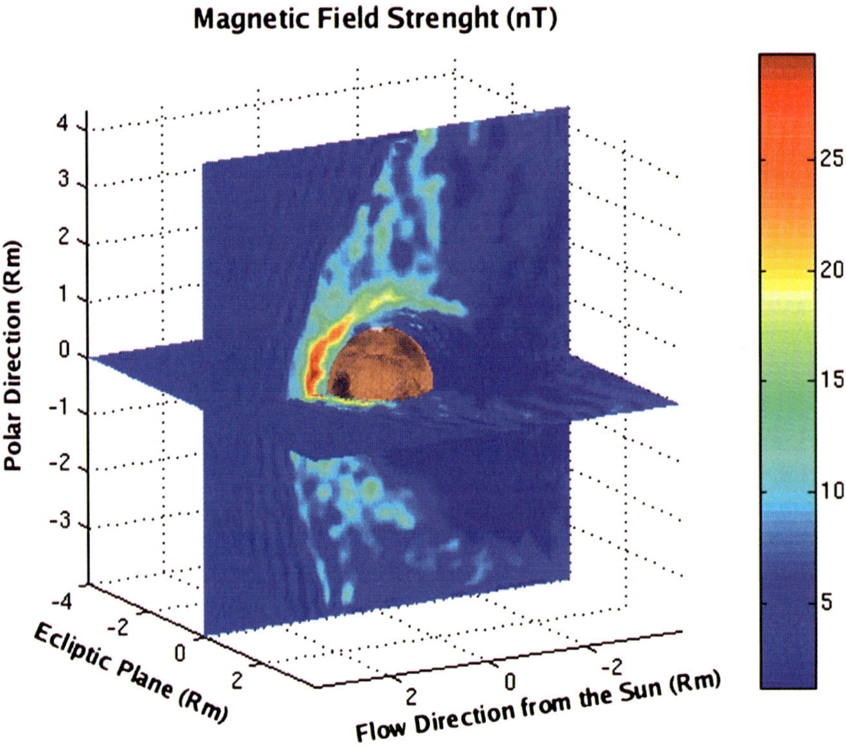

Figure 2. A set of cuts through the Mars simulations showing the typical magnetic field features seen in the simulations. These include the asymmetric shock in the poles and the presence of the magnetic pile up boundary.

3a,b,c illustrates what was seen in the simulations. Figure 2 shows the magnitude of the magnetic field. It is hard to discern some features in this perspective but one see the asymmetric shock structure. The bow shock stands off further than the nominal 1.5 R_m as reported by Vignes *et al.* (2000) and Trotignon *et al.* (2006). The terminator shock crossings are seen to be between 2 and 3 R_m in the North Pole region and slightly more than that in the South Pole. The electromagnetic wave activity seen in the South Pole region makes determination of these features difficult. It would not be surprising to find the crossing further out than the data given the time dependence of the results and the resolution of the simulation.

In Figure 3a one sees the depletion of the solar wind protons near the planet and in the tail. Considerable structure in the barrier is also observed. Further, the depletion is stronger in the South Pole region. Figure 3b and 3c show the exospheric/ionospheric ions. The peak densities in the subsolar region are consistent with the chemistry calculations upon which the chemistry in the code is based. It can be seen that the ions are being lifted from the polar region on the north. However, it appears that the stronger magnetic barrier in the North Pole region is reducing the

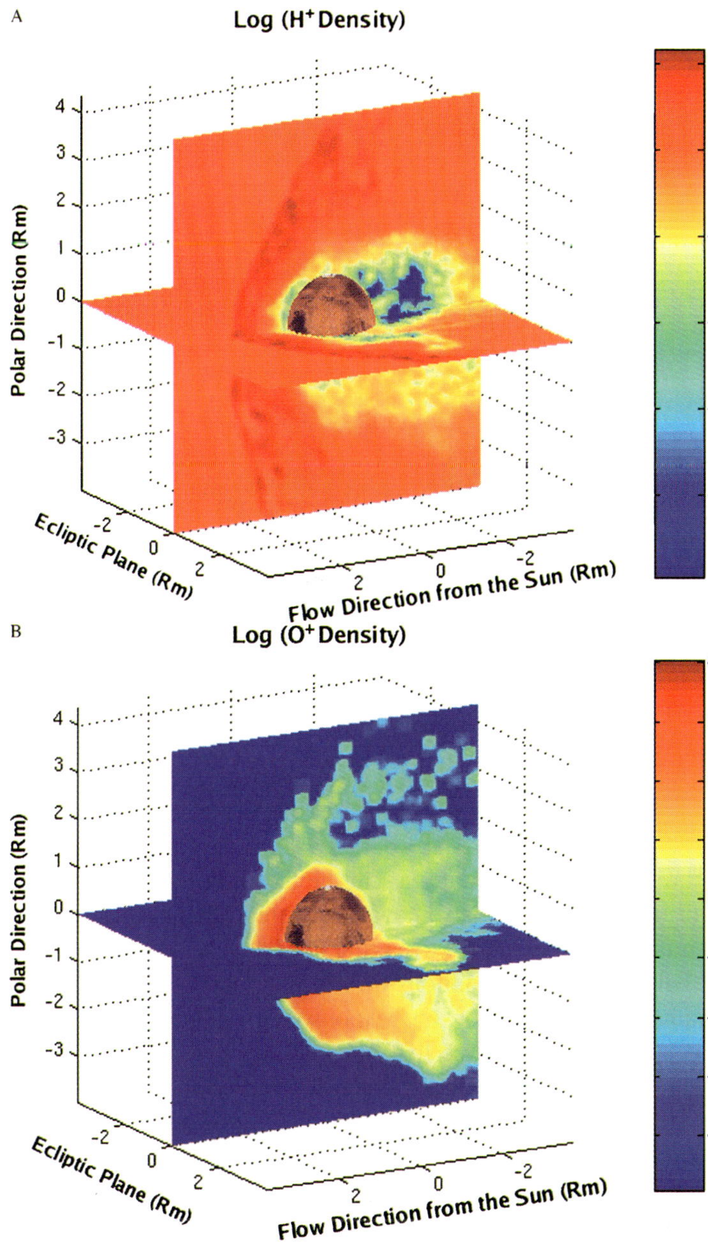

Figure 3. (a) The solar wind proton density. Note the drop out near the planet as well as the structured nature of the regions. (b) Plots of cuts of the O^+ ions as they are picked up primarily from the polar region. (c) Cuts of the density of O_2^+ during one of the simulations. Again notice the preference for pole ward ion pickup consistent with the convection electric field.

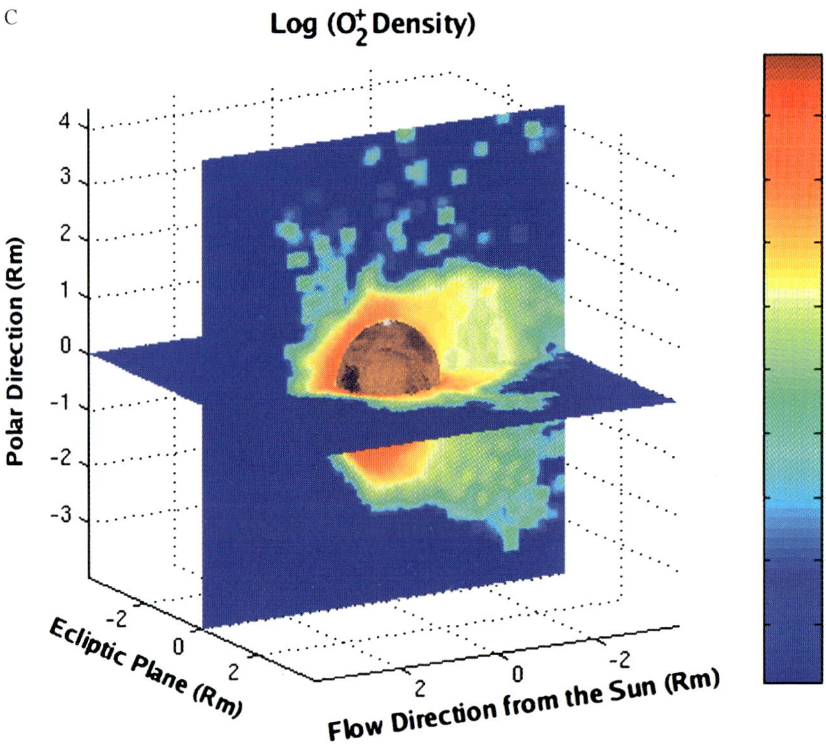

Figure 3. Continued.

ability to lift off the lower altitude ions. In the South Pole region one sees considerable loss of ions. This result is somewhat different than earlier simulations with conducting boundary conditions. Given that Phobos-2 was basically in the ecliptic plane, and that the MHD simulations do not have the necessary physics for the large gyroradius affects that allow the response seen in the poles to the convection electric field it was not a surprise that our loss rates were higher than either the Phobos-2 data or the MHD simulations reported at the time. This will be discussed further in a later section of this paper.

In order to discuss the rates seen it is important to understand how the data was collected. Figure 4 illustrated the computational box through which the pickup ions flux was measured. It starts at the center of the planet in the sunward direction and extends to 2 R_m in the tail. It also extends 2 R_m in all other directions. Snap shots in time were collected and a mean particle velocity normal to the surfaces of each panel was calculated so that the total flux through each panel could be estimated. This is how the rates reported in this paper were determined. Figure 5 shows the time evolution of the O^+ and O_2^+ loss rates. One can see that by slightly over 500 seconds the rates have flattened out. The simulations to be presented and discussed were carried out to 560 seconds.

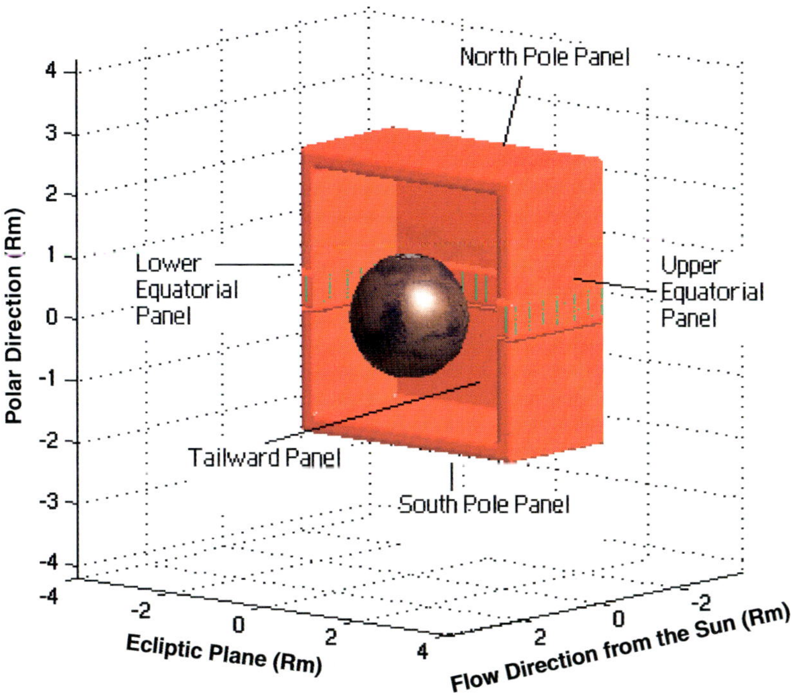

Figure 4. Schematic of the collection regions of the pickup ions. The hashed region is the region where the lost ions are collected and compared to the Phobos-2 Aspera data.

The new ion loss rates using the new simulation approach and a nominal solar maximum EUV fluxes are very different from the earlier rates. First, it is seen that O^+ now has a loss rate much larger than O_2^+. Interestingly, Ma *et al.* (2002, 2004) reported a reversal in the loss rates when they improved their spatial resolution with the atomic oxygen now showing a greater loss than the molecular oxygen ions. The drag term had two obvious affects on the pickup of the ions: a dramatic drop in the O_2^+ loss rates from the very early simulations using fixed profiles and more symmetric density profiles near the planet. In these new simulations the O^+ loss rate is about 5.0×10^{25} ions s^{-1} and the O_2^+ is 1.3×10^{25} ions s^{-1}. These simulations can be compared to those of Ma *et al.* (2004) where the O^+ loss rate was found to be 2.4×10^{24} ions s^{-1} and the O_2^+ loss rate was 3.2×10^{23} ions s^{-1}. These new results are closer to the MHD results but still significantly higher.

3.3. Sensitivity Tests

It seems incumbent upon someone performing simulations of this complexity to examine how sensitive the results are to variations in assumptions. Several tests were performed to determine how sensitive the simulations were to assumptions about the planetary boundary conditions, the neutral density profile, and the electron

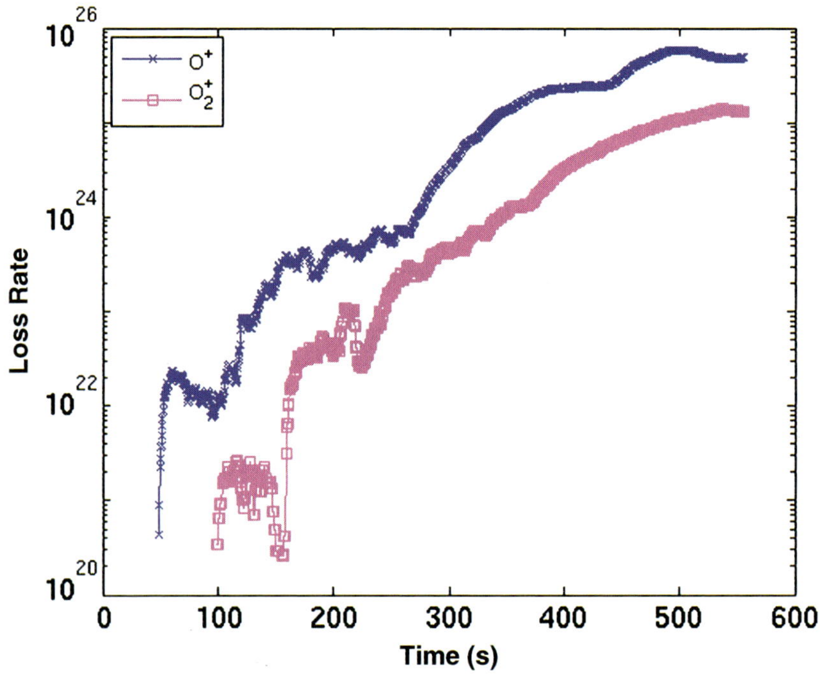

Figure 5. The loss rates for O^+ and O_2^+ from the simulation using the nominal EUV flux and the nominal Parker spiral.

temperature model, the ∇P_e term. The planetary boundary condition was tested by changing it from the original conduction surface to a surface that absorbed both the plasma and the electromagnetic fields. To examine the reality that a neutral density profile would have an azimuthal dependence from noon to midnight, the nominal neutral profile was given a cosine fall off with the midnight profile down to 10% of the nominal profile. To examine the role of the ∇P_e term, it was turned off.

The reasons for these tests were varied. In the case of the planetary boundary it was noticed that the ionospheric plasma tended to want to move radially outward. While the conducting boundary did very well for simple solar wind flow with the planet and did agree reasonably well with Phobos-2 data (cf. Brecht *et al.*, 1993a), the presence of an ionosphere caused us to examine what would happen if the boundary condition was changed. In the neutral density case the single profile for all angles was a very simple assumption. Further, the neutral profile was not modified by the chemistry it was assumed to be a constant. In reality, one could expect the chemistry to vary from the subsolar point and perhaps not in expected ways depending on erosion in this region as compared to the flanks, and cooling on the night side. The testing of the ∇P_e term was driven by one of the assumptions in the simulations. The code solves for an electron temperature and produces profiles very consistent with the data measured by MGS. However, as one goes lower into the atmosphere/ionosphere the electron temperature is governed by chemical processes that are not modeled in these simulations. In these lower altitude regions a temperature profile consistent with Shinagawa and Cravens (1989) is placed in the code. Given that the density gradients and temperature gradients are steepest there and only the density is self-consistently produce, the effect of the electron pressure on the results was of interest.

Changing the planetary boundary condition to an absorbing boundary electromagnetically in addition to absorbing plasma resulted in a change in the ion loss rates. The O^+ loss rate dropped from 6.5×10^{25} ions s^{-1} to 5.0×10^{25} ions s^{-1}. The O_2^+ loss rate increased from 3.1×10^{24} ions s^{-1} to 1.3×10^{25} ions s^{-1}. But, the most compelling reason for using the newer boundary condition was the fact that the ionosphere ceased to advect outward from the planet in the upstream direction. Clearly the presence of an ionosphere made the conducting boundary condition redundant.

The results of changing the neutral density profile as a function of solar zenith angle, SZA, and removing the ∇P_e term are shown in Figure 6a. In the no ∇P_e case the O_2^+ rate increased slightly to 1.4×10^{25} ions s^{-1} from approximately 1.3×10^{25} ions s^{-1}, while the O^+ rate increased from 5.0×10^{25} ions s^{-1} to 8.0×10^{25} ions s^{-1}, indicating that the ∇P_e term made a difference in the loss rates, see Figure 6b for the relative percentages of change. Given that the O_2^+ rate did not change much but the O^+ rate did one can speculate that the lower altitude temperature profile used in the simulations did not have a significant influence. At higher altitudes (above 500 km) the electron temperatures calculated by the temperature equation clearly appear to make a difference via the ∇P_e term as seen by the oxygen pickup.

The change in the neutral density profile resulted in a decrease in the O^+ loss rate of approximately 50% and a decrease in the O_2^+ loss rate by 80%, see Figure 6b. This result suggests that in order to perform simulations with more fidelity, the neutral density profile must have at least solar zenith angle dependence. Perhaps the neutral density profiles must include some erosion based on chemistry. Note that our tests of the chemistry were not in complete photo chemical equilibrium and perhaps this was due to maintaining a constant neutral density profiles.

In summary it is felt that the results presented are sensitive to the models placed in the code. The significance of this will be addressed in Section 4 of this paper, where comparisons will be made to other simulations of the Mars loss rate and some of the data.

3.4. Variations of EUV Flux

The major focus of this work was on the nominal EUV flux to be found during the current solar epoch and the complexity of adding a realistic ionosphere with time dependent production. However, it was of interest to see if simply increasing the EUV flux made a significant difference. Therefore, in addition to the nominal solar maximum EUV flux simulations there were multiple simulations performed with

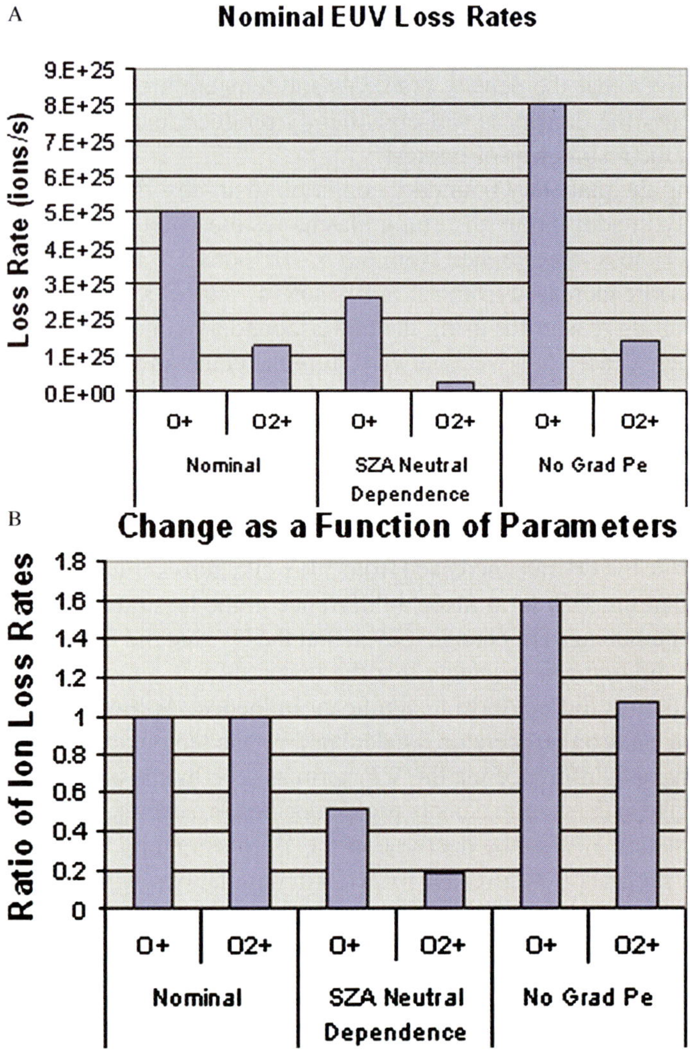

Figure 6. (a) Variations in loss rate as various models in the simulations are changed or removed. (b) Percentage changes as various models in the simulations are changed or removed.

higher EUV fluxes consistent with estimates of (Zhang *et al.* 1993). The simulations were performed with a EUV flux 3 times nominal, and 6 times nominal. The higher EUV cases showed a distinctive change in character. Recall that the neutral profiles for the planet were not modified and neither were the solar wind parameters. Simply the EUV flux was changed thus modifying the chemistry with the enhanced flux. Figure 7 shows that the O^+ loss rates have increased by a factor of roughly six (6) in the 3 EUV case (3.1×10^{26} ions s^{-1}) and a factor of 7 in the 6 EUV case (3.6×10^{26} ions s^{-1}). The O_2^+ the loss rate decreased by a factor of roughly 1.5 for

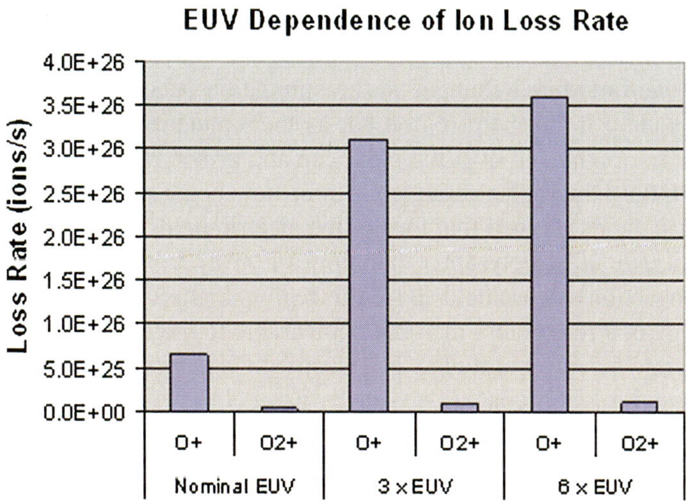

Figure 7. Comparison of the loss rates for each oxygen ion species as a function of EUV flux.

the 3 EUV case (8.4×10^{24} ions s^{-1}), and 1.2 for the 6 EUV case (1.1×10^{25} ions s^{-1}).

Simply increasing the EUV flux shows a very nonlinear result. The overall results show that just changing the EUV flux will lead to very high loss rates of O^+ without introducing solar wind and neutral density levels compatible with earlier epochs (see Lammer *et al.*, 2003). The lack of continued increase of the loss rate as the 3 EUV flux was changed to 6 EUV, is very likely do to using low neutral density profiles consistent with the nominal EUV case. However, given the O_2^+ results one might expect that an extended neutral ionosphere/exosphere coupled with enhanced ionization at higher altitudes might actually serve to protect the planet from erosion by the solar wind. It simply is not clear how the nonlinear interaction of these variables will affect the over all rate. Yet, without increasing the solar wind velocity or IMF field, and without a more extensive atmosphere as expected in the earlier epochs, the loss rates exceed those needed to meet the 2:1 ratio discussed by Lammer *et al.* (2003) and Fox (1997).

4. Discussion

4.1. Observations Based on Simulation Results

One of the larger surprises produced by these simulations was that the preferred direction of loss is in the southern hemisphere with a northern directed convection electric field. It had been thought that the northward directed convection electric field would lift ions from the northern hemisphere and push the ions back into the planet on the southern hemisphere. Previous simulations had indicated a preference to pick up primarily from the northern pole region. However, when the boundary conditions were modified changes in the simulation occurred. One continues to sees rapid pickup in northern region but as the simulations were run longer the magnetic barrier extended over the polar cap and protected the lower ionospheric regions with only the higher altitude O^+ ions now being picked up. Examination of the electric fields showed that locally the electric fields were very small below the magnetic pile up region rather than upward (away from the planet) consistent with the convection electric field above this region. In the southern region, one sees electric fields of strength less than but comparable to the convection electric field penetrating into the lower altitudes. The electric field has a significant anti-sunward component and this component is basically parallel to the curved magnetic fields. So in addition to the tendency for the pick up ions to spiral out of the southern region via the weakened convection field component, there is a very significant field aligned acceleration of the plasma into the tail region.

Based on the asymmetry shown in the pickup ion physics with the presence of the convection electric field, one would not be surprised if the estimates of Lundin *et al.* (1992) for the global loss rate might be low. They were left with little else to do but assume that the fluxes they measured were symmetric when making estimates of the total loss rate. And it seems clear from all of the hybrid simulations that there is a preference to lose ions in regions other than the tail. To see if the results reported in this paper were even reasonable a sanity check was undertaken.

There is a band marked on the collection panels shown in Figure 4. This band is our estimate of the extent of the Phobos orbit assuming a 7 degree inclination to the ecliptic plane at the radius of the collection box; well inside the Phobos-2 circular orbit. The extent north/south is roughly 0.5 R_m in each direction. The ratio between this Phobos collection bands and the total box area is slightly more than 8, although the schematic does not make it seem to be the case. The flux through the banded area was computed by summing up all of the particles in the box that is roughly 600 km thick. In the case of our nominal 1 EUV case the total O^+ loss rate was found to be 2.75×10^{24} ions s^{-1} in the banded region which is a proxy for the ASPERA region during the Phobos-2 orbit. Multiplying by the area ratios one obtains an estimate of the total oxygen ion loss rate of 2.25×10^{25} ions s^{-1}. This result is comparable to the $3. \times 10^{25}$ ions s^{-1} reported by Lundin *et al.* (1992) and shown in Table II. It suggests that these simulations are in fact reasonable particularly given the very asymmetric and inhomogeneous way the ions are found to be picked up by the solar wind interaction with Mars.

Figure 8 shows the loss rate by panel. These panels correspond to those marked on Figure 4. One notices some loss out of the tail region. A lot more out of the polar region, and surprisingly a great deal out of one of the side panels in the direction of the magnetic field line. Given the large gyroradii of the pickup oxygen ions it would not be surprising to find some of these ions gyrating back into the tail region of the planet, especially those found on the South Pole panel and the high equatorial

TABLE II

Ion pick up and Ion loss rates from various researchers.

Loss process	Loss rate (s^{-1})	Authors	Year
Ion pick up: O^+	3.0×10^{25}	Lundin *et al.*	1992
Ion pick up: O^+	1.0×10^{25}	Lammer and Bauer	1991
Ion pick up: O^+	6.0×10^{24}	Luhmann *et al.*	1992
Ion pick up: O^+	8.5×10^{24}	Lichtenegger and Dubinin	1998
Ion pick up: O^+	3.2×10^{24}	Lammer *et al.*	2003
Ion loss: O^+, $O_2{}^+$	1.0×10^{26}	Fox	1997
Ion pick up: O^+, $O_2{}^+$	2.7×10^{25}	Liu *et al.*	1999
Ion pick up: O^+, $O_2{}^+$	2.7×10^{25}	Ma *et al.*	2002
Ion pick up: O^+, $O_2{}^+$	2.4×10^{24}; 3.2×10^{23}	Ma *et al*	2004
Ion loss: O^+, $O_2{}^+$	$3.5 \times 10^{25} - 7.0 \times 10^{27}$	Hodges Jr.	2000
Ion pick up: O^+, $O_2{}^+$	2.37×10^{24}; 7.0×10^{22}	Modolo *et al.*	2005
Ion Pick up: O^+	1.55×10^{25}	Kallio and Janhunen	2002

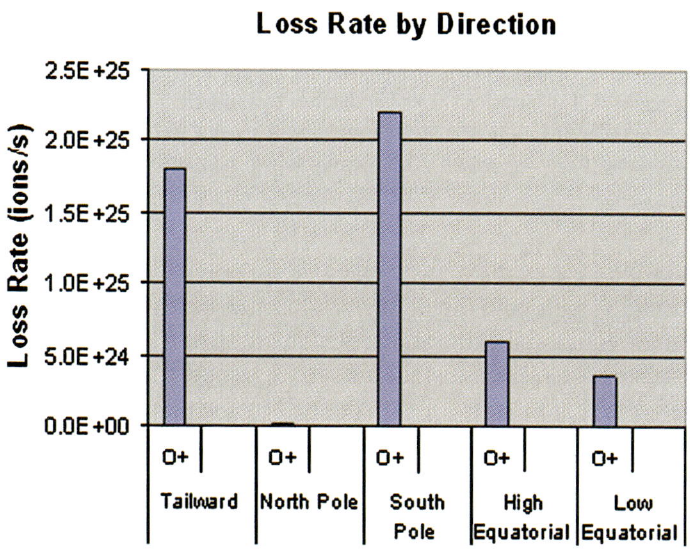

Figure 8. Loss rate broken down by exit planes. The equatorial bins are perpendicular to the equatorial plane with the magnetic field entering or leaving the plane.

panel. At the minimum these results suggest how difficult it is to estimate the actual ion loss rates from Mars via spacecraft instruments because of the very asymmetric and time dependent way the pickup ions are lost from Mars.

Figure 9 shows a slightly different way of looking at the geometry of the ion pickup. In this case the ion loss rate is examined for ions entering the collection

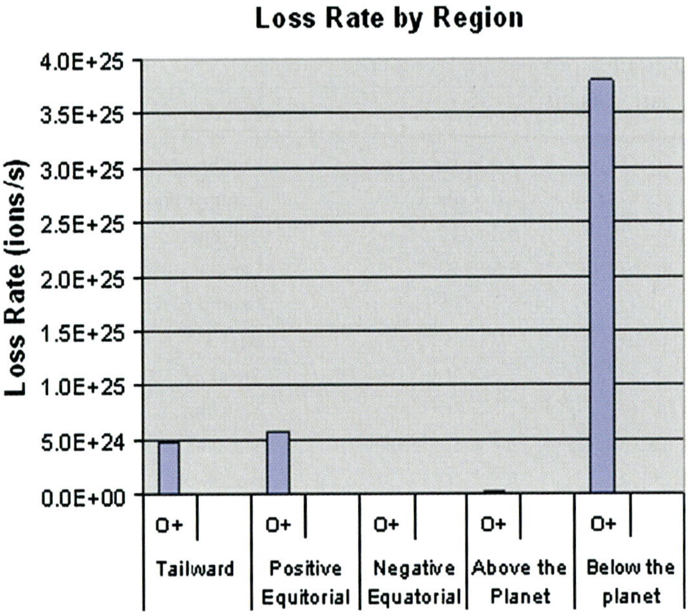

Figure 9. Another break down of ion loss by region. Here above the planet means any ions above 1 R_m from the center of the planet and below the planet means any ions lost into the region 1 R_m below the center of the planet. The tailward values are the ions lost into the tail of the plane but within ± 1 R_m of the center in polar direction. The positive and negative equatorial directions are to the sides of the planet in equatorial plane but within ± 1 R_m of the center in polar direction.

box 1 R_m above he planet the planet center (north pole direction), and 1 R_m below the planet center (south pole direction). Within the shadow of the planet the ions picked up on the equatorial plane and in the tail are shown separately. One sees that most of the pickup ions leave via the southern region (3.8×10^{25} ions s^{-1}), almost none via the north region. In the region defined by the planet radius one sees loses of 4.8×10^{24} ions s^{-1} and 5.8×10^{24} ions s^{-1} in the positive equatorial direction (which is basically along the magnetic field direction). In the anti-magnetic field direction the loss rate is very small.

It is worth repeating that the simulations appear to have reached a reasonable steady state with regard to the ion loss rate as measured through the collection box. This time is roughly 500–600 seconds. A steady state with regard to leaving the entire simulation grid will take longer. However, it was felt that given the purpose of this research (estimate the loss rate from Mars and study the sensitivity of the simulations to various models), it was not necessary to run further at this time.

The fact that the simulations seem to be sensitive to the SZA dependence of the neutral profile has a variety of consequences for future simulation efforts. In fact, these results suggest not too surprisingly that the neutral profile in general must be reconsidered. It was discussed earlier that the chemistry did not reach a photo-chemical equilibrium. It is suspected that the neutral profile would need to reflect the erosion caused by the photo-ionization especially at high altitudes where diffusion is much slower than the loss. This of course does not include the fact that impact ionization is also included in the simulations which would further erode the neutral profiles.

In addition to considering the possibility of eroding the profile on the dayside, it was very clear that a more accurate SZA dependent neutral profile is required to better estimate the loss rates of pickup ions from Mars. It also means that for cases where earlier epochs (higher EUV levels) are to be simulated careful consideration of the neutral profiles for these cases must be undertaken.

Finally, the result of removing the ∇P_e term from the calculation illustrates the importance of the temperature model to these calculations. Given that mainly the oxygen ions are affected by this test one can surmise that region being most affected is above 300–500 km. In this region a temperature equation is solved in the simulations. Bößwetter *et al.* (2004) indicate in their work that the level of temperature they choose for their simulation affects the location of the shock. This again suggests that the electron temperature profile associated with the density gradients is important. It would seem based on the results presented that the electron pressure gradient sets up to impede the loss of ions from the upper ionosphere/exosphere.

4.2. Comparison to Other Simulations

Table II has been created after the table found in Lammer *et al.* (2003). It shows clearly the range of measurement, estimates, and simulation results for the escape of oxygen ions. It is noted that most of the results are below that reported by Lundin *et al.* (1992). The issue is then how can the simulations in this work be so different from the MHD (Liu *et al.*, 1999; Ma *et al.*, 2002, 2004) results and the previous hybrid simulations (Kallio and Janhunen, 2002; Modolo *et al.*, 2005)? We believe that the answer to the comparisons with the MHD calculations is relatively straight forward. The simulations reported here used virtually the same chemistry and neutral profiles as did the MHD simulations. However, as mentioned earlier the real difference lies in the degrees of freedom with regard to plasma motion permitted in the MHD equations vs. the hybrid equations. In the MHD equations the plasma is frozen to the field lines it is born on. This means that the only escape mechanism is down the tail region of the planet as the field lines advect around the planet and thru the exosphere and ionosphere. Indeed when the physical resolution of the simulations was increased in Ma *et al.* (2004), the pickup rate dropped. Their results showed excellent agreement with the Viking data deep in the ionosphere. They state that their higher resolution allowed them to do a better job of calculating the O^+ flux and O_2^+ flux. However, these better results led to lowering the escape rates below those quoted by Lundin *et al.* (1989). The simulations reported here suggest that their results are reasonable given the mechanism of ion pickup available to the MHD code, there is no finite larmor radius effects to lift the ions out of the atmosphere and minimize the chances of recombination as well as drag with the neutrals. Further, unless they have a resistive model in the code the MHD simulations will not produce parallel electric fields.

An examination of the results reported by this paper, as well as, those reported by Kallio and Janhunen (2002) and Modolo *et al.* (2005), shows a large disparity in results. As seen in Table II, Modolo *et al.* (2005) results are an order of magnitude less than those reported by Lundin *et al.* (1992). The results of Kallio and Janhunen (2002) are down by a factor of two. The results reported here are slightly lower than Lundin *et al.* (1992) by roughly 33%. There are differences in the numerical algorithms used by each group in constructing their hybrid codes. These differences may play some role in the differing results. However, these differences would need considerable effort to quantify.

It is clear the differences are not due to resolution as Modolo *et al.* (2005) have cell sizes of 300 km, and Kallio and Janhunen (2002) have minimum cell sizes of 175 km. None of the hybrid simulations have resolution as fine as Ma *et al.* (2004). So where could the differences be? It seems that they may reside in how the ionosphere is created by each group. In Kallio and Janhunen (2002) the strategy is to use the planet as an emitter of ions. Their simulations emit ions at a specified rate, thus having a strong affect on the subsequent pickup rate. But, the purpose of their efforts was only tangentially directed toward pick up rates and was focused on other issues. In the case of Modolo *et al.* (2005) the ionosphere is loaded with a hydrogen profile and an atomic oxygen profile. The paper does not specify anything about molecular oxygen although they do quote an O_2^+ pickup rate. There appears to be no carbon dioxide profiles. The atomic oxygen profile is stated to be modeled after the work by Kim (1998). In examining their paper closely it was noticed that their atomic oxygen profiles were not what we thought. Figure 10 illustrates the findings.

In Figure 10 the carbon dioxide profiles and atomic oxygen profiles used in this work are plotted as solid and dashed lines respectively. The line denoted by boxes is the oxygen profile from Kransopolsky (1993a, b). The line denoted by stars is our interpretation of the atomic oxygen profile presented by Modolo *et al.* (2005). The point is that each group has used a different neutral profile and there is almost an order of magnitude difference in parts of the respective profiles. The differences increase at lower altitudes, but in the 200 to 350 km altitude the difference is still significant. The differences in the neutral profiles coupled with differing chemistry, strongly suggest that the differences in the results are as much to do with the models chosen by each groups research as it is some numerical issue. Given the differences seen, it is not surprising that the work being reported here produces higher ion loss rates from Mars' exosphere/ionosphere.

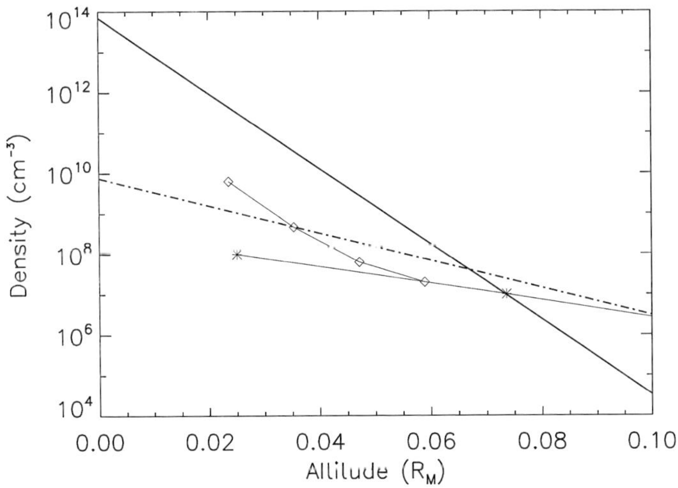

Figure 10. A plot of the various neutral profiles used in hybrid simulations of Mars ion pickup. The solid line is the CO_2 profile used in this paper. The dashed line is the O profile used in this paper. The diamond line is Kransopolsky's profile for O. The star line is our best estimate of the O profile used by Modolo *et al.* (2005).

5. Summary

The purpose of these simulations was to estimate the ion loss rate from Mars using a kinetic simulation. Further, it was clear that implementing the complex chemistry and physics in the region around the planet would require a variety of assumptions. One of the goals was to establish the sensitivity of the results to various assumptions. Both goals have been accomplished. The major conclusion from this research is that while the rates seem consistent, there is much more work to be done.

The results of these simulations produce rates significantly higher than other hybrid simulations, as well as, the most recent MHD simulations. It seems very clear that this is caused by the different approaches used by various hybrid simulation groups to loading the planetary atmosphere/ionosphere/exosphere. Further, the simulations indicate considerable sensitivity to how the models are loaded, and whether or not certain models are included. The result of this effort clearly indicates that further work on this problem is required. All of this suggests that this very difficult problem must be addressed very carefully.

To date, the simulations of a solar maximum level of EUV flux, such as encountered during the Phobos-2 mission, show that nominal loss rates are slightly below the level needed to explain the water loss from Mars. However, estimates of the loss rate as compared to the ASPERA measurements of Lundin *et al.* (1992) are very consistent, suggesting that the simulation approach is on the right track.

The enhanced EUV flux simulations show loss rates in excess of the threshold of 1×10^{26} ions s^{-1}, Fox (1997). It seems that the ion loss rate will not respond linearly to increases in the EUV flux and probably not linearly to extended neutral profiles and higher solar wind velocities and IMF levels consistent with earlier epochs (Lammer *et al.*, 2003). However, the results of the simulations using higher EUV flux suggest that H_2O loss via ion pickup may have occurred in earlier epochs of the solar system evolution.

The simulations presented here do not represent the final or best simulations that can be done. Future research plans include changing to a spherical geometry with much higher resolution in the radial direction. Such simulations will allow more accurate handling of the ionospheric profiles. Further, it is felt that boundary locations will be more accurately represented with the increased resolution. Finally, the better resolution means that the chemistry will be more accurately handled and that coupled with greater ease of implementing planetary boundary conditions should allow for a more accurate calculation of the ion pickup by the solar wind.

The future work will address these issues as well as the issue of what are the "correct" neutral models to use. Once additional confidence with regards to the results is achieved, then the issue of crustal fields will be addressed. One can speculate that the crustal fields will impede the loss rates. If the convection electric field is directed northward, the southern crustal fields may have a significant affect on the loss rate because of the preference for the ions to be lost in the southern hemisphere with a northward directed convection electric field and the potential for very complex magnetic and electric field configurations in this region.

Appendix I

The Darwin approximation consists of introducing a Helmholtz decomposition of the electric field $\mathbf{E}(\mathbf{E} = \mathbf{E}_L + \mathbf{E}_T)$, $\mathbf{E}_L$: longitudinal part, $\mathbf{E}_T$: solenoidal part, with $\nabla \times \mathbf{E}_L = 0$ and $\nabla \cdot \mathbf{E}_T = 0$) and in neglecting only $\frac{\partial \mathbf{E}_T}{\partial \mathbf{t}}$ in Ampère's law. In the hybrid formalism the displacement current $\frac{\partial \mathbf{E}}{\partial \mathbf{t}}$ is neglected (and not only the solenoidal part).

Acknowledgments

SHB was supported by NASA contracts NASW-00001 and NASW-99017. SAL was supported by NASA grant NNG05GA04G. The authors would like to acknowledge many useful discussions with Prof. A. Nagy and some very insightful comments by one of the referees.

References

Acuna, M. *et al.*: 1998, *Science* **279**, 1676.

Birdsall, C. K., and Langdon, A. B.: 1985, *Plasma Physics via Computer Simulations*, McGraw-Hill Book Co., USA.

Brecht, S. H., and Thomas, V. A.: 1988, *Comp. Phys. Comm.* **48**, 135.
Brecht, S. H.: 1990, *Geophys. Res. Lett.* **17**, 1243.
Brecht, S. H., and Ferrante, J. R.: 1991, *J. Geophys. Res.* **96**, 11209.
Brecht, S. H., Ferrante, J. R., and Luhmann, J. G.: 1993a, *J. Geophys. Res.* **98**, 1345.
Brecht, S. H., and Ferrante, J. R.: 1993b, *EOS Transactions, Am. Geophys. Union* **74**, 54.
Brecht, S. H.: 1995a, *Adv. In Space Research: Physics of Collisionless Shocks*, (ed.), C. T. Russell, **15,** 415.
Brecht, S. H.: 1995b, *Geophys. Res. Lett.* **22**, 1181.
Brecht, S. H.: 1997a, *J. Geophys. Res.* **102**, 11287.
Brecht, S. H.: 1997b, *J. Geophys. Res.* **102**, 4743.
Brecht, S. H., Luhmann, J. G., and Larson, D. J.: 2000, *J. Geophys. Res.* **105**, 13119.
Brecht, S. H., Ledvina, S. A., and Luhmann, J. G.: 2003, Global hybrid simulations of the solar wind interaction and the Mars ion loss rate. Fall AGU.
Bößwetter, A., Bagdonat, T., Motschmann, U., and Sauer, K.: 2004, *Annales Geophysicae* **22**, 4363.
Cravens, T. E., Hoppe, A., Ledvina, S. A., and McKenna-Lawlor, S.: 2002, *J. Geophys. Res.* doi:10.1029/2001JA000125.
Fox, J. L.: 1997, *Geophys. Res. Lett.* **24**, 2901.
Harned, D. S.: 1982, *J. Comp. Phys.* **47**, 452.
Hodges, R. R. Jr.: 2000, *J. Geophys. Res.* **105**, 6971.
Kallio, E., and Janhunen, P.: 2002, *J. Geophys. Res.* **107**, doi:10.1029/2001JA000090.
Kim, J., Nagy, A. F., Fox, J. L., and Cravens, T. E.: 1998, *J. Geophys. Res.* **103**, 29339.
Kransopolsky, V. A.: 1993a, *Icarus* **101**, 33.
Kransopolsky, V. A.: 1993b, *Icarus* **101**, 313.
Lammer, H., and Bauer, S. J.: 1991, *J. Geophys. Res.* **96**, 1819.
Lammer, H., Stumptner, W., and Bauer, S. J.: 1996, *Geophys. Res. Lett.* **23**, 3353.
Lammer, H., Lichtenegger, H. I. M., Kolb, C., Ribas, I., Guinan, E. F., Abart, R. *et al.*: 2003, *Icarus* **165**, 9, doi:10.1016/S0019-1035(03)00170-2.
Ledvina, S. A., Brecht, S. H., and Luhmann, J. G.: 2004, *Geophys. Res. Lett.* **31**, L17S10.
Lichtenegger, H. I. M., and Dubinin, E. M.: 1998, *Earth and Planets Space* **50**, 445.
Liu, Y., Nagy, A. F., Gombosi, T. I., DeZeeuw, D. L., and Powell, K. G.: 2001, *Adv. Space Res.* **27**, 1837.
Luhmann, J. G., Johnson, R. E., and Zhang, M. G. H.: 1992, *Geophys. Res. Lett.* **19**, 2141.
Lundin, R., Zakharov, A., Pellinen, R., Barabash, S. W., Borg, H., Dubinin, E. M. et al: 1990, *Geophys. Res. Lett.* **17**, 873.
Lundin, R., Zakharov, A., Pellinen, R., Hultquist, B., Borg, H., Dubinin, E. M., *et al.*: 1989, *Nature* **341**, 609.
Ma, Y. A., Nagy, A. F., Hansen, K. C., and DeZeeuw, D. L.: 2002, *J. Geophys. Res.* **107**, 1282, doi:10.1029/2002JA009293.
Ma, Y., Nagy, A. F., Sokolov, I. V., and Hansen, K. C.: 2004, *J. Geophys. Res.* **109**, 10.1029/2003JA010367.
Matthews, A. P.: 1994, *J. Comp. Phys.* **112**, 102.
Mazelle, C., Winterhalter, D., Sauer, K., Trotignon, J. G., Acuña, M. H., Baumgärtel, K. *et al*.: 2004, *Space Sci. Rev.* **111,** 115.
Mitchner, M., and Kruger, C. H., Jr.: 1973, Partially Ionized Gases, John Wiley & Sons, NY, USA.
Modolo, R., Chanteur, G. M., Dubinin, E., and Matthews, A. P.: 2005, *Annales Geophysicae* **23**, 433.
Nagy, A. F., Winterhalter, D., Sauer, K., Cravens, T. E., Brecht, S. H., Mazelle, C., *et al*.: 2004, *Space Sci. Rev.* **111,** 33.
NRL Plasma Formulary, *NRL/PU/6790-04-477*, Ed. J. D. Huba, 2004.
Sauer, K., Roatsch, T., Motschmann, U., Schwingenschuh, K., Lundin, R., Rosenbauer, H., et al.: 1992, *J. Geophys. Res.* **97**, 6227.

Shinagawa, H., and Cravens, T. E.: 1989, *J. Geophys. Res.* **94**, 6506.

Strangeway, R.: 1996, *J. Geophys. Res.* **101**, 2279.

Trotignon J. G., Mazelle, C., Bertucci, C., and Acuña, M. H.: 2006, *Plan. Space Sci.* **54**, 357–369, doi: 10.1016/j.pss.2006.01.003.

Zhang, M. H. G., Luhmann, J. G., Nagy, A. F., Spreiter, J. R., and Stahara, S. S.: 1993, *J. Geophys. Res*, **98**, 3311.

Vignes, D., Mazelle, C., Reme, H., Acuna, M. H., Connnerney, J. E. P., Lin, R. P. *et al* .: 2000, *Geophys. Res. Lett.* **27**, 49.

ENERGISATION OF O^+ AND O_2^+ IONS AT MARS: AN ANALYSIS OF A 3-D QUASI-NEUTRAL HYBRID MODEL SIMULATION

E. KALLIO[1,*], A. FEDOROV[2], S. BARABASH[3], P. JANHUNEN[1,4], H. KOSKINEN[1,4], W. SCHMIDT[1], R. LUNDIN[3], H. GUNELL[3], M. HOLMSTRÖM[3], Y. FUTAANA[3,6], M. YAMAUCHI[3], A. GRIGORIEV[3], J. D. WINNINGHAM[5], R. FRAHM[5] and J. R. SHARBER[5]

[1]*Finnish Meteorological Institute, Box 503 FIN-00101 Helsinki, Finland*
[2]*Centre d'Etude Spatiale des Rayonnements, BP-4346, F-31028 Toulouse, France*
[3]*Swedish Institute of Space Physics, Box 812, S-98 128, Kiruna, Sweden*
[4]*University of Helsinki, Department of Physical Sciences, P.O.Box 64, FIN-00014 Helsinki, Finland*
[5]*Southwest Research Institute, San Antonio, TX 7228-0510, USA*
[6]*Institute of Space and Astronautical Science, Japan Aerospace Exploration Agency, Yoshinodai 3-1-1, Sagamihara 229-8510, Japan*
*(*Author for correspondence: E-mail: Esa.Kallio@fmi.fi)*

(Received 29 March 2006; Accepted in final form 2 November 2006)

Abstract. We have studied the loss of O^+ and O_2^+ ions at Mars with a numerical model. In our quasi-neutral hybrid model ions (H^+, He^{++}, O^+, O_2^+) are treated as particles while electrons form a massless charge-neutralising fluid. The employed model version does not include the Martian magnetic field resulting from the crustal magnetic anomalies. In this study we focus the Martian nightside where the ASPERA instrument on the Phobos-2 spacecraft and recently the ASPERA-3 instruments on the Mars Express spacecraft have measured the proprieties of escaping atomic and molecular ions, in particular O^+ and O_2^+ ions. We study the ion velocity distribution and how the escaping planetary ions are distributed in the tail. We also create similar types of energy-spectrograms from the simulation as were obtained from ASPERA-3 ion measurements. We found that the properties of the simulated escaping planetary ions have many qualitative and quantitative similarities with the observations made by ASPERA instruments. The general agreement with the observations suggest that acceleration of the planetary ions by the convective electric field associated with the flowing plasma is the key acceleration mechanism for the escaping ions observed at Mars.

Keywords: Mars, Mars-solar wind interaction, Mars Express, ASPERA-3 instrument, ion escape

1. Introduction

Mars does not have a strong global intrinsic magnetic field and, therefore, the solar wind can flow close to the planet where the density of atmospheric neutrals is high. Because of the direct interaction between the atmosphere/exosphere and the solar wind some of the particles ionised from the neutral exosphere and atmosphere can be accelerated by electromagnetic forces and escape the system. This

Space Science Reviews (2006) 126: 39–62
DOI: 10.1007/s11214-006-9120-z

non-thermal escape is one of the loss mechanisms for atmospheric particles. The role of non-thermal escape processes at Mars at present and at times in the past has received attention when the evolution of the Martian atmosphere has been studied (see e.g. Lammer *et al.*, 2006, and ref. therein).

The Mars Express (MEX) spacecraft has observed Mars and the properties of near-Mars space since December 2003. So far the most comprehensive data set for studying ion escape at Mars has been provided by the MEX Analyzer of Space Plasmas and Energetic Atoms (ASPERA-3) experiment (Barabash *et al.*, 2004). This experiment contains four individual instruments one of which, the Ion Mass Analyser (IMA), can distinguish escaping planetary ions (O^+, O_2^+, CO_2^+) from the solar wind ions (H^+, He^{++}). IMA has measured planetary ions near Mars on the dayside (Lundin *et al.*, 2004) and on the nightside (e.g., Fedorov *et al.*, 2006).

Energisation of planetary ions at Mars has been a subject of intense investigations since the Phobos-2 mission in 1989 when distinction between heavy planetary ions and the lighter solar wind ions was performed. Various modelling approaches have been used to interpret Phobos-2 observations, such as a test particle simulation (e.g. Kallio and Koskinen, 1996), a gas dynamic (GD) model (e.g. Kallio *et al.*, 1994), a magnetohydrodynamic (MHD) model (e.g., Liu *et al.*, 1999) and a quasi-neutral hybrid (QNH) model (e.g., Brecht *et al.*, 1993). In this paper we continue to study the ion escape at Mars with a three dimensional (3-D) QNH model. So far the model results have been compared with ASPERA/Phobos-2 ion measurements and Phobos-2 magnetic field measurements (Kallio and Janhunen, 2002), and recently with IMA/ASPERA-3/Mars Express observations (Kallio *et al.*, 2006). In the QNH-IMA/ASPERA-3 comparison, macroscopic plasma parameters based on QNH simulation were compared with IMA energy spectra (Kallio *et al.*, 2006). In this paper we study the properties of escaping O^+ and O_2^+ ions in more detail by analysing the ion velocity distribution functions and by generating energy spectrograms in a similar format as used to display IMA measurements.

The paper is organised by first presenting some basic features of the QNH model and the particular methods used in this paper are introduced. Then the density of O^+ and O_2^+ ions in the tail region and the properties of the magnetic field near Mars based on a QNH model run are presented. Several analyses of the properties of the planetary ions on the nightside at three planes with x = constant ($x = -1.1R_M$, $-2.1R_M$ and $-3.1R_M$) are presented: (1) spatial distribution on each plane, (2) the particle flux, (3) the total outflow rate and its temporal variations, (4) an example of a 3-D O^+ ion velocity distribution function, (5) the direction of the velocity of escaping ions compared with the undisturbed velocity of the solar wind, (6) acceleration of the planetary ions versus the distance from the centre of the tail and, finally, (7) IMA-type of time energy spectrograms generated at the three planes along and perpendicular to the interplanetary magnetic field, IMF.

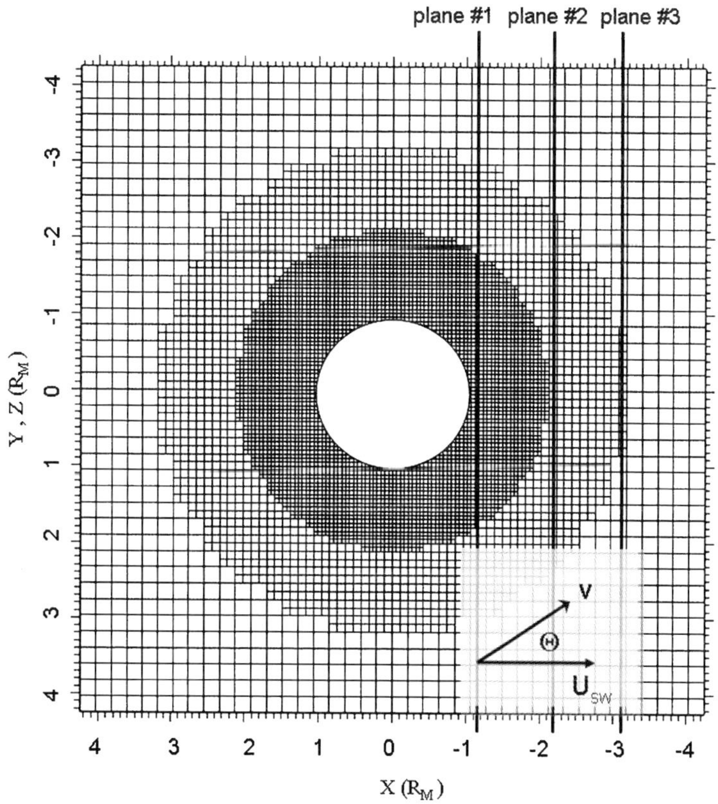

Figure 1. The structure of the grid on the XY and XZ planes used in the simulation in this paper. The velocity and position of O^+ and O_2^+ ions were recorded when they hit the $x = -1.1R_M$, $-2.1R_M$ and $3.1R_M$ planes, labelled in Figure 1 as plane 1, 2 and 3, respectively. The recorded ions were divided in four direction groups depending on the angle Θ between the direction of the velocity of the ion ($\mathbf{v}_i/|\mathbf{v}_i|$) and the direction of the undisturbed solar wind ($\mathbf{U}_{sw}/|\mathbf{U}_{sw}|$): $0° < \Theta < 22.5°$ (direction. No. 1), $22.5° < \Theta < 45°$ (direction No. 2), $45° < \Theta < 67.5°$ (direction. No. 3) and $67.5° < \Theta < 90°$ (direction. No. 4).

2. QNH Model

The description of the 3-D QNH model can be found elsewhere (see Kallio and Janhunen, 2002; Kallio *et al.*, 2006) and here we briefly list some basic features of the model that are of special importance for the present study.

The size of the simulation box is $-4.2R_M < x, y, z < 4.2R_M$ ($R_M = 3393$ km, the radius of Mars) in Mars-centered Solar Orbital (MSO) coordinates (Figure 1). The simulation box contains three cell sizes, 720 km ($\sim 0.2R_M$), 360 km ($\sim 0.1R_M$) and 180 km ($\sim 0.05R_M$). The grid is refined in a spherically symmetric pattern both in dayside and nightside. The inner obstacle in the simulation box is a sphere of radius $r = r_{\text{obstacle}} = 3600$ km that approximates the position of the exobase above which ion-neutral collisions are not frequent. An ion is removed from the situation if it moves inside the obstacle or out of the simulation box. The total number of ions in the simulation is ~11.2 million, the average number of particles per a cell is 30, and the time step, dt, is 0.02 s. The solar wind density(n_{sw}), velocity ($\mathbf{U}_{sw}$) and the interplanetary magnetic field ($\mathbf{B}_{sw}$) are $n_{sw} = n(H^+) = 3$ cm^{-3}, $\mathbf{U}_{sw} = \mathbf{U}(H^+) = [-450, 0, 0]$ km s^{-1}, and $\mathbf{B}_{sw} = [\cos(55°), -\sin(55°), 0]$ $1.12\,\text{nT} = [0.64, -0.92, 0]$ nT. These parameters were adopted in order to use the same upstream parameters than in our previous study (Kallio *et al.*, 2006). The solar wind contains also 4% of He^{++} ions ($n(He^{++}) = 0.04 \times n(H^+)$) which have the same velocity as the solar wind H^+ ions. The model does not contain a Martian crustal magnetic field. The simulation has to run up to $t \sim 300$ s in order to reach a (quasi) stationary state. In the present study the simulation was continued up to $t = 800$ s in order to collect enough ions to the x = const. planes and to study temporal variations.

The model uses so called ion splitting and joining technique. In the analyzed run one ion is split to two if the number of ions becomes much smaller than 30 and three ions are joined to two if the total number of ions within a cell becomes much higher than 30. These methods were chosen to constrain the number of ions in the computer memory and to keep the simulation computationally feasible. These techniques conserve the total energy and the total momentum of ions but the disadvantage of the joining method is that it artificially changes the velocity distribution function. Therefore, one should avoid making strong conclusions from every detail that can be seen in the 3D ion velocity distribution function. We are working to remove this artifact from our forthcoming model versions.

The velocity and energy of the escaping planetary ions are studied in detail by introducing three planes with x = const. into the simulation box and recording the position and the velocity of O^+ and O_2^+ ions when the escaping planetary ions, that is, $v_x < 0$ ions, cross these planes. The planes represent therefore imaginary detectors that collect escaping ions from the 2π angular space. It is also worth noting that, in this study escaping ions are collected by three planes, that is, they are not ions collected inside to a specific volume dV. As a consequence of the ion collection procedure, the number of collected ions which have high $|v_x|$ is larger than the number of collected ions which have small $|v_x|$ although the number of low velocity ions within dV would be same as the velocity of high velocity ions. Also only the ions which have $v_x < 0$ were recorded (escaping ions). Therefore in this study the point where the ion 3-D velocity function was analyzed was far in the tail ($x = -3.1R_M$) where most ions move antisunward ($v_x < 0$).

The ions are collected during the time interval 400 s $< t <$ 734 s from the beginning of the simulation, or until the number of collected ions at a plane exceeded 10^5. The collected ions were divided in four groups depending on the angle, Θ, between the direction of the ion velocity($\mathbf{v}_i$) and the direction of the undisturbed solar wind ($\mathbf{U}_{sw}$): $\Theta = \arccos[(\mathbf{v}_i/|\mathbf{v}_i|)\cdot(\mathbf{U}_{sw}/|\mathbf{U}_{sw}|)]$, (see Figure 1). Therefore, $\Theta = 0°(90°)$ corresponds to ions that move exactly parallel (perpendicular) to the undisturbed solar wind.

Ions are treated as particles accelerated by the Lorentz force. The model contains four ion species: H^+, He^{++}, O^+ and O_2^+. These ions have three sources. The protons are mainly solar wind protons launched into the simulation box at the front face $x = 4.2R_M$. Protons are also generated from the hydrogen corona with scale height 2.61×10^4 km (from Barabash *et al.*, 2002, Table I), but with a low total ion photoionization production rate 1.8×10^{24} s^{-1}. H^+ ions from the charge exchange and electron impact ionization are not included in the simulation. In this study we do not distinguish solar wind protons from protons originating from the neutral hydrogen corona. Alpha particles are only launched into the simulation box from the face at $x = 4.2R_M$. The velocity distribution of the solar wind H^+ and He^{++} ions was assumed to be Maxwellian.

Atomic oxygen ions are produced by two sources: (1) the neutral exosphere with a probability that depends linearly on the neutral density and (2) the obstacle boundary that models the exopause. The neutral scale height is 1.78×10^4 km (from Barabash *et al.*, 2002, Table I). The molecular oxygen ions are emitted from the obstacle boundary without a neutral exosphere. The total ion production rates from the neutral corona, q_{corona}, and the total ion emission rate from the obstacle boundary, q_{iono}, were as in our previous study (Kallio *et al.*, 2006): $q_{\text{corona}}(O^+) = 2.7 \times 10^{23}$ s^{-1}, $q_{\text{iono}}(O^+) = 1.4 \times 10^{25}$ s^{-1} and $q_{\text{iono}}(O_2^+) = 2 \times 10^{25}$ s^{-1}. The used probability function, f_{exobase}, to generate the position of a planetary O^+ or O_2^+ ion on the surface of the obstacle had a solar zenith (SZA) dependence on the dayside but not on the nightside: $f_{\text{exobase}} \sim \cos(\text{SZA}) + 0.1$ on the dayside (i.e. at SZA $< 90°$) and 0.1 on the night side (i.e. at SZA $> 90°$). The velocity distribution function of the planetary ions formed in the neutral corona and emitted from the exobase was Maxwellian. The temperature of ions from the neutral corona and the ions from the exobase was chosen to be 6.5×10^3 K and 10^5 K, respectively. The average total ion loss rates from the simulation box are 1.38×10^{25} O^+ ions s^{-1} and 1.38×10^{24} O_2^+ ions s^{-1}, implying that almost all of the O_2^+ ions emitted from the obstacle boundary return back to the obstacle.

3. Results

In this section we present an overview of the properties of plasma and magnetic field based on the analysed run, the properties of the plasma on the three analysed planes and simulated energy spectrograms.

3.1. Overview of the Run

Figure 2 shows the density of O^+ and O_2^+ ions on the XY-plane ($z = 0$) and on the XZ-plane ($y = 0$) at the time $t = 675$ s. The highest densities are near the planet and an ionotail is formed behind the planet. A notable feature on the XY plane is

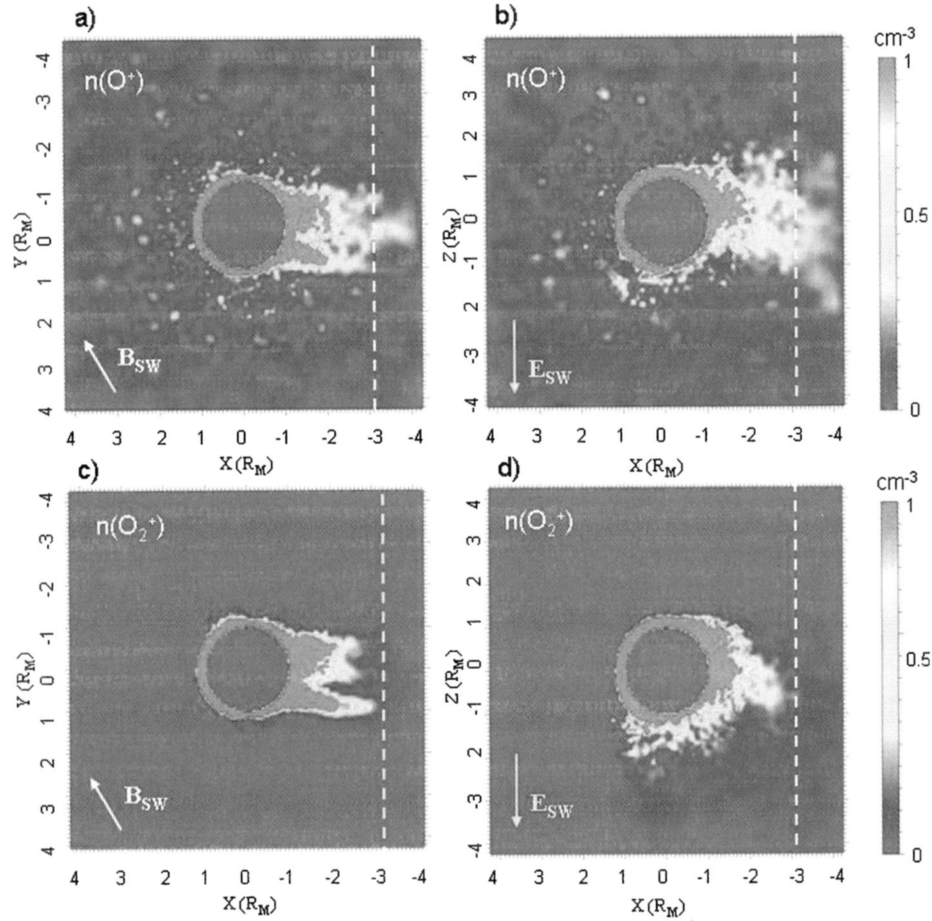

Figure 2. The density of O^+ and O_2^+ ions on the XY and XZ planes in the analysed run. Notice that the IMF is in the XY plane (see $\mathbf{B}_{sw}$ vector in a and b) and that the convective electric field $\mathbf{E}_{sw}(= -\mathbf{U}_{sw} \times \mathbf{B}_{sw})$ is in the XZ-plane. The XY plane is, therefore, a cut through the magnetic tail lobes while the XZ-plane is near the cross tail current sheet. The white dashed vertical lines at $x = 3.1 R_M$ show the lines along which the properties of the escaping planetary ions are analyzed later in this paper in Figures 10–12.

the formation of "tail-ray type" or filamentary structures behind the planet. Furthermore, only a slight dawn-dusk asymmetry, i.e. asymmetry between the $y > 0$ and $y < 0$ hemispheres, can be seen. The dawn-dusk asymmetry is caused by the non-zero IMF x component. In contrast, on the XZ plane there is a clear asymmetry between the $z < 0$ hemisphere and the $z > 0$ hemisphere. The former (latter) hemisphere is the one where the convective electric field in the undisturbed solar wind points away from (towards) the planet, respectively. In this paper as in our previous studies these hemispheres are referred to as the $+E_{sw}$ and $-E_{sw}$ hemisphere. The convective electric field $\mathbf{E}$ $(= -\mathbf{U}_e \times \mathbf{B}$, where $\mathbf{U}_e$ is the electron bulk velocity and **B** is the magnetic field) accelerates planetary ions away from the planet near the surface in the $+E_{sw}$ hemisphere, resulting in "erosion" of planetary ions in that hemisphere and non-zero planetary ion densities relatively far away from the planet in the $+E_{sw}$ hemisphere. Note that the localised O^+ ion enhancements in the solar wind seen in Figure 2a and b are ions originating from the oxygen corona and that the number of O^+ ions in the simulation is not optimised to resolve accurately the properties of that ion population overall in the simulation box.

Figure 3 illustrates how the density of O^+ ions shown in Figure 2 is associated with the morphology of the magnetic field. In Figure 3a–b the grey colour shows the density of O^+ ions on the $y = 0$ and $z = 0$ planes and the red vectors show the direction of the magnetic field ($\mathbf{B}/|\mathbf{B}|$). Figure 3a provides a 3D view of the direction of the magnetic field on the two planes while Figure 3b gives a view along the z axis and Figure 3c along the y axis. Note that the B-vectors are out of the planes. Also note the formation of the magnetic tail lobes in the nightside with the magnetic field pointing away from (towards) the sun at $y < 0$ ($y > 0$) hemisphere (Figure 3b). Furthermore, the magnetic field is "piled up" against Mars, it being tangential to Mars near the planet. The XY plane is therefore a cut through the magnetic tail lobes (Figure 3b) while the XZ plane is a cut near the cross tail current sheet (Figure 3c). Furthermore, Figure 3b and c show the density of O^+ ions in the magnetotail where the magnetic field is highly draped and different than in the solar wind.

3.2. Values at the x = Constant Planes

3.2.1. *Spatial Distribution, Temporal Variations and the Particle Flux*

The position of 5000 O^+ and O_2^+ ions that hit the three analysis planes are given in Figure 4. Note that on the $x = 1.1R_M$ plane the hits of O^+ ions are clustered around $\sim[-1.1, 0, 1]R_M$ and $\sim[-1.1, 0, -1]R_M$, that is, near the so called magnetic "poles" at $[0, 0, \pm 1]R_M$. The spatial distribution is axially asymmetric in all three planes but the asymmetry is relatively small at $x = -3.1R_M$. However, a clear asymmetry with respect to the direction of $\mathbf{E}_{sw}$ can be seen at all three planes in the O_2^+ ions. One reason for the clearer asymmetry in Figure 4a than in Figure 4b can be the fact that the plotted 5000 O^+ ions are originating both from the oxygen neutral corona and from the ionosphere while the 5000 O_2^+ ions are coming only from the ionosphere. Another reason for the differences between the atomic and molecular oxygen ions is the different mass of ions and, consequently, different ion gyroradius. In fact, that is the only difference between O^+ and O_2^+ ions originating from the exobase because these two ion species are emitted in the same way from the model exobase (see Section 2). Note also that escaping O_2^+ ions are concentrated close to the $y = 0$ plane, i.e., near the cross tail current sheet.

The QNH model contains a finite number of ions and the plasma and field parameters never fully reach stationary values. The non-stationary nature can be

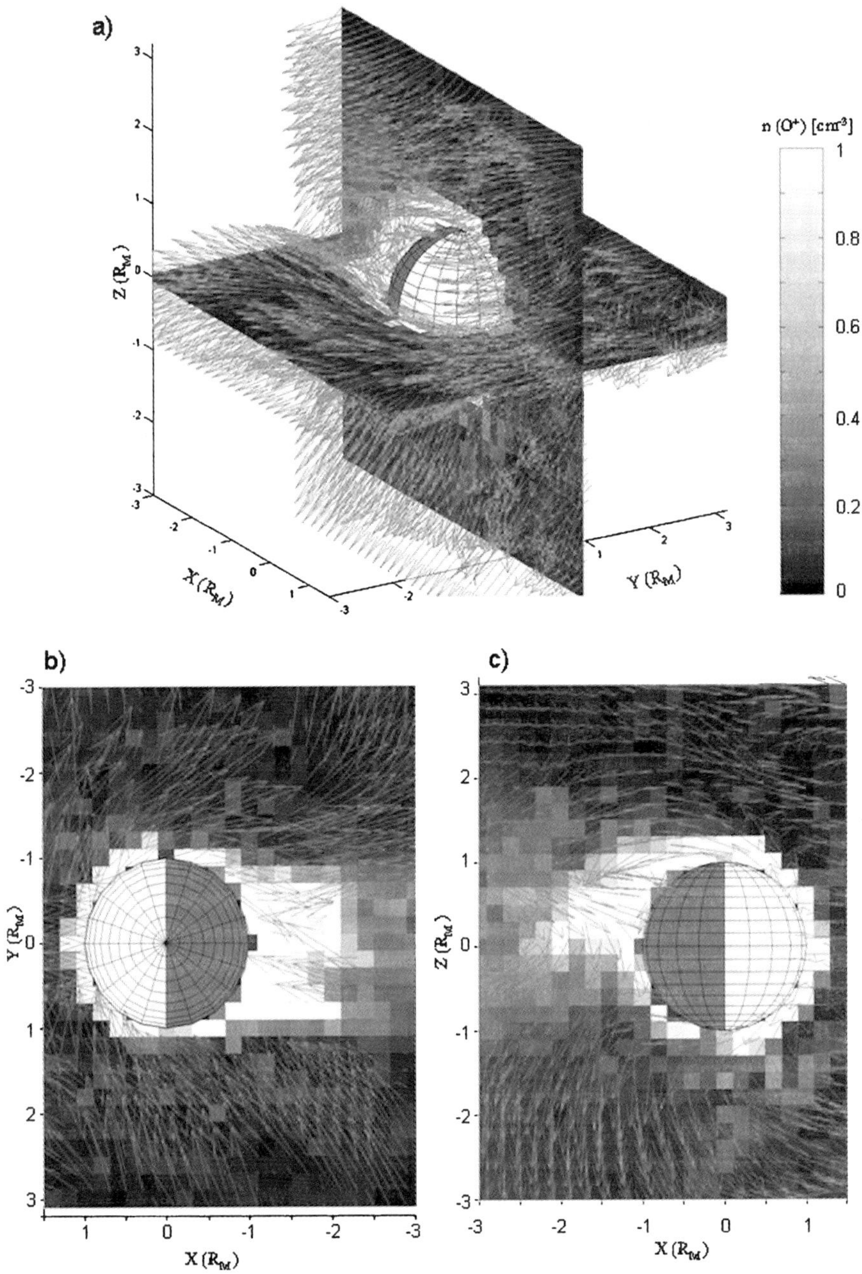

Figure 3. The direction of the magnetic field vectors (red arrows) on the XZ and XY planes. The density of the O^+ ions is shown on these planes by a grey scale for comparison. The direction of the magnetic field and the density of O^+ ions are calculated by interpolating their values from the original grid (see Figure 1) to the $dx = dy = dz = 0.2R_M$ grid. The three-dimensional isometric view is shown in (a) and detailed two-dimension presentations of the XY and YZ planes are shown in (b) and (c), respectively.

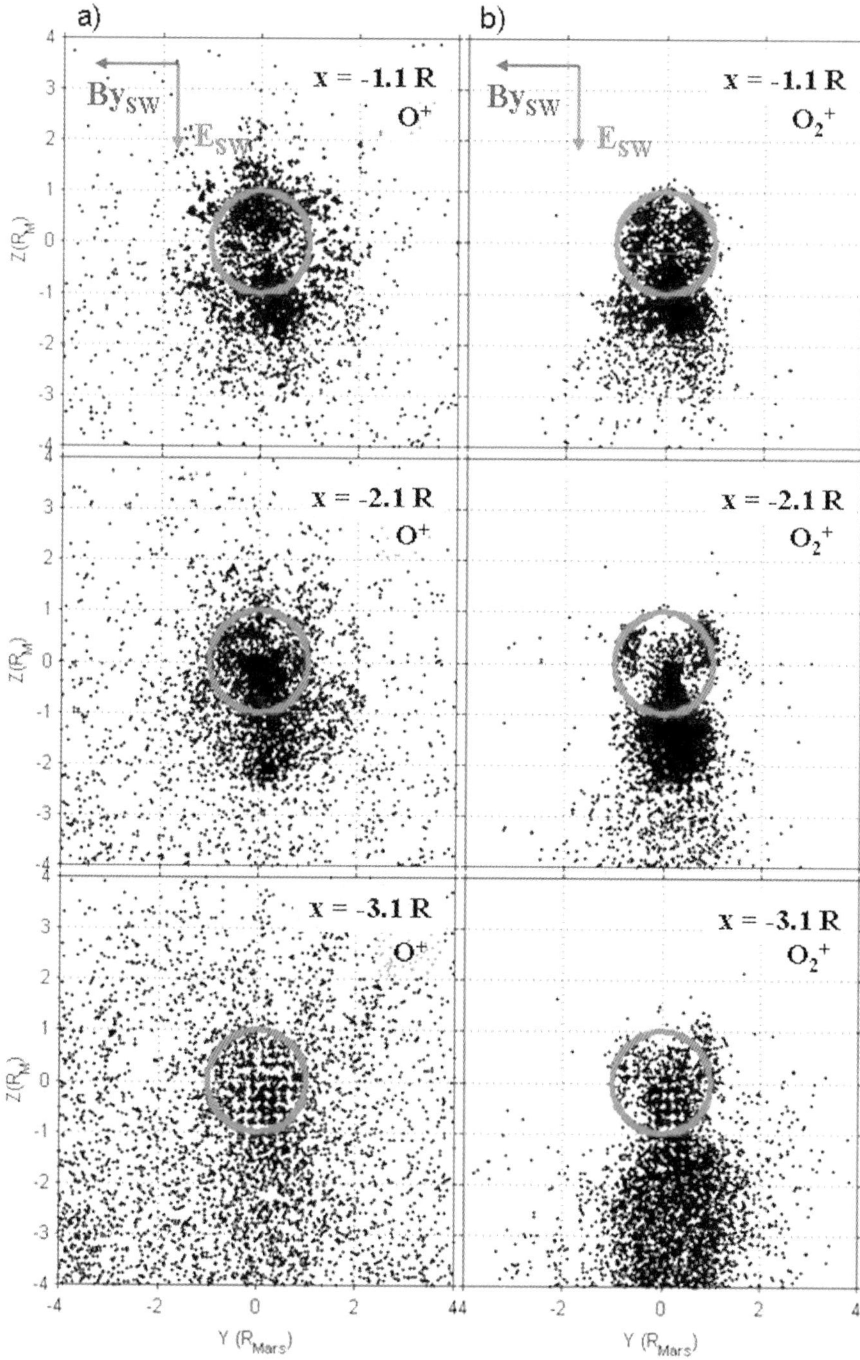

Figure 4. The position of 5000 O^+ ions that had $v_x < 0$ and that hit (column a) the $x = -1.1R_M$, $x = -2.1R_M$, and $x = -3.1R_M$ planes, and 5000 O_2^+ ions that hit (column b) the $x = -1.1R_M$, $x = -2.1R_M$, and $x = -3.1R_M$ planes.

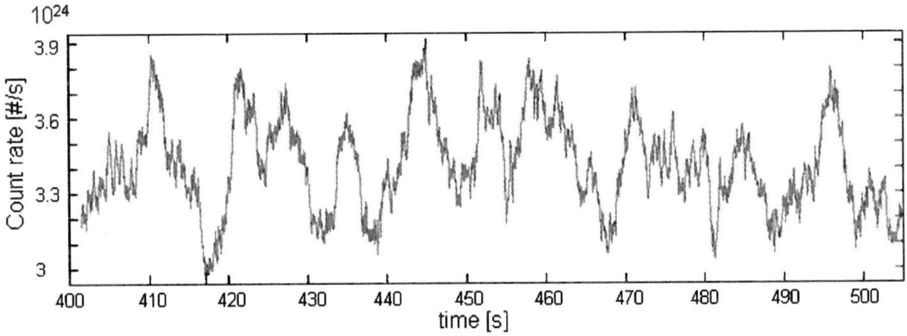

Figure 5. The average outflow rate of O^+ ions through the $x = -3.1R_M$ plane 3 during 400 s $< t <$ 505 s. The count rates (s^{-1}) at a given time are derived by calculating 150 point running mean of the instantaneous ($dt = 0.02$ s) outflow rate.

seen in Figure 5 that shows the average outflow rate of O^+ ions through plane 3 during the time interval 400 s $< t <$ 505 s. The average outflow rate through plane 3 is $\sim 3.4 \times 10^{24}$ s^{-1}, but fluctuations over 10% can be seen in the 150 time step running mean values. One reason for the fluctuations is the finite number of ions that cross the planes at each individual time step dt (=0.02 s). On the other hand, as noted in our previous study (Kallio *et al.*, 2006), the model can result in density enhancements or plasma "clouds" that are generated near the surface of Mars and thereafter move tailward. For example, in Figure 5 a periodicity of about 17 s ($\sim$0.06 Hz) can be found by Fourier analysis, which can also be identified by visual inspection. For a comparison, the gyroperiod of H^+, He^{++}, O^+ and O_2^+ ions in the solar wind (IMF = 1.12 nT) is 60 s, 120 s, 940 s and 1870 s, respectively. In fact, H^+, He^{++}, O^+ ions and O_2^+ ions had to be in the magnetic field of 4 nT, 8 nT, 62 nT and 123 nT, respectively, in order to have the gyroperid of $\sim$17 s. In the simulation several tens of nT magnetic field can be found only near Mars while a few nT magnetic field can be obtained at $x \sim -3.1R_M$ (see, Kallio *et al.*, 2006, Figure 7). This suggests that if the fluctuation is associated with the gyromotion of O^+ or O_2^+ ions, the fluctuations may be originating near Mars while H^+ and He^{++} ions may generate $\sim$17 s fluctuations far in the tail. Figure 5 illustrates that "snapshot" values based on the QNH model run, i.e. values at a given time t, can vary from one time to another.

Figure 6 shows the average particle flux, j_x [s^{-1} cm^{-2}], of the O^+ and O_2^+ ions at the three planes. The flux is derived by using ions that hit the planes during the time period 400 s $< t <$ 734 s. Note that the maximum particle flux in both ion species is found at all planes within the optical shadow or close to the limb. Moreover, the maximum flux is found near the $y = 0$ line, that is, near the cross tail current sheet. In both ion species a $+E_{sw}$ hemisphere/$-E_{sw}$ hemisphere asymmetry exists, j_x being higher on the $+E_{sw}$ hemisphere ($z < 0$) than on the opposite hemisphere. Furthermore, there is a slight dawn/dusk asymmetry caused by the non-zero IMF x-component.

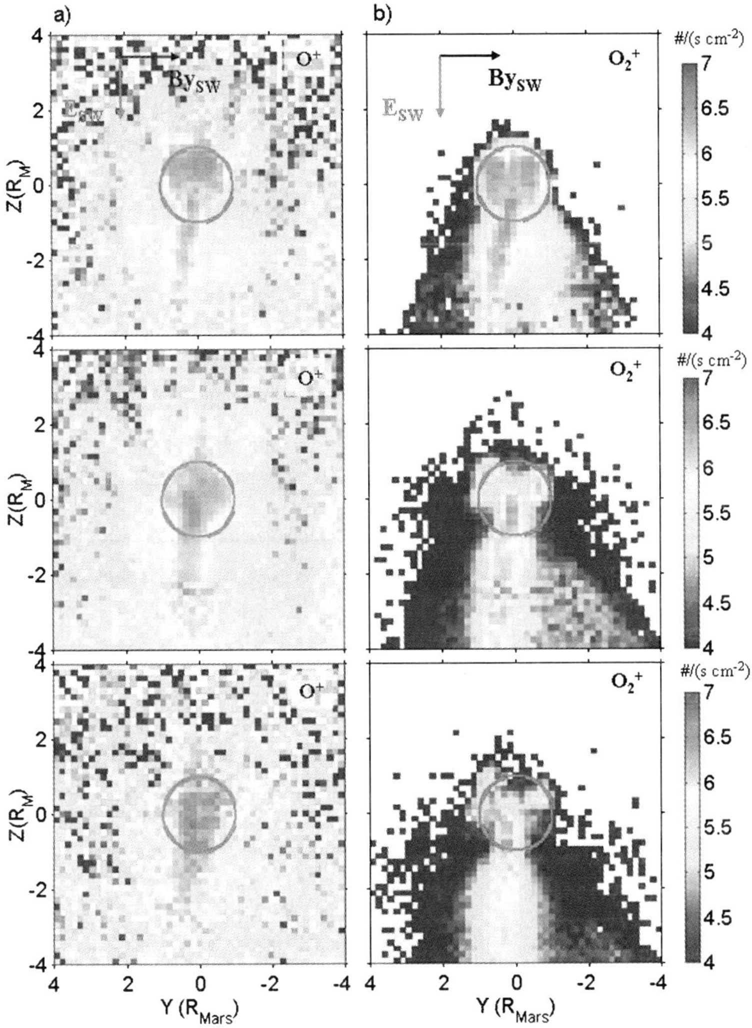

Figure 6. The particle flux of escaping (i.e. $v_x < 0$) along the x-axis, j_x [s^{-1} cm^{-2}], of O^+ ions (column a) and O_2^+ ions (column b) at the three x = const. planes ($x = -1.1R_M$, $-2.1R_M$ and $-3.1R_M$). The highest fluxes are situated near the cross tail current sheet on the $+E_{sw}$ hemisphere ($z < 0$). The red vectors in the top panels show the direction of the convective electric field in the solar wind, $\mathbf{E}_{sw}$, and the direction of the IMF on the YZ plane, $\mathbf{B}_{y\,sw}$.

3.2.2. *Velocity Vectors and Energisation*

In this section we study in detail the velocity distribution of the escaping ions as well as their energy. Figure 7 gives the angle Θ, that is, the angle between the solar wind direction and the direction of O^+ and O_2^+ at the three planes. As already shown in Figure 1, $\Theta = 0°$ corresponds ions that move exactly parallel to the undisturbed solar wind and $\Theta = 90°$ ions that move exactly perpendicular to the undisturbed solar wind U_{sw}. Figure 7 illustrates that predominantly antisunward

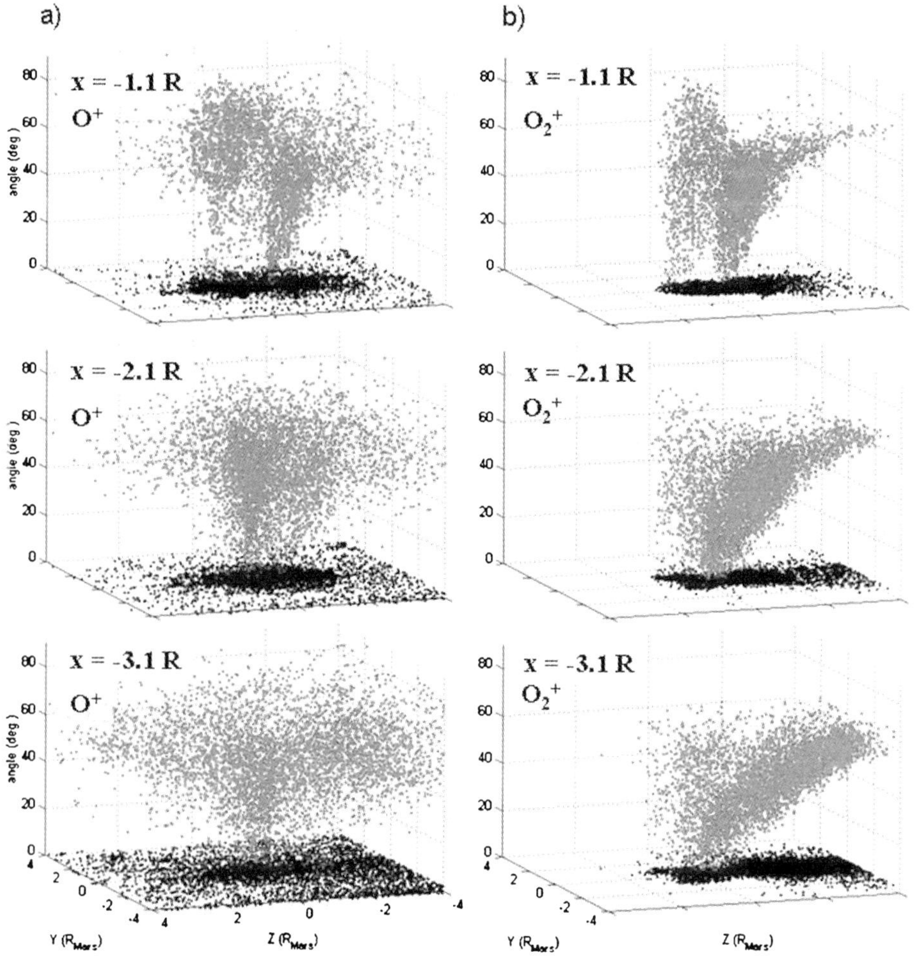

Figure 7. The angle between the O^+ and O_2^+ ions and the $\mathbf{U}_{sw}$ at the three planes for $v_x < 0$ ions. The red points are (y, z, Θ) values where y and z are the values of the y-axis and the z-axis, and Θ , and $\Theta = \arccos[(\mathbf{v}_i/|\mathbf{v}_i|)\cdot(\mathbf{U}_{sw}/|\mathbf{U}_{sw}|)]$ when an ion passes through the plane. The black dots show the points $[y, z, 0]$, i.e. the position of the hits on the YZ-plane. Column a shows plots for the three planes with O^+ and (b) shows plots for the three planes with O_2^+. In all six plots the horizontal axis are Z and Y and the vertical axis is Θ. Note that ions near the horizontal $\Theta = 0°$ plane are moving almost to the same direction than the undisturbed solar wind. Furthermore, the optical shadow is the region within $(y^2 + z^2)^{-1/2} < 1R_M$ and the x axis crosses the three planes at $[Y, Z] = [0, 0]$.

($\Theta \sim 0°$) moving ions can be found within or close to the optical shadow. At the $x = -1.1R_M$ plane, the ion velocities can differ notably from the antisunward direction, i.e., $\Theta \gg 0°$, near the optical shadow corresponding to the convergence of the planetary ions into the tail. Note that the velocity of O_2^+ ions becomes more and more perpendicular (Θ increases) the further the ion resides from the x-axis. Recall that an additional source of O^+ ions is the neutral oxygen corona which

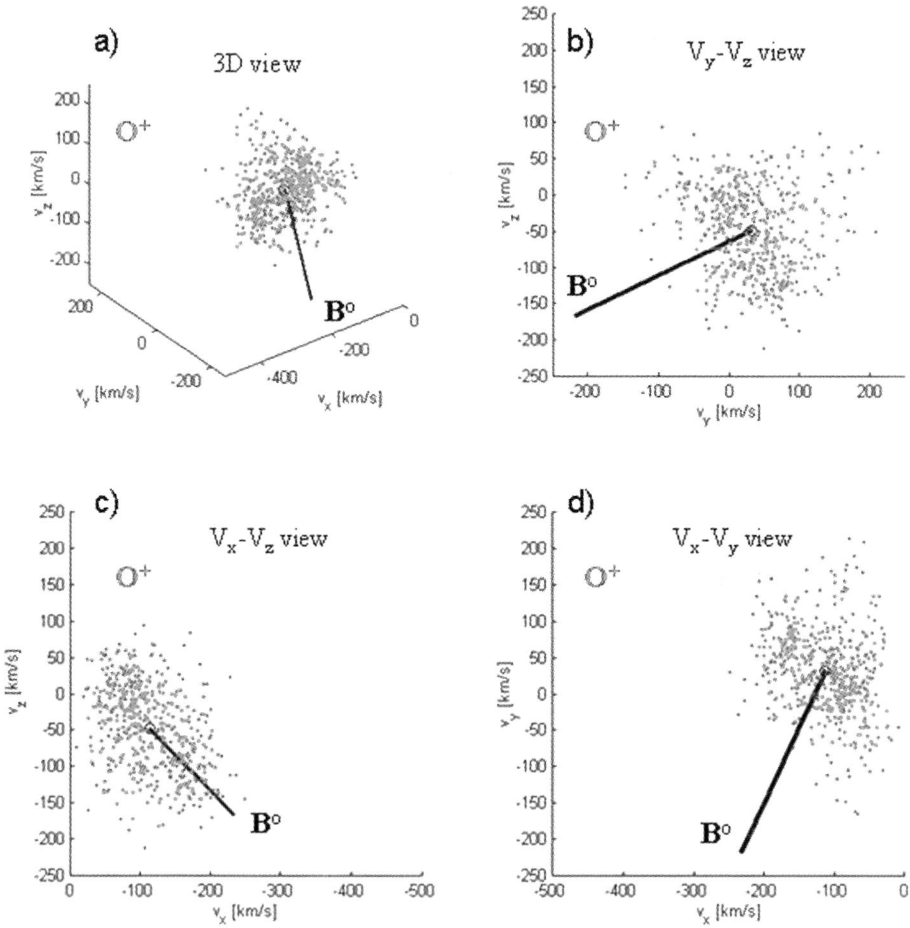

Figure 8. The velocity $\mathbf{v} = [v_x, v_y, v_z]$ km s^{-1} of escaping (i.e. $v_x < 0$) O^+ ions around the point $[x, y, z] = [-2.1, 0, 0]R_M$ is viewed from four directions. The black solid line shows the direction of the magnetic field, $\mathbf{B}^\circ(= \mathbf{B}_{sw}/|\mathbf{B}_{sw}|)$. This O^+ velocity is shown in three dimensions(a) and the dimensional projections in the V_y–V_z plane(b), V_x–V_z plane(c), and in the V_x–V_y plane(d).

causes larger spread in the Θ angle distribution than for O_2^+ ions (which originate only from the ionosphere).

Figure 8 shows in detail the velocity of the O^+ ions that hit plane 2 ($x = -2.1R_M$) near $[y, z] = [0, 0]$, i.e. around the x-axis at the centre of the tail. At this point the magnetic field is $\mathbf{B} = [B_x, B_y, B_z] = [-1.4, -3.0, -1.4]$ nT and the bulk velocity $\mathbf{U}(O^+) = [-77, 17, -40]$ km s^{-1}. At this point the O^+ ions flow away from Mars (antisunward) but they also posses a notable velocity component in the direction of the $\mathbf{E}_{sw}$. The direction of $\mathbf{B}$ at the point is also indicated in Figure 8, but no clear organisation of the O^+ velocities with respect to the direction of $\mathbf{B}$ can be seen. Note that every dot corresponds to an ion passing through the plane and that the distribution of the velocity points in a plot like in Figure 8 would not

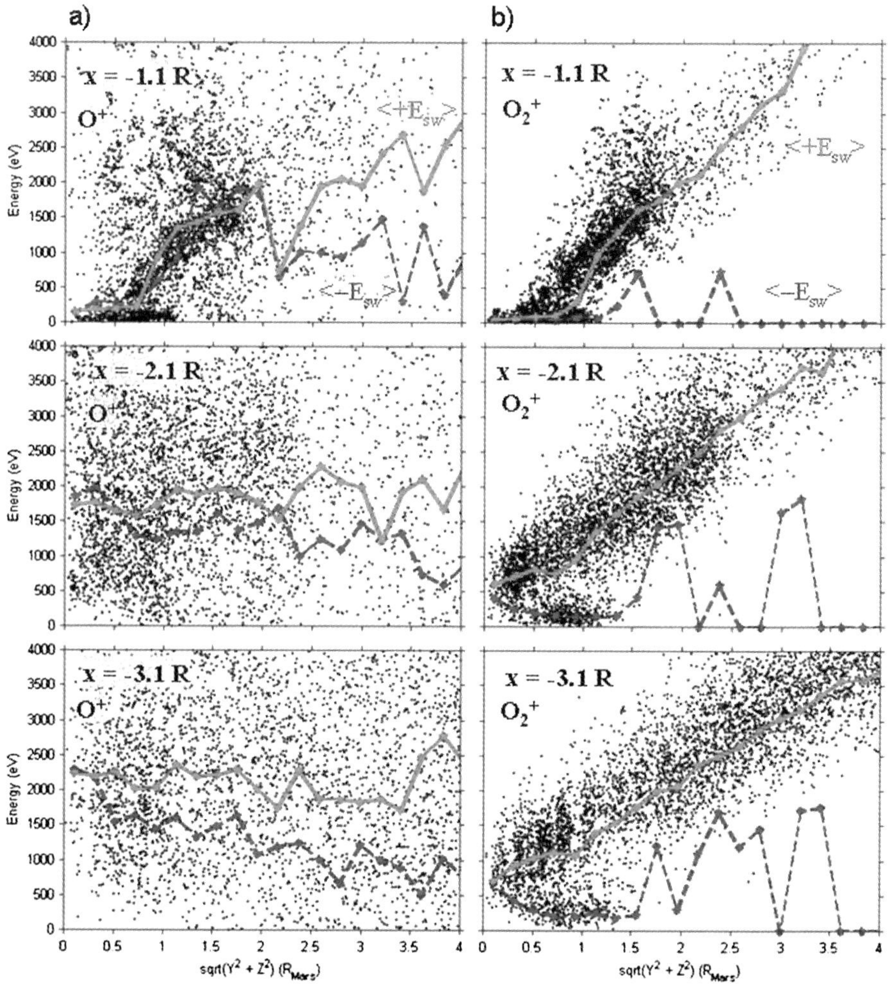

Figure 9. The energy of escaping $v_x < 0$ (a) O^+ ions and (b) O_2^+ ions at the three planes versus the distance from the x-axis. Every black dot corresponds to an ion passing through the plane. The red(blue) line is the average energy $\langle +E_{sw} \rangle$ ($\langle -E_{sw} \rangle$) of ions based on ions collected in the $+E_{sw}$ ($-E_{sw}$) hemisphere. Note the linear increase of the energy of O_2^+ ions in the $+E_{sw}$ hemisphere shown at all three planes.

be Maxwellian even if the velocity distribution functions were Maxwellian (see Section 2).

The energy of the ions versus the distance from the x-axis is given in Figure 9. The red solid lines show the average energy of ions where only ions from the $+E_{sw}$ hemisphere, that is, at $z < 0$, are taken into account. The average energy of ions located on the opposite $-E_{sw}$ hemisphere at $z > 0$ are shown by blue dashed lines. In all panels the average energy is higher on the $+E_{sw}$ side than on the $-E_{sw}$ side. Also, the energy of O_2^+ ions increases almost linearly with the increasing distance from the x-axis. If one approximates the increase by a straight line that passes through the points $[(y^2 + z^2)^{-1/2}$ (in R_M), E (in keV)] = [0.5, 0] and [3.5, 4] at plane 1, the points [0, 0] and [3.5, 4] at plane 2 and the points [0, 0.5] and [4, 4] at plane 3, the value of $d\langle E_{\rm kin}\rangle/(e \times d\rho)(\rho = (y^2+z^2)^{-1/2}$, e is the unit charge, $\langle E_{\rm kin}\rangle$ is the average kinetic energy) is 0.39 mV m^{-1}, 0.34 mV m^{-1} and 0.26 mV m^{-1}, respectively. Such an increase of the kinetic energy would be obtained if the ions were accelerated away from the x axis by the electric field $E_\rho = d\langle E_{\rm kin}\rangle/(e \times d\rho)$. In the model the electric field in the magnetotail is a non-axially symmetric 3D vector but it is worth noting that the values of E_ρ are close to the value of $E_{\rm sw}$ (= 0.41 mV m^{-1} = 450 km s^{-1} × 1.12 sin(55°) nT). The decrease of E_ρ from a plane to another results from the fact that the energy of the ions within the centre of the tail have increased more rapidly than the energy of the ions far away from the x axis. It is finally worth noting that there are two O_2^+ ion populations near the optical shadow at $\rho = 1R_M$: fast ($E \sim$ keV) ions in the $+E_{\rm sw}$ hemisphere and low ($E \sim$ few hundred eV) in the $-E_{\rm sw}$ hemisphere.

3.2.3. *Simulated Energy Spectrograms*

In this section we study in detail the energy of the escaping ions in the magnetic lobes and near the cross tail current sheet by generating IMA-type energy spectrograms. Figure 10 gives the density of O^+ and O_2^+ ions, the magnetic field, and the simulated energy spectrum on the $x = -3.1R_M$ plane along the y-axis.

The ion density and the magnetic field in Figure 10a and c are derived at $t = 675$ s. The energy spectrum in Figure 10b and d gives the particle flux j_x (# s^{-1} sr^{-1} cm^{-2}) calculated by collecting ions on plane 3 that have $|z| < 0.2R_M$ and dividing the number of hits by the energy intervals $[E_i, E_{(i+1)}]$ where $E_i = 3\text{eV} \times (1 + 0.08)^i$ ($i = 0, 1, \ldots, 95$). These energy intervals were chosen to mimic the energy steps used in IMA measurements that have an energy resolution of $dE/E \sim 0.08$ (see Barabash *et al.*, 2004, for the details of the IMA instrument). The energy spectra were also calculated separately in the four directions shown in Figure 1: direction No. 1 ($0° < \Theta < 22.5°$), direction No. 2 ($22.5° < \Theta < 45°$), direction No. 3 ($45° < \Theta < 67.5°$) and direction No. 4 ($67.5° < \Theta < 90°$). These Θ intervals were chosen because IMA has a field of view (FOV) of 4.5° × 22.5°. Note that the simulated energy spectra do not represent any individual IMA measurement because the simulated FOV is not identical to the FOV of the IMA instrument, and also because the direction of FOV of IMA depends on the MEX orientation. In the simulation, the directions of the FOVs are fixed in the MSO frame. Moreover, MEX never crosses the tail along the y or z-axis, resulting in more complicated energy spectra than those presented here.

The magnetic field in Figure 10a and c shows that when an imaginary spacecraft moves along the y-axis, the maximum magnetic field associated with the magnetic tail lobes is observed at $|y| \sim 1R_M$, and that the cross tail current sheet is crossed near $y \sim 0$. There is a slight asymmetry resulting from the positive IMF x-component that has boosted the magnetic field in the magnetic tail lobe where

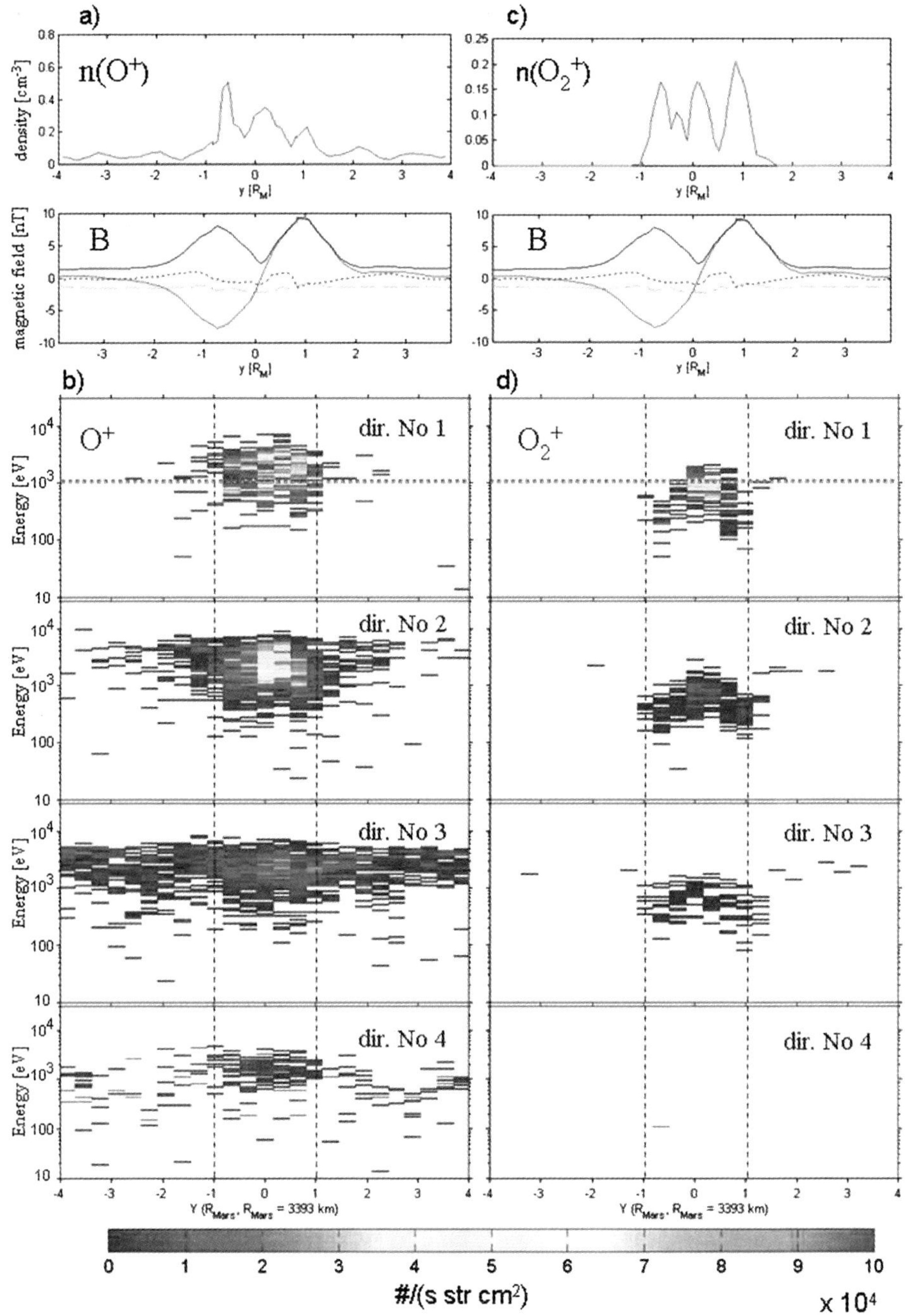
a)
n(O+)
density [cm-3]
y [RM]
B
magnetic field [nT]
c)
n(O2+)
b)
O+
dir. No 1
dir. No 2
dir. No 3
dir. No 4
Energy [eV]
Y (RMars, RMars = 3393 km)
d)
O2+
#/(s str cm2)
x 104

B_x is positive, compared with the tail lobe where B_x is negative. More details of the properties of the magnetic field can be found in our previous work (Kallio *et al.*, 2006) when the macroscopic parameters for the same case than presented in this work was analyzed. Also note that there are three local maxima in $n(O^+)$ and $n(O_2^+)$, one maximum being associated with ions in the cross tail current sheet, the two other maxima being associated with the escaping ions within the magnetic tail lobes. The properties of $n(O^+)$ on the XY plane, as well as the position of the analysis line, can be seen in Figure 2a and c.

As seen in Figure 10b and d the highest particle fluxes are located within the optical shadow near the cross tail current sheet. The energy dispersion of the fluxes show that the average energy within the optical shadow is higher at the centre of the optical shadow ("inverted U shape"). One possible reason for differences between O^+ and O_2^+ energy spectra is the fact that O^+ ions are originating both from the neutral corona and from the exobase while O_2^+ ions are originating from the exobase only. The other possible source for the differences can be associated with the different mass of the atomic and molecular oxygen ions. The properties of $n(O_2^+)$ on the XY plane, as well as the position of the analyses line, can be seen in Figure 2b and d.

Figure 11 presents similar simulated parameters as these given in Figure 10, but now they are calculated along the z-axis. In this case, the tail is crossed near the cross tail current sheet and the magnetic field is weaker than in the previous case when an imaginary spacecraft crossed the magnetic tail lobes (Figure 11a and c, bottom panels). The increase of the average energy with increasing distance from the x-axis on the $+E_{sw}$ hemisphere ($z < 0$) is clearly seen in the O_2^+ energy spectra. The highest O^+ count rates can be found within the optical shadow or near it, as was also the case in Figure 10. It is worth of noting that it is not obvious that such a correlation should exist in the Martian tail because the convective electric field depends on the magnitude and the direction of the magnetic field and the bulk velocity of all ion species. These are fully 3D parameters in the Martian tail, their values depending on how the solar wind is decelerated, accelerated and deviated around Mars.

←

Figure 10. Simulated plasma and field parameters along the y-axis at $x = -3.1R_M$ plane, i.e., values through the magnetic tail lobes and through the cross tail current sheet. The parameters are: The density of O^+ ions (Figure 10a, top panel), the density of O_2^+ ions (Figure 10c, top panel), the magnetic field (Figure 10a and c, bottom panels; the red solid line: B_x, the green dashed line: B_y, the blue dotted line: B_z, black solid line: $|B|$) and energy spectrograms for O^+ ions (Figure 10b) and O_2^+ ions (Figure 10d) calculated for four directions dir. No. 1 (No. 4) looking ions move predominantly parallel (perpendicular) to the direction of the undisturbed solar wind. The optical shadow is the region between the vertical dashed lines. The units of the particle density, the magnetic field, and the energy spectra are cm^{-3}, nT, and (s^{-1} cm^{-2} sr^{-1}). Note the different scales in $n(O^+)$ and $n(O_2^+)$. The horizontal red dotted lines in (b) and (d) at dir. No. 1 show the energy of H^+ ions in the undisturbed solar wind, 1060 eV ($U = 450$ km s^{-1}).

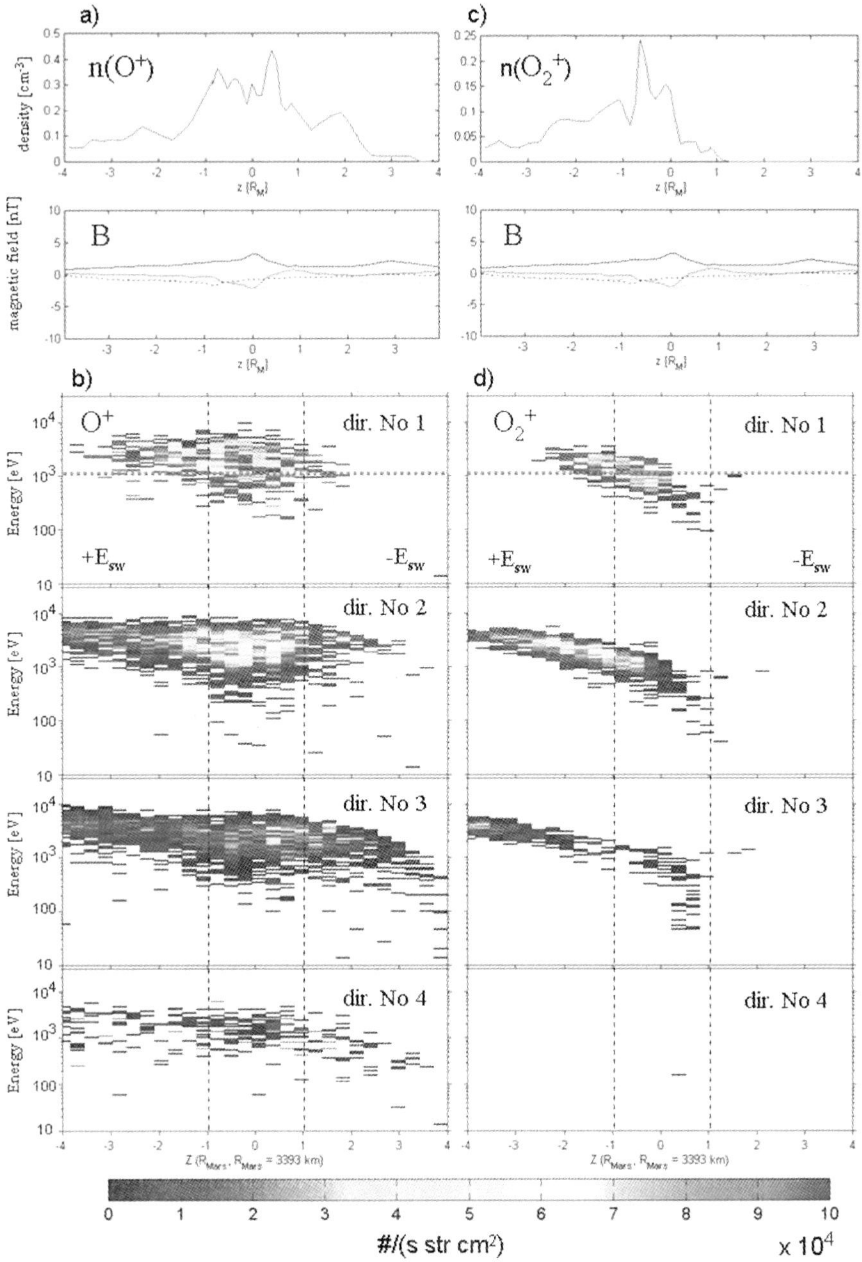

Figure 11. Simulated plasma and field parameters along the z-axis at $x = -3.1R_M$, i.e. when the tail is crossed near the cross tail current sheet. See Figure 10 for the description of the parameters. The $+E_{\rm sw}$ ($-E_{\rm sw}$) text labels are added to show the hemisphere where the direction of the convective electric field in the undisturbed solar wind points away from (toward) Mars. Note the different scales in $n(O^+)$ and $n(O_2^+)$. The horizontal red dotted lines in (b) and (d) at dir. No. 1 show the energy of H^+ ions in the undisturbed solar wind, 1060 eV ($U = 450$ km s^{-1}).

Figure 12 shows the simulated energy spectrograms at the three planes along the z-axis and along the y-axis. The highest fluxes are observed in the antisunward direction (dir. No. 1) near or within the optical shadow, and the highest flux values correspond to a few keV energy. A clear increase of the average energy of O_2^+ ions can be seen in Figure 12a–c when z is decreased, i.e. when one moves toward the $+E_{SW}$ hemisphere (Figure 12a–c; panels on the right hand side). A similar increase, although not as intense as in O_2^+, is observed in the O^+ energy spectra which contains ions both from the ionosphere and from the neutral oxygen corona (Figure 12a–c; panels on the left hand side). It is worth noting that O_2^+ ions are concentrated within the optical shadow or close to the limb when the path is along the y-axis (Figure 12d–f; panels on the right hand side). Notice also that the energy of O^+ and O_2^+ ions increase notably near the optical shadow at plane 1 when moving along the z axis (the current sheet) away from the x axis. In that case the energy of the escaping ions is in a wide energy range from a few tens of eV's to a few kilo eV's (Figure 12a). Finally, the flux of planetary ions seems to be higher near points $[-1.1, 0, \pm 1]R_M$ than points $[-1.1, \pm 1, 0]R_M$. This effect probably results from the $j \times B$ "slingshot" force caused by the magnetic tension at the highly draped magnetic field lines near the so called magnetic poles, that in the analyzed IMF case are at $[0, 0, \pm]R_M$ (see Tanaka, 1993, for more discussions about the $j \times B$ slingshot force). In the analyzed case the $j \times B$ slingshot force can accelerate planetary ions effectively near the XZ plane near the magnetic poles.

4. Discussion

4.1. Similarities with Observations

In this paper several properties of the escaping O^+ and O_2^+ ions in the Martian tail have been analysed. The main motivation for this study is to interpret ASPERA-3/Mars Express ion observations, especially the observations of escaping planetary ions in the tail. In a very detailed IMA data – QNH model comparison the energy spectra should be generated by using the positions of MEX, the attitudes of IMA instrument, and the accurate IMA FOV but this is beyond the present study.

Figures 2–12 represented some parameters based on the analysed run. The following similarities between the simulated values and the observations made by ASPERA-3/Mars Express and ASPERA/Phobos-2 (measurements in 1989) are worth noting.

4.1.1. *About keV Planetary Ions at the Centre of the Tail*

ASPERA/Phobos-2 ion measurements near Mars in 1989 showed that there was a beam of ~keV planetary ions in the centre of the Martian tail (see, for example, Kallio *et al.*, 1995). Also ASPERA-3/MEX has measured ~1 keV planetary ions in the nightside (see, for example, Kallio *et al.*, 2006, Figure 1). The present

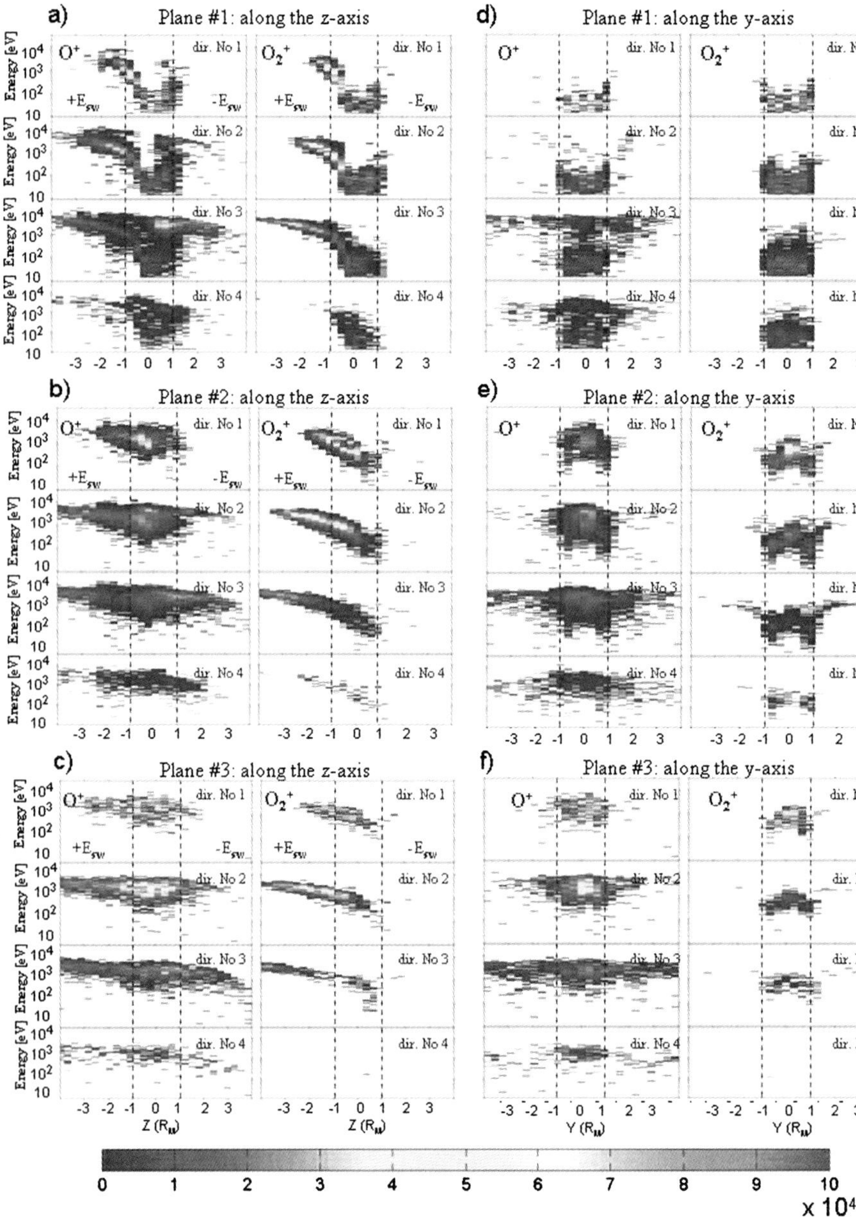

Figure 12. Comparison of the simulated energy spectrograms at the three planes calculated along the z-axis (Figure 12a–c) and along the y-axis (Figure 12d–f). In all figures the energy is in $\log_{10}$ scale from 10 eV to 30 keV. Figure 12c is the same as Figure 11b and d. Figure 12f is the same as Figure 10b and d. The vertical dashed lines indicate the region of the optical shadow. Note that moving along the z-axis corresponds to the crossing the tail near the cross-tail current sheet while moving along the y-axis corresponds to crossing through the magnetic tail lobes. The counts that can be seen in dir. No. 1 (dir. No. 4) are planetary ions moving predominantly parallel (perpendicular) to the direction of the undisturbed solar wind.

study as well as our previous study (Kallio and Janhunen, 2002) showed that our self-consistent QNH model produces a few keV planetary ions to the tail (see Figures 10–12). In addition, the model resulted in energy dispersion and "inverted U" type of energy spectra (see Figure 10b and d) being, at least qualitatively, in agreement with energy spectra observed by ASPERA/Phobos-2 (see, for example, Kallio and Koskinen, 1999, Figure 9b, the bottom panel). In fact planetary ions with energies near 1 keV have also been produced in test particle simulations (Kallio and Koskinen, 1999).

4.1.2. *Energisation of Planetary Ions Outside of the Optical Shadow*

ASPERA-3 ion observations suggest that the energy of the planetary ions can increase with increasing distance from the planet and that there are events when the increase of energy with altitude is linear in the magnetosheath (Dubinin *et al.*, 2006). The observed linear increase of the energy can be obtained if the ions are accelerated by the electric field whose magnitude is about the magnitude of the convective field in the undisturbed solar wind (Dubinin *et al.*, 2006). It is therefore notable that in this study the QNH model was found to result in a linear energy increase of planetary ions with increasing distance from the x-axis (see Figure 9). Furthermore, the magnitude of the derived electric field, E_ρ, was found to be close the value of the $|\mathbf{E}_{\mathrm{sw}}|$ (see Section 3.2.2). It is worth of noting that it is not obvious that such a clear correlation between E_ρ and $|\mathbf{E}_{\mathrm{sw}}|$ should exists in the Martian tail because the convective electric field depends on the magnitude and the direction of the magnetic field, B, and the bulk velocity of all ions species, $U(H^+)$, $U(O^+)$ and $U(O_2^+)$. These B and U fields are fully 3-D in the Martian tail their values depending on how the solar wind was decelerated, accelerated and deviated around Mars.

4.1.3. *Spatial Distribution of the Escaping Ions in the Tail*

A statistical study of the spatial distribution of the escaping planetary ions in the Martian tail suggests that the planetary ions are escaping asymmetrically with respect to the direction of the undisturbed convective electric field, $\mathbf{E}_{\mathrm{sw}}$ (Fedorov *et al.*, 2006). More ions were found to be lost on the $+E_{\mathrm{sw}}$ hemisphere than in the $-E_{\mathrm{sw}}$ hemisphere (Fedorov *et al.*, 2006). Similar type of $+E_{\mathrm{sw}}/-E_{\mathrm{sw}}$ asymmetry can be also seen in the QNH model simulation (see Figure 6).

It should also be noted that the "clustering" of the escaping planetary ions in the tail (see Figure 4, top and middle panels) have some similarities with the earlier test particle simulations that was developed to interpret ASPERA/Phobos-2 observations (Kallio and Koskinen, 1999). Although a comparison of the test particle simulations and the present study is not straightforward because of different input parameters and different collection surfaces for ions (in the test particle simulation ions were collected on the 2.8 R_M sphere and the upstream conditions were different from the present study), in both studies, the escaping ions formed localised groups or clusters in the tail (c.f. Figures 4a, d, e, 7, and 8 and Kallio and Koskinen,

1999, and Figure 4 in this paper) and about ~1 keV escaping planetary ions at the cross tail current sheet.

4.1.4. *Temporal Variations*

The ASPERA-3 Electron Spectrometer (ELS) has observed electron oscillations with frequency peaks which are typically in the range 0.01–0.02 Hz (Winningham *et al.*, 2006). In addition, plasma and magnetic field fluctuations have been observed by the MGS spacecraft (see Espley *et al.*, 2004). It is therefore interesting, that as seen in Figure 5 and as noted in our previous study (Kallio *et al.*, 2006), fluctuations with about 10–20s periods can be found in the properties of the escaping planetary ions in the tail which were produced by the QNH model simulations. These fluctuations might result from statistical fluctuations caused by the finite number of ions in the simulation box, but they may also be a manifestation of instabilities generated near the ionosphere where the flow of the solar wind meets the planetary ions. In fact, an instability has been found to take place in 2-D QNH model simulations made for Venus which resulted in density fluctuations in the $+E_{sw}$ hemisphere (Terada *et al.*, 2002).

4.2. Miscellaneous Remarks

An interesting issue is the question: "Are the escaping planetary ions organised in the tail with respect to the direction of the magnetic field"? That question is also related to the question of the magnetisation of the escaping ions. We have studied this question by calculating the value $T(\Theta_i, \varphi_i) = \text{sum}[w_i(\mathbf{v}_i - \mathbf{U})\cdot\mathbf{n}(\Theta_i, \varphi_i)]^2$, which was developed in this study to provide a rough measure for the spreading of the distribution at different directions. Here $\mathbf{n}$ is the unit vectors whose direction is determined by the polar angles (Θ, φ), $\mathbf{U}$ is the bulk velocity, w_i takes into account the weight of an ion and the sum is over all velocities $\mathbf{v}_i$ that hit the $x = \text{const.}$ planes in a given grid (dy, dz). For example, ions gyrating around the magnetic field would result to $\min(T)$ in the directions where n is along $\pm B$ and to $\max(T)$ in the direction where n is perpendicular to B. We studied whether a correlation exists between the direction of $\mathbf{n}_{\max}$ (that gives $\max(T)$ value) or $\mathbf{n}_{\min}$ (that corresponds $\min(T)$ value) and the direction of $\mathbf{B}$ at the analysed position. In the case shown in Figure 8, no clear correlation between $\mathbf{n}_{\max}$ nor $\mathbf{n}_{\min}$ with the direction of $\mathbf{B}$ could be found. One reason may be that the point $[-2.0, 0, 0]R_M$ used in Figure 8 is at the centre of the cross tail current sheet where the magnetic field is weak. At other locations, where the magnetic field is stronger than in the case shown in Figure 8, $\mathbf{n}_{\max}$ was found to occasionally be close to the direction of $\mathbf{B}$ (figures not shown). However, determination of the direction of $\mathbf{B}$ from the shape of the 3-D velocity distribution function of the escaping planetary ions was not found to be an accurate method, at least, in the relatively weak IMF case used for this study. From the observational point of view, such a correlation would have been useful in order to determine the direction of IMF for the case where there is no magnetometer available, as is the case for the Mars Express spacecraft.

Finally, it is worth noting that the QNH model version used in this work does not contain a Martian crustal magnetic field. It has been proposed that the acceleration of the planetary ions near the Martian magnetic anomalies may have many similarities with the acceleration processes associated with auroras in the terrestrial magnetosphere (Lundin *et al.*, 2006a,b). In this work the acceleration related to the magnetic anomalies are therefore not included and the presented properties of the escaping planetary ions result purely from acceleration caused by the convective electric field.

Summary

We have studied the properties of escaping O^+ and O_2^+ ions in the Martian tail with a self-consistent 3-D quasi-neutral hybrid model that does not contain crustal magnetic field. Our analysis shows that the model can reproduce qualitatively, and in many cases also quantitatively, many properties of ions observed by ASPERA/Phobos-2 in 1989 and ASPERA-3/Mars Express in 2004–2006, especially, (1) the acceleration of the planetary ions with an electric field $E_\rho \sim |\mathbf{E}_{\mathrm{sw}}|$ in the magnetosheath, (2) keV-class planetary ions within the optical shadow at $x \sim -3R_M$, (3) an "inverted U" shape type of energy spectra within the optical shadow, (4) $+E_{\mathrm{sw}}/-E_{\mathrm{sw}}$ hemisphere asymmetry, (5) different ion outflow along a line that goes through the magnetic tail lobes than along a line that goes near the cross tail current sheet and (6) that the highest particle flux comes from the same direction than the undisturbed solar wind.

References

Barabash, S., Holmström, M., Lukyanov, A., and Kallio, E.: 2002, *J. Geophys. Res.* **A10**, 1280, JA000326.

Barabash, S., *et al.*: 2004, *ESA publication SP-1240*, 121.

Brecht, S. H., Ferrante, J. F., and Luhmann, J. G.: *J. Geophys. Res.* **98**, 1345.

Dubinin, E., and 39 co-authors: 2006, *Icarus* **182**(2), 337, doi:10.1016/j.icarus.2005.05.022.

Espley, J. R., Cloutier, P. A., Brain, D. A., Crider, D. H., and Acuña, M. H.: 2004, *J. Geophys. Res.* 109, doi:10.1029/2003JA010193.

Fedorov, A., and 44 co-authors: 2006, *Icarus* **182**(2), 329, doi:10.1016/j.icarus.2005.09.021.

Kallio, E., Koskinen, H., Barabash, S., Nairn, C. M. C., and Schwingenschuh, K.: 1995, *Geophys. Res. Lett.* **22**, 2449.

Kallio, E., Koskinen, H., Barabash, S., Lundin, R., Norberg, O., and Luhmann, J. G.: 1994, *J. Geophys. Res.* **99**(A12), 23,547.

Kallio, E., and Koskinen, H.: 1999, *J. Geophys. Res.* **104**, 557.

Kallio, E., and Janhunen, P.: 2002, *J. Geophys. Res.* **107**, A3.

Kallio, E., and 46 co-authors: 2006, *Icarus* **182**(2), 350, doi:10.1016/j.icarus.2005.09.018.

Lammer, H., Lichtenegger, H. I. M., Biernat, H. K., Erkaev, N. V., Arshukova, I. L., Kolb, C., *et al.*: 2006, *Planetary and Space Science*, in press.

Liu, Y., Nagy, A. F., Groth, C. P. T., DeZeeuw, D. L., Gombosi, T. I., and Powell, K. G.: 1999, *J. Geophys. Res.* **26**, 2689.

Lundin, R., and 44 co-authors: 2004, *Science* **305**(5692), 1933, DOI: 10.1126/science.1101860.

Lundin, R., and 22 co-authors: 2006, *Science* **311**(5763), 980, DOI: 10.1126/science.1122071.

Lundin, R., and 43 co-authors: 2006, *Icarus* **182**(2), 308, doi:10.1016/j.icarus.2005.10.035.

Tanaka, T.: 1993, *J. Geophys. Res.* **98**(A10), 17,251.

Terada, N., Machida, S., and Shinagawa, H.: 2002, *J. Geophys. Res.* **107**(A12), 1471, doi:10.1029/2001JA009224.

Winningham, J. D., and 44 co-authors: 2006, *Icarus* **182**(2), 360, doi:10.1016/j.icarus.2005.10.033.

MARS GLOBAL MHD PREDICTIONS OF MAGNETIC CONNECTIVITY BETWEEN THE DAYSIDE IONOSPHERE AND THE MAGNETOSPHERIC FLANKS

MICHAEL W. LIEMOHN[1,*], YINGJUAN MA[1], RUDY A. FRAHM[2], XIAOHUA FANG[1], JANET U. KOZYRA[1], ANDREW F. NAGY[1], J. DAVID WINNINGHAM[2], JAMES R. SHARBER[2], STAS BARABASH[3] and RICKARD LUNDIN[3]

[1]*Center for Planetary Sciences, AOSS Department, University of Michigan, Ann Arbor, MI*
[2]*Southwest Research Institute, San Antonio, TX*
[3]*Swedish Institute of Physics, Kiruna, Sweden*
*(*Author for correspondence: E-mail: liemohn@umich.edu)*

(Received 10 May 2006; Accepted in final form 13 November 2006)

Abstract. Atmospheric photoelectrons have been observed well above the ionosphere of Mars by the ASPERA-3 ELS instrument on Mars Express. To systematically interpret these observations, field lines from two global MHD simulations were analyzed for connectivity to the dayside ionosphere (allowing photoelectron escape). It is found that there is a hollow cylinder behind the planet from 1–2 R_M away from the Mars-Sun line that has a high probability of containing magnetic field lines with connectivity to the dayside ionosphere. These results are in complete agreement with the ELS statistics. It is concluded that the high-altitude photoelectrons are the result of direct magnetic connectivity to the dayside at the moment of the measurement, and no extra trapping or bouncing mechanisms are needed to explain the data.

Keywords: Mars, atmospheric photoelectrons, Mars Express, numerical modeling

1. Introduction

Frahm *et al.* (2006, this issue) have reported observations of atmospheric photoelectrons well above the ionosphere of Mars by the ASPERA-3 electron spectrometer (ELS) instrument (Barabash *et al.*, 2004) on Mars Express (MEX) (Chicarro *et al.*, 2004). These high-altitude measurements reveal the characteristic photoelectron primary production peaks in the 20–30 eV energy range in a small angular width looking back toward the planet. These electrons are created in the dayside ionosphere, yet they are seen thousands of kilometers above the planet, well into the flanks of Mars' induced magnetosphere. Frahm *et al.* (2006) believe that this electron signature (the 20–30 eV number flux peaks) sufficiently identifies them as atmospheric photoelectrons because nothing else is known to create such an energy spectrum. They are not observed everywhere, but rather in limited spatial regions around the planet, namely a slowly-expanding cone of 1–2 R_M in radius on the flanks and just behind Mars (Frahm *et al.*, this issue). Anecdotally, the observation

Space Science Reviews (2006) 126: 63–76
DOI: 10.1007/s11214-006-9116-8

of these photoelectrons has been shown to be associated with direct magnetic connection between the measurement location and the dayside ionosphere (Liemohn *et al*., 2006). The electrons were thought to simply flow out along an open or draped field line from the dayside ionosphere directly to MEX.

The question still remains of whether this conclusion from a few case studies is the dominant mechanism for allowing the observation of atmospheric photoelectrons high above the ionosphere of Mars. That is, direct magnetic connection with the dayside ionosphere is one scenario, but other possibilities exist. One such process is trapping on closed field lines associated with the strong crustal magnetic field anomalies. Such trapping regularly occurs at Earth (e.g., Swartz *et al*., 1975; Lejeune and Wörmser, 1976; Khazanov and Liemohn, 1995). Downward-flowing flux spikes have been observed by Mars Global Surveyor across the nightside of Mars, occurring at the "cusps" of the strong crustal field regions (Mitchell *et al*., 2001). These precipitating electrons are presumably of solar wind origin, but they could also be photoelectrons, trapped during the day and dumped into the ionosphere sometime during the night. At Earth, trapped photoelectrons slowly precipitate throughout the night, contributing to the thermal balance of the ionospheric F layer (e.g., Nagy and Banks, 1970; Swartz *et al*., 1975; Khazanov *et al*., 1998, 2000). The MEX observations of high-altitude photoelectrons could be related to trapping within mini-magnetospheres. Another scenario is that these atmospheric photoelectrons are bouncing along draped interplanetary magnetic field (IMF) lines, and that there are some preferential locations where they can be observed flowing away from the planet and others where they should be seen flowing towards Mars.

The present study addresses this question to determine the true transport scenario for these electrons. Specifically, this study addresses the first hypothesis of direct magnetic connection between the dayside ionosphere and the locations where MEX statistically observes these particles. The magnetohydrodynamic (MHD) model of Ma *et al*. (2004) is used for this investigation, extracting many field lines from two simulation results to examine the magnetic topology around Mars. In brief, it is found that direct magnetic connection is fully capable of explaining the Mars Express high-altitude photoelectron observations, and more complicated methods of transport are not necessary.

2. Numerical Approach

Ma *et al*. (2002, 2004) provide a detailed description of the MHD model employed in the present study. Briefly, it solves the non-ideal MHD equations in dimensionless conservative form, using multiple continuity equations for the various ion species (H^+, O^+, O_2^+, and CO_2^+), and single-fluid versions of the momentum and energy equations. The solution, obtained on a non-uniform spherical grid, uses the second-order accurate numerical scheme of Powell *et al.* (1999). Thermospheric parameters are taken from the model of Bougher *et al.* (2001), and the strong crustal fields are included with the spherical harmonic model of Arkani-Hamed (2001, 2002). A nominal Parker spiral interplanetary magnetic field and average solar wind flow conditions are assumed for the upstream conditions in the simulations, in agreement with the typical values reported by Luhmann and Brace (1991). The MHD model is run until a steady-state solution is achieved. This is justified because of the fast transit times of photoelectrons in the Mars space environment. For example, a 25 eV electron has a speed of nearly 3000 km/s, and can therefore, if directed upward, leave the Mars ionosphere in much less than a second and travel many Mars radii distance away from the planet in just a few seconds. The temporal change in the global magnetic topology can therefore be neglected for this application. Note that these are the same conditions used in the simulations discussed by Ma *et al.* (2004) and Liemohn *et al.* (2006), and additional details of the configuration set up are discussed there.

Results from two simulations are examined in this study. The only difference between these two simulations is the subsolar longitude of Mars: 0° and 180°. Because the strong crustal field region is concentrated around 180° east longitude (e.g., Acuña *et al.*, 1998; Connerney *et al.*, 2001), these two choices represent the extremes of the influence of the crustal fields on the magnetic topology of near-Mars space. The 0° subsolar longitude case has the strongest crustal fields on the nightside, and therefore their effect on magnetic field connection to the dayside ionosphere is minimized. In the 180° subsolar longitude case, the effect is maximized, and the existence of open magnetic field lines (connected to the planet and to the IMF) is likely.

Ma *et al.* (2004) has already presented and discussed the general magnetic topology of the two MHD simulation results to be examined in this study. To summarize, IMF draping dominates the magnetic field configuration in both cases, but the strong crustal field sources create localized regions of closed field lines (mini-magnetospheres). Some of these field lines, especially those on the nightside, can extend thousands of kilometers above Mars. On the dayside, the closed field line regions are, in general, contained by the IMF, and the magnetic pile-up region around Mars (e.g., Luhmann and Brace, 1991; Crider *et al.*, 2002; Vennerstrom *et al.*, 2003) is pressed outward in some places, but only by a few hundred kilometers at most (Ma *et al.*, 2004).

In order to examine the magnetic connection between the dayside ionosphere and the MEX observation locations of high-altitude photoelectrons, many magnetic field lines were extracted from each of these simulations. The field line traces began from a grid in the $x = 0$ (terminator) plane. This starting point grid extended from 200 km altitude out to 6600 km (near 2 R_M) altitude, with an extraction every 200 km altitude in this range. This was done every 10° around the planet in the y–z terminator plane. Thus, there were 33 altitude starting points at 36 clock angle locations, resulting in 1188 field lines extracted from each simulation.

The field lines are designated as connected to the dayside ionosphere when they cross below an altitude of 300 km at daytime local times (i.e., 06–18 LT). This is roughly the altitude when the ionospheric density is 1% of the peak value (see Figure 13.10 of Schunk and Nagy (2000), taken from Hanson *et al.* (1977)). Selection of this altitude means that this study cannot easily differentiate between open magnetic field lines and draped magnetic field lines that pass below this threshold. This is unfortunate, but this information the nature of the MHD results already made this distinction difficult. That is, the assumed grid resolution in the MHD simulations lowers the fidelity of the field line tracing through the ionosphere. This is because the inner boundary condition for the magnetic field is specified at the average crustal field vector for each horizontal cell face at 100 km altitude (the lower boundary of the MHD model). The cells are roughly 100 km wide; therefore, resolving small-scale structures in the crustal fields, such as cusp-like features, is problematic. However, it is precisely at these cusps where reconnection between the crustal and interplanetary fields might occur. Resolution of the issue of open or draped field line topology is not necessary for the present study, however. The only thing that matters is magnetic connection to the dayside ionosphere, and a 300 km altitude sphere has been chosen for this designation. Note that this means the entire ring of extracted field lines beginning at 200 km altitude in the terminator plane are designated as having dayside ionospheric connection.

3. MHD Modeling Results

Figure 1 shows a few example 3-D field line traces extracted from the MHD results with 180E longitude at local noon. In Figures 1a and 1b are plotted lines that start somewhere along a semicircle in the northern hemispheric terminator plane at 1000 km, seen from the front and from the side. The color of the field lines shows the local magnetic field magnitude. Figures 1c and 1d are similar plots, but for field lines starting at 2000 km altitude in the terminator plane. The plotting of each field line stops when it reaches either 300 km altitude or the boundary of the plotting domain (into Mars' atmosphere to 300 km altitude, to the side to $Y = \pm 3\ R_M$, behind the planet to $X = -5\ R_M$, or up or down to $Z = 0$ or $+1.5\ R_M$).

It is seen that most of the lines in Figures 1a and 1b (1000 km starting altitude) connect to the dayside ionosphere, while very few of those in Figures 1c and 1d (2000 km starting altitude) show such a connection. As noted above, it is unclear whether these field lines are connected to the planetary crustal field sources, or are simply draped IMF field lines. For the northern hemisphere, shown here, it is assumed that most are draped IMF lines because of the weak crustal sources at northern latitudes. Note that for the connected field lines, B is seen to vary much more along the field line than for the unconnected field lines.

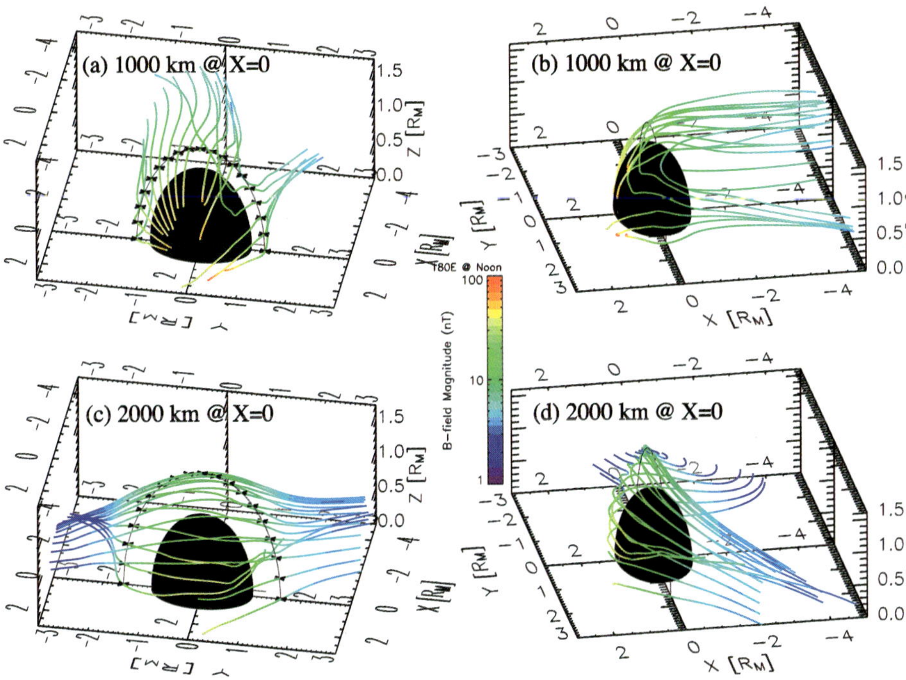

Figure 1. Field lines around Mars extracted from an MHD simulation, taken every 10° in a ring in the terminator plane ($X = 0$) at an altitude of (a and b) 1000 km and (c and d) 2000 km. The view in (a) and (c) is from the Sun, slightly above and to the right of the Sun-Mars line, while the view in (b) and (d) is from the dusk side, slightly above and to the left of the $+y$ axis. The color shows the magnetic field magnitude along the field lines (log scale), the thin black circle shows the ring of positions from which the extractions began, and the black crosses in (a) and (c) show the starting points of the field line traces. The black hemisphere in the middle of each panel represents Mars.

For comparison with the MEX data, it is useful to identify where the field lines with dayside ionospheric connectivity go in their paths around Mars. Figure 2 is an illustration of this, showing the locations of the dayside-ionosphere-connected field lines as they pass through the $x = 0, -2$, and -5 R_M planes. The two columns show the dayside-connected field line locations for the 2 simulations (180E at noon on the left, 0E at noon on the right).

Several prominent features are worth noting in Figure 2. First, many of the field lines pass through a ring 1–2 R_M from the $-x$ axis. Across the northern hemisphere, these are most likely draped IMF field lines passing through the inner edge of the magnetic pile-up region in front of Mars. In the southern hemisphere, the abundance of field lines in this ring is much larger in Figures 2a, 2c, and 2e (180E at local noon) than in Figures 2b, 2d, and 2f (0E at local noon). This implies that many of them in this $z < 0$ region are open field lines, i.e., the field lines that are connected to one of the planet's crustal sources and to the IMF.

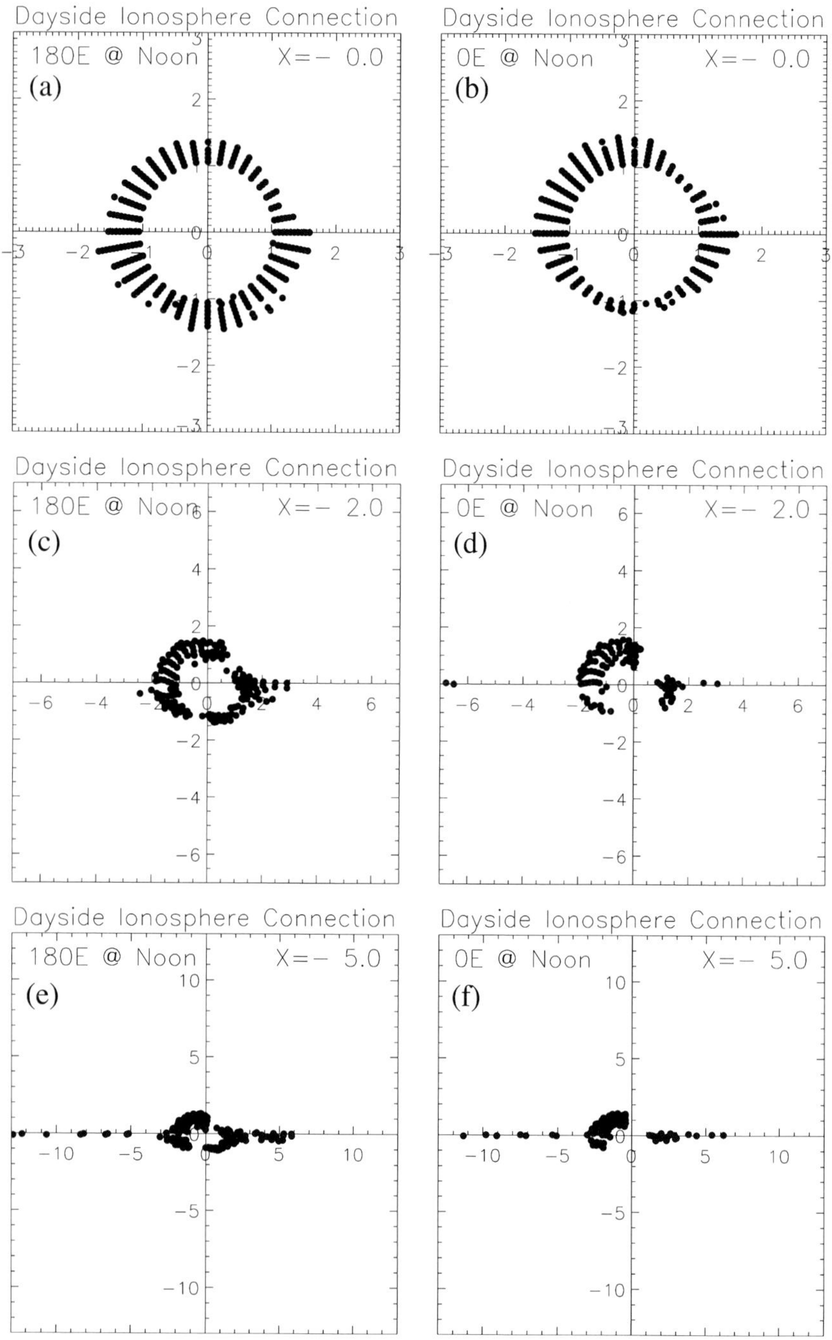

Figure 2. The y–z plane at (a and b) $X = 0$, (c and d) $X = -2$, and (e and f) $X = -5\ R_M$ showing points of magnetic field connectivity to the dayside ionosphere for the simulation with subsolar longitude of (left column) 180 E and (right column) 0 E (strong crustal field region toward and away from the Sun, respectively).

There is the possibility that the center of this cylinder of connectivity is also connected to the dayside ionosphere. The selection of the terminator plane for the beginning of the field line traces means that the patterns in Figure 2 could be reflecting the region of high magnetic field intensity and not necessarily the entire region of dayside ionospheric connection. This is most likely not the case, however. The field line extractions include a ring of starting points at 200 km altitude, and none of these field lines in either of the simulations escapes beyond the ionosphere of Mars. Therefore, if there were field lines from very close to the $-X$ axis with dayside ionospheric connection, they would have to pass through a portion of the nightside ionosphere/thermosphere, where the numerous Coulomb collisions would seriously degrade the flux intensities. This would make a high-altitude observation difficult and unlikely. To confirm this, additional field line extractions were conducted with starting points distributed in the $X = -2\,R_M$ plane. The resulting patterns of dayside ionospheric connection are very similar to those presented in Figure 2. While it could be that the MHD model is incorrectly tracking the exact field line paths at low altitudes, it is believed that the field lines within the cylinder are not populated by atmospheric photoelectrons.

Another feature is the presence of several field lines along the $+y$ axis in both sets of plots. These are draped field lines that extend into the magnetosheath, yet still have a connection to the dayside ionosphere. These field lines, in general, have slightly higher closest approach altitudes within the ionosphere, but still pass within 300 km altitude of the planet.

A final feature of these plots is the presence of field lines far out on the $-y$ axis in Figures 2d–2f. These are locations beyond the bow shock, but on field lines that form an S shape in the x–y plane, crossing the bow shock on the dawn side, continuing back a bit in the magnetosheath, and then sweeping sunward again to drape around the dayside ionosphere. They cross the terminator plane 3 times in their trace around Mars. Two examples of such field lines are shown in Figure 3.

The locations of y–z plane crossings at various x distances can be quantified for direct comparison with the statistical results of Frahm *et al.* (this issue). To do this, a cylindrical grid was defined, transforming the y–z plane location into polar coordinates, measured as ρ (distance from the x axis) and ø (counterclockwise angle measured from the $+y$ axis). The cylindrical grid was defined by dividing the x distance and the radial distance ρ into 0.1 R_M-wide bins, and dividing the ø angle into 24 equal bins around its 2π extent. If one or more field lines with dayside ionospheric connectivity passed through a particular bin, that bin was assigned a value of 1. If not, the bin was given a zero value. These values were then averaged around each ø ring, producing an array of fractions in the x-ρ plane of the amount of dayside ionospheric connectivity for that cylindrical location.

The results of this exercise are shown in Figure 4, for the cases of 180E and 0E longitude at local noon (Figures 4a and 4b, respectively). The color scale is logarithmic in order to highlight the smaller fractions in the magnetosheath and upstream solar wind. Of course, the dayside ionosphere has fractions at or near

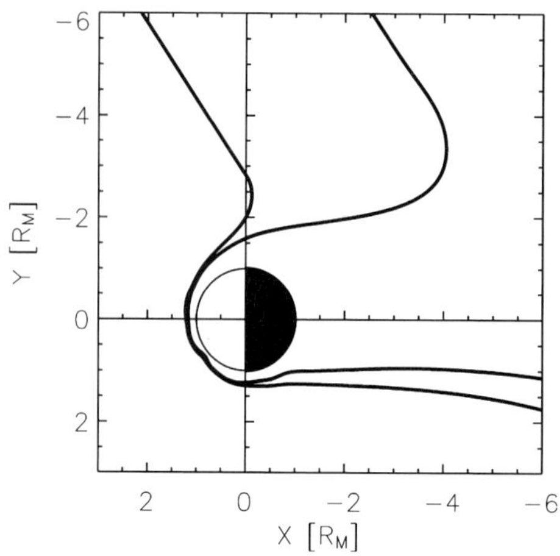

Figure 3. Two examples of S-shaped field lines that cross the $x = 0$ terminator plane three times.

unity, as expected. However, it is seen that there is a channel (actually, a cylinder) of high fractions behind Mars located 1 to 2 R_M away from the x axis. The fractions are very high, with values above 0.5 even 4 R_M and 2 R_M downtail (in Figures 4a and 4b, respectively). At larger ρ values, in the magnetosheath, the fractions can still be as high as 0.1 in some places. Even in the solar wind upstream of the bow shock, there are non-zero fractions present in both simulation results (from the S-shaped field lines in the x–y plane).

These plots in Figure 4 clearly demark the regions of high probability for a spacecraft to be located on a field line with direct magnetic connection to the dayside ionosphere, and therefore for the electron detector onboard the satellite to measure atmospheric photoelectrons far away from their source region. A comparison with high-altitude measurements of atmospheric photoelectrons is now possible.

4. Comparison with High-Altitude Photoelectron Observations

One such spacecraft with a good orbit and appropriate detector is MEX. The eV-energy-range electron detector on MEX, ELS, consists of a collimator followed by a standard top-hat electrostatic analyzer, with a micro channel plate and anode ring below this. The anode is divided into 16 sectors of 22.5°, making the field of view 360° with a 4° collimator window. The energy range extends from below an electron volt to ~20 keV, with a $\Delta E/E$ of 0.08.

At high altitudes, ELS is oriented on the MEX satellite in such a way that anode sector 3 is in the Sun-looking direction. Thus, Frahm *et al.* (this issue) chose to examine the data in this sector in their statistical analysis of high-altitude photoelectron observations. By looking for the primary production peaks from the photoionization of CO_2 by 30.4 nm solar photons (in the 20–30 eV energy

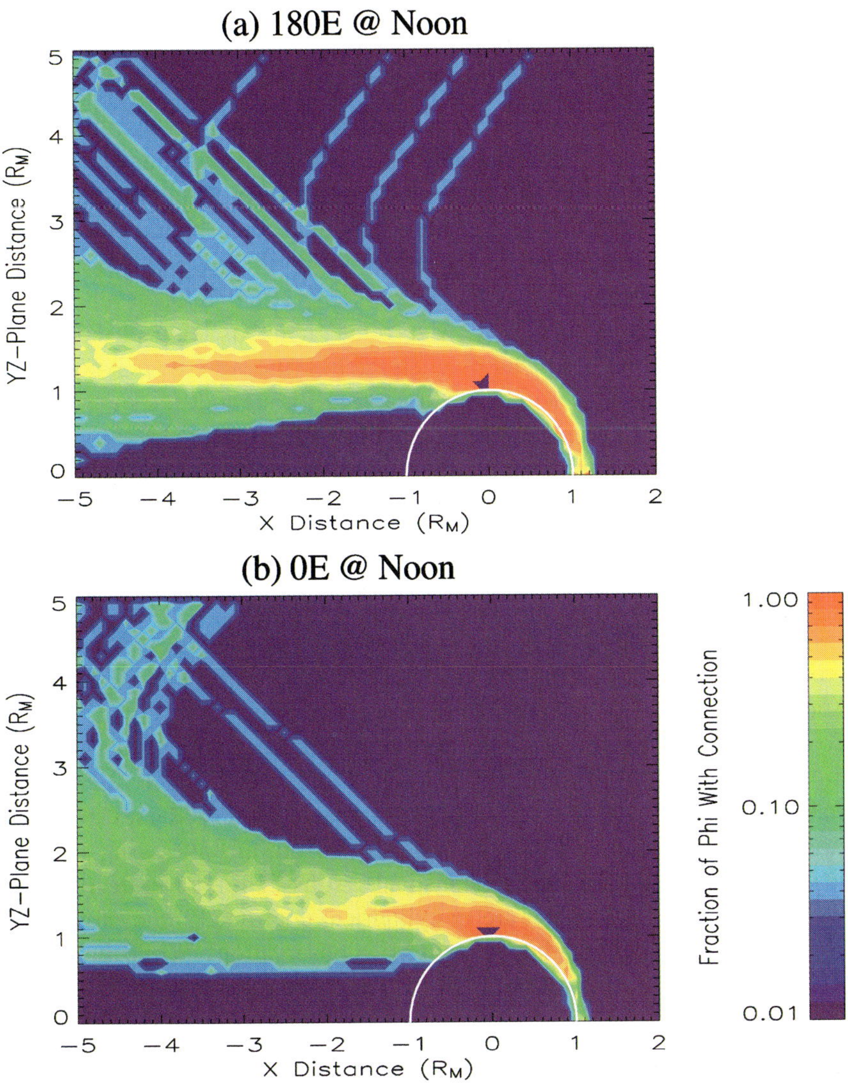

Figure 4. Fraction of field lines that have magnetic connectivity to the dayside ionosphere as a function of x distance and cylindrical distance from the x axis for the simulation with subsolar longitude of (a) 180E and (b) 0E. Note the logarithmic color scale.

range), they were able to determine if a given energy spectrum from this anode sector contained an atmospheric photoelectron signature. Data were considered throughout most of 2004, and so the analysis included tens of thousands of energy spectra across a wide range of near-Mars space.

Figure 5a presents a plot of the fraction of observations that contained atmospheric photoelectrons in sector 3, shown in cylindrical coordinates, like Figure 4.

It is seen that the dayside ionosphere has very high fractions, but there are also high fractions, approaching unity, in a channel (cylinder) roughly 1 to 2 R_M from the x axis extending downtail. This result is in excellent agreement with the MHD simulation results of direct dayside ionospheric connection shown in Figure 4.

Also shown in Figure 5a is the region of ELS data coverage included in the Frahm *et al.* (this issue) study. While the sampling is not uniform within this region, most $x - \rho$ grid points within this region have tens (and often hundreds) of sector 3 spectra included in the analysis.

The "excellent" comparison with Figure 4 is qualitative, however. Two main reasons account for this caveat. Firstly, Mars rotates about its axis each day, exposing a different configuration of crustal magnetic field sources to the magnetic pile-up region. Secondly, the solar wind is also varying throughout the day. Therefore, neither of the plots in Figure 4 is truly analogous to the statistical results. While the latter effect requires additional simulations to investigate, two points lend validity to this study without such additional numerical experiments. The first is the choice of average solar wind and IMF conditions for these simulations, which should yield a "typical" solar wind interaction with Mars. The second is that the results are averaged in azimuth in the y–z plane, which means that different IMF clock angles will not affect the results. That is, the locations of dayside ionospheric connectivity will rotate around in the y–z plane with changes in the IMF By and Bz components, but such rotation will not change the ø-averaged fractions.

To crudely take into account the former factor (accounting for Mars' daily rotation), the results from the 2 simulations, which represent the 2 extremes of crustal field influence on the magnetic field topology around Mars, can be simply averaged together. Figure 5b presents the results from this averaging step. These results are now comparable to the statistical results of Frahm *et al.* (this issue), shown in Figure 5a. Both plots are in cylindrical coordinates and plotted with a linear color scale to highlight the region of MEX observations. While the agreement is not perfect, the existence of the high-fraction channel extending into the Mars magnetotail roughly 1 to 2 R_M from the x axis is clearly visible in both plots.

One major difference between the two plots in Figure 5 is at and just behind the terminator ($x = 0$ to -1), where the observation-based fractions are low but the MHD-based fractions are high. While the true reason for this discrepancy is unknown, the likely explanation is an observational bias due to the use of sector 3 for the automatic identification routine. In this region, MEX sometimes (on about half of the orbits) changes its orientation relative to the Sun and Mars so that the cameras face the planet surface. In addition, the magnetic field line is often not parallel to the x axis in this region, but rather curved, and so the field-aligned flows are directed into sector 4 or 5 rather than sector 3. Because of the dramatic magnetic field decrease between the photoelectron source region and this flank region, the photoelectron source cone is probably 20° or less in pitch angle (i.e., smaller than a single sector width). So, the photoelectron stream could simply be striking a different anode sector. Another complication is that even when sector 3 is

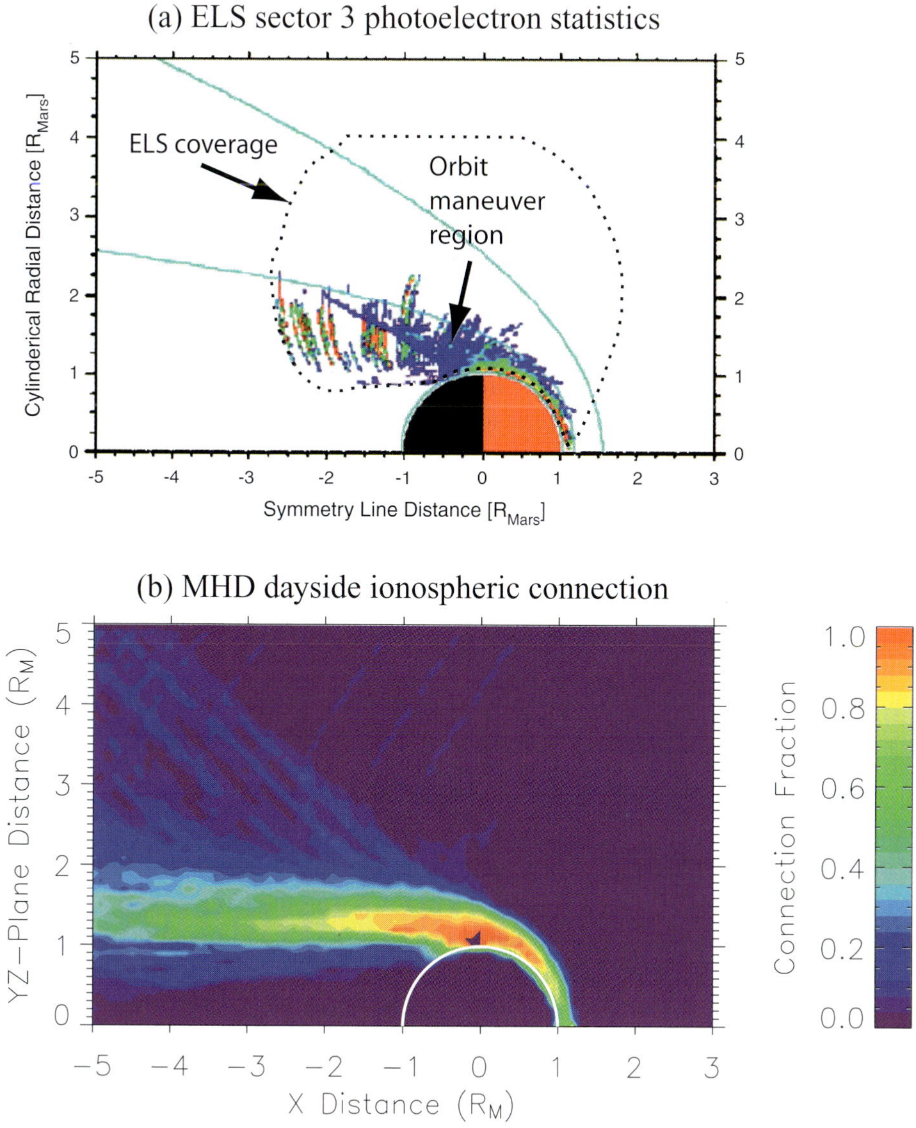

Figure 5. Measures of dayside ionospheric connectivity from the ELS instrument and the MHD model results, as a function of x distance and cylindrical distance from the x axis. (a) Fraction of ELS spectra containing atmospheric photoelectrons in sector 3 (from Frahm *et al.*, this issue). The black dotted line shows the region of ELS coverage included in the statistical analysis, and the two blue curves show the statistical locations of the magnetopause and bow shock. (b) Fraction of MHD field lines with connectivity, averaging the fractions from the 2 simulations. Note the linear color scale, which is the same for both plots.

Sun-looking, the ELS detector plane is parallel to the x–y plane, and any tilt of the magnetic field out of the x–y plane reduces the observed pitch angle extent seen by ELS. The first pitch angles to be lost because of this tilt are those near 0° and 180°, i.e., the source cone where the photoelectrons are located. As seen in Figure 1, the field lines between $x = 0$ and $x = -1$ R_M are often not parallel to the x axis, as they are farther downtail, but rather they are pointed in many different directions. So, this effect also acts to reduce the chances of observation in this window of the x axis. Thus, sector 3 is an unreliable sector for photoelectron detection in this particular spatial region near Mars, and other sectors must be examined, depending on the spacecraft and magnetic field orientation. That is, sector 3 simply might not have seen the atmospheric photoelectrons in this spatial region.

On the dayside, the observations are near the source region of the atmospheric photoelectrons. Therefore, the pitch angle distribution is nearly isotropic and the photoelectrons often appear in many (or all) ELS sectors. So, even though the field lines point in all directions here as well, the observation fraction for sector 3 is still high.

A final note on the data-model comparisons is that while there are many ELS sector 3 energy spectra recorded while MEX was in the magnetosheath and unshocked solar wind. That is, in Figure 5a, compare the ELS coverage region encompassed by the black dotted line with the statistical locations of the magnetopause and bow shock shown by the blue lines. The occurrence fractions of atmospheric photoelectron observations are essentially zero in these regions. Again, this is because the statistics are based on an examination of ELS sector 3 observations, which (at high altitudes) requires a magnetic field parallel to the x axis (or an isotropic distribution) for any chance of detection. Therefore, it is not surprising that the ELS statistics have fractions at or near zero in these regions. Note that the Frahm *et al.* (this issue) analysis of sector 3 photoelectron measurements is a pilot study and will be followed up with a more detailed investigation of additional sectors.

5. Discussion and Conclusions

In this study, the question has been systematically addressed of how atmospheric photoelectrons can be seen with such high probability thousands of kilometers away the dayside ionosphere of Mars. To do this, the near-Mars magnetic topology was simulated with an MHD model, and extracted many field lines from the results. These field lines were checked for connection to the dayside ionosphere, and a similar "statistical analysis" of the simulation results of atmospheric photoelectron probabilities was conducted. The resulting maps are in excellent agreement with the observation-based statistics of the electron spectrometer on Mars Express, and the main discrepancies between the data and the model results can be readily explained.

The observation of atmospheric photoelectrons far from the dayside ionosphere can be a powerful tool for interpreting the solar wind interaction with Mars. It reveals direct magnetic connection with the dayside ionosphere and upper atmosphere, and therefore is an indicator of where to look for very-low-energy (i.e., thermal) escaping planetary ions (those streaming along the field line). In fact, atmospheric photoelectron intensities should be closely related to ionospheric temperatures (photoelectron energy deposition is a major heat source) and therefore the flux of these electrons could be used as a proxy for the flux of escaping ions. This issue of photoelectron flux at high altitudes has not been addressed in this study; only the magnetic connection to the dayside ionosphere is being examined. Liemohn *et al.* (2006) calculated such fluxes and discuss many of the processes affecting these fluxes. Another feature of high-altitude photoelectron observations is that their location will vary with changes in the solar wind and IMF, and therefore they can be used to deduce the upstream conditions, even in the absence of a direct measurement.

To address the question posed in the Introduction, the answer is that no "fancy" trapping or bouncing mechanisms are needed to get photoelectrons to the high altitudes where ELS observes them. All that is needed is a simple and direct magnetic connection between the observation location and the dayside ionosphere of Mars.

Acknowledgments

The authors would like to thank support for this research by NASA under grants NASW-00003, NAG5-10887, NNG04G055G, and NAG5-13332, by the NSF under grant ATM-0455729. We also wish to thank the Swedish National Space Board for their support of the main PI institute and we are indebted to ESA for their courage in embarking on the Mars Express program.

References

Acuña, M. H., *et al.*: 1998, *Science* **279**, 1676.

Arkani-Hamed, J.: 2001, *J. Geophys. Res.* **106**, 23,197.

Arkani-Hamed, J.: 2002, *J. Geophys. Res.* **107**(E10), 5083, doi: 10.1029/2001JE001835.

Barabash, S., *et al.*: 2004, in Wilson, A. (ed.), *Mars Express: The Scientific Payload*, European Space Agency Publications Division, European Space Research and Technology Centre, Noordwijk, The Netherlands, SP-1240, p. 121.

Chicarro, A., Martin, P., and Trautner, R.: 2004, in Wilson, A. (ed.), *Mars Express: The Scientific Payload*, European Space Agency Publications Division, European Space Research and Technology Centre, Noordwijk, The Netherlands, SP-1240, p. 3.

Connerney, J. E. P., Acuña, M. H., Wasilewski, P. J., Kleteschka, G., Ness, N. F., Rème, H., *et al.*: 2001, *Geophys. Res. Lett.* **28**, 4015.

Crider, D. H., *et al.*: 2002, *Geophys. Res. Lett.* **29**(8), 1170, doi: 10.1029/2001GL013860.

Frahm, R. A., *et al.*: 2006, *Icarus*, in press.

Frahm. R., *et al.*: *Space Sci. Rev.*, this issue, doi: 10.1007/s11214-006-9119-5.

Hanson, W. B., Sanatani, S., and Zuccaro, D. R.: 1977, *J. Geophys. Res.* **82**, 4351.

Khazanov, G. V., and Liemohn, M. W.: 1995, *J. Geophys. Res.* **100**, 9669.
Khazanov, G. V., Liemohn, M. W., Kozyra, J. U., and Moore, T. E.: 1998, *J. Geophys. Res.* **103**, 23, 485.
Khazanov, G. V., Liemohn, M. W., Kozyra, J. U., and Gallagher, D. L.: 2000, *J. Atmos. Solar-Terr. Physics* **62**, 947.
Lejeune, J., and Wörmser, F.: 1992, *J. Geophys. Res.* **97**, 159.
Liemohn, M. W., *et al.*: 2006, *Icarus* **182**, 383.
Luhmann, J. G., and Brace, L. H.: 1991, *Rev. Geophys.* **29**, 121.
Ma, Y., Nagy, A. F., Hansen, K. C., DeZeeuw, D. L., and Gombosi, T. I.: 2002, *J. Geophys. Res.* **107**(A10), 1282, doi: 10.1029/2002JA009293.
Ma, Y., Nagy, A. F., Sokolov, I. V., and Hansen, K. C.: 2004, *J. Geophys. Res.* **109**, A07211, doi: 10.1029/2003JA010367.
Mitchell, D. L., *et al.*: 2001, *J. Geophys. Res.* **106**(E10), 23,419.
Nagy, A. F., and Banks, P. M.: 1970, *J. Geophys. Res.* **75**, 6260.
Powell, K. G., Roe, P. L., Linde, T. J., Gombosi, T. I., and De Zeeuw, D. L.: 1999, *J. Comp. Phys.* **153**, 284.
Schunk, R. M., and Nagy, A. F.: 2000, *Ionospheres*, Cambridge University Press, New York.
Swartz, W. E., Bailey, G. J., and Moffett, R. J.: 1975, *Planet. Space Sci.* **23**, 589.
Vennerstrom, S., Olsen, N., Purucker, M., Acuña, M. H., and Cain, J. C.: 2003, *Geophys. Res. Lett.* **30**(7), 1369, doi: 10.1029/ 2003GL016883.

MARS GLOBAL SURVEYOR MEASUREMENTS OF THE MARTIAN SOLAR WIND INTERACTION

D. A. BRAIN
University of California, Berkeley Space Sciences Laboratory, Berkeley, CA 94720
(E-mail: brain@ssl.berkeley.edu)

(Received 8 August 2006; Accepted in final form 17 November 2006)

Abstract. The solar wind at Mars interacts with the extended atmosphere and small-scale crustal magnetic fields. This interaction shares elements with a variety of solar system bodies, and has direct bearing on studies of the long-term evolution of the Martian atmosphere, the structure of the upper atmosphere, and fundamental plasma processes. The magnetometer (MAG) and electron reflectometer (ER) on Mars Global Surveyor (MGS) continue to make many contributions toward understanding the plasma environment, thanks in large part to a spacecraft orbit that had low periapsis, had good coverage of the interaction region, and has been long-lived in its mapping orbit. The crustal magnetic fields discovered using MGS data perturb plasma boundaries on timescales associated with Mars' rotation and enable a complex magnetic field topology near the planet. Every portion of the plasma environment has been sampled by MGS, confirming previous measurements and making new discoveries in each region. The entire system is highly variable, and responds to changes in solar EUV flux, upstream pressure, IMF direction, and the orientation of Mars with respect to the Sun and solar wind flow. New insights from MGS should come from future analysis of new and existing data, as well as multi-spacecraft observations.

Keywords: Mars, MGS, magnetosphere, solar wind interaction

1. Introduction

The Martian interaction with the solar wind provides an interesting contrast to the plasma interactions at other solar system bodies. The solar wind obstacle is a combination of a global atmospheric obstacle (like those at Venus or comets) punctuated by many smaller-scale obstacles formed by strong crustal magnetic fields (similar, perhaps, to Earth or the Moon). The supersonic solar wind evolves in density, temperature, and the strength of its entrained Interplanetary Magnetic Field (IMF) as it expands into the solar system, so that the incident plasma at Mars has properties intermediate between those experienced by the inner and outer planets.

In addition to being of general interest, the plasma environment influences at least three "big picture" science issues. First, studies of the Martian solar wind interaction provide important contributions toward understanding the long-term evolution of the Martian climate since the end of the late heavy bombardment. A variety of lines of evidence suggest that the Martian atmosphere has been substantially altered

Space Science Reviews (2006) 126: 77–112
DOI: 10.1007/s11214-006-9122-x

over time (see Jakosky and Phillips, 2001). Escape of atmospheric particles to space is known to occur in the present epoch (e.g. Lundin *et al.*, 1989; Carlsson *et al.*, 2006), and likely has been the most efficient loss process over the last 3.5 billion years or more (Brain and Jakosky, 1998). Of the variety of physical processes collectively termed 'escape to space', all ion loss processes are directly influenced by the solar wind plasma and magnetic field, as is the loss of neutrals via 'sputtering' by pickup ions (and the upper atmospheric reservoir for escaping neutrals). Second, the solar wind provides a boundary condition for the current state of the upper atmosphere, and therefore plays a role in determining its structure, composition, chemistry, and dynamics. Solar wind charged particles (as well as neutrals formed in the solar wind via charge exchange) have access to the thermosphere at low altitudes (Mitchell *et al.*, 2001a), and can contribute to atmospheric energy deposition and ionization. Sharp contrasts in structure and composition can develop near crustal field boundaries (Gurnett *et al.*, 2005), driving dynamics. The Martian upper atmosphere would be remarkably different without the plasma interaction, which is therefore a necessary component in its understanding. Finally, Mars offers a natural laboratory for exploration of fundamental plasma processes observed at Earth and elsewhere in the solar system and universe. Processes such as particle acceleration, magnetic reconnection or merging, and the generation of instabilities in the form of plasma waves and shocks all occur at Mars in plasma conditions that differ significantly from those observed elsewhere. Mars has the potential to provide a useful end-member data point on how these processes operate.

The main features of the Martian global plasma interaction are summarized in cartoon form in Figure 1. Solar wind ions (indicated in blue) and the associated interplanetary magnetic field (IMF) interact with the extended Martian atmosphere (indicated in orange) and ionosphere. A variety of different plasma regimes and boundaries form as a result, and can be distinguished using spacecraft particle and field measurements. The solar wind transitions from supersonic to subsonic as it crosses the bow shock into the hotter, denser, more turbulent magnetosheath. Some solar wind plasma is reflected from the shock into the foreshock region. Few or no solar wind protons are observed downstream from a boundary sometimes called the magnetic pile-up boundary (or MPB) and its tailward extension (however the shocked IMF and solar wind electrons are found downstream from this boundary, presenting a challenge in interpretation). Below the MPB, the photoelectron boundary (PEB) separates the planetary ionosphere from the magnetic pileup region (MPR). A two-lobed induced magnetotail forms on the night side, with a current sheet carrying planetary ions between the two lobes. Crustal remnant magnetic fields perturb the global interaction at low altitudes.

What is known about the Martian solar wind interaction has been derived primarily from measurements made by spacecraft missions to Mars over the past 40 years. Historical spacecraft measurements are described in several review articles (e.g. Luhmann *et al.*, 1992; Barabash and Lundin, 2006); the main contributions are summarized here. Equipped with a magnetometer, the Mariner 4 spacecraft made

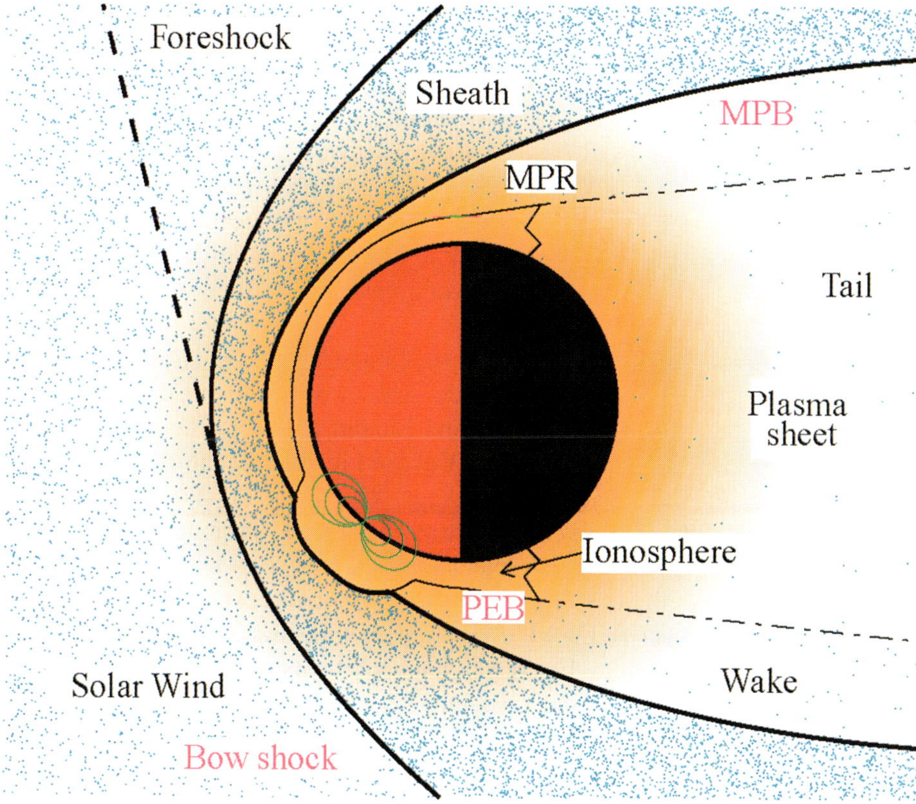

Figure 1. Cartoon of the global Martian solar wind interaction. Orange shading indicates density of planetary neutrals. Blue indicates relative density of solar wind ions in different plasma regions (labeled in black), separated by different plasma boundaries (labeled in magenta). Boundary names in this figure are those specific to MGS literature.

the first measurements of a non-terrestrial bow shock on two flybys in July 1965 (Smith, 1969). From 1971 to 1974 the Soviet Mars missions (2, 3, 5) also measured the bow shock and the underlying sheath using magnetometers and ion and electron instruments (e.g. Bogdanov and Vaisberg, 1975; Dolginov *et al.*, 1976). Additionally, they made the first measurements of the "ion cushion", identified in Figure 1 as the MPR. There were no more measurements of the plasma interaction at Mars until the arrival of the Phobos 2 spacecraft in 1989, though the Viking 1 and 2 Landers in 1976 measured *in situ* vertical density and temperature profiles in the ionosphere (Hanson *et al.*, 1977). Phobos 2 measured escape products from the Martian atmosphere, and provided a wealth of useful information about the Martian wake, tail, sheath, and upstream regions (e.g. Lundin *et al.*, 1989; Riedler *et al.*, 1989; Rosenbauer *et al.*, 1989; Pedersen *et al.*, 1991; Verigin *et al.*, 1991, 1993; Dubinin *et al.*, 1993, 1994, 1996). Mars Global Surveyor (MGS), discussed in this review, discovered strong crustal magnetic fields that interact directly with the shocked solar wind, and placed a new upper limit on the strength of any lingering Martian dynamo (Acuña *et al.*, 1998, 2001). The Mars Express (MEX) spacecraft, in orbit since late 2003, has made the first low-altitude measurements of planetary heavy ions (Lundin *et al.*, 2004), the first measurements of Energetic Neutral Atoms (ENAs) at Mars (Futaana *et al.*, 2006), and has discovered aurora in the crustal magnetic fields (Bertaux *et al.*, 2005). At the time of this review there is considerable opportunity for new discovery. Both MGS and MEX continue to operate and make new measurements, Venus Express is making measurements at Venus that can be directly compared to MEX data, and several new spacecraft missions are being considered or proposed.

Previous review papers have been published on many of the spacecraft missions and results described above. These include reviews of Phobos results (Zakharov, 1992), Mars Express results (Barabash and Lundin, 2006), subsets of the MGS results (Crider, 2004; Bertucci *et al.*, 2005a), and reviews of the Martian system that are not specific to any one spacecraft (e.g. Nagy *et al.*, 2004; Luhmann *et al.*, 1992). The purpose of this review is to illustrate the unique contributions of the MGS mission to the study of the Martian interaction with the solar wind, incorporating recent results and highlighting opportunities for future discoveries. In the sections that follow we will describe the MGS instrument suite and orbit (Section 2), and MGS contributions related to the crustal magnetic fields (Section 3), global solar wind interaction (Section 4), and variability (Section 5). We follow with a brief summary (Section 6) and directions for the future.

2. MGS Measurements

2.1. Instrumentation

MGS carries three instruments capable of returning information about the solar wind interaction with the upper atmosphere: a magnetometer (MAG), an electron reflectometer (ER), and a radio science investigation (RS).

MAG consists of two triaxial fluxgate magnetometers mounted on the spacecraft solar panels. MAG returns full vector magnetic field measurements every 0.75–3 s, and successive vector field differences 24 times as often. The instrument has dynamic range of 0.005–65536 nT. The instrument was calibrated in-flight to remove spacecraft-generated magnetic fields, and is accurate to ~ 1 nT (Acuña *et al.*, 2001). As of early 2006, MAG has returned more than 4.2 billion vectors from the Martian system. Further details about the instrument can be found in Acuña *et al.* (1992, 1998). Previous spacecraft to carry magnetometers to Mars include Phobos, Mars 2, 3, and 5, and Mariner 4.

ER is a top-hat electrostatic analyzer designed to measure fluxes of superthermal electrons in a planar slice through the 3D distribution. Full 3D electron distributions are not measured because the MGS spacecraft is three-axis stabilized. Directional information in energy channels ranging from 10 eV to 20 keV is obtained every 2–8 seconds from 16 sectors measuring 14° by 22.5°. Omni-directional energy spectra with 25% energy resolution are recorded every 12–48 seconds. More information on the MGS ER can be found in Acuña *et al.* (1992); Mitchell *et al.* (2001a). The Phobos and Mars 5 spacecraft carried instruments capable of measuring electrons prior to the arrival of MGS, and MEX also carries an electron sensor.

The MGS radio science (RS) investigation returns information relevant to the solar wind interaction in the form of upper atmospheric electron density profiles derived from radio occultations (Tyler *et al.*, 2001; Hinson *et al.*, 1999). Details about the RS instrument can be found in Tyler *et al.* (1992). The MGS RS investigation has returned many more profiles than previous missions, combined. RS results will not be discussed in detail in this review. Major results include identification of different ionospheric scale heights in the vicinity of crustal magnetic sources (Krymskii *et al.*, 2004), identification of "anomalous" electron density profiles in the vicinity of crustal magnetic fields (Withers *et al.*, 2005), measurement of enhancements in ionospheric densities due to solar flares (Mendillo *et al.*, 2006), and measurement of simultaneous variability in the ionospheres of Mars and Earth (Mendillo *et al.*, 2003). Ionospheric measurements at Mars prior to MGS are summarized in Mendillo *et al.* (2003).

2.2. Orbit

In some respects, the many contributions of the MAG/ER to the study of the Martian solar wind interaction were made possible by the unique orbit of the MGS spacecraft. The mission had two main phases – premapping and mapping, also described in Albee *et al.* (2001).

The first phase, premapping, lasted from 13 September 1997 through late January 1999. During this time period MGS had an elliptical orbit that precessed in local time and gradually circularized, with periapsis as low as ∼101 km and apoapsis as high as ∼16 Mars radii (R_M). MGS was actively aerobraking in the Martian atmosphere during some of these orbits, and was in a "holding" orbit at other times as it precessed to the local time for the mapping orbit. Figure 2 shows the orbital coverage of low altitude MAG data as a function of altitude, planetary latitude, solar zenith angle, and local time. Overall, the data coverage is unprecedented at all altitudes. Data coverage is uniform in planetary longitude. There is good altitude coverage as a function of latitude, though much of the lowest altitude data north of 60° S were recorded when MGS was in sunlight. Coverage is poor near the dawn terminator, and the subsolar and anti-solar points. Further, it is apparent from the figure that solar zenith angle coverage is convolved with both latitude and local time, making it difficult to exclusively identify trends in the observations as being associated with one parameter. Finally, most of the local time coverage occurred when MGS had periapsis in the northern hemisphere.

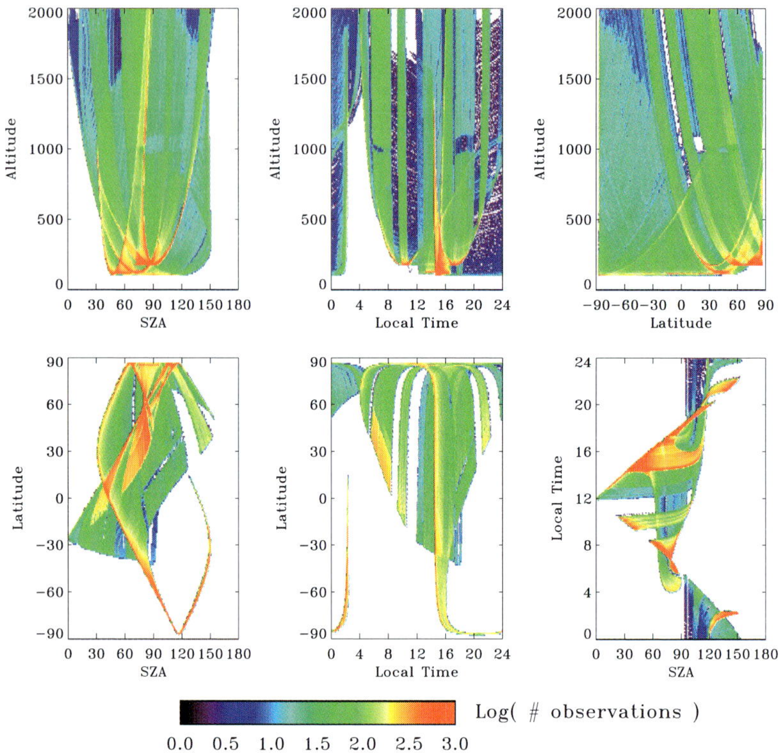

Figure 2. Data coverage for MGS MAG full vector measurements recorded during premapping orbits from 13 September 1997 through 28 January 1999. Panels show data density for times when the spacecraft was below 2000 km altitudes, as a function of altitude, solar zenith angle, local time, and latitude.

Since mid-1999, MGS has been in a nearly circular mapping orbit with fixed local time near 2am/2pm. The spacecraft altitude ranges from ~369–441 km, and since the periapsis latitude is very near −90° latitude, altitude and latitude covary. Figure 3 shows the analogue of Figure 2 for the mapping orbits. All parameters are convolved during mapping, though the seasonal orientation of Mars with respect to the Sun allows some solar zenith angle coverage at certain altitudes, latitudes, and local times.

Three features of the MGS orbit enabled it to make important contributions not possible using earlier spacecraft:

1. *Low altitude* – The low periapsis altitude of the premapping orbits allowed one of the most significant discoveries of the MGS mission – the detection of Mars' crustal magnetic fields. Though Phobos approached to within ~800 km of the planet, low enough to measure the strongest crustal field signatures over a small region of the southern hemisphere (Brain *et al.*, 2003), unambiguous association

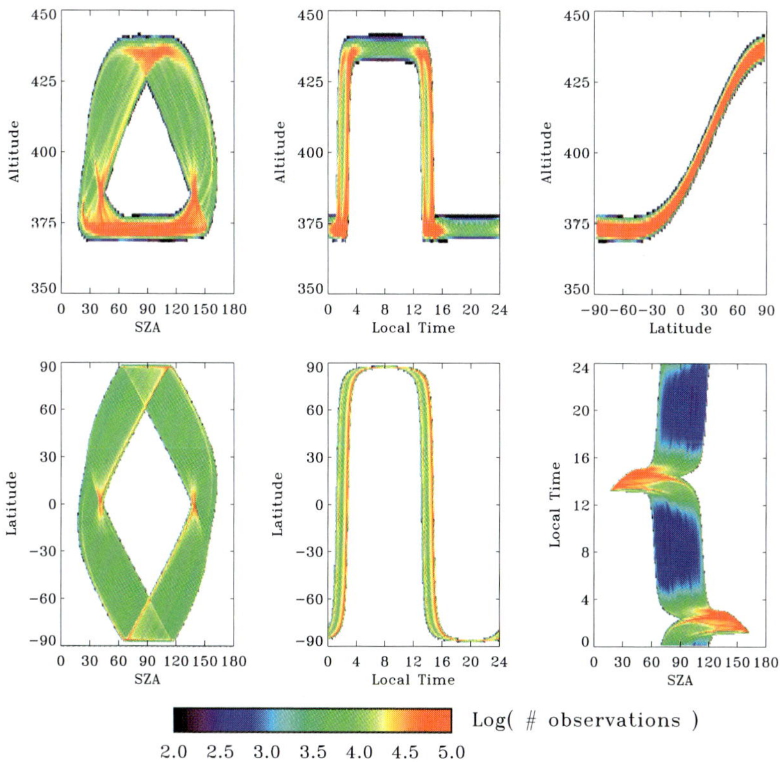

Figure 3. Data coverage for MGS MAG during mapping orbits from 2 June 1999 and 31 March 2004, in the same format as Figure 2.

of these signatures with crustal fields would have been very difficult without supporting data from lower altitudes.

2. *Global coverage* – The long duration of the premapping phase of the MGS mission, made necessary by a possible problem with a hinge on one of the solar panels (Albee *et al.*, 2001), was a boon for the MAG/ER experiment. MGS achieved much better coverage of the global interaction region than it would have otherwise, and much better coverage than any previous spacecraft to visit Mars. Each of the regions identified in Figure 1 was visited at a variety of local times and solar zenith angles by MGS.
3. *Long-lived mapping orbit* – MGS has been in its mapping orbit for more than three Martian years, making repeated measurements of a small slice of the global interaction region. The long baseline of observations enables investigation of the many factors that control variability in this slice over timescales ranging in length from hours to a solar cycle.

Each of the three science consequences of the MGS orbit listed above is described in more detail in the following sections.

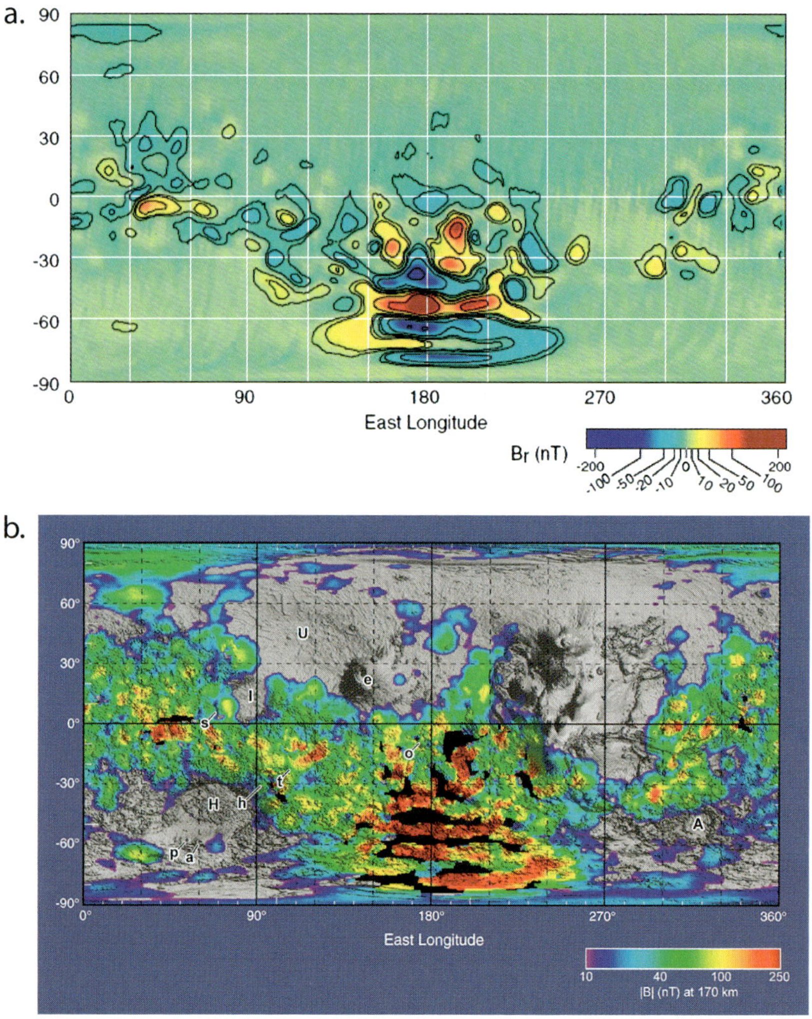

Figure 4. Martian crustal magnetic field maps based on MGS mapping data in eclipse. (a) Radial field component typically measured by MAG at ~400 km altitude from Connerney *et al.* (2001). (b) Field strength at 170 km altitudes inferred calculated from the shape of ER angular distributions from Mitchell *et al.* (2006).

3. Crustal Fields

The low periapsis of MGS allowed the discovery of crustal magnetic fields over much of the surface. Figure 4 shows that crustal fields are strongest in the heavily cratered (and therefore older) southern hemisphere, except in the locations of large impact basins such as Hellas and Argyre, which are largely devoid of enhanced magnetic fields (Acuña *et al.*, 1999). The large strength of the crustal fields measured at spacecraft altitude imply that there are large volumes of coherently magnetized material in the outer layers of the Martian crust. The most likely scenario for their formation is in the presence of a global dynamo magnetic field that has since ceased (Acuña *et al.*, 1998). Even in the younger, sparsely cratered northern hemisphere there is evidence from both MAG and electron data for weaker crustal magnetic fields in some locations (Lillis *et al.*, 2004; Connerney *et al.*, 2005; Mitchell *et al.*, 2006; Brain *et al.*, 2003), with implications for the formation mechanism of the north-south dichotomy at Mars.

For reference, the largest crustal magnetic field strength measured at Mars was ~1600 nT near 100 km altitude. The draped IMF at Mars typically reaches strengths of 30–60 nT. At the ~400 km mapping orbit of MGS, crustal fields measure as much as 200 nT, compared to 26000 nT and 10 nT from Earth's global field and anomalies, respectively. The Martian crustal fields are also much stronger than lunar anomalies, which measure ~30 nT at ~20 km altitudes.

The large strength of the crustal fields also has unanticipated implications for the Martian solar wind interaction. Maps of crustal magnetic fields are created using data from the Martian nightside, in shadow, where the contributions from the draped IMF are minimized (Connerney *et al.*, 2001). However, the signature of crustal fields can be measured at all solar zenith angles at Mars, and to considerable altitude. Using pre-mapping MAG data above individual regions of the surface, Brain *et al.* (2003) qualitatively determined the typical altitude to which crustal magnetic fields can be distinguished in observations. Crustal fields extend above 120 km altitude (near the ionospheric main peak) over ~70% of the surface, and even extend above 1000 km altitudes over the strongest southern source.

The large region of influence of crustal magnetic fields adds several layers of complexity to the study of the Martian solar wind interaction. Even if all other sources of variability were held constant, the Martian interaction would be highly variable simply because the planet's rotation would change the orientation of crustal fields with respect to the solar wind. Many consequences of the crustal fields for the plasma interaction, locally and globally, have been considered. All of these effects are related to one of two influences that crustal fields have on the system: the upward perturbation of plasma boundaries, and the modification of magnetic field topology. Each effect is discussed further below.

3.1. Crustal Influences on Plasma Boundaries

Magnetic pressure from crustal magnetic fields can be comparable to or even far exceed ionospheric thermal pressure above selected regions of the Martian surface. This additional pressure contribution locally raises the altitude of the solar wind obstacle. The cartoon in Figure 5a illustrates that the theoretical pressure balance obstacle to the solar wind can exceed altitudes of 1200 km in some locations. This is far higher than the obstacle in the northern hemisphere, where the crustal

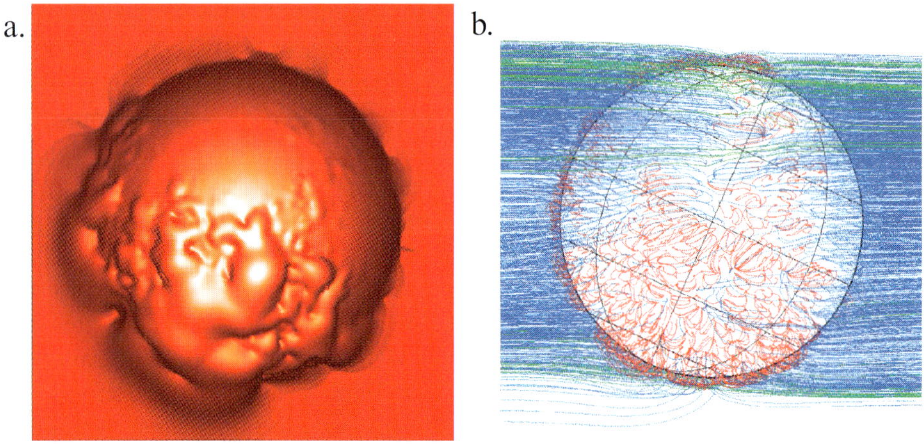

Figure 5. Cartoon showing: (a) the Martian pressure balance obstacle and (b) magnetic field topology. (a) The shape of the Martian solar wind obstacle is derived from a calculation of pressure balance between upstream solar wind dynamic pressure and a combination of ionospheric thermal pressure and magnetic pressure from crustal fields. (b) The magnetic field topology results from field line tracing in a vacuum superposition of a crustal field model with a uniform background magnetic field. Field lines are colored according to their topology: closed (red), open (blue), or draped (green). Mars has the same orientation in both panels. From Brain (2002).

magnetic pressure is small compared to thermal pressure contributions. Therefore, the Martian obstacle should be qualitatively similar to Earth's magnetopause in some locations, and similar to the Venus obstacle in others.

Crustal fields are known to perturb two boundaries in MGS data. The photoelectron boundary reported by (Mitchell *et al.*, 2000) is located at higher altitudes over regions of strong horizontal crustal magnetic field Mitchell *et al.* (2001b). Crustal fields prevent solar wind electrons from accessing the ionosphere in these locations, locally shielding the atmosphere so that ionospheric photoelectrons signatures are detectable by ER. Consequently, ionization processes associated with the solar wind (electron impact and charge exchange) can not occur under these protective bubbles of field, and global fluxes of escaping ions may be reduced.

At higher altitudes the MPB is also perturbed or modified by the presence of strong crustal magnetic fields. Crider *et al.* (2002) used MGS premapping orbits to show that the MPB occurs at higher altitudes in the southern hemisphere, where crustal fields are strongest (see Figure 6). Subsequently, Brain *et al.* (2005a) demonstrated using mapping orbits that, like the PEB, the MPB is higher in particular over strong horizontal crustal fields. Hall MHD model results reported by Harnett and Winglee (2003) suggest that the MPB is different above crustal fields, with more similarities to a magnetopause.

There has been no detected influence of the crustal fields on the location of the bow shock (Vignes *et al.*, 2002). One might expect a connection (e.g. Acuña *et al.*,

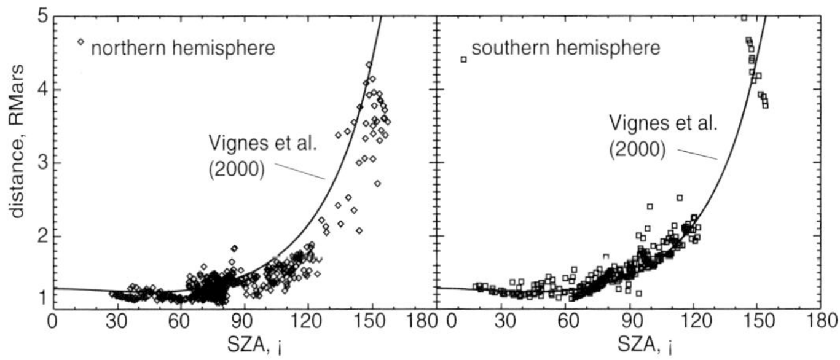

Figure 6. MPB crossings from MGS premapping data are plotted as a function of solar zenith angle (SZA) in the northern hemisphere (left) and southern hemisphere (right). The best-fit shape for the MPB calculated by Vignes *et al.* (2000) using all MPB crossings is shown in both panels for reference. From Crider *et al.* (2004).

1998), since the bow shock location is effectively determined by the location of the solar wind obstacle, which is surely perturbed by crustal fields. However, the bow shock occurs at much higher altitudes than the PEB or MPB, so that local effects from crustal fields may be of minor importance compared to the larger variations in bow shock location caused by external influences such as the direction of the IMF or changes in solar wind pressure (Dubinin *et al.*, 1998; Vignes *et al.*, 2002).

In addition to these local effects, crustal fields may also have global influence on the plasma interaction. There is some indication from statistical analysis of MGS mapping data that the altitude of the magnetosheath on the entire day side (even far from crustal fields) is raised during southern summer at Mars, when the strongest southern crustal fields approach subsolar latitudes (Brain *et al.*, 2005a). The interaction of the shocked solar wind with crustal fields also likely creates current systems in the ionosphere over the entire dayside (Luhmann *et al.*, 2002), creating a global ionospheric dynamo (Withers *et al.*, 2005). Finally, on the Martian nightside, the width of the induced magnetotail calculated from model fits to the dayside MPB crossings was found to be wider when strong crustal sources were beneath the MPB on the dayside (Verigin *et al.*, 2004). Additional results on the degree to which crustal fields perturb plasma boundaries near Mars are becoming available from Mars Express (see Fraenz *et al.* and Dubinin *et al.*, this issue).

3.2. Crustal Influences on Topology

In the absence of crustal sources the IMF would provide the only source of magnetic field at Mars (induced ionospheric magnetic fields ultimately originate from the solar wind). In this Venus-like case, all magnetic field lines, regardless of how they are configured in the Martian system, have both "ends" in the passing IMF. Crustal fields make possible a system with far greater complexity. Field lines near Mars may have one of three different topologies – closed field lines connected at both ends to Mars (i.e. crustal field lines), unconnected field lines connected at both ends to the IMF, and open field lines connecting Mars to the IMF. Open field lines provide an additional opportunity for direct particle exchange between the solar wind and the upper atmosphere of Mars. Changes in topology via reconnection or merging enable the trapping of solar wind plasma in crustal magnetic field "umbrellas", and release of confined ionospheric plasma to the solar wind.

Several groups have considered the Martian field topology and its implications. The cartoon in Figure 5b shows magnetic field lines predicted by a simple linear superposition of the Cain *et al.* (2003) crustal field model with a uniform background "IMF". All three field topologies are present, and magnetic cusps are predicted above some of the strong crustal field regions. More closed field lines are predicted in the southern hemisphere than in the north. The topology of magnetic field near Mars is in many ways more similar to that of the Sun than any other solar system body. More sophisticated vacuum superpositions of crustal field models with an external field have been performed by Luhmann *et al.* (2002) and Brain (2002), with similar qualitative results. Global simulations that include crustal fields predict the presence of open and closed field lines for different orientations of Mars with respect to the Sun and solar wind (e.g. Ma *et al.*, 2002, 2004; Harnett and Winglee, 2005), and there is opportunity to compare these predictions to observations.

Other groups have used MGS data to identify locations having different field topologies. This has proved difficult using *in situ* observations. Krymskii *et al.* (2002a) used maps of the nightside magnetic field orientation created by Connerney *et al.* (2001), and identified locations of likely solar wind energy deposition in magnetic cusps through comparison with idealized dipoles having different orientations. As they note, the search for cusps is complicated by the fact that the orientation of a magnetic field line with respect to the surface does not necessarily dictate its topology. One can conceive of horizontal field lines that connect to cusps of open (and radial) field at lower altitudes, and radial field lines that are part of closed loops. The electron energy spectrum was used by Mitchell *et al.* (2001a) to indirectly identify closed field lines on the night side. Observations for which the measured electron fluxes were consistent with instrumental background, termed 'plasma voids', were inferred to be made on closed field lines where electron source processes are negligible. Plasma voids are observed with regularity in regions of strong horizontal crustal magnetic field (see Figure 7). Plasma voids are punctuated in MGS data by 'flux spikes' observed on radially oriented crustal field lines, where electron fluxes exceed those observed on the night side far from crustal fields. Flux spikes were taken as indicators of open magnetic field lines. Most recently, Brain *et al.* (2004) have used the shape of electron pitch-angle distributions to infer the topology of field lines visited by MGS. They find that the pitch angle distributions have characteristic shapes that can be associated with different topologies. For example, plasma voids and trapped distributions occur on closed field lines. One-sided loss cones indicate field lines where a portion of the most field-aligned incident electron flux has been partially absorbed by the atmosphere below the spacecraft; such

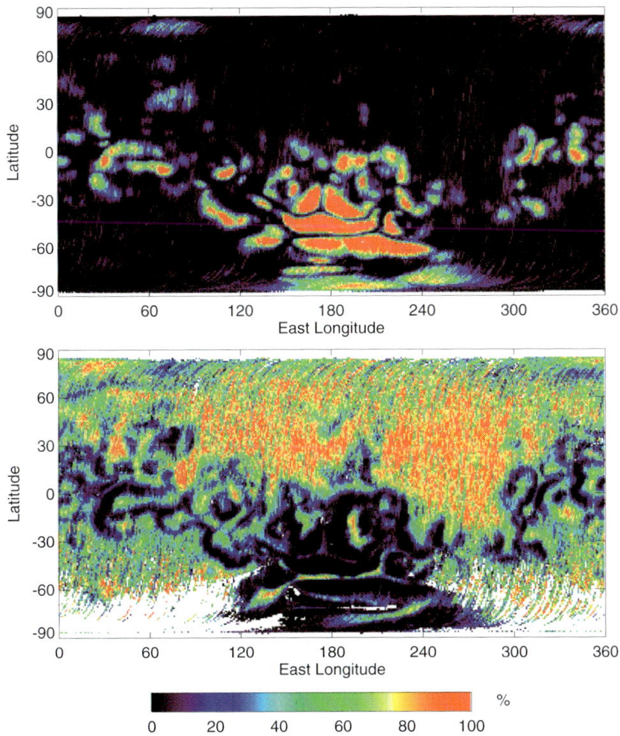

Figure 7. Geographic probability maps showing the location of plasma voids (top) and one-sided loss cones in ER mapping pitch angle distributions recorded in eclipse. From Brain *et al.*, JGR, to be submitted.

distributions are found on open and draped field lines. This method has been used to produce maps of magnetic field topology for Mars (Figure 7), but the analysis is somewhat hampered by the 2-dimensional nature of the ER observations (which results in incomplete coverage in pitch-angle space for many observations), and has only been successfully applied to a fraction of the MGS dataset.

One specific consequence of the complex topology at Mars is the exchange of energy and particles between the upper atmosphere and solar wind (Acuña *et al.*, 1998). The neutral thermospheric scale height is larger above regions of vertical magnetic field (Krymskii *et al.*, 2002a), presumably because the solar wind heats the neutral atmosphere on open field lines. Cusps of open field may also allow ion outflow to occur, analogous to Earth's cusps (Lundin *et al.*, 2005; Ergun *et al.*, 2006). The radar sounder (MARSIS) on MEX has observed near-vertical ionization layers in the Martian atmosphere close to regions of strong radial magnetic field, as might be formed by solar wind particle and energy deposition (Nielsen *et al.*, 2006). And energetic electron distributions of the type reported by Lundin *et al.* (2005); Brain *et al.* (2005b) are predicted to create localized patches of ionization on the night side (Fillingim *et al.*, 2006). Finally, electrons reflected and backscattered from the Martian nightside atmosphere along open field lines have even be used to probe neutral thermospheric densities (and map crustal magnetic field strengths) below the MGS mapping altitude (Lillis *et al.*, 2004, 2005; Mitchell *et al.*, 2006).

4. Global Interaction

Prior to MGS, Phobos was the only spacecraft to cover the Martian plasma interaction globally, and this coverage was both sparse (there were only 4 elliptical orbits and ~100 circular orbits) and incomplete (periapsis did not go below 800 km). The elliptical premapping orbits of MGS allowed it to visit all plasma regimes in the interaction region (albeit with fewer plasma instruments), from upstream of the bow shock to the ionosphere near the main peak. It covered most of these regions over a range of solar zenith angles and local times, making new discoveries and confirming past measurements. In the following section we review highlights of the MGS contributions toward understanding the global plasma interaction.

4.1. Global Views

MGS observations have been used to examine the interaction in a global sense by visualizing the entire system with data. The structure of magnetic fields near Mars has been illustrated in a number of ways by different investigators. MGS confirmed that field magnitude is greatest at low altitudes and solar zenith angles, and RMS (root mean square) is greatest in the sheath between the bow shock and the MPB at low solar zenith angles (see Figure 8a and b) (Brain *et al.*, 2003) . The field is draped on the day side, flares away from the planet with increasing solar zenith angle, and stretches into a two-lobed magnetotail on the night side. The flaring angle of the draped magnetic field has been treated more quantitatively by Crider *et al.* (2001), who found that the flaring of the IMF is less pronounced (but more variable) in the ionosphere than above it. Crider *et al.* (2004) showed that the average measured magnetic field as a function of location in cylindrical coordinates closely resembles the predictions of a simple gasdynamic model where the best-fit MPB was taken as the solar wind obstacle.

MAG data have also been analyzed in order to visualize the properties of electromagnetic plasma waves throughout the Martian system. Properties such as wave frequency, polarization, ellipticity, and propagation direction have been mapped in cylindrical coordinates in the Martian sheath, MPR, and tail by Espley *et al.* (2004a). Wave power at the local gyrofrequency has been mapped in the upstream region by Brain *et al.* (2002). Different wave properties dominate in different regions. Both whistler waves and waves at the local gyrofrequency are detected upstream (Brain *et al.*, 2002; Mazelle *et al.*, 2004), while in the sheath the observed wave properties are consistent with mirror mode waves on the day side and oscillations associated

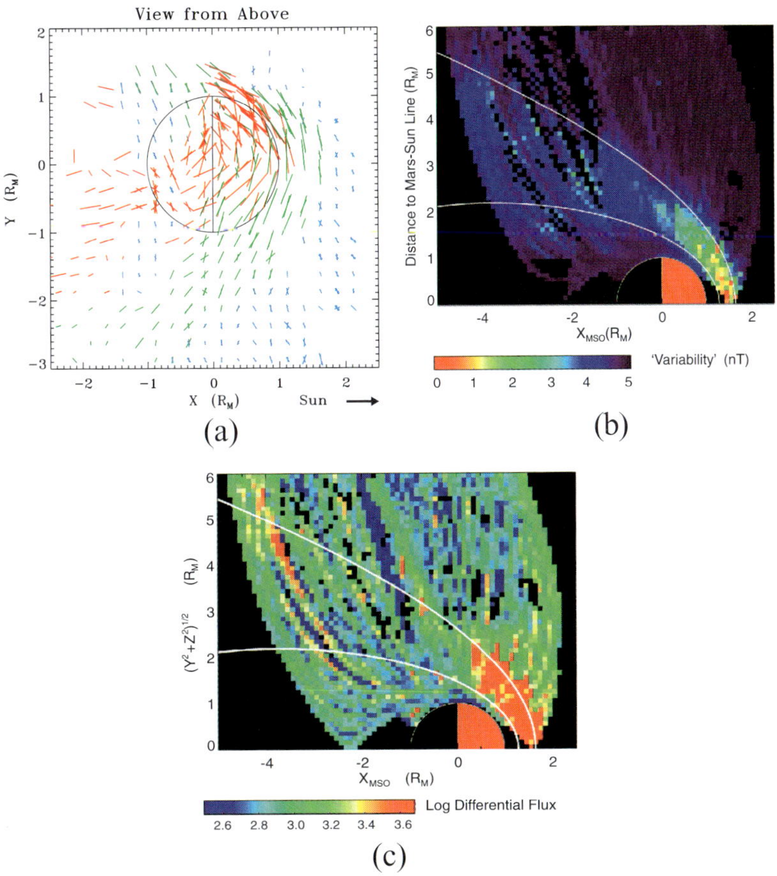

Figure 8. Global views in MSO cylindrical coordinates of the interaction region using MGS premapping observations. (a) Projection of median magnetic field vectors onto the $x - y$ plane, with 20 nT = 1 R_M. Vectors are colored according to whether they are outside the shock (blue), in the sheath(green), or below the MPB (red) as determined from the fits of Vignes *et al.* (2000). From Brain *et al.* (2003). (b) RMS noise in MAG data on 30 second timescales. From Brain *et al.* (2003). (c) Differential electron flux (#/s cm^2 ster eV) at 300 eV.

with the interaction of cold pickup ions with solar wind protons on the night side (Bertucci *et al.*, 2004; Espley *et al.*, 2004a). A mixture of wave modes is observed in the MPR (Espley *et al.*, 2004a), including fast mode MHD waves (Bertucci *et al.*, 2004). A mixture of wave modes is also likely to be responsible for observations of magnetic field fluctuations in the ionosphere, including a magnetosonic mode predicted by kinetic theory (Espley *et al.*, 2004b). Though a variety of wave properties and modes have been inferred from MAG data, coupling of these data with particle observations or more complete plasma data (e.g. electron oscillations reported by Winningham *et al.* (2006)) would be beneficial to understanding the instabilities responsible for each type of plasma wave at Mars.

Finally, electron data may also be used to provide a global view of the Martian plasma interaction. Figure 8 shows that the average electron flux at 300 eV is highest in the sheath, at low solar zenith angles. Below the MPB the flux drops abruptly, consistent with the electron signatures observed at the MPB in MGS data (see Section 4.4).

4.2. Upstream and Foreshock

The upstream and foreshock regions of Mars were visited early in the premapping mission phase, while the MGS orbit periapsis altitude was still high. The typical magnetic field upstream from the shock is 2–4 nT and conforms to the expected Parker spiral configuration of 56° (Crider *et al.*, 2001; Brain *et al.*, 2003). As mentioned above, analysis of oscillations in MAG data confirmed the presence of plasma waves near the proton gyrofrequency (Brain *et al.*, 2002; Mazelle *et al.*, 2004), which are also seen as oscillations in electron data (Mazelle *et al.*, 2004). These waves have been interpreted as standing waves resulting from the interaction of solar wind ions with planetary ions (Mazelle *et al.*, 2004), and as resulting from direct pickup of planetary hydrogen (Russell *et al.*, 1990). MAG data also contained evidence for previously undetected whistler waves in the Martian foreshock (Brain *et al.*, 2002), which have properties at Mars consistent with expectations based on observations at other solar system bodies (see Orlowski and Russell, 1995). Finally, hot diamagnetic cavities upstream of the Martian shock have been reported from MGS data (Figure 9), analogous to hot flow anomalies at Earth believed to result from the interaction of solar wind discontinuities with the bow shock (Øieroset *et al.*, 2001). Upstream phenomena have been reviewed in detail by Mazelle *et al.* (2004); Bertucci *et al.* (2005a).

4.3. Boundary Shapes

MGS greatly increased the number of crossings of plasma boundaries, allowing quantitative fits to idealized shapes for the bow shock and MPB and comparison with fits based on previous measurements. Vignes *et al.* (2000) calculated a bow shock shape based on MGS crossings (shown in Figure 1), and Trotignon *et al.* (2006) have calculated a shape combining crossings detected in MGS MAG data and Phobos plasma wave data. These fits are similar to previous fits, summarized in Trotignon *et al.* (2006), though are more accurate due to the greater number of crossings and better solar zenith angle coverage. The Martian bow shock is located at altitudes of ~2000 km near the subsolar point, and ~5500 km near the terminator, and appears to be insensitive to solar cycle (Vignes *et al.*, 2000). However, the shape of the bow shock is asymmetric with respect to IMF direction (Dubinin *et al.*, 1998; Vignes *et al.*, 2002), and in general appears to be highly variable. The use of a "best-fit" bow shock shape then is only a first step toward understanding the shape and

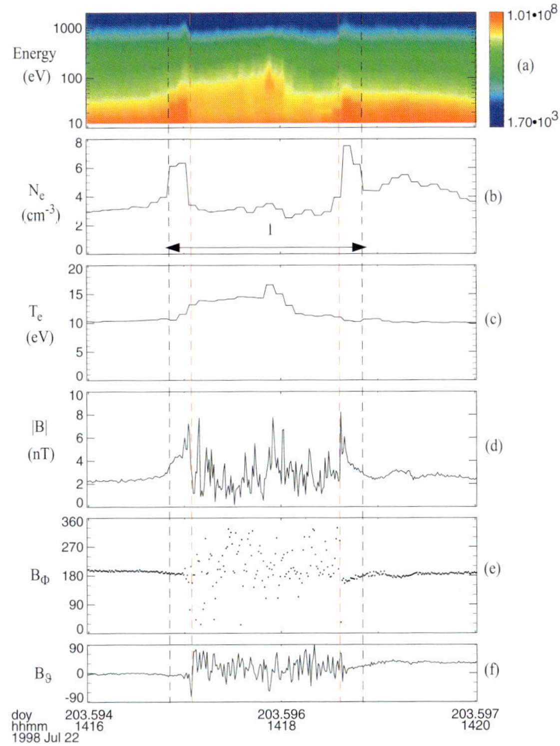

Figure 9. Hot diamagnetic cavity upstream of the Martian shock, with regions of high density and magnetic field surrounding a central hot, turbulent plasma with the same density as the undisturbed solar wind. This event is associated with an IMF rotation. From Øieroset *et al.* (2001).

location of the shock. Future studies might account for a number of controlling factors in order to parameterize the shape of the bow shock as a function of location and external conditions. With more than 700 crossings (Trotignon *et al.*, 2006), not including those already returned by MEX, such a study will soon be possible.

MGS increased the number of recorded MPB crossings at Mars from 41 (from Phobos) to nearly 900. The shape has been fit using Phobos data (Trotignon *et al.*, 1996), MGS data (Vignes *et al.*, 2000), a combination of Phobos and MGS data (Trotignon *et al.*, 2006), and MEX data (Dubinin *et al.*, this issue). The four modeled shapes are in rough agreement. A peculiar feature of the Vignes *et al.* (2000) fit (shown in Figure 1) is that the MPB has higher altitudes near the subsolar point than at moderate solar zenith angles. This result is almost certainly not physical and simply results from the assumed shape for the boundary (an ellipsoid offset from the center of Mars), coupled with the lack of coverage by MGS of low solar zenith angles (see Section 2.2 and Figure 2). From the model fits, the MPB is situated at ~850 km altitudes near the subsolar point, and ~1500 km near the terminator. Similar to the bow shock, the MPB location is highly variable, and the variability

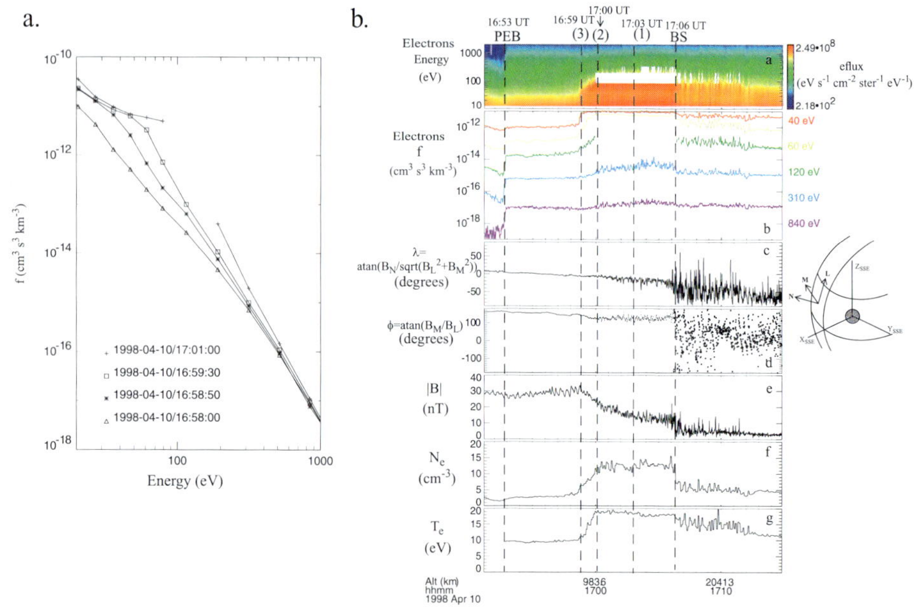

Figure 10. Signatures of the MPB in MGS data. (a) Superthermal electron fluxes decrease across the MPB. (b) Changes in electron fluxes, field strength, turbulence, and orientation are observed across the MPB, along with changes in electron density and temperature. Magnetic field is shown in a minimum variance coordinate system. From Øieroset *et al.* (2004).

increases with solar zenith angle. Factors that control the location of the MPB are discussed in Section 5.2.

The PEB observed by MGS is known to be highly variable in its location (discussed further in Section 5.2). Using a combination of premapping and mapping MGS data, the PEB could also be fit to a model shape. Such an effort, incorporating the many thousands of crossings in MGS premapping and mapping observations, should be undertaken in the future.

4.4. MPB Signatures and Physics

Previous spacecraft to visit Mars have crossed the MPB, and have referred to it by many names (planetopause, ion composition boundary, mantle boundary, protonopause, magnetopause, etc.). The large number of crossings by MGS has enabled several new insights into the signatures and underlying physics responsible for this boundary. The signatures of the MPB in MGS MAG/ER data (crossing from upstream to downstream) include: an increase in field magnitude, a decrease in field fluctuations, an increase in the field 'draping', and a decrease in superthermal electron fluxes (see Figure 10). These signatures have been used in a number of papers to study the MPB shape (Vignes *et al.*, 2000; Trotignon *et al.*, 2006), its variability (Crider *et al.*, 2002, 2003; Verigin *et al.*, 2004; Brain *et al.*, 2005a), and its similarity in characteristics and structure to boundaries observed at other planets (Bertucci *et al.*, 2005b). Additionally, MGS data show that the dominant ULF waves differ on either side of the MPB (Bertucci *et al.*, 2004).

Despite the many different names and plasma signatures associated with this boundary, it seems clear that it results from the interaction of the shocked solar wind with planetary heavy ions (see discussion in Nagy *et al.*, 2004). Comparisons of models to data suggest that ionization of the exosphere (via electron impact and charge exchange) play a role in creating the signatures observed by MGS (Crider *et al.*, 2000; Chen *et al.*, 2001). Observation of the same boundary by the Phobos instruments allowed a more complete set of identifying signatures to be constructed, including a change in the ion population from solar wind dominated to planetary dominated (e.g. Breus *et al.*, 1991; Dubinin *et al.*, 1996). Continued measurements and comparison to simulations will help to identify the detailed physics responsible for forming and maintaining the MPB. MEX data are already providing important new information (see other papers in this issue).

A "big picture" question about the MPB is whether this apparently common feature of plasma interactions with atmospheres has an analog at magnetized planets. It has been suggested that the MPB has similarities in structure and behavior to the plasma depletion layer upstream of Earth's magnetopause Øieroset *et al.* (2004). Further, the MPB appears to be the inner boundary for solar wind protons, similar to a magnetopause. Of all bodies in the solar system, the question may best be answered through observations at Mars, which exhibits features of both a Venus-like atmospheric interaction (Cloutier *et al.*, 1999) and an Earth-like magnetospheric interaction near crustal sources (see, for example Krymskii *et al.*, 2000).

The interested reader is referred to reviews by Bertucci *et al.* (2005a); Nagy *et al.* (2004) for further information on the MPB.

4.5. Ionosphere

The only *in situ* sampling of the ionosphere prior to MGS was made by the Viking Landers during their descent. The ionosphere is detected by the ER instrument on MGS using electron energy spectra (Mitchell *et al.*, 2000). Below the PEB, where contributions from solar wind-like electrons are relatively weak, ER measures features attributable to photoemission of oxygen. The transition from a regime dominated by solar wind electrons to one dominated by photoelectrons (shown in Figure 11), was seen at altitudes ranging from 180–800 km in the northern hemisphere at high solar zenith angles. Multiple crossings, evident in some orbits, indicate detached ionospheric clouds or surface waves (Mitchell *et al.*, 2001a).

In addition to determination of the ionosphere's upper boundary, there has been some progress in measuring the Martian ionosphere using MGS data. Vignes *et al.* (2004) studied flux ropes identified in the Martian ionosphere at high latitudes in

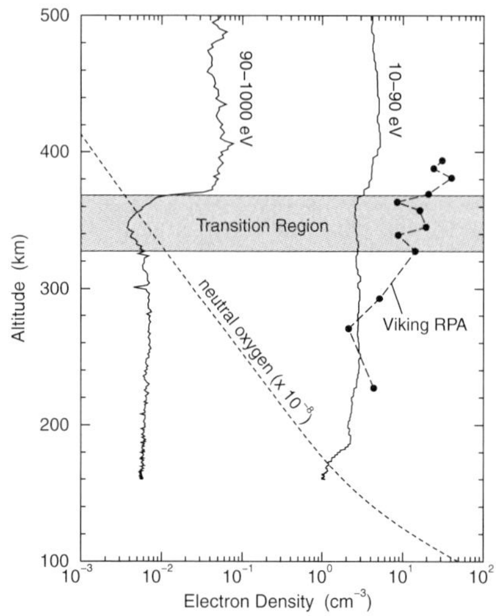

Figure 11. Altitude profiles of electron density for two different energy ranges measured by MGS ER. The PEB is located at a sharp transition in 90–1000 eV density. From Mitchell *et al.* (2000).

the northern hemisphere (Cloutier *et al.*, 1999). This preliminary study suggests that flux ropes are observed more often at Venus than Mars (where the ionosphere is often magnetized), and never near crustal fields in the southern hemisphere. The orientation of the draped IMF in the ionosphere, discussed in Section 4.1, has been quantified (Crider *et al.*, 2001). And several studies have analyzed ionospheric profiles measured using radio science, including influences from crustal fields and solar X-ray flux (Krymskii *et al.*, 2002a, 2004, 2002b; Ness *et al.*, 2000; Withers *et al.*, 2005; Mendillo *et al.*, 2003, 2006). In the future, photoelectron fluxes in MGS data might be studied as a function of solar zenith angle and external conditions to learn more about the distribution of ionospheric electrons.

4.6. Wake

The wake and tail are particularly important for studies of present day atmospheric escape, since much of the escaping ion flux passes through these regions. MGS lacks ion measurements, and studies of the properties of these regions have necessarily focused on nightside magnetic field structure and superthermal electron distributions (each of which may indirectly provide clues about escaping particles and processes). The central wake and tail was also explored by Phobos, and its thickness and field orientation were used as support for both intrinsic and induced Martian obstacles to the solar wind (Riedler *et al.*, 1989; Russell *et al.*, 1995; Axford, 1991; Möhlmann *et al.*, 1991; Dubinin *et al.*, 1994). Comparison of MGS premapping MPB crossings to a model for the boundary shape by Verigin *et al.* (2004) suggests

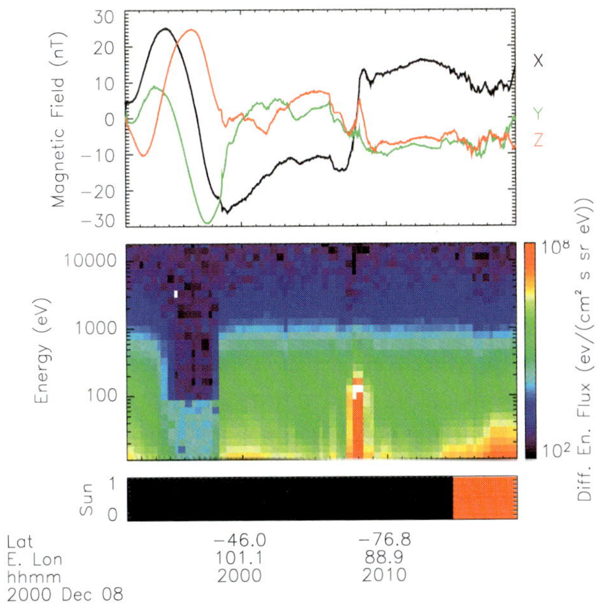

Figure 12. Nightside current sheet crossing by MGS. The magnetic field in MSO coordinates (top) has a sharp reversal near 20:08, accompanied by an increase in 10–200 eV electron energy fluxes (middle) Crustal fields are evident at the beginning of the time period. From Halekas *et al.* (2006).

that crustal magnetic fields act to make the tail boundary (the nightside extension of the MPB) thicker by up to 1000 km, so that both intrinsic and induced fields influence the tail structure. MGS mapping data revealed a magnetic flux asymmetry between the two lobes of the induced magnetotail (Ferguson *et al.*, 2005). More detailed study of nightside current sheet crossings (inferred from field reversals) shows that the thin current sheets measured at the 400 km mapping altitude of MGS (shown in Figure 12) have locations and variability consistent with reconnection of the draped IMF to crustal magnetic fields (Halekas *et al.*, 2006). And Mitchell *et al.* (2001a) showed that electron fluxes at energies less than ~400 eV are lower in the tail.

Measurements made by MEX in the past two years prompted re-examination of the MGS tail measurements for evidence of particle acceleration. First, UV auroral emission detected by MEX was reported by Bertaux *et al.* (2005) in a region of radial nightside crustal magnetic field. MGS observations during this time show that this observation occurred during the passage of a SEP event through the Martian system, before the arrival of the CME shock (Brain *et al.*, 2005b), suggesting a possible link between auroral emission at Mars and SEP events, as has been suggested for Venus (Phillips *et al.*, 1986). Secondly, Lundin *et al.* (2005) reported nightside ion and electron spectra peaked in energy, indicative of a field-aligned auroral-like acceleration process. Examination of MGS mapping data at much lower altitude revealed thousands of auroral-like peaked electron energy spectra (Brain *et al.*, 2005b), and showed that these spectra occurred unambiguously above regions of radial crustal magnetic field lines, on the edges of closed field regions. The MGS data also revealed that in some locations the observations of auroral-like electron distributions is influenced by IMF direction, Martian season, and (weakly) by solar wind pressure. Each of these dependencies indicates an external influence on the conditions required for observation of accelerated electrons. The upward-accelerated ions seen by Mars Express show that atmospheric escape may occur out of cusps of crustal magnetic field on the Martian night side (Lundin *et al.*, 2005). Further study of particle acceleration signatures, in tandem with MEX UV and particle observations, should prove useful in uncovering how the particles are accelerated and their effect on the Martian atmosphere.

5. Variability

The preceding sections of this review have focused in large part on MGS contributions toward describing the steady-state Martian solar wind interaction. However, the particles and fields environment is highly variable. MGS mapping data are particularly useful for exploring variability in the Martian system, since they have been collected in one small region of this interaction over a period now in excess of six years. MGS data have shown how the Sun and solar wind influence plasma boundaries, magnetic fields, and field topology near Mars, and that unexplained asymmetries exist in the data. MGS results on variability are discussed below.

5.1. Proxies

There is no upstream solar wind monitor at Mars, and no monitor of solar EUV. Therefore, proxy information derived from MGS and Earth-based observations must be employed in order study external influences on the *in situ* MGS measurements. To date, proxies for solar EUV flux, upstream solar wind pressure, and the clock angle of the IMF have been constructed. Figure 13 shows a timeseries for each of these proxies during the MGS mapping orbit. The proxy for the solar EUV flux at Mars has been inferred from the F10.7 radio flux measured at Earth, extrapolated from 1 AU to the heliocentric distance of Mars, and time-shifted to account for the difference in solar longitude of Mars and Earth (Mitchell *et al.*, 2000).

Magnetic pressure on the day side, far from crustal fields, is assumed to be proportional to upstream solar wind dynamic pressure in a proxy developed by Crider *et al.* (2003). This method compares favorably with extrapolation of Earth-based

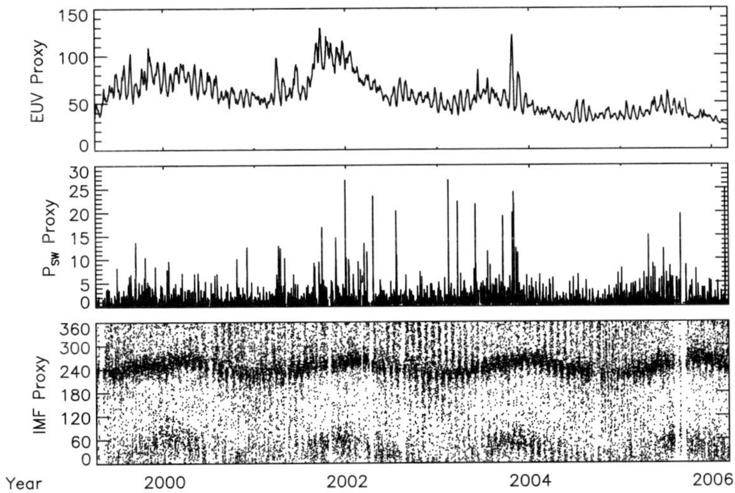

Figure 13. Three proxy datasets for Mars: EUV flux in solar flux units (top); solar wind pressure in units of nPa (middle); and IMF draping direction in degrees (bottom).

measurements to Mars' orbital distance. Pressure values are computed on an orbit-by-orbit basis, and assume that sudden changes in pressure do not occur during each two-hour orbit period. Further, the MPB is pushed below the mapping altitude of MGS on ~20% of the mapping orbits, so that magnetic pressure in the sheath, rather than the pileup region, is used to calculate the proxy (Brain *et al.*, 2005a). In the sheath, however, the thermal plasma pressure may constitute a larger fraction of the total pressure, so that use of magnetic pressure alone provides an underestimate of the upstream solar wind pressure. For these reasons pressure proxy information is best-used in statistical studies that seek to separate high pressure time periods from low pressure time periods. In addition to the MGS-based pressure proxy, Vennerstrom *et al.* (2003) have extrapolated ACE data to the heliocentric distance of Mars during time periods when Mars and Earth were magnetically aligned.

The orientation of the IMF upstream from Mars has been estimated in three different ways. For pre-mapping data, (Crider *et al.*, 2004) determined the IMF draping direction from field vectors recorded immediately downstream from the bow shock, where field amplitudes are large enough that the determination is not overly sensitive to spacecraft-generated magnetic fields. For mapping data, Brain *et al.* (2006) used the configuration of the draped IMF on the dayside, far from crustal fields, as indicative of the clock angle of the upstream IMF. Like the proxy for solar wind pressure, both these methods calculate a proxy on an orbit-by-orbit basis, and assume that external conditions do not change during each orbit. Vennerstrom *et al.* (2003) showed that, for time periods when Earth and Mars are magnetically aligned, ACE data provide an adequate estimate of the IMF orientation on timescales associated with solar wind sector changes.

TABLE I

Drivers affecting variability in the location of plasma boundaries at Mars, and references to analyses of MGS data.

	Bow shock	MPB	PEB
Solar wind pressure	?	Yes[a]	Yes[b]
IMF direction	Yes[c]	Yes[d]	?
EUV	?	?	Yes[e]
Martian season	?	Yes/?[d]	?
Crustal fields	No/?[c]	Yes[a]	Yes[b]

[a]Crider *et al.* (2003).
[b]Mitchell *et al.* (2001b).
[c]Vignes *et al.* (2002).
[d]Brain *et al.* (2005a).
[e]Mitchell *et al.* (2000).
[f]Crider *et al.* (2002).

5.2. Boundaries

The response of the location of plasma boundaries near Mars to different influences has been well-studied. Table I shows whether the bow shock, MPB, and PEB have been demonstrated to vary in response to five drivers using MGS data. High solar wind pressure compresses both the MPB and the PEB, and likely similarly affects the bow shock (Crider *et al.*, 2003; Mitchell *et al.*, 2001b). The IMF direction controls the direction of the solar wind convection electric field ($\mathbf{E}_{SW} = -\mathbf{v}_{SW} \times \mathbf{B}$), which in turn affects the motion of charged particles in the planetary interaction region. The bow shock and MPB have both been shown to have asymmetric shapes (or, less likely, to change size) determined by the IMF orientation (Vignes *et al.*, 2002; Brain *et al.*, 2005b). Mass loading of the flow by planetary heavy ions is thought to influence the global interaction at Mars similar to comets, particularly near the MPB. It has not yet been demonstrated whether mass loading is directly responsible for the observed variability through creation of a hemispherically asymmetric obstacle to the flow, or whether simple particle motion controls the observed asymmetries. Solar EUV flux, initially shown to have little influence on the PEB location (Mitchell *et al.*, 2000), has more recently been demonstrated to raise the altitude of the PEB during observations made when the EUV flux is high (Mitchell *et al.*, 2001b). Seasonal effects have been observed in the location of the MPB (Brain *et al.*, 2005b), but this influence is thought to be caused indirectly by crustal fields, which are strongest at southern mid-latitudes and therefore approach closer to the subsolar point during southern summer, raising the altitude of the MPB. Finally, the effects of crustal fields have been reported for the location of the PEB

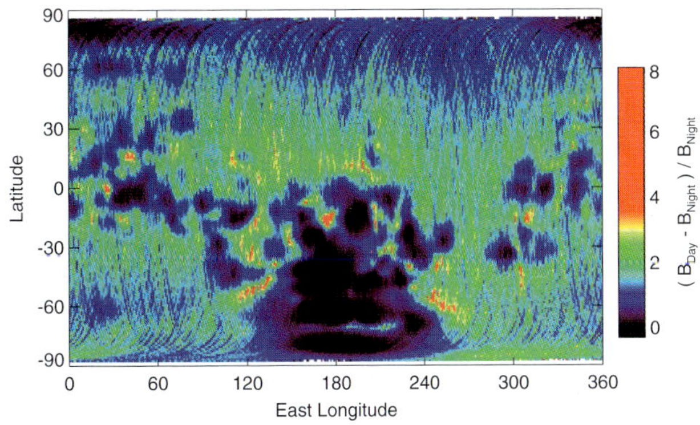

Figure 14. Geographic map of compression in the magnetic field magnitude, expressed as $\Delta B/B_{\mathrm{night}}$, where $\Delta B = B_{\mathrm{day}} - B_{\mathrm{night}}$.

and MPB (Mitchell *et al.*, 2001a; Crider *et al.*, 2002), but have not been definitively measured for the more distant bow shock (Vignes *et al.*, 2002).

5.3. Field and Topology

Variability has been detected in MGS MAG measurements of field amplitude and orientation throughout the Martian system. One example is the day-night variability shown in Figure 14. The figure shows the increase in field magnitude on the day side in many geographic locations, relative to the field magnitude at each location on the night side. On average, at mid-latitudes the dayside field is a factor of two or more higher than on the night side, with large increases in certain regions near crustal sources. This excess field on the dayside likely has two sources: the draped IMF and current-generated magnetic fields. One unexplained feature of this map is a ~(20%) reduction in average field strengths on the day side relative to the night side in some regions of strong horizontal crustal magnetic field. Other variability in magnetic field demonstrated using MGS data includes the influence of upstream pressure and IMF direction on the night side (Ferguson *et al.*, 2005; Brain *et al.*, 2006). These analyses show that external influences have not been entirely removed from the data also used to construct models for the crustal magnetic fields (Purucker *et al.*, 2000; Arkani-Hamed, 2001; Cain *et al.*, 2003; Langlais *et al.*, 2004).

Field topology, determined from ER and MAG measurements exhibits variability on both the night side and day side with IMF direction and upstream pressure. Figure 15 shows geographic probability maps of the likelihood of observing one-sided loss cone distributions in ER pitch angle data for two sets of MGS orbits. For orbits where the IMF on the Martian day side is roughly westward, there is a low probability of observing distributions associated with open field lines in the region centered near (210°E, 45°S). For eastward IMF, however, open field lines are often

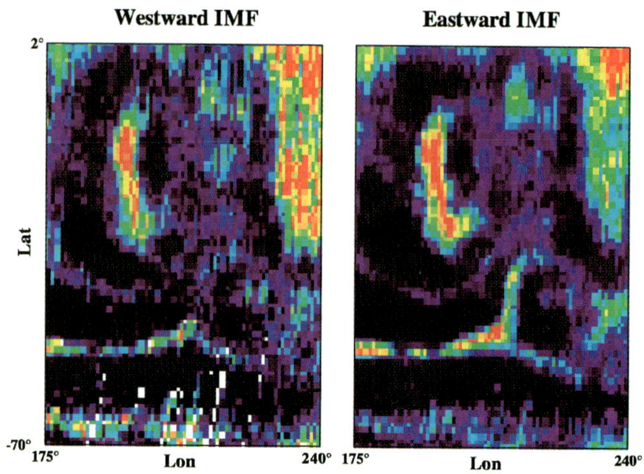

Figure 15. Geographic probability maps of one-sided loss cone distributions, analogous to Figure 7b, separated according to IMF direction.

observed, indicating that IMF direction in part determines whether a crustal field cusp at 400 km is open or closed. Other differences between the two maps are evident in the figure, and may be associated with topological changes governed by the IMF orientation. This change in topology could be achieved by magnetic reconnection, or by large scale motion of closed and open field regions – both governed by IMF direction. Variability in topology has also been measured for upstream pressure variations (Brain *et al.*, 2004).

5.4. Asymmetries

A number of unusual field asymmetries have been discovered in MAG/ER data, and their origin has not been fully resolved. First, Krymskii *et al.* (2002a) discovered that the total magnetic flux calculated from a map of the median nightside magnetic field (Connerney *et al.*, 2001) is non-zero, with significant additional flux toward the planet. The map was constructed from observations made over a long time period, so that the different directions of the IMF should have largely averaged out of the map. This observation of non-zero flux may simply result from the fact that the magnetic field map used in the calculation was not made over a closed surface; instead Mars rotated underneath the spacecraft situated at fixed local time as statistics were accumulated in each bin. Therefore, a negative net flux at the 2am orbit may be compensated by a net positive flux at other local times. Ferguson *et al.* (2005) also observed an asymmetry in the sunward component of the nightside magnetic field, after a spherical harmonic crustal field model was subtracted (see Figure 16). Further, the asymmetry between the number of observations with B toward the Sun/planet vs. away grew with upstream solar wind pressure. Brain *et al.*

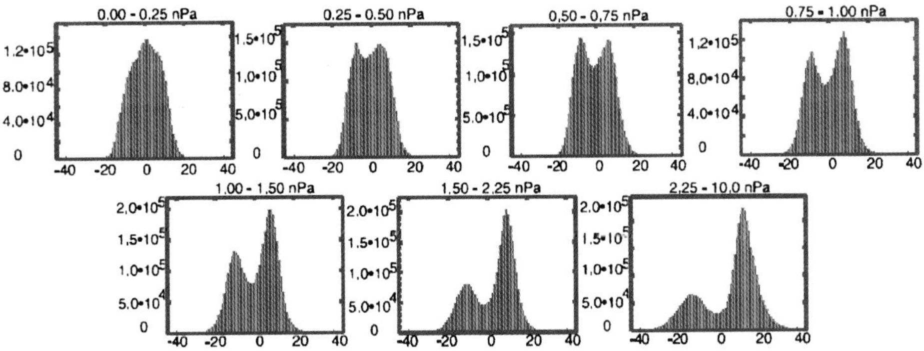

Figure 16. Histograms of the sunward component of magnetic field on the Martian night side, after the Cain *et al.* (2003) crustal magnetic field model has been subtracted, separated according to upstream pressure. From Ferguson *et al.* (2005).

(2006) showed an asymmetry in the nightside radial field component controlled by IMF direction. This asymmetry might be explained by an asymmetrically shaped MPB and tail, however there is a peculiar long-wavelength dependence of the asymmetry on planetary longitude which remains unexplained. Finally, Brain *et al.* (2006) showed an asymmetry in the draping directions of the magnetic field on the day side in the northern hemisphere, where draping directions cluster in a direction pointed toward the subsolar point when Mars is in one sector of the solar wind, but not the other. This asymmetry might be explained by an asymmetry in "weather-vaning" of the draped field in the ionosphere for one IMF direction, coupled with the asymmetrically shaped MPB mentioned above. "Weathervaning" refers to the antisolar draping of the low-altitude portion of magnetic field lines as they are embedded in the ionosphere, and was observed extensively at Venus (Law and Cloutier, 1995) and reported at Mars (Cloutier *et al.*, 1999).

5.5. SEP Effects

A more extreme source of variability has been observed at Mars in the form of Solar Energetic Particle (SEP) events. SEP events are associated with Coronal Mass Expansions (CMEs) from the Sun, and charged particles can be accelerated to energies of hundreds of MeV near the Sun or at the shock front of the CME as it expands and propagates into the solar system. The influence of SEP events, many of which have also been observed at Earth, has been detected in MAG/ER data and related to effects measured by other spacecraft instruments in the upper atmosphere. A particularly large event occurred at Earth on 28 October 2003, and is referred to as the Halloween 2003 event. During this event, MGS observed compression of the Martian system and an increase of field strengths on the day side (Crider *et al.*, 2005). Solar wind access to low altitudes (determined from electron

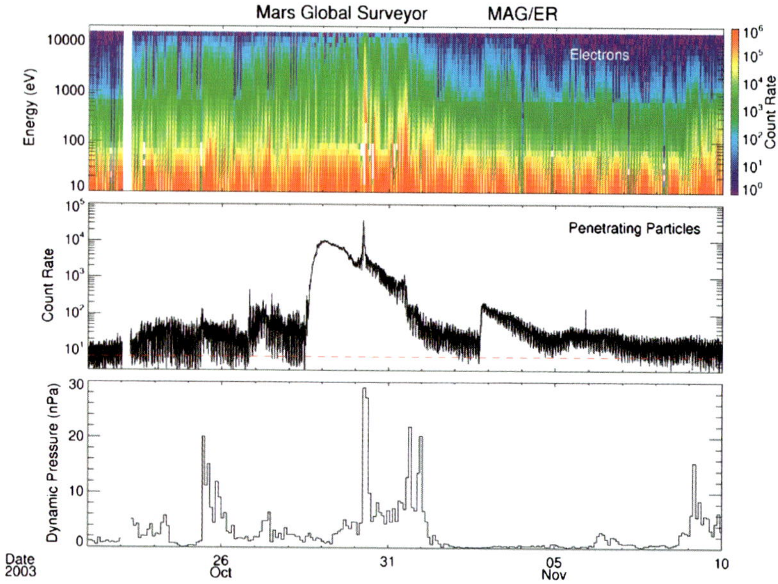

Figure 17. The Halloween 2003 solar events at Mars. (Top) Flux of energetic (10–20 keV) electrons measured by ER, after subtraction of SEPs and GCRs. (Middle) Count rate of SEP and GCR particles penetrating the ER instrument. (Bottom) Solar wind pressure proxy. (Figure courtesy of D. Mitchell).

measurements) increased (Crider *et al.*, 2005), and waves near the gyrofrequency of pickup hydrogen and oxygen were observed on the typically quiet night side (Espley *et al.*, 2005). Each of these observations suggests that atmospheric escape rates are elevated during solar storms, which may be more reminiscent of conditions early in Martian history. Recently, MARSIS radar observations have been shown to contain evidence for additional ionization layers in the atmosphere during the times of SEP events (Morgan *et al.*, 2006).

Figure 17 shows a timeseries of background countrates from the ER during a large SEP event at Mars, and the solar wind pressure proxy during this time period. The three highest energy channels of the ER instrument (10–20 keV) typically measure background, which is dominated at quiet times by the Galactic Cosmic Ray (GCR) flux. However, the ER is sensitive to solar energetic protons with energies of 10's MeV, which directly penetrate the instrument housing and strike the instrument anode to be recorded as counts. The countrate of penetrating particles (SEPs and GCRs) is independent of the ER energy channel, since these particles do not pass through the instrument optics and are therefore unaffected by the energy to which the instrument is tuned. For large penetrating particle fluxes, then, the three highest energy channels have nearly equal countrates when the signal is dominated by penetrating particles, and unequal countrates when there are significant numbers of 10–20 keV electrons present. The event is apparent in ER data as a rise in background countrate in all energy channels equally on 28–29 October, followed by a spike in the ER countrate and associated increase in solar wind pressure on 30 October, followed by a steady decline during which there are large temporal variations in the ER backgrounds as well as a consistently lower countrates in the highest energy channel. These three phases are interpreted as an initial period where SEPS accelerated in front of the CME shock encounter the Martian system, followed by the passage of the shock by Mars, and a timeperiod where both SEPs and energetic electrons are measured by ER. This event is only one of many tens of examples of SEP events detected by ER at Mars during the mapping phase of the mission, and we estimate that ~4% of MGS observations occur during SEP events.

6. Summary

The Martian plasma interaction has similarities to the atmospheric interaction at Venus and comets, and to a magnetospheric interaction at Earth or on small scales at the Moon. The main features of the Martian solar wind interaction region are well-known, including a bow shock, magnetosheath, MPB (or alternatively named boundary), MPR, ionosphere, wake, tail, and plasma sheet. In addition, crustal magnetic fields are an important part of the interaction. The solar wind interaction has bearing on the problems of atmospheric escape and climate evolution at Mars, the structure and variability of the upper atmosphere, and on fundamental plasma processes such as reconnection and particle acceleration. A number of spacecraft have made relevant measurements at Mars, dating back more than forty years.

MGS carries a vector magnetic field instrument and an electrostatic analyzer, dedicated to study of the plasma environment and intrinsic magnetic field at Mars. Magnetometers and electron measurements had both been made at Mars before, but the unprecedented orbit of MGS enabled many new discoveries. The spacecraft went much lower than any previous spacecraft carrying plasma instruments, covered the global interaction region better than any previous spacecraft during its elliptical aerobraking period, and has made mapping observations from ~400 km and fixed local time for more than six years. This orbit allowed discovery of crustal sources, characterization of different regions of the interaction, and determination of variability in response to many different drivers and on many timescales.

Crustal fields are sufficiently strong that they extend upward into the plasma interaction and modify it. Because they contribute magnetic pressure that helps the ionosphere divert the solar wind, they perturb the locations of boundaries, and may fundamentally change their nature. Crustal fields also enable new magnetic topologies, including closed field lines that shield portions of the atmosphere from the solar wind, and open field lines that allow access of the solar wind to the lower ionosphere (and particle escape).

MGS has made new discoveries and contributions in each region of the solar wind interaction. The entire system has been visualized using MAG/ER data in various forms. Hot diamagnetic cavities have been observed upstream and whistler waves have been detected in the foreshock. The shape of the bow shock and MPB have been determined with greater accuracy. New defining features of the MPB have been defined and compared to similar boundaries at other planets, and the role of ionization processes in creating the boundary has been explored. Flux ropes have been observed in the ionosphere, and the upper boundary to the ionosphere has been characterized as a function of location and external drivers. Evidence for electron acceleration in cusps of crustal magnetic field has been reported, and low-altitude current sheets have been reported, possibly resulting from reconnection of the draped IMF with crustal fields.

Variability has been observed throughout the system. The bow shock, MPB, and PEB have all been observed to respond to external drivers, including solar wind pressure, IMF direction, EUV fluxes, crustal fields, and Mars' orientation with respect to the Sun and solar wind. MGS data have proved capable of supplying proxy information about the upstream solar wind pressure and IMF orientation. This proxy information is useful for organizing both MGS and MEX observations. The bow shock and MPB are both asymmetrically shaped, with the asymmetry apparently controlled by the IMF direction. Magnetic field magnitude, orientation, and topology all respond to conditions in the upstream solar wind. Asymmetries have been observed in magnetic field measurements which have not yet been completely explained. SEP events noticeably disturb the Martian system, compressing the interaction region after the arrival of a CME shock, increasing atmospheric escape fluxes, and depositing energy in the upper atmosphere.

6.1. Looking Forward

Though the many contributions from MGS have been summarized in this review, it is likely that many more will follow. Three different avenues of research should bring new results from MGS. First, the existing MGS premapping and mapping data have not yet been fully mined for the information they carry. Promising areas include further analysis of the ER angular electron distributions, investigation of high time resolution vector magnetic field data for waves and discontinuities, and investigation of the detailed physical processes (such as reconnection and particle acceleration) likely evident in mapping orbit data.

Second, MAG/ER continues to make measurements from the MGS mapping orbit, and may do so for several more years. New data would undoubtedly promote new discoveries. Additionally, a few more years of observations would enable investigation of the dataset over an entire solar cycle at Mars, and reduce uncertainties in statistical analyses (for example, investigation of mapping orbits recorded during periods of high solar wind pressure, only).

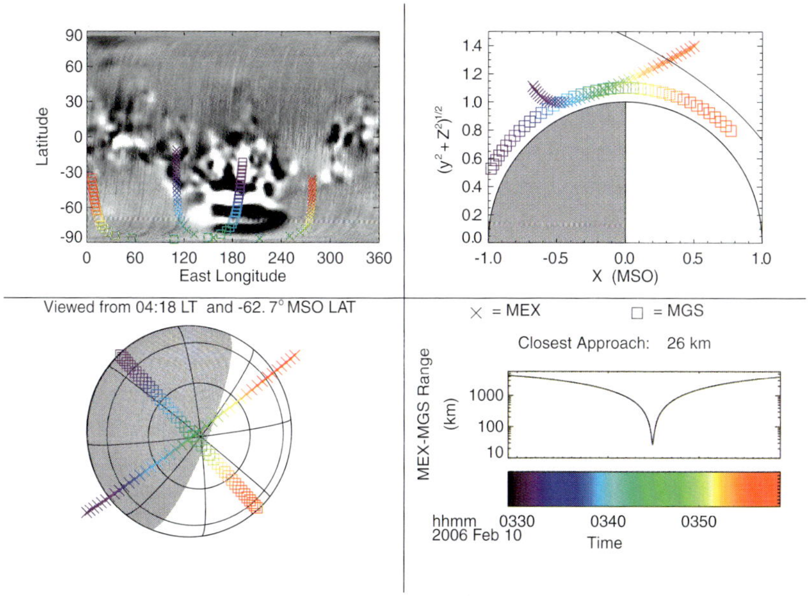

Figure 18. Orbit geometry for MGS and MEX during a close spacecraft pass. Trajectories are colored according to time (bottom right), with × denoting MEX position and square denoting MGS position. Trajectories are shown in rectangular planetary coordinates (top left), MSO cylindrical coordinates (top right), and in satellite projection (bottom left).

Finally, simultaneous measurements by other spacecraft may enhance the scientific return of MAG/ER, and vice versa. Since early 2004, MGS and MEX have been making *in situ* measurements of the plasma environment that are complementary (in terms of both instrument and orbit). Each of MGS and MEX data in tandem provides an opportunity to mitigate the shortcomings of each dataset, and increase our overall understanding of the Martian solar wind interaction and atmospheric escape. Close passes of spacecraft (conjunctions) are one particularly powerful means of increasing the utility of measurements, as evidenced by the Cluster mission at Earth. At Mars, conjunctions might be used to obtain more complete simultaneous and/or co-located plasma measurements, which can be used to study a variety of phenomena, including measurements of auroral-like particle acceleration near crustal fields and the three-dimensional motion and shape of plasma boundaries.

Approximately forty conjunctions (instances with instantaneous spacecraft separation smaller than 400 km) of MGS and MEX have already been identified between January 2004 and February 2006. The closest pass was 27 km, near the South Pole (Figure 18). Conjunctions occur both at mid-latitudes (when the surface-projected orbit tracks of the two spacecraft nearly overlap), and at the poles. These conjunctions will be explored further in the coming months, including intercomparison of MGS and MEX electron data, the addition of MGS magnetic field and MES ion data, and the inclusion of solar wind proxy information to establish context. Other configurations of MGS and MEX may also prove useful, including times when they are on the same flux tube (for spatial evolution of particle distributions), times when they pass through the same region of space separated by a delay (for time evolution of plasma populations in certain regions), and times when they are on opposite sides of plasma boundaries (to determine boundary shapes and motion).

Comparison of MGS observations simultaneous with those of other spacecraft is already proving quite useful in understanding the system. Measurements from the Mars Express instruments ASPERA-3 (e.g. Fedorov *et al.*, 2006) and MARSIS (e.g. Morgan *et al.*, 2006) have been compared with MGS-derived solar wind proxies and observations. Continued study of simultaneous MGS-MEX observations will provide leverage needed to better understand the interaction of the solar wind with the Martian atmosphere.

Acknowledgement

D. Brain acknowledges the useful discussions and input of D. Mitchell, J. Halekas, D. Crider, and J. Luhmann in preparing this manuscript.

References

Acuña, M. H., Connerney, J. E. P., Wasilewski, P., Lin, R., Anderson, K., Carlson, C. W., *et al.*: 1992, *J. Geophys. Res.* **97**, 7799.

Acuña, M. H., Connerney, J. E. P., Wasilewski, P., Lin, R., Anderson, K., Carlson, C. W., *et al.*: 1998, *Science* **279**, 1676.

Acuña, M. H., Connerney, J. E. P., Ness, N., Lin, R., Mitchell, D., Carlson, C. W., *et al.*: 1999, *Science* **284**, 790.

Acuña, M. H., Connerney, J. E. P., Ness, N., Lin, R., Mitchell, D., Carlson, C. W., *et al.*: 2001, *J. Geophys. Res.* **106**(E10), 23,403.

Albee, A. L., Arvidson, R.E., Palluconi, F., and Thorpe, T.: 2001, *J. Geophys. Res.* **106**(E10), 23,291.

Arkani-Hamed, J.: 2001, *J. Geophys. Res.* **106**(E10), 23,197.

Axford, W. I.: 1991, *Planet Space Sci.* **39**(7), 167.

Barabash, S., and Lundin, R.: 2006, *Icarus* **182**(2), doi: 10.1016/j.icarus.2006.02.015.

Bertaux, J. L., Leblanc, F., Witasse, O., Quemerais, E., Lilensten, J., Stern, S. A., *et al.*: 2005, *Nature* **435**, doi: 10.1038/nature03603.

Bertucci, C., Mazelle, C., Crider, D. H., Mitchell, D. L., Sauer, K., Acuña, M. H., *et al.*: 2004, *Adv. Space Res.* **33**, 1938, doi: 10.1016/j.asr.2003.04.054.

Bertucci, C., Mazelle, C., and Acuña, M. H.: 2005a, *J. Atm. Terr. Phys.* **67**(17–18), 1797, doi: 10.1016/j.jastp.2005.04.007.

Bertucci, C., Mazelle, C., Acuña, M. H., Russell, C. T., and Slavin, J. A.: 2005b, *J. Geophys. Res.* **110**, A01209, doi: 10.1029/2004JA010592.

Bogdanov, A. V., and Vaisberg, O. L.: 1975, *J. Geophys. Res.* **80**, 487.

Brain, D. A.: 2002, PhD thesis, University of Colorado.

Brain, D. A., and Jakosky, B. M.: 1998, *J. Geophys. Res.* **103**(E10), 22,689.

Brain, D. A., Bagenal, F., Acuña, M. H., Connerney, J. E. P., Crider, D. H., Mazelle, C., *et al.*: 2002, *J. Geophys. Res.* **107**(A6), 1076, doi: 10.1029/2000JA000416.

Brain, D. A., Bagenal, F., Acuña, M. H., and Connerney, J. E. P.: 2003, *J. Geophys. Res.* **108**(A12), 1424, doi: 10.1029/2002JA009482.

Brain, D. A., Mitchell, D. L., Lillis, R., and Lin, R.: 2004, AGU Fall Meeting Abstracts pp SA13A–1119 (abstract).

Brain, D. A., Halekas, J. S., Lillis, R. J., Mitchell, D. L., Lin, R. P., and Crider, D. H.: 2005a, *Geophys. Res. Lett.* **32**(18), L18203, doi: 10.1029/2005GL023126.

Brain, D. A., Halekas, J. S., Peticolas, L. M., Lin, R. P., Luhmann, J. G., Mitchell, D. L., *et al.*: 2005b, *Geophys. Res. Lett.* **33**(1), L01201, doi: 10.1029/2005GL024782.

Brain, D. A., Mitchell, D. L., and Halekas, J. S.: 2006, *Icarus* **182**(2), 464, doi: 10.1016/j.icarus.2005.09.023.

Breus, T. K., Krymskii, A. M., Lundin, R., Dubinin, E. M., Luhmann, J., Yeroshenko, Y., *et al.*: 1991, *J. Geophys. Res.* **96**(A7), 11,165.

Cain, J. C., Ferguson, B. B., and Mozzoni, D.: 2003, *J. Geophys. Res.* **108**(E2), 5008, doi: 10.1029/2000JE001487.

Carlsson, E., Fedorov, A., Barabash, S., Budnik, E., Grigoriev, A., Gunell, H., *et al.*: 2006, *Icarus* **182**(2), 320, doi: 10.1016/j.icarus.2005.09.020.

Chen, Y., Cloutier, P. A., Crider, D. H., Mazelle, C., and Rème, H.: 2001, *J. Geophys. Res.* **106**(A12), 29,387.

Cloutier, P. A., Law, C. C., Crider, D. H., Walker, P. W., Chen, Y., Acuña, M. H., *et al.*: 1999, *Geophys. Res. Lett.* **26**(17), 2685.

Connerney, J. E. P., Acuña, M. H., Wasilewski, P. J., Kletetschka, G., Ness, N. F., Rème, H., *et al.*: 2001, *Geophys. Res. Lett.* **28**(21), 4015.

Connerney, J. E. P., Acuña, M. H., Ness, N. F., Kletetschka, G., Mitchell, D. L., Lin, R. P., *et al.*: 2005, *Proc. Nat. Acad. Sci.* **102**(42), 14,970, doi: 10.1073/pnas.0507469102.

Crider, D.: 2004, *Adv. Space Res.* **33**, 152, doi: 10.1016/j.asr.2003.04.013.

Crider, D., Cloutier, P., Law, C., Walker, P., Chen, Y., Acuña, M., *et al.*: 2000, *Geophys. Res. Lett.* **27**(1), 45.

Crider, D., Acuña, M., Connerney, J., Mitchell, D., Lin, R., Cloutier, P., *et al.*: 2001, *Adv. Space Res.* **27**(11), 1831.

Crider, D., Brain, D. A., Acuña, M., Vignes, D., Mazelle, C., and Bertucci, C.: 2004, *Space Sci. Rev.* **111**, 203.

Crider, D., Espley, J., Brain, D. A., Mitchell, J. E. P., Connerney, D. L., and Acuña, M.: 2005, *J. Geophys. Res.* **110**(A9), A09S21, doi: 10.1029/2004JA010881.

Crider, D. H., Acuña, M. H., Connerney, J. E. P., Vignes, D., Ness, N. F., Krymskii, A. M., *et al.*: 2002, *Geophys. Res. Lett.* **29**, 11.

Crider, D. H., Vignes, D., Krymskii, A. M., Breus, T. K., Ness, N. F., *et al.*: 2003, *J. Geophys. Res.* **108**(A12), 1461, doi: 10.1029/2003JA009875.

Dolginov, S. S., Yeroshenko, Y. G., and Zhuzgov, L. N.: 1976, *J. Geophys. Res.* **81**, 3353.

Dubinin, E., Fraenz, M., Woch, J., Roussos, E., Barabash, S., Lundin, R., *et al.*: *Space Sci.Rev.*, this issue, doi: 10.1007/s11214-006-9039-4.

Dubinin, E., Lundin, R., Norberg, O., and Pissarenko, N.: 1993, *J. Geophys. Res.* **98**(A11), 3991.

Dubinin, E., Lundin, R., and Schwingenschuh, K.: 1994, *J. Geophys. Res.* **99**(A11), 21,233.

Dubinin, E., Sauer, K., Lundin, R., Norberg, O., Trotignon, J. G., Schwingenschuh, K., *et al.*: 1996, *J. Geophys. Res.* **99**(A12), 27,061.

Dubinin, E., Sauer, K., Delva, M., and Tanaka, T.: 1998, *Earth Planets Space* **50**, 873.

Ergun, R. E., Andersson, L., Peterson, W. K., Brain, D., Delory, G. T., Mitchell, D. L., *et al.*: 2006, *Geophys. Res. Lett.* **33**, L14103, doi: 10.1029/2006GL025785.

Espley, J., Cloutier, P. A., Brain, D. A., Crider, D. H., and Acuña, M. H.: 2004a, *J. Geophys. Res.* **109**, A07213, doi: 10.1029/2003JA010193.

Espley, J., Delory, G. T., and Cloutier, P. A.: 2004b, *J. Geophys. Res.* **111**, E06S22, doi: 10.1029/2005JE002587.

Espley, J., Cloutier, P. A., Crider, D. H., Brain, D. A., and Acuña, M. H.: 2005, *J. Geophys. Res.* **110**, A09S33, doi: 10.1029/2004JA010935.

Fedorov, A., Budnik, E., Sauvaud, J. A., Mazelle, C., Barabash, S., Lundin, R., *et al.*: 2006, *Icarus* **182**(2), 329, doi: 10.1016/j.icarus.2005.09.021.

Ferguson, B., Cain, J. C., Crider, D. H., Brain, D. A., and Harnett, E.: 2005, *Geophys. Res. Lett.* **32**(16), L16105, doi: 10.1029/2004GL021964.

Fillingim, M. O., Peticolas, L. M., Lillis, R. J., Brain, D. A., Halekas, J. S., Mitchell, D. L., *et al.*: 2006, European Geophysical Union 2006 meeting, pp EGU06–A–09,238 (abstract).

Fraenz, M., Dubinin, E., Roussos, E., Woch, J., Winningham, J. D., Frahm, R., *et al.*: *Space Sci. Rev.*, this issue, doi: 10.1007/s11214-006-9115-9.

Futaana, Y., Barabash, S., Grigoriev, A., Holmström, M., Kallio, E., Brandt, P. C., *et al.*: 2006, *Icarus* **182**(2), 424, doi: 10.1016/j.icarus.2005.09.019.

Gurnett, D. A., Kirchner, D. L., Huff, R. L., Morgan, D. D., Persoon, A. M., Averkamp, T. F., *et al.*: 2005, *Science* **310**(5756), doi: 10.1126/science.1121868.

Halekas, J. S., Brain, D. A., Lillis, R. J., Fillingim, M. O., Mitchell, D. L., and Lin, R. P.: 2006, *Geophys. Res. Lett.* **33**, L13101, doi: 10.1029/2006GL026229.

Hanson, W. B., Santani, S., and Zuccaro, D. R.: 1977, *J. Geophys. Res.* **82**(28), 4351.

Harnett, E. M. and Winglee, R. M.: 2003, *Geophys. Res. Lett.* **30**(20), 2074, doi: 10.1029/2003GL017852.

Harnett, E. M., and Winglee, R. M.: 2005, *J. Geophys. Res.* **110**(A7), A07226, doi: 10.1029/2003JA010315.

Hinson, D. P., Simpson, R. A., Twicken, J. D., Tyler, G. L., Flasar, F. M.: 1999, *J. Geophys. Res.* **104**(E11), 26,297.

Jakosky B. M., Phillips R. J. (2001) Mars volatile and climate history. *Nature* **412**(6843), 237.

Krymskii, A. M., Breus, T. K., Ness, N. F., and Acuña, M. H.: 2000, *Space Sci. Rev.* **92**(A9), 535.

Krymskii, A. M., Breus, T. K., Ness, N. F., Acuña, M. H., Connerney, J. E. P., Crider, D. H., *et al.*: 2002a, *J. Geophys. Res.* **107**(A9), 1245, doi: 10.1029/2001JA000239.

Krymskii, A. M., Breus, T. K., Ness, N. F., Hinson, D. P., and Bojkov, D. I.: 2002b, *J. Geophys. Res.* **108**(A12), 1431, doi: 10.1029/2002JA009662.

Krymskii, A. M., Ness, N. F., Crider, D. H., Breus, T. K., Acuna, M. H., and Hinson, D. P.: 2004, *J. Geophys. Res.* **109**, A11306, doi: 10.1029/2004JA010420.

Langlais, B., Purucker, M. E., and Mandea, M.: 2004, *J. Geophys. Res.* **109**(E18), 2008, doi: 10.1029/2003JE002048.

Law, C. C., and Cloutier, P. A.: 1995, *J. Geophys. Res.* **100**(A12), 23,973.

Lillis, R. J., Mitchell, D. L., Lin, R. P., Connerney, J. E. P., and Acuña, M.H.: 2004, *Geophys. Res. Lett.* **31**(15), L15702, doi: 10.1029/2004GL020189.

Lillis, R. J., Engel, J. H., Mitchell, D. L., Brain, D. A., Lin, R. P., Bougher, S. W., *et al.*: 2005, *Geophys. Res. Lett.* **32**(23), L23204, doi: 10.1029/2005GL024337.

Luhmann, J. G., Russell, C. T., Brace, L. H., and Vaisberg, O. L.: 1992, *Mars, Univ. Arizona press, chap The Intrinsic Magnetic Field and Solar-Wind Interaction of Mars*, p. 1090.

Luhmann, J. G., Acuña, M. H., Purucker, M., Russell, C. T., and Lyon, J. G.: 2002, *Planet Space Sci.* **50**, 489.

Lundin, R., Borg, H., Hultqvist, B., Zakharov, A., and Pellinen, R.: 1989, *Nature* **341**, 609.

Lundin, R., Barabash, S., Andersson, H., Holmström, M., Grigoriev, A., Yamauchi, M., *et al.*: 2004, *Science* **305**, 1933.

Lundin, R., Winningham, D., Barabash, S., Frahm, M., Holmström, R., Sauvaud, J. A., *et al.*: 2005, *Science* **311**(5763), L23204, doi: 10.1126/science.1122071.

Ma, Y., Nagy, A. F., Hansen, K. C., DeZeeuw, D. L., and Gombosi, T. I.: 2002, *J. Geophys. Res.* **107**(A10), 1282, doi: 10.1029/2002JA009293.

Ma, Y., Nagy, A. F., Sokolov, I. V., and Hansen, K. C.: 2004, *J. Geophys. Res.* **109**, A07211, doi: 10.1029/2003JA010367.

Mazelle, C., Winterhalter, D., Sauer, K., Trotignon, J. G., Acuña, M. H., Baumgärtel, K., *et al.*: 2004, *Space Sci. Rev.* **111**, 115.

Mendillo, M., Smith, S., Wroten, J., Rishbeth, H., and Hinson, D.: 2003, *J. Geophys. Res.* **108**(A12), a432, doi: 10.1029/2003JA009961.

Mendillo, M., Withers, P., Hinson, D., Rishbeth, H., and Reinisch, B.: 2006, *Science* **311**(5764), 1135–1138, doi: 10.1126/science.1122099.

Mitchell, D. L., Lin, R. P., Rème, H., Crider, D. H., Cloutier, P. A., Connerney, J. E. P., *et al.*: 2000, *Geophys. Res. Lett.* **27**(13), 1871.

Mitchell, D. L., Lin, R. P., Mazelle, C., Rème, H., Cloutier, P. A., Connerney, J. E. P., *et al.*: 2001a, *J. Geophys. Res.* **106**(E10), 23,419.

Mitchell, D. L., Lin, R. P., Rème, H., Cloutier, P. A., Connerney, J. E. P., Acuña, M. H., *et al.*: 2001b, Probing Mars' crustal magnetic field with the MGS Electron Reflectometer. AGU Spring Meeting Abstracts, pp GP22A–07.

Mitchell, D. L., Lillis, R. J., Lin, R. P., Connerney, J. E. P., and Acuña, M. H.: 2006, *J. Geophys. Res.* in press.

Möhlmann, D., Riedler, W., Rustenbach, J., Schwingenschuh, K., Kurths, J., Motschmann, U., *et al.*: 1991, *Planet Space Sci.* **39**(7), 83.

Morgan, D. D., Gurnett, D. A., Kirchner, D. L., Huff, R. L., Brain, D. A., Boynton, W. V., *et al.*: 2006, *Geophys. Res. Lett.* **33**(13), L13202, doi: 10.1029/2006GL026637.

Nagy, A. F., Winterhalter, D., Sauer, K., Cravens, T. E., Brecht, S., Mazelle, C., *et al.*: 2004, *Space Sci. Rev.* **111**(1), 33, doi: 10.1023/B:SPAC.0000032718.47512.92.

Ness, N. F., Acuña, M. H., Connerney, J. E. P., Kliore, A. J., Breus, T. K., Krymskii, A. M., *et al.*: 2000, *J. Geophys. Res.* **105**(A7), 15,991.

Nielsen, E., Wang, X. D., Gurnett, D. A., Kirchner, D. L., Huff, R., Orosei, R., *et al.*: 2006, *J. Geophys. Res.* in press.

Øieroset, M., Mitchell, D. L., Phan, T. D., and Lin, R. P.: 2001, *Geophys. Res. Lett.* **28**(5), 887.

Øieroset, M., Mitchell, D. L., Phan, T. D., Lin, R. P., Crider, D. H., and Acuña, M. H.: 2004, *Space Sci. Rev.* **111**(1), 185.

Orlowski, D. S., and Russell, C. T.: 1995, *Adv. Space Res.* **15**(8/9), 37.

Pedersen, A., Nairn, C., Grard, R., and Schwingenschuh, K.: 1991, *J. Geophys. Res.* **96**(A7), 11,243.

Phillips, J. L., Luhmann, J. G., and Stewart, A. I. F.: 1986, *Geophys. Res. Lett.* **13**(A4), 1047.

Purucker, M., Ravat, D., Frey, H., Voorhies, C., Sabaka, T., and Acuña, M. H.: 2000, *Geophys. Res. Lett* **27**, 2249.

Riedler, W., Möhlman, D., Oraevsky, V. N., Schwingenschuh, K., Yeroshenko, Y., Rustenbach, J., *et al.*: 1989, *Nature* **341**, 604.

Rosenbauer, H., Shutte, N., Apáthy, I., Galeev, A., Gringauz, K., Grünwaldt, H., *et al.*: 1989, *Nature* **341**, 612.

Russell, C. T., Luhmann, J. G., Schwingenschuh, K., Riedler, W., and Yeroshenko, Y.: 1990, *Geophys. Res. Lett.* **17**(6), 897.

Russell, C. T., Mulligan, T., Delva, M., Zhang, T. L., and Schwingenschuh, K.: 1995, *Planet Space Sci.* **43**(7), 875.

Smith, E. J.: 1969, *Advanced Space Experiments, American Astronautical Society*, chap Planetary Magnetic Field Experiments, Tarzana, CA.

Trotignon, J. G., Dubinin, E., Grard, R., Barabash, S., and Lundin, R.: 1996, *J. Geophys. Res.* **101**(A11), 24,965.

Trotignon, J. G., Mazelle, C., Bertucci, C., and Acuña, M. H.: 2006, *Planet Space Sci.* **54**, 357, doi: 10.1016/j.pss.2006.01.003.

Tyler, G. L., Balmino, G., Hinson, D. P., Sjogren, W. L., Smith, D. E., Woo, R. T., *et al.*: 1992, *J. Geophys. Res.* **97**(E5), 7759.

Tyler, G. L., Balmino, G., Hinson, D. P., Sjogren, W. L., Smith, D. E., Simpson, R. A., *et al.*: 2001, *J. Geophys. Res.* **106**(E10), 23,327.

Vennerstrom, S., Olsen, N., Purucker, M., Acuña, M. H., and Cain, J. C.: 2003, *Geophys. Res. Lett.* **30**(7), 1369, doi: 10.1029/2003GL016883.

Verigin, M., Vignes, D., Crider, D., Slavin, J., Acuña, M., Kotova, G., *et al.*: 2004, *Adv. Space Res.* **33**, 2222, doi: 10.1016/S0273-1177(03)00522-2.

Verigin, M. I., Shutte, N. M., Galeev, A. A., Gringauz, K. I., Kotova, G. A., Remizov, A. P., *et al.*: 1991, *Planet Space Sci.* **39**(1–2), 131.

Verigin, M. I., Gringauz, K. I., Kotova, G. A., Remizov, A. P., Shutte, N. M., Rosenbauer, H., *et al.*: 1993, *J. Geophys. Res.* **98**(A2), 1303.

Vignes, D., Acuña, M. H., Connerney, J. E. P., Crider, D. H., Rème, H., and Mazelle, C.: 2000, *Geophys. Res. Lett.* **27**(1), 49.

Vignes, D., Acuña, M. H., Connerney, J. E. P., Crider, D. H., Rème, H., and Mazelle, C.: 2002, *Geophys. Res. Lett.* **29**(9), 1328, doi: 10.1029/2001GL014513.

Vignes, D., Acuña, M. H., Connerney, J. E. P., Crider, D. H., Rème, H., and Mazelle, C.: 2004, *Space Sci. Rev.* **111**(1), 223, doi: 10.1023/B:SPAC.0000032716.21619.f2.

Winningham, J. D., Frahm, R. A., Sharber, J. R., Coates, A. J., Linder, D., Soobiah, Y., *et al.*: 2006, *Icarus* **182**(2), 360, doi: 10.1016/j.icarus.2005.10.033.

Withers, P., Mendillo, M., Rishbeth, H., Hinson, D. P., and Arkani-Hamed, J.: 2005, *Geophys. Res. Lett.* **32**, L16204, doi: 10.1029/2005GL023483.

Zakharov, A. V.: 1992, in: Luhmann, J. G., and Tatrallyay, Pepin R. O. (eds.), *Venus and Mars: Atmospheres, Ionospheres, and Solar Wind Interactions*, AGU, Washington, DC, Geophys. Monogr. Ser., vol. 66, p. 327.

THE ANALYZER OF SPACE PLASMAS AND ENERGETIC ATOMS (ASPERA-3) FOR THE MARS EXPRESS MISSION

S. BARABASH[1,*], R. LUNDIN[1], H. ANDERSSON[1], K. BRINKFELDT[1], A. GRIGORIEV[1], H. GUNELL[1], M. HOLMSTRÖM[1], M. YAMAUCHI[1], K. ASAMURA[2], P. BOCHSLER[3], P. WURZ[3], R. CERULLI-IRELLI[4], A. MURA[4], A. MILILLO[4], M. MAGGI[4], S. ORSINI[4], A. J. COATES[5], D. R. LINDER[5], D. O. KATARIA[5], C. C. CURTIS[6], K. C. HSIEH[6], B. R. SANDEL[6], R. A. FRAHM[7], J. R. SHARBER[7], J. D. WINNINGHAM[7], M. GRANDE[8], E. KALLIO[9], H. KOSKINEN[9,16], P. RIIHELÄ[9], W. SCHMIDT[9], T. SÄLES[9], J. U. KOZYRA[10], N. KRUPP[11], J. WOCH[11], S. LIVI[7], J. G. LUHMANN[12], S. McKENNA-LAWLOR[13], E. C. ROELOF[14], D. J. WILLIAMS[14], J.-A. SAUVAUD[15], A. FEDOROV[15] and J.-J. THOCAVEN[15]

[1]*Swedish Institute of Space Physics, Kiruna, Sweden*
[2]*Institute of Space and Austranautic Studies, Sagamichara, Japan*
[3]*University of Bern, Physikalisches Institut, Switzerland*
[4]*Instituto di Fisica dello Spazio Interplanetari, Rome, Italy*
[5]*Mullard Space Science Laboratory, University College London, Surrey, UK*
[6]*University of Arizona, Tucson, USA*
[7]*Southwest Research Institute, San Antonio, USA*
[8]*Rutherford Appleton Laboratory, Oxfordshire, UK*
[9]*Finnish Meteorological Institute, Helsinki, Finland*
[10]*Space Physics Research Laboratory, University of Michigan, Ann Arbor, USA*
[11]*Max-Planck-Institut für Sonnensystemforschung, Katlenburg-Lindau, Germany*
[12]*Space Science Laboratory/University of California in Berkeley, Berkley, USA*
[13]*Space Technology Ltd., National University of Ireland, Maynooth, Co. Kildare, Ireland*
[14]*Applied Physics Laboratory/John Hopkins University, Laurel, USA*
[15]*Centre d'Etude Spatiale des Rayonnements, Toulouse, France*
[16]*University of Helsinki, Helsinki, Finland*
(**Author for correspondence: E-mail: stas@irf.se*)

(Received 23 August 2006; Accepted in final form 17 November 2006)

Abstract. The general scientific objective of the ASPERA-3 experiment is to study the solar wind – atmosphere interaction and to characterize the plasma and neutral gas environment within the space near Mars through the use of energetic neutral atom (ENA) imaging and measuring local ion and electron plasma. The ASPERA-3 instrument comprises four sensors: two ENA sensors, one electron spectrometer, and one ion spectrometer. The Neutral Particle Imager (NPI) provides measurements of the integral ENA flux (0.1–60 keV) with no mass and energy resolution, but high angular resolution. The measurement principle is based on registering products (secondary ions, sputtered neutrals, reflected neutrals) of the ENA interaction with a graphite-coated surface. The Neutral Particle Detector (NPD) provides measurements of the ENA flux, resolving velocity (the hydrogen energy range is 0.1–10 keV) and mass (H and O) with a coarse angular resolution. The measurement principle is based on the surface reflection technique. The Electron Spectrometer (ELS) is a standard top-hat electrostatic analyzer in a very compact design which covers the energy range 0.01–20 keV. These three sensors are located on a scanning platform which provides scanning through 180° of rotation. The instrument also contains an ion mass analyzer (IMA). Mechanically IMA is a separate unit connected by a cable to

Space Science Reviews (2006) 126: 113–164
DOI: 10.1007/s11214-006-9124-8

the ASPERA-3 main unit. IMA provides ion measurements in the energy range 0.01–36 keV/charge for the main ion components H^+, He^{++}, He^+, O^+, and the group of molecular ions 20–80 amu/q. ASPERA-3 also was its own DC/DC converters and digital processing unit (DPU).

Keywords: space instrumentation, Mars solar wind – interaction

1. Introduction

The purpose of the Analyzer of Space Plasma and Energetic Atoms (ASPERA-3) experiment on-board the European Space Agency Mars Express mission is to investigate the solar wind – Mars interaction through the use of ion, electron, and energetic neutral atom (ENA) measurements. The Mars Express mission was designed to fulfill the requirement of remotely sensing the planet and thus is not ideally suitable for space plasma investigations. The main drawback is the absence of a magnetometer onboard the spacecraft. Therefore, the ASPERA-3 investigation focused its objectives in relation to planetological issues rather than plasma processes. The main focus of the ASPERA-3 investigation was given to studies of the solar wind – Mars interaction impact on the atmosphere and atmospheric evolution. In the introduction we describe how near-Mars space is coupled to the atmosphere, then review the ENA environment of the planet and explain how ENAs can be used to diagnose plasma – gas interactions. In the following sections we review the instrument, its individual sensors, subsystems, accommodation on the spacecraft, and briefly describe the science operation concept.

1.1. Impact of the Solar Wind – Mars Interaction on the Atmosphere

Because of Mars lacks an intrinsic magnetic field, the following basic features characterize the near-planet environment:

(1) the upper atmosphere/ionosphere is a subject of solar wind induced escape in the form of pick-up ions and bulk plasma escape;
(2) the solar wind deposits matter, momentum, and energy into the Martian upper atmosphere;
(3) the atmosphere is a subject of sputtering which is caused by planetary ions either extracted from the ionosphere or resulted from the exospheric neutrals ionization, picked-up by the solar wind, and accelerated back to the atmosphere;
(4) the main domains and boundaries are the collisionless bow shock, the magnetospheath, a boundary which excludes the solar wind (induced magnetosphere boundary or magnetic pile-up boundary), the photoelectron boundary separating the region dominated by the ionospheric photoelectrons;
(5) the local crustal magnetizations affect the solar wind interaction pattern.

Without the magnetic cavity of a magnetosphere to shield the upper atmosphere from the on-coming solar wind, Mars is a subject to comet-like atmosphere erosion processes and solar-wind-induced current systems. The solar wind induced atmospheric escape occurs through three main channels; ion pick – up, bulk thermal plasma removal, and sputtering. While the first channel is extensively investigated both theoretically and experimentally, the two other remain poorly constrained because up to now the relevant plasma investigations have not been conducted. However, bulk plasma escape and sputtering are most efficient at removing heavy molecules such as CO_2. The atmospheric carbon dioxide is critical for understand ing the water evolution on Mars. It is the thick CO_2 atmosphere thought to be present 3–4 billions years ago that provided the strong green house effect needed to rise the atmosphere temperature to the level at which water could exist in the liquid form. Apart from the solar wind induced escape mechanics, in operation at Mars now are the Jeans escape process (affect mainly hydrogen and deuterium) and non-thermal photochemical escape processes. The relative contribution of all these processes to the overall atmospheric evolution is still unclear because all estimations are based mainly on models, which, because of complex feedbacks are still quite limited. Figure 1 based on the review by Lammer *et al.* (2003) provides a summary of the escape mechanisms at Mars and illustrates their relative importance. Actual measurements by the PHOBOS-2 spacecraft determined that the scavenging of planetary ions amounts to $(0.5–3) \times 10^{25}$ ions/s (Lundin *et al.*, 1991; Verigin *et al.*, 1991). The initial results from the Mars Express mission confirm the existence of the effective particle loss induced by the solar wind interaction and even show that the solar wind may reach much lower altitudes (ca. 330 km) than was previously thought (Lundin *et al.*, 2004). It was also established that the escaping flux consists approximately of an equal amount of O_2^+ and O^+ and some admixture of CO_2^+ (ca. 20%) (Carlsson *et al.*, 2006). The newly reported escape rates measured by ASPERA-3 (Barabash *et al.*, 2007) are a factor of 100 lower than the PHOBOS-2 numbers, a puzzling finding still to be explained.

The role of the Martian crustal magnetic anomalies (Acuña *et al.*, 1998) in the overall ion escape is still unclear and the effect on the thermosphere is still to be investigated in detail. As was shown by the Mars Global Surveyor (MGS) mission (Crider *et al.*, 2002) the magnetic anomalies do affect the global pattern of the solar wind interaction by “lifting” the magnetic pile-up boundary and screening the ionosphere from the access of magnetospheath electrons. The analysis of the MGS MAG/ER (Magnetometer and Electron Reflectrometer) data indicated that the precipitating solar wind particles can heat the Martian thermosphere and that the crustal magnetic fields can control the sites where precipitations occur (Krymskii *et al.*, 2003). However, the recent Mars Express results indicate that there are no well-pronounced effects on ion fluxes (Nilsson *et al.*, this issue) (interpretation of the Mars Express particle data is complicated by the absence of magnetic field measurements). Nevertheless, Mars Express observed a short burst of CO Cameron band emissions (strongest lines) interpreted as Martian aurora (Bertaux *et al.*, 2005)

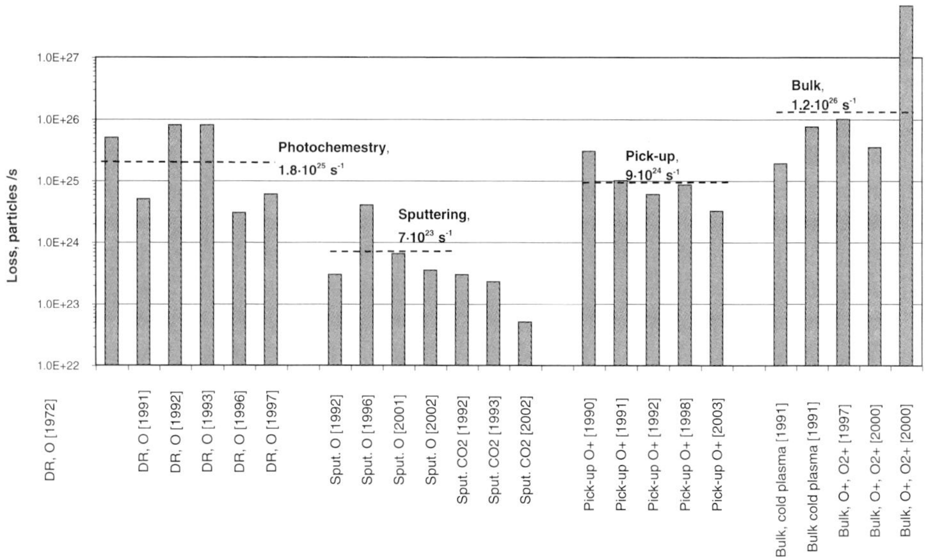

Figure 1. Non-thermal escape processes at Mars (adopted from review by Lammer *et al.*, 2003). DR: dissociative recombination; Sput.: sputtering. The horizontal dashed lines show log-averaged values. All available data, i.e., measurements, models, and estimations, are shown.

in the vicinity of a strong magnetic anomaly, indicating particle precipitation in the area. In addition, Lundin *et al.* (2006) reported signatures in Martian electron spectra above magnetic anomalies which are typical of those observed in terrestrial auroral electron spectra. The bulk escape might be the strongest among ion loss mechanisms driven by solar wind interactions (Figure 1). Ionospheric profiles observed by the Viking 1 and 2 landers indicate that the solar wind may erode the Martian ionosphere (Fox, 1997). Ionospheric bubbles or clouds triggered by plasma instabilities at the solar wind-ionopause transition region were observed on Venus and may also contribute to the loss of heavy ions as an additional escape process from the Martian ionosphere as shown in hybrid code simulations (Terada *et al.*, 2002). However, very limited experimental data are available to constrain the associated escape. Nairn *et al.* (1991) reported "tail rays," cold electron enhancements in the Martian tail with densities 10–65 cm^{-3} observed in association with broadband wave activity from a few Hz to several kHz. The associated ion outflow was estimated to $1.9–7.5 \times 10^{25}$ ions/s.

The atmospheric loss due to sputtering is the mechanism experimentally constrained the least, but is the key mechanism for determining the carbon dioxide inventory in the last Noachian Martian atmosphere because sputtering is the most effective way to remove CO_2. In light of the non-detection of carbonates on the surface (Bibring *et al.*, 2005), the current sputtering rate becomes the most important parameter which is used to constrain the amount of the carbon dioxide in the ancient Martian atmosphere.

Another problem of the solar wind – atmosphere coupling that has not been experimentally explored concerns the energetic consequences for the Martian atmosphere of the lack of a Martian dipole field. The kinetic and test-particle models of the Mars-solar wind interaction (Brecht, 1997; Kallio *et al.*, 1997) have suggested that solar wind absorption by the Martian atmosphere may be an important energy source for the upper atmosphere. The ENAs generated as a product of the solar – wind interaction further enhance the deposition of solar wind energy (Kallio and Barabash, 2000). According to the models, some of the solar wind ions (mainly protons and alphas) directly impact Mars' upper atmosphere near its exobase (at $\sim$180 km altitude) because their gyroradii are too large to behave as a deflected "fluid" in the subsolar magnetosheath (Kallio and Janhunen, 2001) or because they are partially thermalized by the bow shock (Kallio *et al.*, 1997). Thus, solar wind energy is "directly" deposited into the upper atmosphere resulting in increasing ionization rates and UV emissions.

1.2. ENA Production and ENA Diagnostics

As a result of the low gravity on Mars, the neutral gas (exosphere) density can reach 10^4–10^6 cm^{-3} in the solar wind interaction region where the main plasma boundaries, the bow shock and the magnetic pile-up boundary (induced magnetosphere boundary), are located. The co-existence of these two components, the solar wind plasma and the planetary neutral gas, results in strong interactions. One of the fundamental collisional interactions is the charge – exchange process

$$A^+(\text{energetic}) + M(\text{cold}) \Rightarrow A(\text{energetic}) + M^+(\text{cold}),$$

that provides ENAs. A global image formed from the directional detection of ENAs yields a picture of this interaction region while analysis of the time variation of the ENA signal and of the ENA energy spectrum provides a means to diagnose the plasma properties exhibited within the interaction region. The main populations of ENAs at Mars are described in Table I and the hydrogen ENA flux morphology is depicted in Figure 2.

The supersonic solar wind upstream of the bow shock can experience charge – exchange with the Martian hydrogen exosphere over very long distances resulting in a narrow (··10°) anti-sunward beam of ENAs called the neutral solar wind which has an energy of the bulk flow of the solar wind. This population was detected by the Low Energy Neutral Atom (LENA) instrument onboard the terrestrial magnetospheric Imager for Magnetopause-to-Aurora Global Exploration (IMAGE) mission (Collier *et al.*, 2001) which was capable of looking directly towards the Sun. This component can also be measured with sensors not fully immune against solar UV photons from vantage points inside the Martian eclipse (Brinkfeld *et al.*, 2006). In this geometry,

TABLE I
ENA at Mars.

Original ion/neutral population	Production mechanism	ENA Energy
Upstream solar wind (Holmström *et al.*, 2002)	Charge – exchange of the undisturbed solar wind on the extended hydrogen exosphere	~1 keV
Shocked solar wind (Holmström *et al.*, 2002; Gunell *et al.*, 2006)	Charge – exchange of the shocked solar wind on the hydrogen corona in the magnetospheath	Few hundreds eV
Accelerated planetary ions (Barabash *et al.*, 2002; Lichtenegger *et al.*, 2002)	Charge – exchange of H^+, O^+, O_2^+, CO_2^+ on the neutral coronae (H, H_2, O)	Few hundreds eV – few keV
Backscattered hydrogen (ENA albedo) (Kallio and Barabash, 2001; Holmström *et al.*, 2002)	Backscattering of the precipitating neutral solar wind and solar wind protons from the upper atmosphere	Few hundreds eV
Sputtered O and CO_2 (Luhmann and Kozyra, 1991; Luhmann *et al.*, 1992)	Sputtered of the atmospheric gasses by precipitating O^+	Few tens eV (low flux)
Phobos torus (Mura *et al.*, 2002)	Charge – exchange of the solar wind (upstream/shocked) on the neutral gas density increase due to possible Phobos torus.	Few hundreds eV

Mars blocks the solar photons but neutral solar wind can reach the umbra region due to thermal spreading and scattering by the upper atmosphere (Kallio *et al.*, 2006).

The shocked solar wind is the strongest source producing ENAs since the protons flowing around the Martian obstacle have the most time to interact with the dense (in the ENA context) hydrogen exosphere. Detailed modeling of the ENA production from this source was performed by Kallio *et al.* (1997), Holmström *et al.* (2002), Mura *et al.* (2002), and Gunell *et al.* (2006). Figure 3 taken from Gunell *et al.* (2006), shows simulated ENA images for several vantage points at different solar zenith angles. The estimated ENA fluxes are well above $10^4\,\mathrm{cm^{-2}\,s^{-1}\,keV^{-1}}$ and the directional fluxes exceed $10^5\,\mathrm{cm^{-2}\,s^{-1}\,sr^{-1}}$ (images for solar zenith angle 80°, 100°, and 120°).

Barabash *et al.* (2002) and Lichtenegger *et al.* (2002) investigated the detailes of the ENA signal associated with oxygen and hydrogen pick-up ions. Using the empirical model of the solar wind plasma flow near Mars developed by Kallio

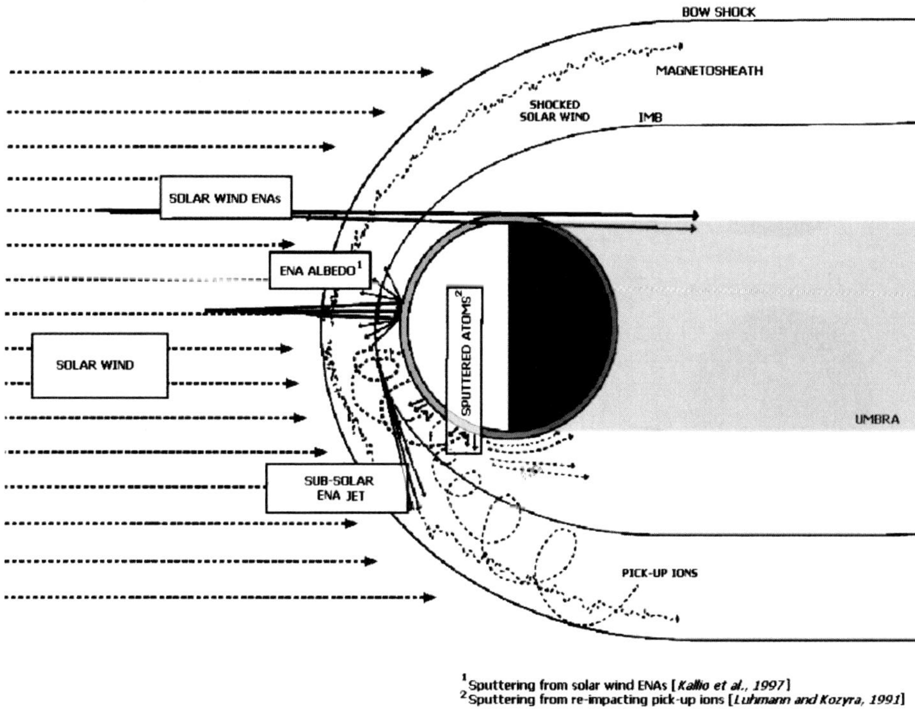

Figure 2. ENA populations at Mars. Not shown are ENAs which originated from the planetary ion charge – exchange (From Brinkfeld *et al.*, 2006).

(1996), Barabash *et al.* (2002) numerically solved the kinetic equation to obtain the global distribution of oxygen ions. This distribution was then converted to the corresponding ENA flux. It was found that the fluxes of oxygen ENAs could reach 10^4 cm^{-2} s^{-1} sr^{-1} eV^{-1} and fully reflect the morphology of the oxygen population. This process provides a technique for determining the instantaneous oxygen escape rate. One of the simulated images for the energy range 0.1–1.65 keV is reproduced in Figure 4. The projection is similar to the one used to display the hydrogen ENA images of Figure 3. The Figure 4 image shows a strong jet of ENAs coming out from the subsolar point and flowing tailward.

ENA images reveal morphological features of the ENA sources such as locations of boundaries and relative sizes. Directionally separated ENA signals can be converted into global distributions of the proton flow and of neutral gas using inversion techniques similar to those developed for conditions at the Earth (Roelof and Scinner, 2000). Holmström *et al.* (2002) showed that the ENA fluxes generated from the shocked solar wind are the most sensitive to the neutral hydrogen distribution, which is controlled by the exobase temperature, and the position of the boundary separating the solar wind and planetary plasmas.

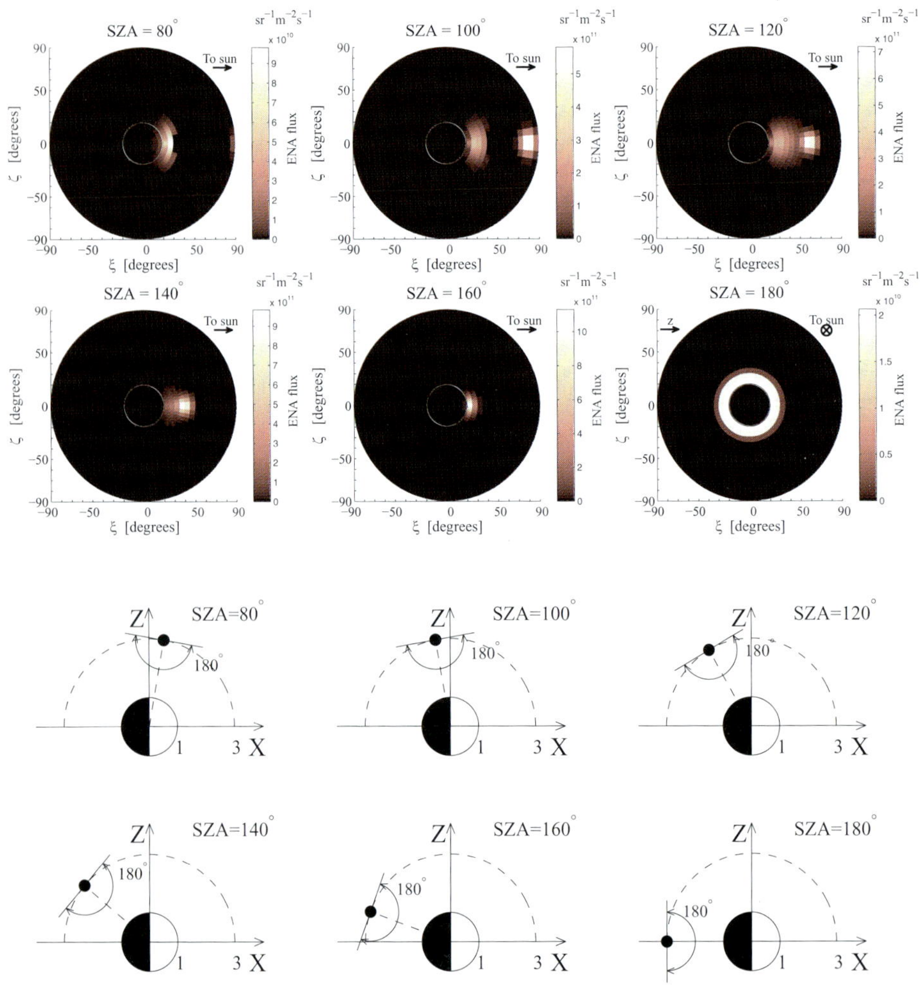

Figure 3. Hydrogen ENA images. The ENA images are from vantage points which are 3Rm from the center of Mars looking towards nadir. The solar zenith angles (SZA) are from 80° to 180°. The horizontal and vertical axes show the angular offsets in degrees, from the centre of the image (from Gunell *et al.*, 2006).

The directionally separated ENA signals of the escaping plasma display globally and instantaneously the size and geometry of the outflowing plasma as well as provide constrains on the total escape rate (Barabash *et al.*, 2002). For example, the non-detection of oxygen ENAs by the ASPERA-3 sensor has constrained the total escape rate of oxygen to be "less than 1 g/s" (Galli *et al.*, this issue), which is close to the 4 g/s obtained by direct measurements using the ion plasma spectrometer (Barabash *et al.*, 2006).

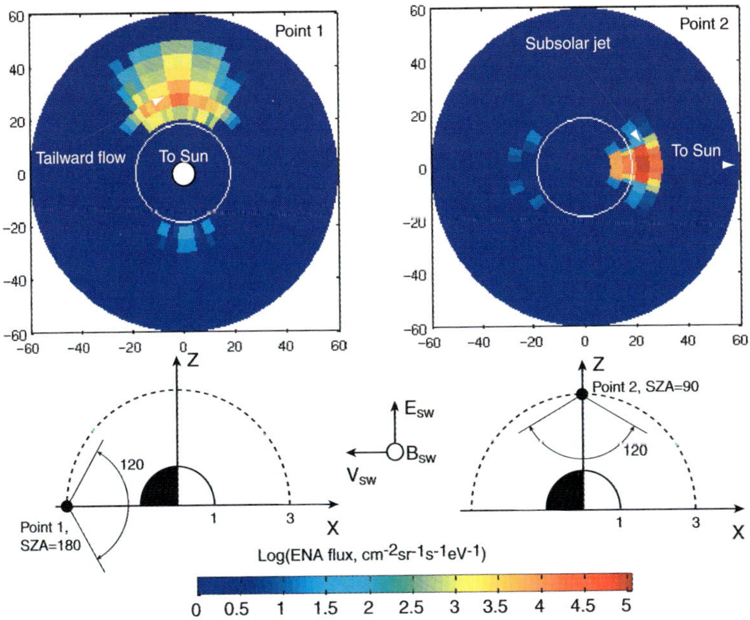

Figure 4. ENA images of pick-up oxygen ions from two vantage points: in the tail and at the pole. The vantage points are in the plane perpendicular to the ecliptic. The polar axis is towards the planetary center. The image projection is similar to those shown in Figure 3. The energy range is 0.1–1.65 keV. The electric and magnetic field vectors in the solar wind are also shown for reference (Adopted from Barabash *et al.*, 2002).

ENA diagnostics is a useful tool in the study of plasma dynamics because temporal and spatial variations are inherently separated. A single directional observation provides measurements of the typical time scale for plasma variations in the interaction region. The global response of the system against internal disturbances can also be studied, for example, variations of the magnetic pile-up boundary during interplanetary shock impact or internal variability related to different kinds of instabilities (Futaana *et al.*, this issue; Grigoriev *et al.*, this issue). Imaging ENA albedo, i.e., ENAs resulted from back scattering and sputtering, can be used to map the precipitation fluxes.

2. Scientific Objectives

In order to study the processes related to the atmospheric impact of the solar wind – Mars interaction on the atmosphere, the ASPERA-3 experiment was designed to measure electrons and ions in the hot plasma energy range as well as to provide remote sensing (diagnostics) of the plasma – neutral gas interaction *via* ENA measurements. The key objectives of the ASPERA-3 experiment are (1) to determine

TABLE II
The ASPERA-3 objectives and requirements.

Scientific objectives	Measurements requirements
Solar wind induced ion escape (pick-up, bulk escape, sputtering)	Mass resolving ion measurements in the energy range few eV – tens keV
	Mass resolving (H/O) ENA measurements in the energy range few eV – tens keV. ENA flux $>10^3$ cm^{-2} s^{-1} keV^{-1}
Momentum, energy, and mass deposition from the solar wind to the upper atmosphere/ionosphere	Mass resolving ion and electron measurements in the energy range few eV – tens keV
	Mass resolving (H/O) ENA measurements in the energy range down to tens eV from the nadir direction. ENA flux $>10^6$ cm^{-2} s^{-1} keV^{-1} (100 eV)
Morphological structure of the Martian interaction region and local plasma characteristics	Mass resolving ion and electron measurements in the energy range few eV – tens keV with 4π coverage
	Measurements of the ENA flux in the energy range tens eV – few keV with 4π coverage.
	ENA flux $>10^4$ cm^{-2} s^{-1} keV^{-1}
	Measurements of the upstream solar wind parameters.
Search for the solar wind – Phobos interactions	Mass resolving ion and electron measurements in the energy range few eV – tens keV
	ENA measurements in the energy range tens eV – few keV with 4π coverage. ENA flux 10^4 cm^{-2} s^{-1} keV^{-1}

as precisely as possible the total ion escape (particles/s) for the major ion species (O, O_2^+, CO_2^+), (2) to study momentum, energy, and mass deposition from the solar wind into the upper atmosphere/ionosphere and its response (sputtering), (3) to investigate the morphological structure of the Martian interaction region and define its local plasma characteristics. A search for the solar wind – Phobos interaction was also undertaken. The ASPERA-3 objectives and associated measurement requirements are listed in Table II.

The scientific objectives of ASPERA-3 are formulated to describe the current solar wind interaction with Mars. The results of the experiment can be projected backward in time in order to determine the reaction of the planet at times past. Thus, results of the ASPERA-3 investigation are instrumental in determining and constraining the role of the solar wind interaction in the overall atmosphere evolution throughout the history of Mars.

3. The Instrument

No instruments with similar scientific objectives and capabilities of ASPERA-3 have been or are planned to be flown to Mars. The only similar experiment, ASPERA-C, was on-board the Mars-96 mission; however, the ASPERA-C did not contain a mass-resolved energy-analyzing ENA detector. The Japanese Nozomi spacecraft, launched to Mars in 1998, was concentrating on plasma measurements, but it did not carry any ENA detectors. Both the Mars-96 and Nozomi missions failed to reach Mars. MGS (launched in 1996 operational until November 2006) carried a magnetometer and an electron spectrometer (energy resolution 25%), but contained no ENA or ion measurement capability.

3.1. Overview

Mechanically the ASPERA-3 instrument consists of two units, the Main Unit (MU) and the Ion Mass Analyzer (IMA) (Figure 5). The MU comprises three sensors Neutral Particle Imager – NPI, Neutral particle Detector – NPD, Electron Spectrometer – ELS and a Digital Processing Unit – DPU, which are all located on a scanner. The MU is also equipped with two solar sensors allowing identification of the Sun direction for automatic reduction of the detector bias during scanning because the ENA sensors are UV sensitive. All mechanical and electrical interfaces are made through the scanner. The total mass of the instrument is 8.2 kg and the power consumption is ca. 16 W. The MU envelope is $359 \times 393 \times 234\,mm^3$ and the IMA envelope is $255 \times 150 \times 150\,mm^3$.

NPI provides measurements of the integral ENA flux with no mass or energy resolution but with $5^\circ \times 11^\circ$ angular resolution. The intrinsic field of view is $9^\circ \times 344^\circ$. The sensor utilizes the particle – surface interaction technique for ENA detection. ENAs incident on a target surface coated by a graphite – containing material (DAG 213, a resin-based graphite dispersion) at a grazing angle of 20° are reflected and/or cause ion sputtering. An MCP (Microchannel Plates) stack detects the reflected particles and sputtered fragments with a discrete anode. The NPI head is a replica of the NPI-MCP sensor developed for the ASPERA – C experiment on the Mars-96 mission (launch failure) and successfully flown on the Swedish microsatellite Astrid launched in 1995 (C:son Brandt *et al.*, 2000). A detailed overview of the NPI development and calibration results including UV response can be found in Brinkfeldt (2005). The NPI also accommodates two solar sensors.

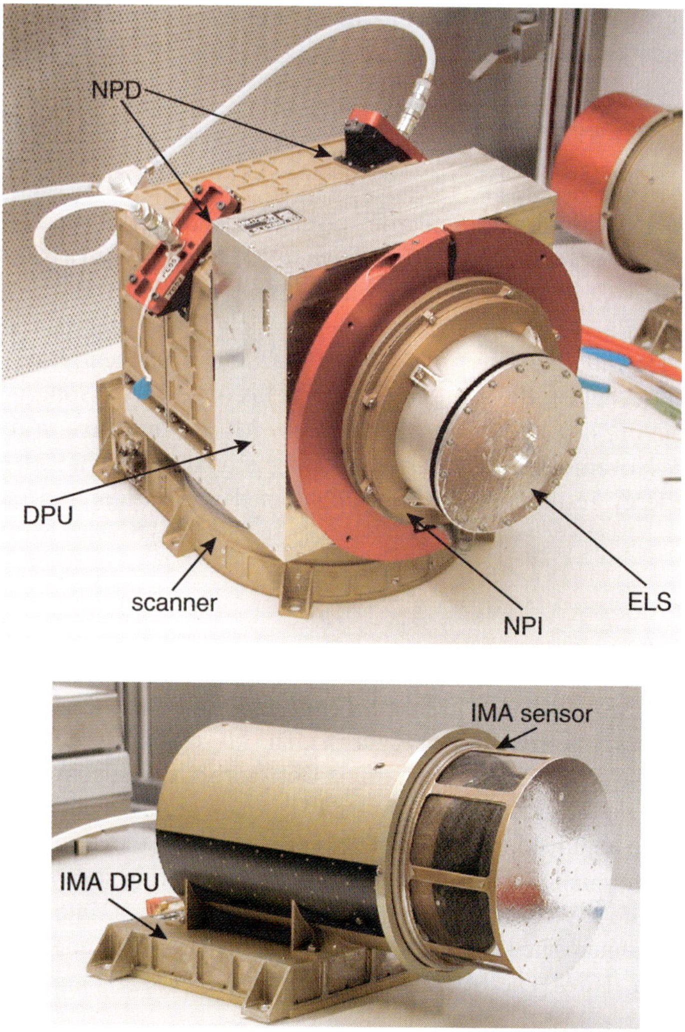

Figure 5. The ASPERA-3 configuration (flight model with red-tagged covers installed).

The NPD provides measurements of the ENA differential flux over the energy range 100 eV–10 keV, resolving H and O with coarse ($5^\circ \times 30^\circ$) angular resolution. The sensor consists of two identical heads, each with the $9^\circ \times 90^\circ$ intrinsic field of view. The measurement technique is based on the principle similar to NPI. ENAs incident on a surface at a grazing angle of 15° are reflected and cause secondary electron emission. The secondary electrons are transported to an MCP assembly, which gives the START signal. The reflected ENAs hit a second surface and again produce secondary electrons used to generate the STOP signal. The NPD time-of-flight (TOF) electronics determines the ENA velocity. Mass analysis is based on the large velocity difference between two main ENA species, hydrogen and oxygen. The pulse – height distribution (PHD) analysis of the STOP signals can also be used to provide a rough determination of the ENA mass.

ELS provides electron measurements in the energy range 0.01–20 keV. The intrinsic full field of view is $4^\circ \times 360^\circ$. The 360° aperture is divided into 16 sectors. The sensor is a standard top – hat electrostatic analyzer in a very compact design. ELS is a reduced and modified version of the Miniaturized Electrostatic Dual-top-hat Spherical Analyzer (MEDUSA) experiment for the Astrid-2 and Munin missions launched in 1998 and 2000 (Norberg *et al.*, 2001).

IMA is an improved version of the Three-dimensional Ion Composition Spectrometer (TICS)/Freja, Ion Mass Imaging Sensor (IMIS)/Mars-96, Ion Mass Imager (IMI)/Nozomi (Norberg *et al.*, 1998), and an exact copy of the Ion Composition Analyzer (ICA) instrument flying on the Rosetta mission. In the ASPERA-3 design, the IMA is a separate unit connected by a cable to the MU. The IMA provides ion measurements in the energy range 0.01–36 keV/q for the main ion components H^+, H_2^+, He^+, O^+, and for the group of molecular ions $20 < M/q < \sim 80$. The IMA instantaneous field of view is $4.6^\circ \times 360^\circ$. Electrostatic sweeping performs elevation ($\pm 45^\circ$) coverage. The IMA sensor is a spherical electrostatic analyzer followed by a circular magnetic velocity separating section. A large diameter MCP with a discrete anode system images a matrix, which is azimuth $\times$ mass.

The three sensors (NPI, NPD, and ELS) are located on a scanning platform. The combination of the 360° field of view of NPI and ELS along with the scanner motion from 0° to 180° could provide the 4π maximum coverage; however, a fraction of the field of view is blocked by the spacecraft body. The scanner also provides the capability of pointing ELS, NPI, and NPD independently of the spacecraft. Table III summarizes the instrument performance.

The following features make ASPERA-3 a truly unique instrument:

(1) two ENA sensors, the first at Mars
(2) combined electron, ion, and ENA sensors in a single package with common data acquisition system
(3) high energy resolution of electron measurements (8%) ensuring resolving photoelectron peaks.

Mars Express has quite a favorable orbit for studies of the solar wind – planet interaction. With its pericenter height of 270 km and apocenter of 11,580 km (first year in-orbit) and 10,050 km (later in the mission), the 86.6° inclination the orbit allows sampling of all plasma domains, and during certain periods, grazes the induced magnetosphere boundary (magnetic pile-up region) for hours.

TABLE III

Performance of the NPI, NPD, ELS, and IMA sensors.

Parameter	NPI	NPD	ELS	IMA
Particles to be measured	ENA	ENA	Electrons	Ions
Energy range, keV per charge	≈0.1–60[a]	0.1–10	0.01–20	0.01–30
Energy resolution, $\Delta E/E$	No	0.5	0.08	0.07
Resolved masses, amu/q	No	1, 16	N/A	1, 2, 4, 8, (16, 32, 44)[b]
Intrinsic field of view	$9^\circ \times 344^\circ$	$9^\circ \times 180^\circ$	$4^\circ \times 360^\circ$	$90^\circ \times 360^\circ$
Angular resolution (FWHM)	$4.6^\circ \times 11.5^\circ$	$5^\circ \times 40^\circ$	$2^\circ \times 22.5^\circ$	$4.5^\circ \times 22.5^\circ$
G-factor/pixel or sector cm^2 sr (NPI, NPD) cm^2 sr eV/eV (ELS, IMA)	2.5×10^{-3} (ε not incl.)	6.2×10^{-3} (ε not incl.)	7×10^{-5}	1.6×10^{-6}
Efficiency, ε, %	~1	1–15	Inc. in G	Inc. in G
Time resolution (one scan), s	32	32	32	196[c]
Mass, kg	0.7	1.3	0.3	2.2
Power, W	0.8	1.5	0.6	3.5

[a]Upper cut-off of the deflector system.
[b]The ions O^+, O_2^+, CO_2^+ (mass/charge = 16, 32, 44) are resolved by peak-fitting technique (Carlsson *et al.*, 2006).
[c]Full energy – elevation scan.

3.2. Neutral Particle Imager (NPI)

The significant components of NPI are identified in Figures 6–8. Figure 6 identifies the main components of NPI in a cross-sectional view while the integrated sensor is shown in Figure 7 . At selected locations during assembly, the critical components of NPI are identified in Figure 8 (the actual NPI shown here is from ASPERA-4 for the Venus Express mission, identical to the ASPERA-3 NPI except for the outer structure surface thermal coating). The charged particles, electrons and ions, are removed by the electrostatic deflection system, which consists of two disks separated by a 3 mm gap. The 5 kV positive potential between the grounded and biased disks results in a strong electric field, which sweeps away all charged particles with energies up to 60 keV. Since the integral ENA flux substantially exceeds the

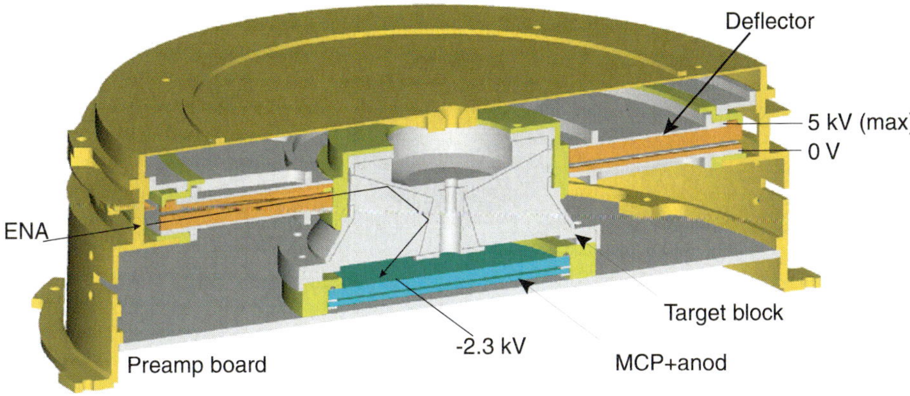

Figure 6. NPI sensor cross-sectional view. The deflector voltage and MCP bias are identified.

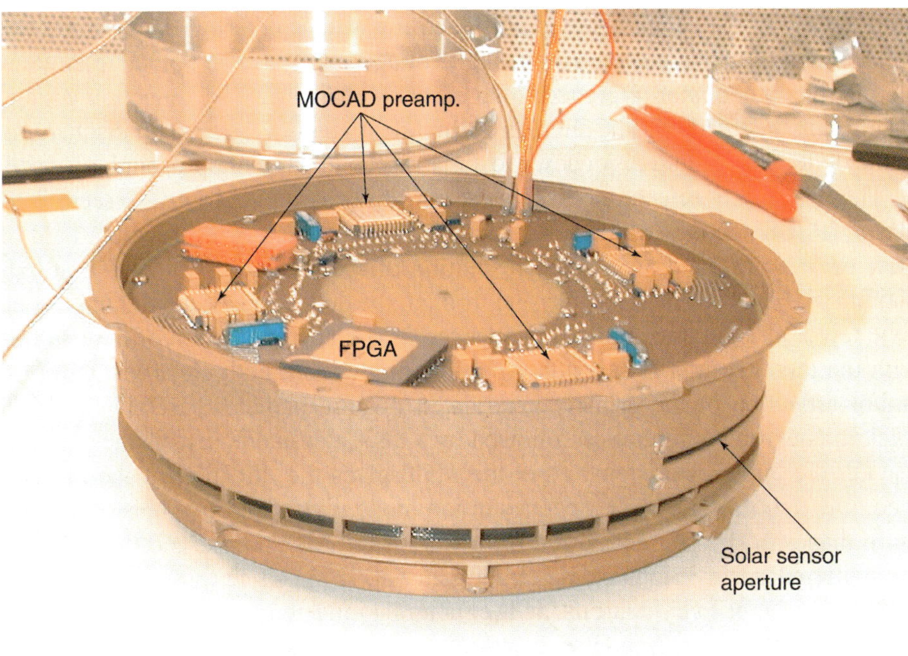

Figure 7. The completed NPI sensor.

charged particle flux for energies greater than 60 keV, this rejection energy is sufficient for satisfactory performance. The disks also collimate the incoming beam over elevation and azimuth angle. Apart from being ON or OFF, the deflection system can be operated in two other modes, alternative mode and sweeping mode.

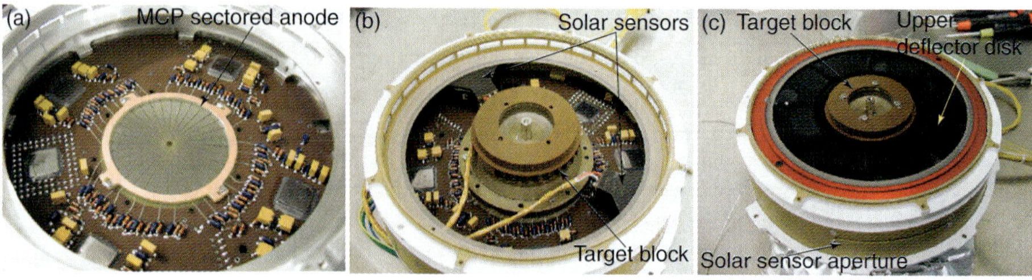

Figure 8. NPI components: the sectored anode (a), the target block installed (b) and the deflection system installed (c).

In the alternative mode, the deflection system is turned on and off, alternating each sampling period; whereas, in sweeping mode, the potential across the deflection system is continually changed to form a voltage sweep. These modes are used to more accurately separate charged and neutral particles entering the system. The deflection system is connected to the high voltage supply *via* an optocoupler. Regulating the optocoupler reference voltage causes changes in the deflection voltage performing the sweeping and alternating. In order to reduce the time for discharging of the deflection system disks down to 1 ms, a second parallel optocoupler is used.

The space between the deflection system disks is divided into 32 sectors by plastic (PEEK) spokes which form 32 azimuthal collimators, each with a field of view of $9^\circ \times 18^\circ$ and an angular resolution of $4.6 \times 11.25^\circ$ (full width at half maximum, FWHM). Neutrals passing through the deflection system hit a 32 sided cone target at a grazing angle of incidence of 20°. The interaction of the neutrals with the target results in secondary particle production, both electrons and ions, and/or reflection of the primary neutrals. A 56 mm diameter PHOTONIS MCP stack in chevron configuration followed by a 32 sector anode detects the particles leaving the target. The signal from the MCP gives the direction of the primary incoming neutral. The MCP operates in ion mode with a negative bias of slightly more than -2.3 kV applied to the front side of the MCP stack, and thus, detects (a) sputtered positive ions of the target material, (b) positive ions resulting from ionizing of the primary neutrals, and (c) neutrals reflected from the target surface. In order to improve the angular resolution and collimate the particles leaving the interaction surface, 32 separating walls are attached to the target forming a star-like structure. This configuration allows the entering particles to experience multiple reflections and reach the MCP. NPI covers 4π in one half scanner cycle (except for the fraction blocked by the spacecraft body) and produces an image of the ENA flux distribution in the form of an azimuth $\times$ elevation matrix once per (32s, 64s, or 128s). The flux in each of the 32 elements is read out once every 62.5 ms and integrated over 1 s by the DPU. Two sectors around the scanner spin axis and looking toward the spacecraft body are blocked, intended to provide monitoring of the MCP assembly dark counts. This space is also used to connect the ELS electrical

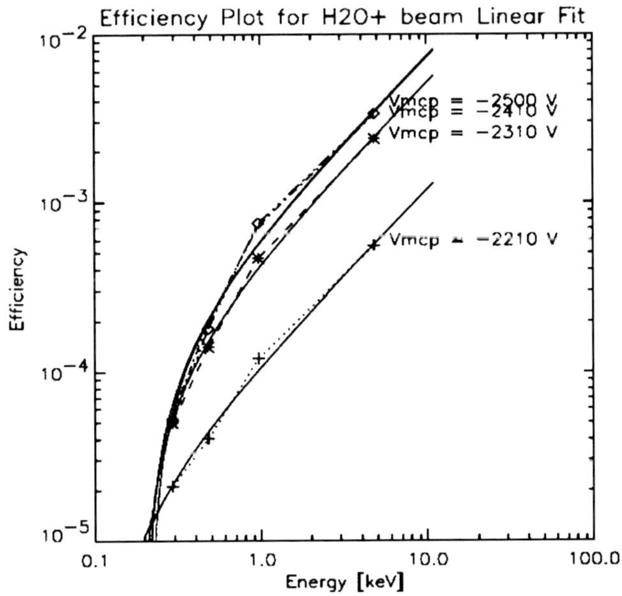

Figure 9. The NPI efficiency of the sector 14 with neighboring sectors mechanically blocked against H_2O beams for different bias voltages and energies.

harness to the DPU. The NPI sensor also accommodates two solar sensors which can be seen in Figures 7 and 8.

NPI calibrations were performed at the Swedish Institute of Space Physics, Kiruna, Sweden jointly with Institute of Space and Astronautic Studies, Sagamichara, Japan as well as at the University of Arizona, Tucson, United States of America. Calibrations of the NPI sensor were preformed to characterize the sensor response, MCP working bias, dark count level, angular response in elevation and azimuth, and efficiency. In addition, comprehensive investigations of the sensor's immunity against UV photons were performed.

The MCP working bias is a bias at which the MCP count rate starts to saturate for a given discriminator level. During calibrations, the measurement of the MCP count rate saturation started at slightly more than an MCP bias of −2.4 kV. However, to optimize the angular resolution, the MCP operational bias was chosen to be −2.3 kV (Brinkfeldt, 2005).

Figure 9 shows the NPI efficiency for different values of the MCP bias and for different particle energies using H_2O^+ beams. The efficiency was measured in sector 14 where the neighboring sectors were mechanically blocked. This single sector was thoroughly investigated in terms of efficiency measurements. The NPI sector 14 efficiency of the solar wind at an energy of 1 keV is around 5×10^{-4} for an operating bias of 2.3 kV. An azimuth scan of all sectors with the particle beam located in the central plane (shown in Figure 10) was used to determine the relative

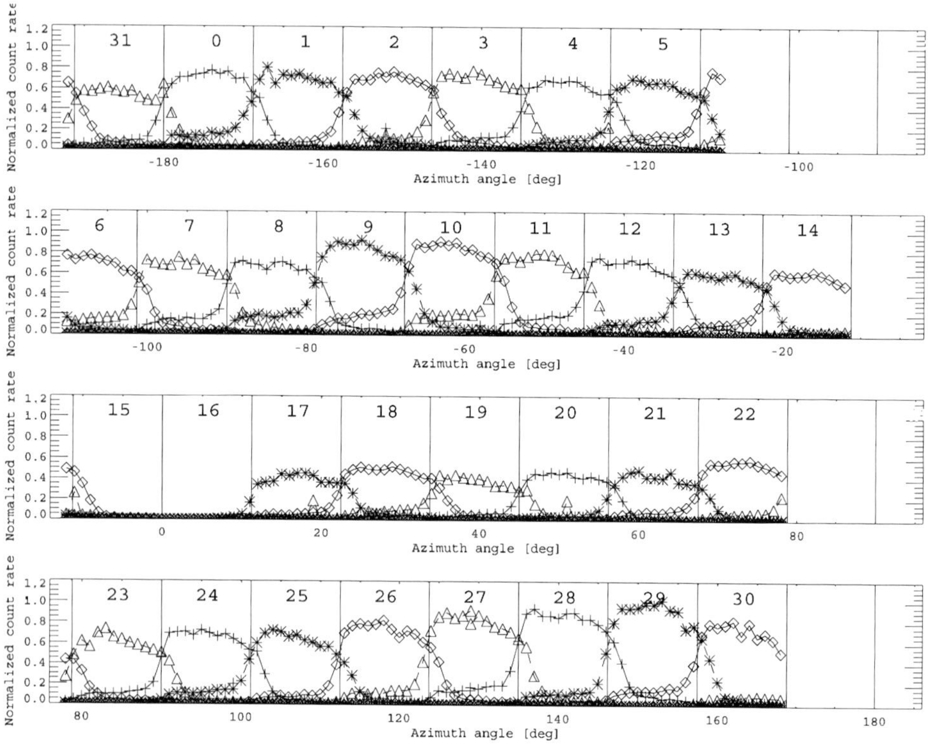

Figure 10. Scan through the central plane of the NPI sectors showing the relative efficiency against a H_2O^+ beam.

response of the other 31 sectors and define inter-calibration factors for each NPI sector. The azimuthal scan shows that there are large differences in the response of different sectors and the defined inter-calibration factors are significant. For example, sectors 17–21 show lower responses than the remaining sectors (sectors 15 and 16 are mechanically blocked and used to monitor MCP background). The sensitivity variation between different sectors is about a factor of 2 and the inter-calibration factors need to be taken into account during ENA image processing. At the nominal operation bias, Figure 11 shows the angular response of sector 14 to a 5 keV beam of protons. The lower plots in Figure 11 show a polynomial fit to the response. The analytical function describing the sensor angular response is included into ENA image inversion software performing ENA image inversion.

The calibration provided a pure geometrical factor of 2.7×10^{-3} cm^2 sr, not including the efficiency. This number is also confirmed by ray-tracing. The equation below can be used to convert from the count rate in sector n, C_n, to the flux F_n

$$F_n = \frac{C_n - C_{\text{dark}}}{G\eta_n\varepsilon}$$

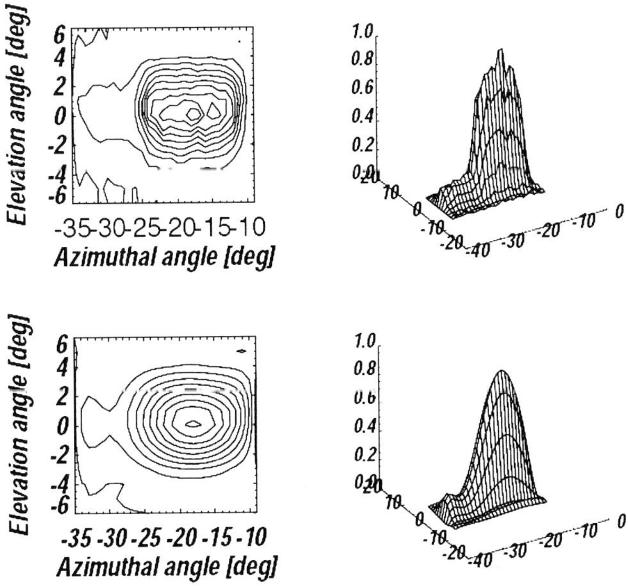

Figure 11. Full angular response of the sector 14 against a proton beam. The lower plot is a polynomial fit.

where C_{dark} is the dark count of sector n (defined during calibrations and monitored in flight through the count rate of the blocked sectors 15 and 16), G is the pure geometrical factor, η_n is the relative sensitivity to sector 14, and ε is the efficiency of sector 14.

An important issue in the NPI design is coating the target block for suppression of UV photon fluxes, which enter the instrument and produce a UV background in the measurements. NPI uses the same coating as in the Prelude in Planetary Particle Imaging (PIPPI) and NPI/ASPERA-C experiments, namely, DAG 213, a resin-based graphite dispersion. This substance is similar to Aquadag, which is a graphite dispersion in water. The coating demonstrated satisfactory performance in the PIPPI experiment flown in the Earth's magnetosphere (C:son Brandt *et al.*, 2000). To verify the expected UV contamination levels the Mars Express NPI spare model (the ASPERA-4 flight model) was calibrated against Lyman- α photons ($\lambda = 121.6$ nm). The calibration philosophy was similar to that of the particle calibration, i.e., the response of one sector (number 4) was fully characterized and then a relative measurement was made for the other sectors.

Using a much simpler set-up (Barabash, 1995), the previous measurement gave the DAG 213 UV rejection efficiency around 10^{-5}.The efficiency of the MCP itself for UV at the Lyman-α wavelength is $\sim$1%. So the combined efficiency of the NPI for Lyman-α photons was expected to be 10^{-7}.

Figure 12a shows the dependence of a single channel count rate (sector 4) on UV beam intensity. The count rate was determined by integrating over angle. The

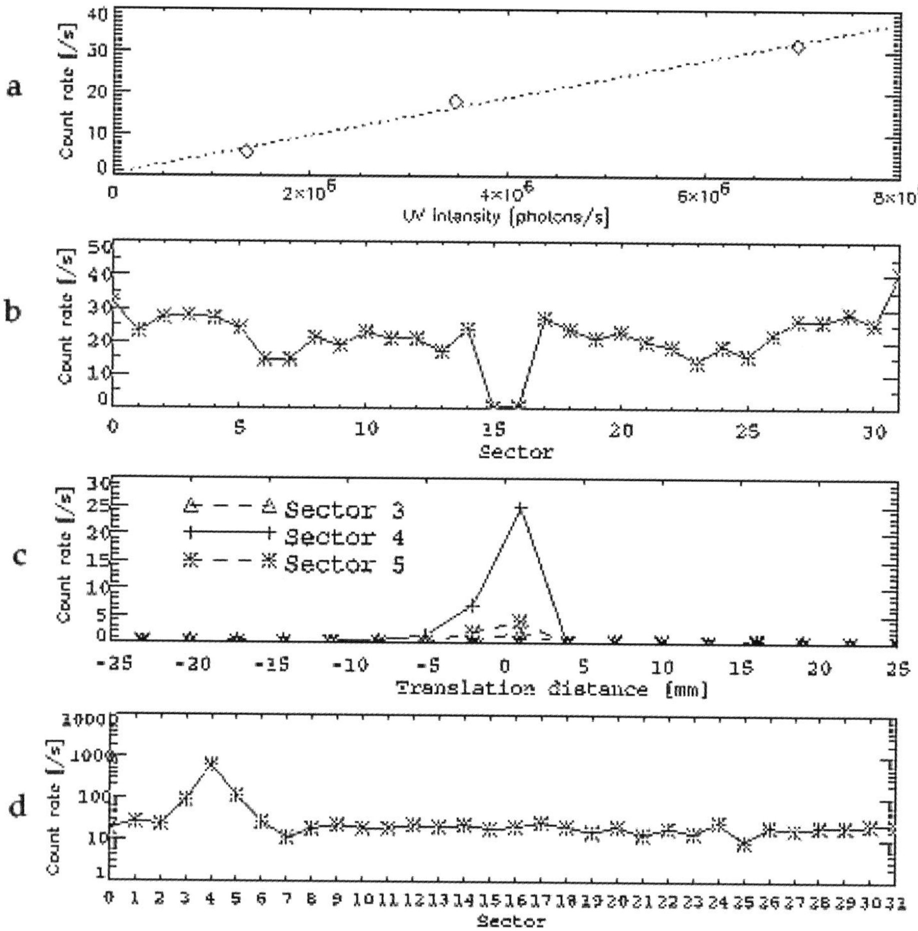

Figure 12. (a) Absolute UV calibration results. NPI count rate (sector 4) as a function of the Lyman – α photon flux. (b) Relative sensitivity of the NPI sectors against UV. (c) NPI sectors 3, 4, and 5 response during a translation perpendicular to the UV beam. (d) The count rate for each sector during the translation of the sensor perpendicular to the UV beam.

parallel UV photon beam was centered about sector 4. Due to limitations in the beam intensity variations, only three points on this curve were determined. The detection efficiency was found to be 5×10^{-6}, close to the previous (less accurate) measurements. The response of the other 31 sectors against Lyman- α photons was measured only at the center of the aperture of each sector, following the approach similar to the particle calibrations. The count rate normalized to the same beam intensity is shown in Figure 12b. Sectors 15 and 16 show zero counts because they are mechanically blocked. In general, the UV sensitivity varies by a factor of 2 between different sectors, similar to the particle measurements. This result indicates that there are gain variations across the MCP surface.

To investigate internal photon reflection from the deflector spokes separating the different sectors, the sensor performed parallel translation with respect to the UV beam, again with the center of the UV beam directed at the center of sector 4 when directly under the UV beam. Figure 12c shows responses from sector 4 (center parallel to the beam) and two neighboring sectors 3 and 5 (each centered 11.25° from parallel) resulting from the translation. Figure 12d shows the integrated count rate for each different sector during the translation. This test showed that internal photon reflections result in a cross-talk between neighboring sectors to a level of 10–15%.

The sensor includes only front-end-electronics (FEE) and interfaces with the DPU *via* an FPGA (Field Programmable Gate Array). High voltage is provided by an external high voltage supply located inside the DPU. The FEE is based on 4 MOCAD (Monolitic Octal Charge Amplifier/Pulse Discriminator) chips each containing 8 channels. The signals from the MCP anode are fed directly to the MOCAD inputs, which generate Transistor–Transistor Logic (TTL) pulses. The discriminator threshold is the same for all 8 channels. The sensitivity may vary between different channels within a chip which may also contribute to the variations in efficiency observed between sectors (Figure 10). The logical pulse from the FEE is fed to the FPGA, which generates the respective address read-out by the DPU. All data accumulations and consecutive data processing are performed by the DPU.

3.3. Neutral Particle Detector (NPD)

The NPD consists of two identical detectors, each of which is a pinhole camera. Figure 13 provides a conceptual view of one detector and also indicates how the azimuth and elevation angles are defined relative to the overall instrument geometry. In each detector the charged particles, electrons and ions, are removed by the deflection system, which consists of two trapezoidal plates separated by a 3.0 mm gap. In the normal operational mode a 10 kV potential (±5 kV) is applied to the plates and the resulting strong electric field sweeps away all charged particles with energies up to 70 keV. The deflector also collimates the incoming beam in elevation angle. The collimated ENA beam emerging from the entrance 3.0 × 4.5 mm pin-hole hits the START surface with a 15° grazing angle and causes secondary electron emission. By a system of collecting grids, the secondary electrons (SE) are transported to one of two MCP assemblies, producing the START signal for TOF electronics. Depending on the azimuth angle, the collection efficiency varies from 80 to 95%. The incident ENAs are reflected from the START surface nearly specularly. Since the charge state equilibrium is established during the interaction with the surface, the emerging beam contains both the neutral and ionized (positive and negative) components. To increase the total efficiency, no further separation by the charge is made. As proven by ion ray-tracing, there is very little disturbance of the reflected atomic ions leaving the START surface with an energy above 80 eV introduced by the START electron optics. Figure 14 shows the results of electron

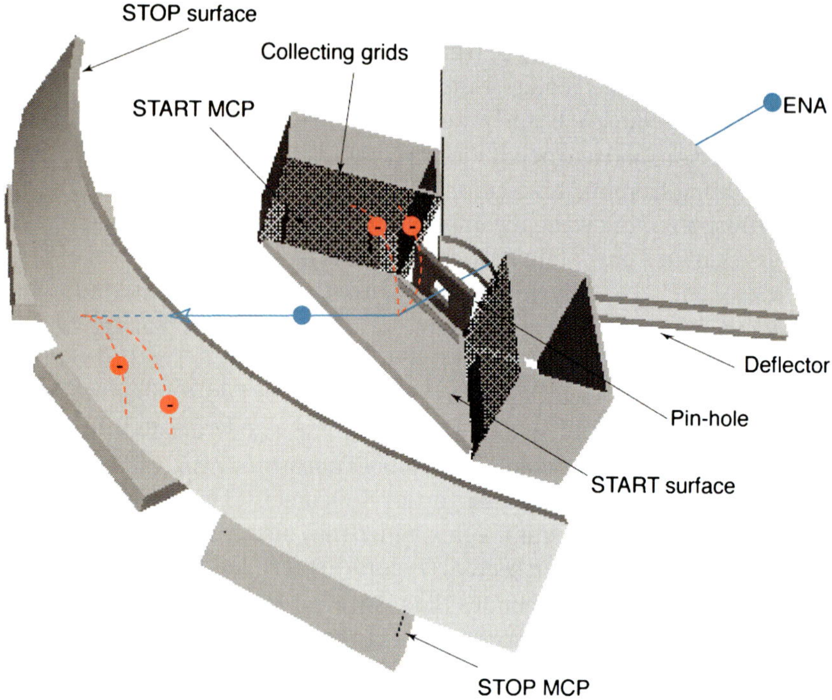

Figure 13. Three-dimensional view of the NPD principal components.

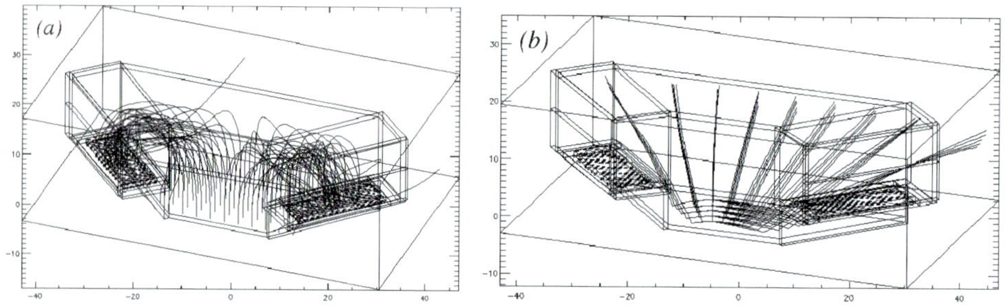

Figure 14. The ray-tracing of the electron (a) and 80 eV ion trajectories (b) in the START NPD assembly electron optics.

and ion ray-tracing in the START electron optics assembly. Therefore particles of all charge states – negative, neutral, and positive – will impact the second surface, the STOP surface. Again secondary electrons will be produced which are detected by one of the three MCP assemblies, giving the STOP signal. The time of flight over a fixed distance of 8 cm defines the particle velocity. Three STOP MCPs also give crude resolution over azimuth to be within the NPD 90° acceptance angle. Since the SE yield depends on mass for the same velocity, the pulse height distribution analysis of the STOP signals can provide the estimation of the ENA mass.

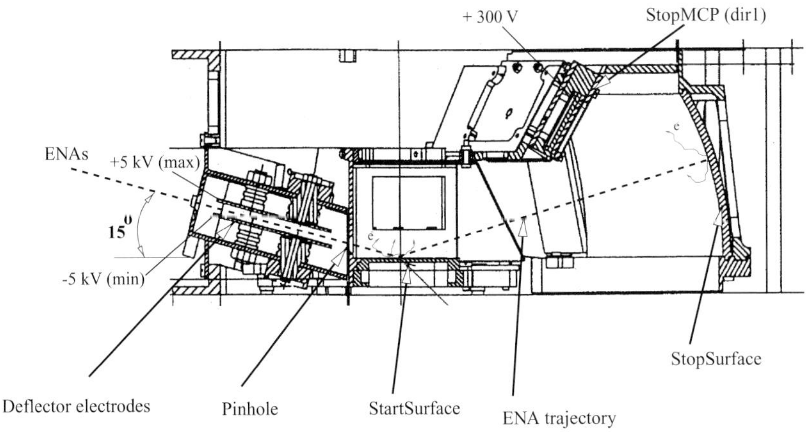

Figure 15. Cross-sectional view of the NPD sensor. The deflector and the STOP MCP front voltages are identified.

Fortunately, the difference in TOF between hydrogen and oxygen is sufficient for direct mass identification within the energy range in question. The accumulation time is commandable, but normally 1 s.

Figure 15 shows a cross-section of the NPD sensor and its main elements. Figure 16 shows different views of the NPD flight model. Two identical sensors are built in a package and installed on the scanner. The NPD package provides 180° coverage, and when combined with the motion of the scanner, 2π coverage (see Figure 5). In order to minimize the sensor mass, all sensor elements are fixed to the outer shell of the mechanical structure. When both sensors are mounted together, electrical screening between units is provided by a kapton film with conductive coating.

The selection of the START and STOP surfaces was the most challenging part of the NPD development. Extensive studies have been performed at University of Bern (Jans, 2000) and Brigham Young University (USA) to optimize the performance of the surfaces which must satisfy a number of requirements, namely, high secondary electron yield, high UV absorption even at grazing angles, high particle reflection coefficient (START surface), low angular scattering, and low photoelectron yield. For the START surface, a multi-layer coating composed of a thin layers of Cr_2O_3, covered by a thicker layer of MgF, and topped with a thin layer of WO_2 was used. The coating was optimized for the absorption of the 121.6 nm line at a 15° incident angle. The reflection coefficient was about 30%, a factor of 2 lower than the uncoated surface. The coating was applied on a titanium substrate polished down to 100 Å roughness.

The STOP surface is graphite (roughness around 100 nm) covered by a MgO layer of about 500 nm. This combination has a very high secondary electron yield, low photoelectron yield, and high UV absorption. Much effort has been expended

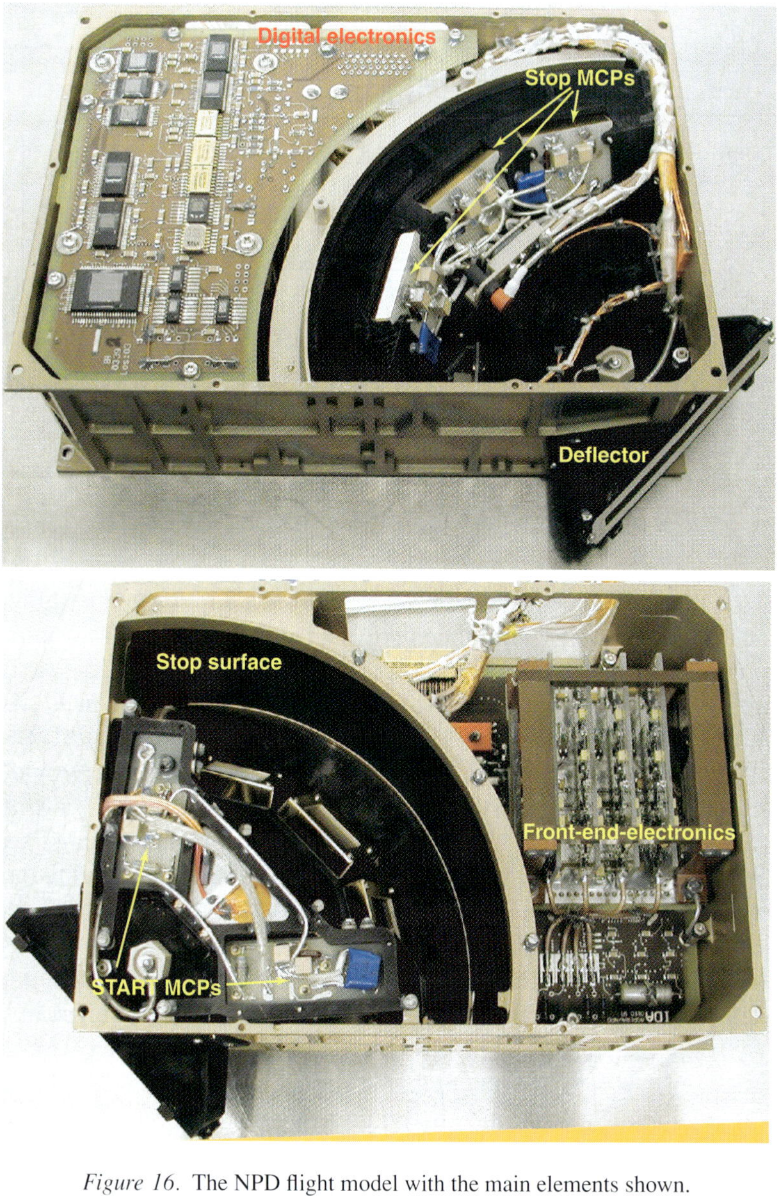

Figure 16. The NPD flight model with the main elements shown.

into increasing the stability of the MgO coating against moisture absorption. It was established that polishing the graphite down to the roughness of around 100 nm improves the surface stability such that no problems related to surface performance (such as possible surface property changes due to increases in the absorption of air humidity during storage and pre-launch operations) are encountered. Thus, the surfaces did not require special maintenance to retain their stability. The importance

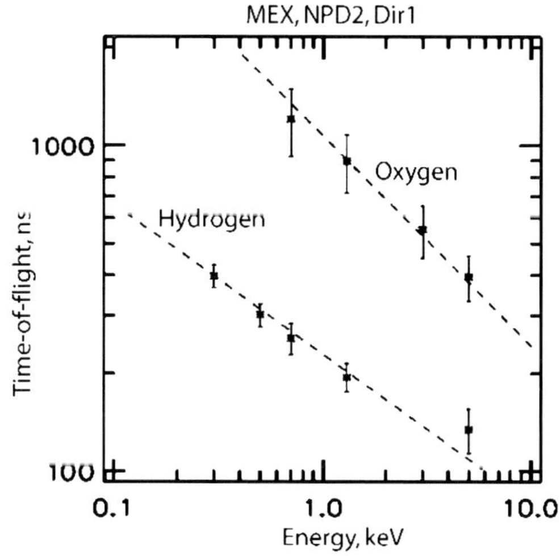

Figure 17. NPD laboratory calibrations: time-of-flight for different masses as a function of energy. The dashed lines give the theoretical dependence corresponding to the 33% energy loss on the START surface.

of achieving this goal was that the surface properties did not change from their calibrated state.

The NPD calibrations included determining the sensor efficiency, total geometrical factor, angular response, and energy resolution. The calibration results fully correspond to the specified performance. Since charge – equilibrium is established in just over a few Å along the parth of the material, a particle interacts with an ion beam can be used for calibrations of ENA sensors. Figure 17 shows the dependence of the measured TOF on the incoming particle energy for protons and H_2O^+. The dashed lines show the theoretical dependence corresponding to the 33% energy loss in the START surface. In the ion source, the produced ionic water molecules break-up during the impact and the residual components carry the same initial velocity as the ion beam, corrected for the energy loss in the target. Therefore, water can be used to calibrate the instrument response to oxygen. Since the TOF for oxygen with an energy below 2 keV is longer than the TOF corresponding to slowest protons at around 100 eV, TOF measurements alone can be used to identify the particle mass (at least in the low energy range).When examining the TOF of both oxygen and hydrogen, it was found that the slopes of the atoms, energy versus TOF are slightly different. The reason for this could be a slight dependence of the reflection properties of the START surface (and thus the effective TOF length) on mass; however, this is only speculation and the true cause remains unknown. Note that, this effect was not observed during calibration of the similar ASPERA-4 NPD.

A 3 keV proton beam was used to generate Figure 18 which shows the NPD angular response over azimuth and elevation for the three STOP sectors, central

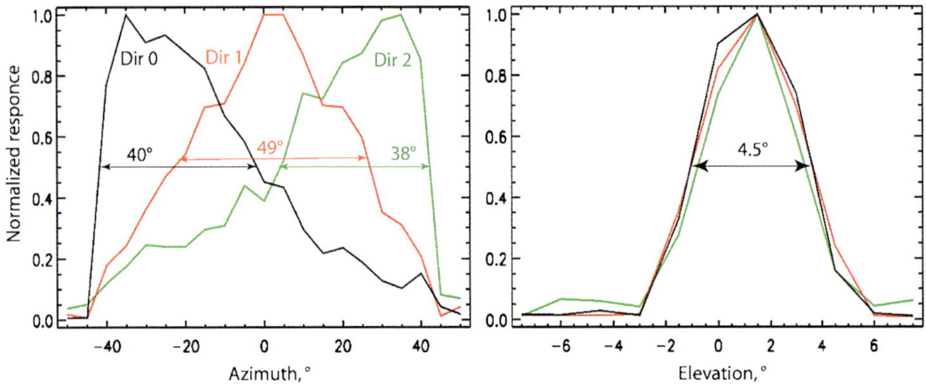

Figure 18. The angular response of the NPD 2 sensor over azimuth and elevation.

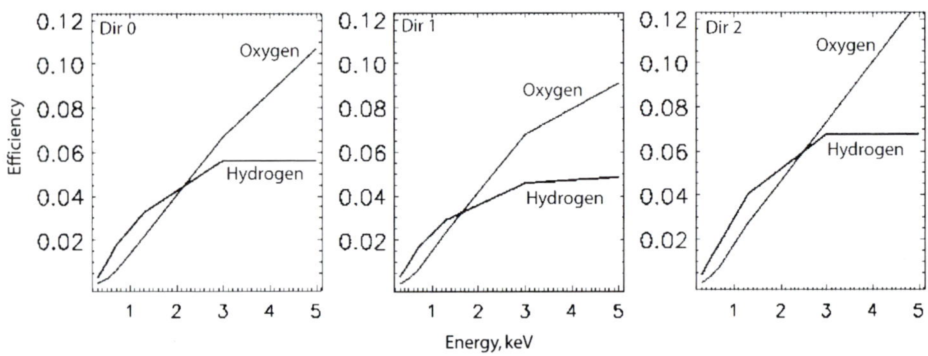

Figure 19. The NPD2 sensor absolute efficiency for oxygen and hydrogen ions (atoms).

(red) and two side (green and black). The FWHM varies over azimuth from 49° for the central sector to 38° and 40° for the side sectors. The FWHM is constant over elevation, 4.5° for all directions. The difference in the azimuthal response for different directions is a geometrical effect; particles from the side direction have a lower probability of reaching the STOP surface. In addition, the STOP and START secondary electron collection efficiency is lower due to fringe field effects.

Figure 19 shows the absolute efficiency of the NPD sensor as a function of energy for proton and H_2O^+ (oxygen) beams. As seen on the graph at 5 keV, the efficiency is more than 10% for oxygen ENAs, and the efficiency reaches 5–6% for hydrogen ENAs. Up to a certain energy, the efficiency increases with energy corresponding to an increase in the secondary electron yield on both START and STOP surfaces. At the energy where the yield exceeds the unity on both surfaces, the efficiency levels out (ca. 3 keV for protons and 6 keV for oxygen). At energies below 1 keV, the efficiency is around a few percent.

Each NPD sensor is an "intelligent" device and its electronics perform a substantial portion of the data pre-processing. The NPD electronics consists of two boards:

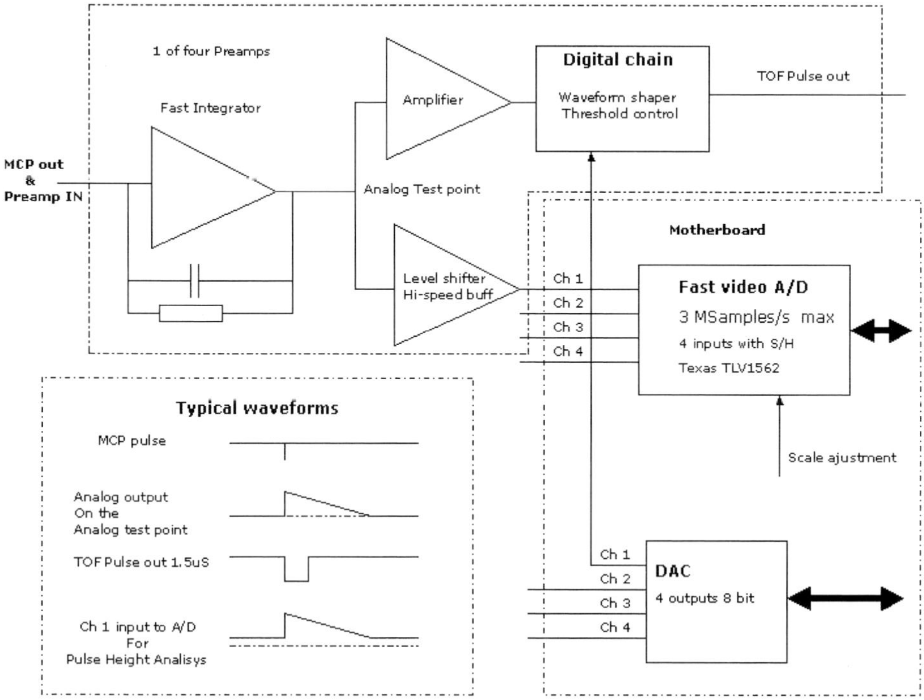

Figure 20. NPD FEE (Front End Electronics) block diagram.

FEE and digital TOF electronics (DigTOF). All high voltages are provided and controlled externally. The FEE block diagram and typical signal shapes are shown in Figure 20. Pulses from four MCP anodes (one START and three STOPs) are fed to fast MOS FET – based discrete charge sensitive preamplifiers (frequency cut-off at 6 GHz) which is followed by operational amplifies. The wave shaper generates a fast logic pulse for the Time-To-Digit converter (TDC) of the DigTOF board and fast video ADC Texas TLV1562 (Analog-To-Digital converters) performs pulse height analysis. The signals are up shifted by 0.8 V to reach the ADC working range of 0.8–3.8 V. Four DACs (Digital-to-Analog converters) provide threshold control. The thresholds are commandable. FEE provides theoretically a TOF resolution of 0.1 ns that is much below what is required for this measurement technique. Even a resolution of 1 ns would be sufficient.

The DigTOF electronics serves a number of tasks including TOF measurements from one START to one out of three STOP signals, serving FEE and initiating analog to digital conversion of STOP pulses, coincidence check and selection of valid TOF STOP pulse height pairs, counting events, buffering three different data types in SRAM (Static RAM, Random Access Memory), and interfacing DigTOF to the DPU. Figure 21 shows the DigTOF block diagram focused on the functionality of the diffirent componets and FPGA subroutines. The occurrence of a START

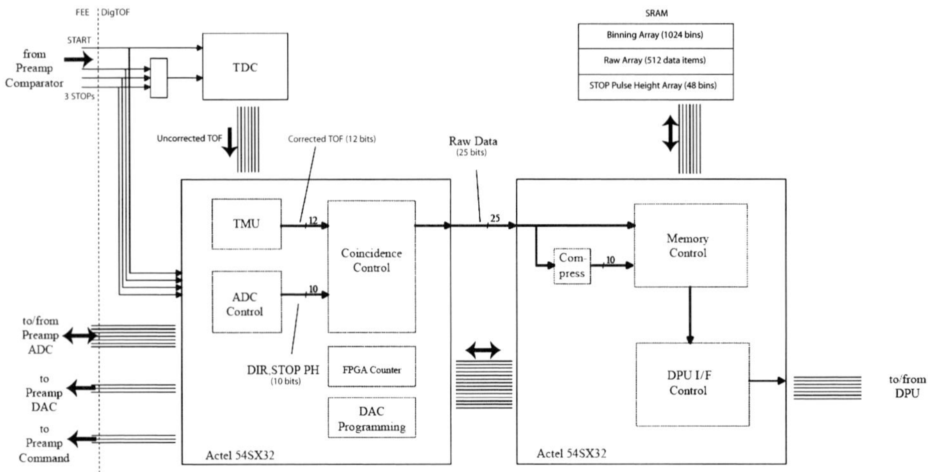

Figure 21. NPD DigTOF (Digital TOF electronics) block diagram.

signal, followed by a STOP signal leads to a TOF measurement in the TDC and generation of a data item with uncorrected TOF information. The obtained time is then corrected to compensate for a possible temperature shift. The calibrations are performed continuously using 6 MHz clock supplied by the DPU and the calibration constants at DigTOF are continuously updated. The TDC control and TOF correction are performed by a TDC Management Unit (TMU), a part of FPGA. The final time data item is a TOF value with 12-bit binary time information. In parallel to this, a sampling process by FEE ADC follows the occurrence of a STOP signal, thus a 10-bit data item with information of both direction and 8-bit STOP pulse height is generated. The coincidence control then checks for a valid coincidence of these two data items and additionally flags the occurrence of more than one START or STOP signal during TOF measurements. This now leads to a 25-bit raw data item that is used for the pending storing process. The memory control has to handle three different memory areas in the SRAM. For the binning array, the raw data is compressed into 10-bit data that represents the bin number inside the array. The respective bin is incremented by one up to the bin depth of 65,536 (16 bit). For the raw data array, incoming raw data items are successively stored until this array is filled completely (512 entries). The STOP counter array is filled in the same way as the binning array, but with the compressed STOP pulse height together with the respective direction. All data arrays are filled in parallel (binning array is excluded, if coincidence level does not fit). Readout and the following initialization of these arrays are performed by using burst read access from the DPU. Besides the STOP counter array, 16-bit event counters and two registers facing the preamplifier board are implemented in FPGAs. One of these registers is used to program FEE DACs, the other for directly commanding FEE. All control, counter, and memory data are

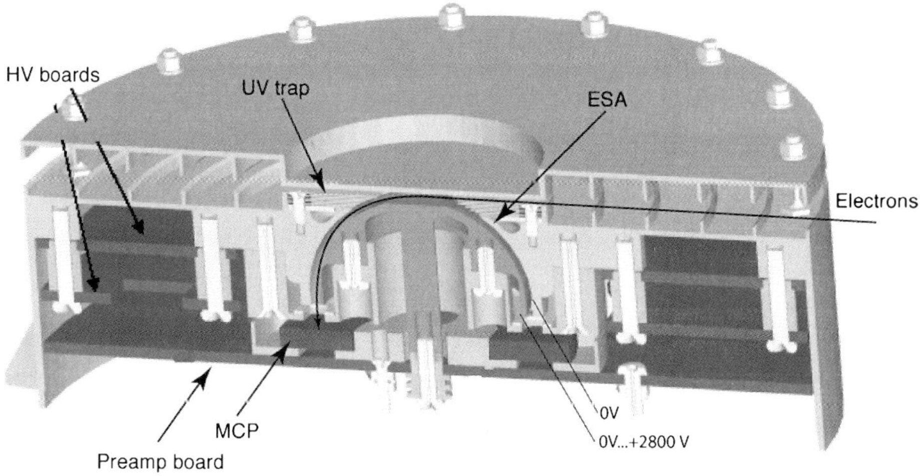

Figure 22. Cut-sectional view of the ELS sensor. The electrostatic analyzer voltages are also indicated.

accessed from the DPU over 16-bit registers; physically the connection to the DPU is a 16-bit bus. DigTOF is realized in two FPGA Actel 54SX32.

3.4. Electron Spectrometer (ELS)

The ELS sensor represents a new generation of ultra-light, low-power, electron sensors (Figures 22 and 23). It is formed by using a collimator system followed by a spherical top-hat electrostatic analyzer (Sablik *et al.*, 1990), achieving a geometric factor of 5.88×10^{-4} cm^2 sr. The top-hat has a 17° opening angle and a 90° turn angle. The radii of the inner and outer hemispheres are 14.9 and 15.9 mm, respectively. Particles enter the aperture at any angle within the "plane" of incidence, which is determined by the collimator to be 4° × 360°. Electrons are then deflected into the spectrometer by applying a positive voltage to the inner hemisphere. The electrons hit an MCP after being filtered in energy by the analyzer plates. There are 16 anodes behind the MCP, each anode defining a 22.5° sector and each connected to a preamplifier. Each ELS preamplifier generates TTL signals, which are then counted by the DPU. ELS deflection plates are stepped in voltage sequences to achieve differential spectral measurements. Electrons with energies up to 20 keV/q can be measured.

The ELS unit has a self-contained, dual range, linear power supply. The first power supply range is from 0 to about 21 V (about 150 eV) and has 4096 possible settings. The second power supply range is from 0 to 2800 V (about 20 keV) and also has 4096 possible settings. The ELS sweep is fully programmable within the constraint that the maximum decay rate is 32 steps/s. On any given step, the deflection plate voltage is held constant during a minimum of 28.125 ms, which

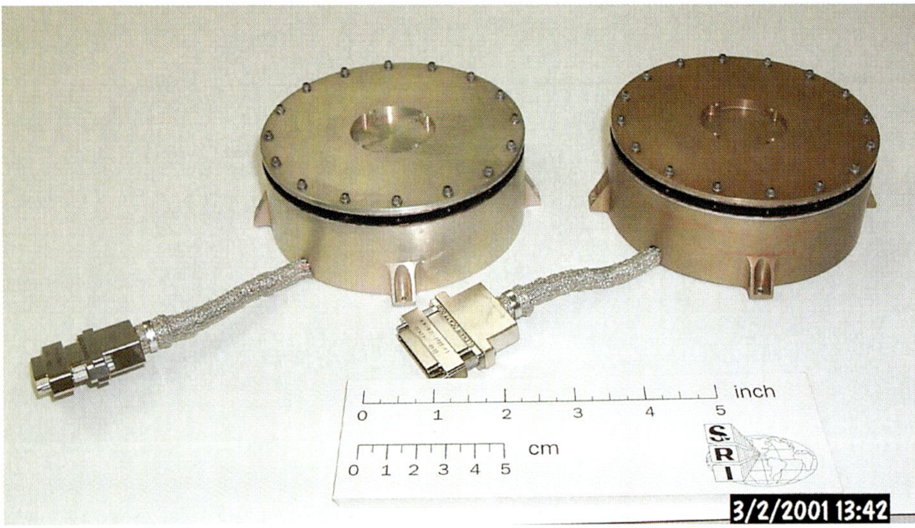

Figure 23. ELS spare (*left*) and flight models (*right*).

is used to accumulate electrons. There is a minimum of 3.125 ms of data latency between energy steps for transition. Thus, the total time per energy step is 31.25 ms.

The ELS FEE (similar to NPI) generates outputs from 16 anodes which are fed through individual amplifiers housed in two MOCAD chips. These amplifiers convert and amplify the MCP pulse into a TTL signal which is sent from the sensor through communication lines (16) to the DPU. The DPU gathers the samples and controls the accumulation time for the step, as well as the latency between steps and the energy sweep. The DPU also controls the instrument settings and determines when to read the monitor outputs. The ELS sensor is mounted on top of the NPI sensor, which is mounted on top of the DPU as an assembly. This assembly is mounted on the ASPERA-3 scan platform (Figure 5) in such a way that the full 4π angular distribution of electrons is measured during each platform scan (the scanner takes 32, 64, or 128 s to complete a sample rotation).

ELS was designed to be solar blind so that it may operate in exposure to direct sunlight. This has been achieved by using two UV reducing mechanisms and one secondary electron suppression technique. UV is minimized through the use of a series of light baffles in the ELS collimator and a series of UV light traps at the entrance to the spherical deflection plates. Secondary electrons are reduced by the addition of a special coating, based on a modified Ebanol-C process, which is included throughout the deflection surface, light trap, and collimator system (Johnstone *et al.*, 1997). To avoid overload of the MCP by photoelectrons, a grid is installed in front of the MCP assemblies, which can be biased by a voltage (256 programmable) in the range from 0 V to −5 V.

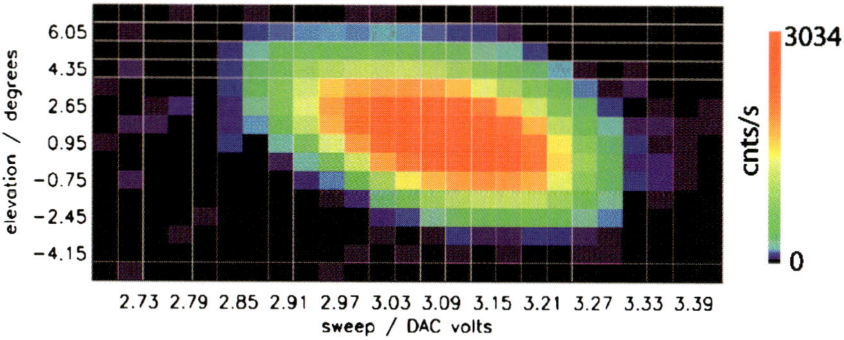

Figure 24. ELS calibrations: Typical spectrograph of the elevation angle versus the applied voltage on the inner hemisphere (100 eV beam, sector 7).

ELS was calibrated at the Mullard Space Science Laboratory (MSSL) in the United Kingdom. The calibration facility (Johnstone *et al.*, 1997), based on the technique described in Marshall *et al.* (1986), provides a wide area photoelectron beam at energies ranging from a few eV to 15 keV with variable beam intensities from a few Hz to several MHz. The system is fully automated facilitating calibration scans over the complete range of polar and azimuth angles at several instrument voltage settings both for the analyzer as well as the MCP. A flexible data acquisition system was integrated into the automation to provide simultaneous measurements from the 16 preamplifier channels, coordinated with the instruments position and voltage settings. Before performing the instrument calibration, a profile of the beam output is recorded at each of the calibration energies by means of a channel multiplier mounted on an X-Y table. During calibrations, the channeltron is mounted as close as possible to the instrument aperture in order to provide a constant reference to the beam intensity.

The instrument has two operational voltage ranges for the energy sweeps as described earlier, and hence, tests were carried out at several energies in both ranges to cover $\pm 180^\circ$ in polar and $\pm 3^\circ$ in azimuth. Figure 24 is a typical plot of the voltage-angle scans carried out over sector 7 using a 100 eV electron beam (similar for other sectors). Such plots were used to calculate the *k*-factor factor (central energy/analyzer voltage) of the analyzer and energy resolution for all 16 sectors. Figure 25 is a plot of the *k*-factor across the 16 anodes. Although the *k*-factor is lower than the design value of 7.5 eV/V, the variation across the anodes is less than 10%, allowing small deviations in the instrument response to be determined from good calibrations. The calibrations also established the initial MCP operational level for flight (2250 V).

Finally, calibration of ELS also tested the UV rejection ratio of the analyzer using a Lyman α UV source (Alsop *et al.*, 1996). Figure 26 shows the rejection efficiency versus analyzer voltage for sector 5 with the screen grid on and off (these

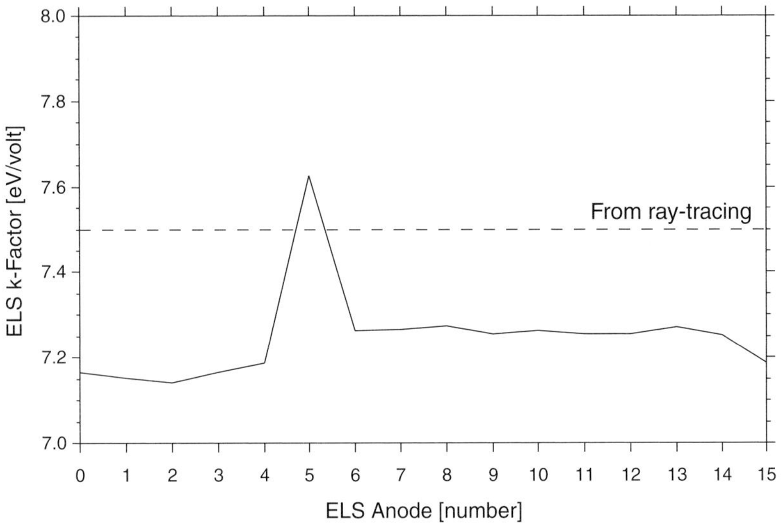

Figure 25. ELS calibrations: variation in *k*-factor (central energy/analyzer voltage) across the 16 anodes.

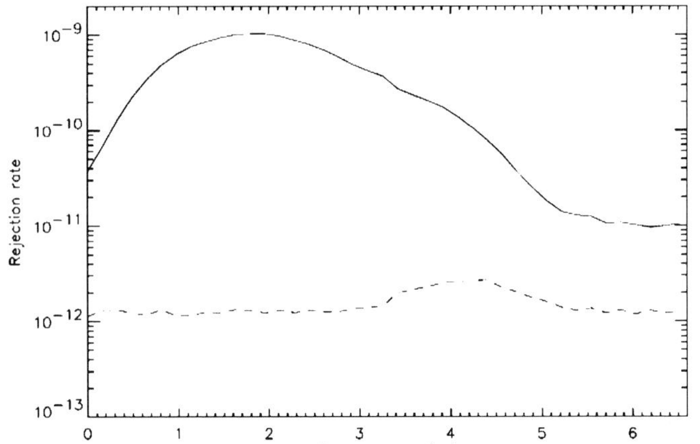

Figure 26. ELS calibrations: UV rejection (summed over all elevations) versus sweep energy for sector 5 (*solid line*: grid off, *dashed line*: grid on).

data are typical for all other sectors). When the grid is on, photoelectrons produced inside the analyzer are effectively repelled, which results in a decrease of the UV related signal by a factor of almost 1000.

ELS is operated in several modes with different sweeps. The most common mode is a survey mode where the ELS sweep steps through 128 of its possible 8192 voltage settings in 4 s to produce a logarithmic energy sweep covering the full ELS energy range (to 20 keV). Adjacent energy steps coincide at about the 50% point

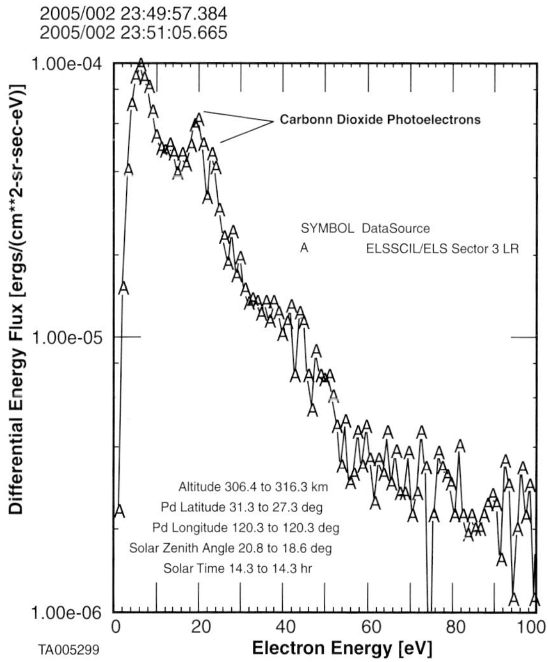

Figure 27. ELS measurement at Mars: photoelectron spectrum from the Mars ionosphere.

of the sector response. In this mode, energy and/or time degradation is selectable, but is seldom used.

ELS is sometimes operated in a linear stepping mode which produces an oversampled spectrum of 128 selected voltage settings every 4 s. Here, the energy width of a sector is usually larger than the center energies between steps. When combined with the knowledge of the instrumental response function, flux values can be adjusted to account for measurement overlap of the same parent function by removing the overlap flux, and thus sharpening the measurement (Early and Long, 2001). Electron measurements from this linear stepping mode are shown in Figure 27, which shows that ELS can resolve of the CO_2 photoelectron peaks, characteristic of the Martian atmosphere.

Since ELS is fully programmable, there are energy modes defined which select randomly sampled energies, and other modes which alter the time for the sweep resolution and number of samples in a sweep. The minimum requirement for programming the ELS energy sweep is that there can be no more than 32 energy steps per second and a change between voltage levels of 10% requires a 3.125 ms dead time for the ELS power supply to settle to the new deflection value (factor of safety is 3 for dead time). ELS will take advantage of its programming capability to best sample Martian phenomena.

3.5. Ion Mass Analyzer (IMA)

IMA is a stand alone instrument with its own DPU and high voltage power supplies (Figure 5). The ASPERA MU only provides the required voltages (+5 V, −5 V, +12 V, and −12 V, +30 V) and serves as a relay for IMA telemetry (TM) packages. The IMA instrument comprises a sensor, high voltage supply, and a DPU with housekeeping electronics which interface the sensor and the ASPERA-3 DPU.

3.5.1. *IMA Sensor*

The IMA sensor consist of four main components: electrostatic deflection system to provide elevation sweep, top-hat electrostatic energy analyzer, permanent magnet-based velocity analyzer, and an MCP detector with a position sensitive anode (Figure 28). Ions enter the sensor through a grid grounded to the spacecraft such that the entrance grid is at the spacecraft potential. Behind the grid is a deflection system which has the purpose of sweeping the elevation acceptance angle between 45° and 135° with respect to the symmetry axis. Particles passing the deflector continue into a spherical 127° top-hat electrostatic analyzer (ESA). The dimensions of the ESA are 45.0 mm (center radius) and 2.2 mm (gap between the two plates). The outer hemisphere of the analyzer is kept at a fixed low voltage (−100 V), while the inner hemisphere is stepped through the high-energy part of the voltage sweep, from −4 kV up to −100 V. During the lower energy portion of the sweep

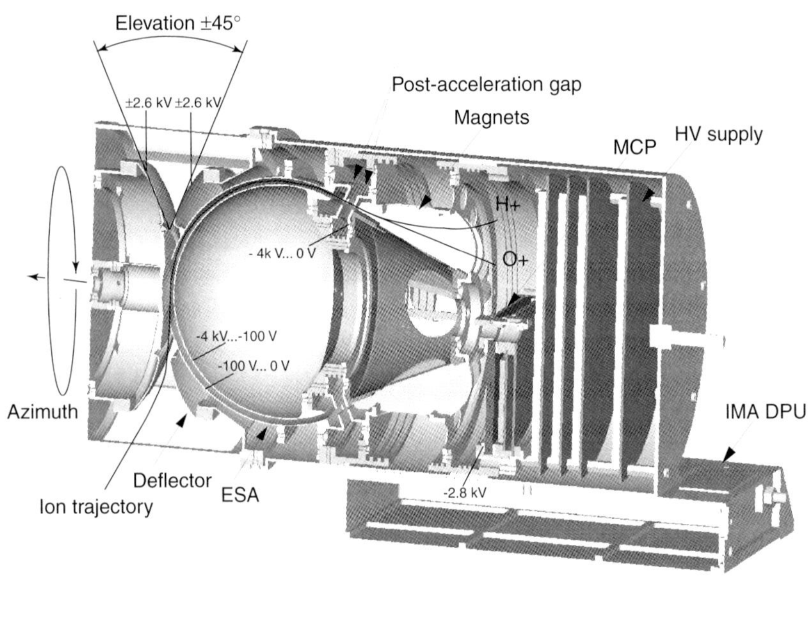

Figure 28. The IMA instrument cross-section. The key voltages are identified.

the situation is reversed, the inner hemisphere is kept at a fixed potential and the potential on the outer hemisphere is stepped up towards ground potential. That allows more accurate (closer energy steps) and precise (lower voltage setting error) measurements in the low energy range. Ions transmitted through the ESA pass through a two-slit electrostatic lens consisting of a 7 mm slit (closer to ESA) and a 1 mm slit (closer to the magnetic section). The size of the slits is a compromise between mass resolution and the instrument sensitivity. The mass analyzer uses a circular magnetic field, which deflects the ions outward, away from the central axis of the analyzer system. The magnetic system consists of 16 radially oriented permanently magnetized neodymium-iron-boron magnets (VACODYM® alloy). To lower the stray magnetic fields, the 16 magnets are matched to have the same magnetization within $\pm 1\%$ accuracy. The magnetic field strength is approximately 1200 G in the center of the section. The entire magnet assembly can be biased by a post-acceleration voltage between 0 and -4 kV. The post-acceleration is used to vary the mass resolution of the instrument. Low post-acceleration increases the mass resolution due to an increasing difference in the gyroradius between different masses, although low energy protons cannot be detected because they impact the analyzer walls due to the small gyroradius. Protons with energy below 1 keV require the maximum post-acceleration to reach the detector.

Particles striking the front of the 100 mm diameter MCP (two stacked plates biased on the front at -2.8 kV and with a 300 V drop between the MCP back and the anode) produce an electron cloud of approximately 10^5–10^6 electrons. The electron cloud is split between rings and anodes of the IMA position sensitive detector. A system of 32 concentrically spaced anode rings behind the MCP measures the radial impact position (representing ion mass), whereas 16 sector anodes measure the azimuthal impact position (representing the ion entrance angle). The 32 ring anodes and the 16 sector anodes are each connected to 6 charge-sensitive MOCAD preamplifiers (8 channels per chip). When the preamplifier TTL output of both a ring and a sector is high, coincidence circuitry implemented in an FPGA resolves the impact coordinates. The coordinates are used by another FPGA to update a 32 by 16 memory position matrix (16-bit deep) with one count every time a valid particle is detected in a given ring-sector (mass-angle) position. Two memories are used to allow for double buffering of data. While data is being accumulated into one memory, the IMA DPU reads the other memory.

Internal working mode of the IMA sensor is always the same. The fastest changing parameter is the particle energy. It sweeps from 36,000 eV down to 10 eV over 96 logarithmically equidistant steps. The sampling time on the each energy step is 125 ms. The "mass image" of 16 azimuthal sectors $\times$ 32 rings (mass) is read-out once per sampling time. After each complete energy sweep the instrument changes the polar angle of the field of view. The polar angle scans from -45° up to $+45^\circ$ over 16 steps. The total time to complete a 3D spectrum is 192 s. This spectrum consists of 32 rings (mass) $\times$ 16 azimuthal sectors $\times$ 96 energy steps $\times$ 16 polar angles.

3.5.2. *IMA Electronics and Data Processing Unit*

The IMA electronics includes high voltage power supplies for biasing of the various electrostatic filters and the MCP assembly, and digital electronics to handle the instrument operations. The high voltage supplies are laid out on two circular boards in the cylindrical sensor part of the instrument (Figure 28). The high voltages are achieved by switched-mode power supplies converting an input voltage of 25–30 V DC to approximately 4500 V DC through the use of transformer ratios and subsequent chains of diode-capacitor voltage multipliers. IMA has two separate high voltage power supplies. One supply is dedicated to keeping the front side of the MCP biased to −2.8 kV, while the other supply generates all other high voltages as fixed "raw" voltages. These raw voltages are then regulated by the use of high voltage optocouplers (Amptek HV601B). The electrostatic entrance filter can be stepped between ±2.6 kV to an accuracy of 1.2 V using a 12-bit DAC and the electrostatic energy filter is stepped between 0 and −4.0 kV also using an 8-bit DAC with a high voltage optocoupler. As described above, high accuracy at low voltage settings is achieved by fixing the −4.5 kV supply at a voltage of about −100 V, and using a −100 V supply with a 12-bit DAC connected to the outer ESA hemisphere to step through the lower energy range. The post-acceleration voltage is also taken from the raw −4.5 kV voltage and set to the desired voltage using an optocoupler.

The IMA DPU is built around a 16-bit processor (MA31750) and performs the following main functions: (a) reading of data from the double-buffered sensor memory and processing of the data, (b) feed the IEEE 1355 digital serial interface to the MU DPU with processed and formatted data, (c) receive commands on the serial interface from the MU DPU, (d) control the high voltage power supplies' settings and monitor voltages and temperatures. Data processing includes three steps (1) summation (if commanded) of the angle bins (azimuth/elevation) and/or mass bins, (2) logarithmical compression of the count values from 16 to 8 bits, (3) loss-less RICE compression of the final spectrum. The mode of the processing is set by a telecommand (TC) or chosen automatically according to telemetry rate limitations. The post-acceleration level is changed by a TC. Only 3 values of the post-acceleration are available: 0 V, −2150 V, and −3650 V. Due to the data compression, the output data packets have variable length, and are buffered until enough compressed and time-stamped packets are available to fill a fixed length packet transmitted to the MU. To improve the compression efficiency a fixed background level can be subtracted onboard. Directions shadowed by the spacecraft can be masked out to avoid sending data which are not valid.

3.5.3. *IMA Calibrations*

IMA calibrations were performed at Centre d'Etude Spatiale des Rayonnements, Toulouse, France (Figure 29). Before calibrations IMA was extensively ray-traced and the model was then verified against the calibrations. That allowed minimizing the calibrations and calibrating only individual sectors. During the calibration the

Figure 29. The IMA instrument in the vacuum chamber during calibrations.

full response function was defined for selected masses. Figure 30 demonstrates the IMA angular responses: the instrument's effective aperture (in cm^2) as a function of the elevation angle for a fixed azimuth at the maximum response (a), and the azimuth response for a fixed elevation at the maximum response (b). Despite the ion optics can provide the azimuth resolution down to 15° (FWHM), the instrument resolution defined by the full field of view divided by 16 sectors is assumed to be 22.5°. As expected the mass resolution strongly depends on the post-acceleration voltage (Figure 31), but even at the maximum post-acceleration value, IMA can resolve amu/charge = 1, 4, 14. For the incoming beam energy of 1.2 keV, the proton peak is observed to disappear below $V_{postacc} = -2000$ V while at sufficiently high post-acceleration voltages, all masses can reach the detector. At the medium post-acceleration, $V_{postacc} = -2000$ V, the mass resolution of IMA is sufficient to resolve all masses relevant to the near-Mars space (Figure 32). The calibrated curves of the constant mass in the energy-mass matrix are in excellent agreement with the observations at Mars. Figure 33 shows the energy-mass matrix for $V_{postacc} = -2000$ V accumulated by the IMA sensor for 1.5 years at Mars along with the theoretical mass lines. The yellow band at about 700 eV is a contamination from the solar wind protons.

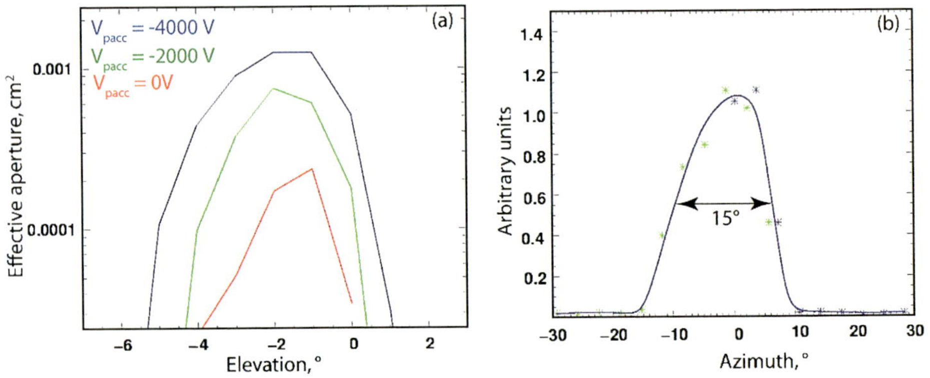

Figure 30. IMA angular response over elevation for different post-accelerations (V_{pacc}) and azimuth.

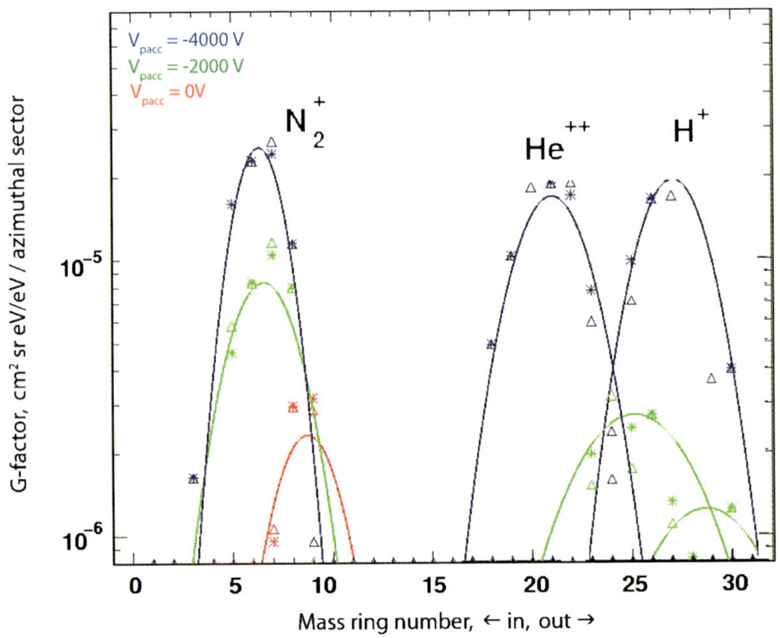

Figure 31. IMA mass resolution for different post-acceleration voltages. The beam energy is 1.2 keV.

3.6. Scanner

The scanner platform was originally developed as part of the ASPERA-C experiment during the Russian Mars-96 project. Some modifications were made for the ASPERA-3 on Mars Express, most of them concerned optimization of the performance during long-term operations and reduction of its mass. The scanner, shown in Figure 34 and technical characteristics are given in Table IV, constitutes the 0° to 180° rotating platform on which the sensors (ELS, NPI, NPD1 and NPD2) as

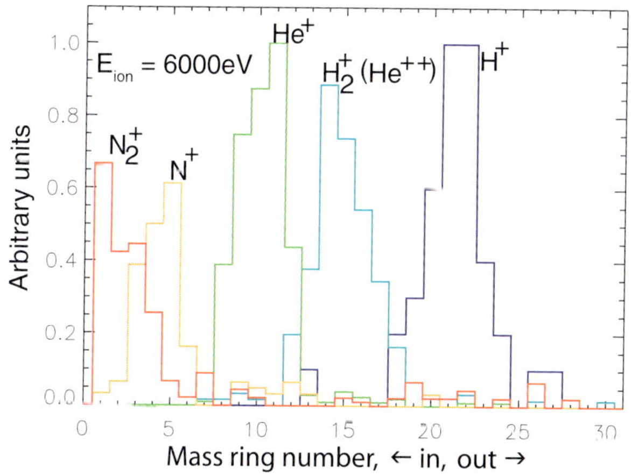

Figure 32. IMA mass resolution for 6 keV particles and $V_{postacc} = -2000$ V.

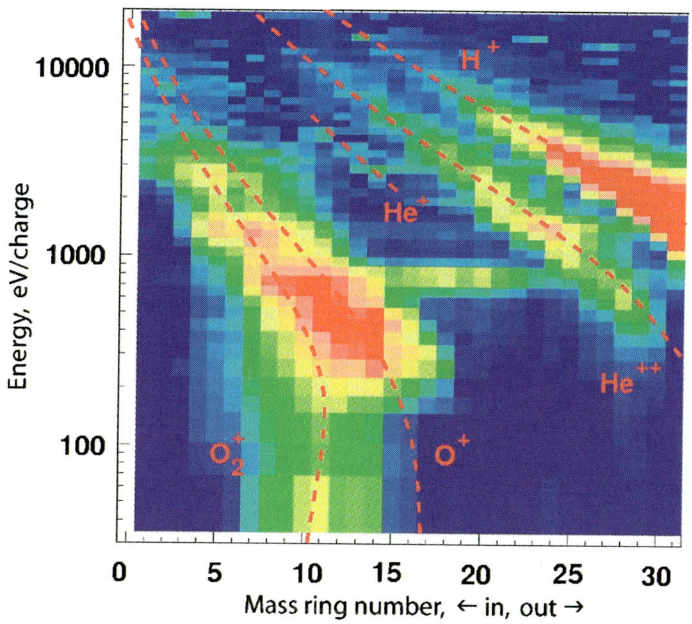

Figure 33. Energy – mass matrix for $V_{postacc} = -2000$ V, accumulated by the IMA sensor for 1.5 years at Mars with the theoretical mass lines. The yellow band at about 700 eV is a contamination from the solar wind protons.

well as the DPU, high voltage power supply, house keeping and DC/DC boards are situated. The scanner provides all necessary mechanical and electrical interfaces between the spacecraft and ASPERA-3. Rotation is accomplished by the use of a worm gear mechanism, which was selected in order to minimize friction, and

TABLE IV
Scanner technical characteristics.

Parameter	Value
Maximum angle of rotation (°)	±100
Angular movement per step (°)	0.0095–0.0190
Angular position feedback resolution (°)	0.05
Angular positioning accuracy (°)	0.2
Operational rotation rate (°/s)	1.5/3.0/6.0
Maximum rotation rate (°/s)	~25.0
Power dissipation (W)	0.5–2.0
Platform load (kg)	3.7
Maximum platform load (kg)	~12
Dimensions (mm)	60 × 254 × 232
Mass (kg)	1.42
Operational lifetime in vacuum (years)	~3
Gear ratio	188

Figure 34. The ASPERA-3 scanner.

obtain a high gear ratio (1:188). The scan platform is made as a plug-in unit for the sensor assembly. Considerable efforts were spent to reduce mass, volume, power consumption, and out-gassing in vacuum as well as to achieve high reliability.

We have not made any particularly investigations to measure the level of microvibrations induced by the scanner motion since the scanner does not have any

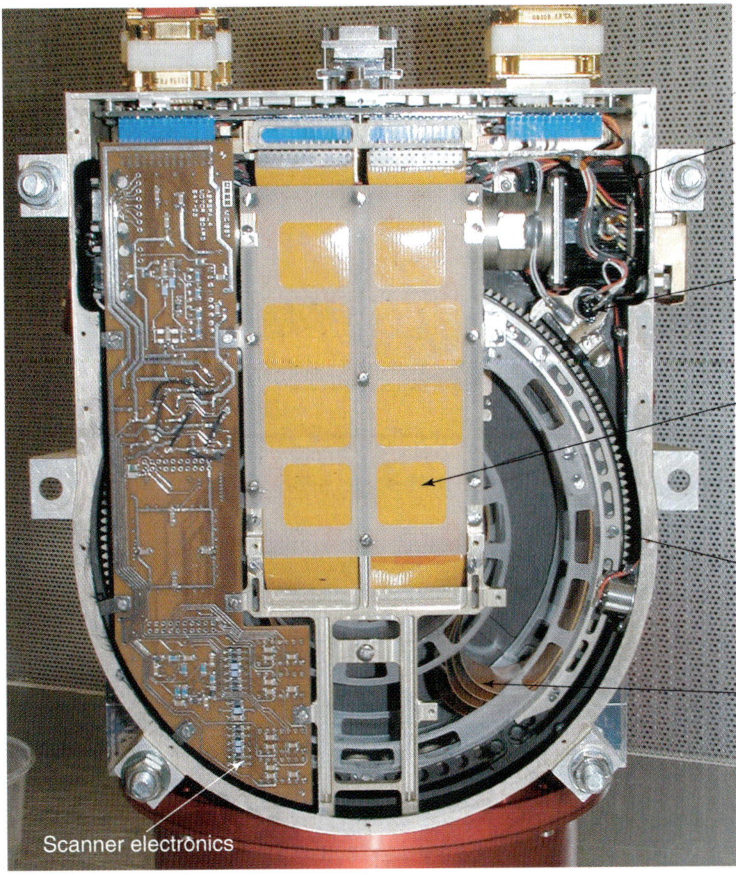

Figure 35. Scanner interior with the main features identified.

parts moving with high frequency. What might be a problem are extremely small changes in the spacecraft attitude caused by the Main Assembly motion and the associated onboard attitude control system reaction. The level of the spacecraft attitude changes as given by the Mars Express gyros readings was in the range 0.3 mrad (maximum over all three axes). To illuminate any possible effects on remote sensing instruments the scanner operations were conducted outside their operations.

3.6.1. *Mechanics*

Figure 35 shows the scanner internal view. The large diameter worm wheel to which the sensor assembly is fixed is rotated by a stepper motor *via* a co-axial worm screw (not visible below the flat cable). The worm wheel is fixed to the structure with a large diameter angular contact ball bearing. During the scanner lifetime tests, several types of bearing balls were tested, including the balls originally mounted in the bearings. The bearing balls which were ultimately used are of ceramic type Si_3N_4, which was found to be the best suited to meet the ASPERA-3 requirements.

Figure 36. Scanner locking mechanism.

The housing and circular sensor platform is manufactured using a high-strength aluminum alloy. The position of the movable parts relative the scanner is given by three magnetic sensors: two end-sensors at 0° and 180°, and one step counter. Because of the long-term operational requirements, no mechanical contact exists with the sensors. A feed-through cable loop consisting of six flat cables with connectors, each cable with 26 conductors (a maximum of 156 connections possible) interfacing through D-SUB connectors, provides electrical interface to the satellite electrical systems for the entire ASPERA-3 instrument.

During launch and other necessary transports (when the instrument is without electrical power), ASPERA-3 was expected to encounter heavy mechanical vibrational loads. A worm gear type of mechanism provides a self-locking behavior without electrical power, and thus, was selected for use on the scanner. A true locking mechanism was also included on the scanner (Figure 36) and prevents unwanted movements of the platform. This locking mechanism consists of a wire which ties together two small levers, locking the square-shaped worm screw axis. By command, the axis is unlocked by applying a voltage to a resistor which burns the wire and then the levers are forced to move apart by the actions of a spring. The release scanner command can be executed only once.

3.6.2. *Motor*

The stepper motor used is a modified P430 from Escap. The two ball bearings within the stepper motor holding the motor shaft were modified. The original ball bearings are replaced by type NMB R-1350ZZ bearings which have a surface treated with NoWear Gamma by SKF, Sweden. The balls are made of Si_3N_4 (NBD200) by Saint Gobain/Cerbec and have a diameter of 2.00018 mm, grade 5. The ball holders are designed and manufactured at IRF, Kiruna, Sweden and are made of Beraloy (Acoflon 100 Mo, 97% PTFE + 3% MoS_2). The ball bearings are mounted by IRF, Kiruna, Sweden in the motor with shims for a correct axial displacement. Also, the ball bearings are mounted without any axial tension in order to release the ball bearings from wear. The motor shaft is isolated from any axial load created by the worm screw with a soft split of the motor shaft and worm screw.

In the original design the scanner had two motors (one redundant) and two sets of driving electronics. The worm gear was directly coupled to both motor shafts. While the operating motor was driving the worm gear, the cold back-up motor was a passive mechanical load. However, during life-time test it was found that the cold back up motor is a potential source of a mechanical failure (jamming). Therefore, only one motor and respective electronics were used in the final design.

3.6.3. *Electronics*

The motor electronics located in the scanner provides motor control and driving. The stepper motor is driven by a classical H-bridge drive system with a motor current control system. The location of the board is shown in Figure 35.

For a smooth stepping of the motor, 16 microsteps per full step and winding are implemented. The 16 microstep levels are set by an ACTEL FPGA to a DAC, both situated on the scanner electronics board and set the current reference value for the motor current control system. In addition to current control, for minimizing mechanical interaction with its environment, the scanner electronics controls the scanner start up sequence by providing a start up ramp. Given a start up command, the scanner electronics ramps up the scanner speed from zero to full speed in 4 steps. In order to increase the torque of the motor, an offset of the current setting can be changed by command. By command, one can also set the coast as well as the ramp current separately, meaning that there can be a higher current (higher torque) during the ramp period than during the coast period.

The following modes of the scanner operation are possible: (1) continuous scanning back-and-forth between 0° and 180° in speed steps of 128, 64 and 32 seconds/180°, (2) continuous back-and-forth scans in steps of predefined by TC degrees and predefined by TC sampling time for each step, (3) positioning in any predefined by TC position. The determination of the scanner position is obtained by counting pulses from the wheel sensor that sits on the motor shaft as it rotates away from one of the two end position sensors. The end sensors reset the pulse counter. The angular positioning accuracy of the scanner pointing direction is 0.2°.

3.7. Main Unit (MU) Electronics

Functionally, the MU electronics includes DC/DC electronics, a high voltage power supply (HVPS), and data processing electronics. The DC/DC electronics is a single board, the HVPS includes two boards in a stack configuration, and the data processing electronics also includes two boards, a DPU board and a housekeeping (HK) board. The later two are connected together with the sensor control electronics, the power supply, and the HVPS *via* a common bus system with 8 address and 16 data lines besides control, analog and power supply lines.

3.7.1. *Digital Processing Unit (DPU)*

The DPU is built around a 16-bit processor (MA31750 from Dynex) with 12 MHz system clock frequency and an FPGA RT54SX32S, which implements memory management, watchdog functions, and the serial spacecraft interface protocol. The software runs inside a 128 kByte RAM, organized in 2 banks each processing two 32 kByte of static memory chips. On power-up, a two times 16 kByte bipolar PROM (HARRIS) (Programmable Read-Only Memory) is activated with a boot loader, which transfers the complete PROM contents into the RAM, changes the program control to the RAM area, and then switches the power to the PROMs off *via* transistor switches to conserve power. A 512 kByte radiation hardened EEPROM (Maxwell) (Electrically Erasable Programmable Read Only Memory) contains additional program code and configuration information, which can be modified from ground. A 2 MByte mass memory RAM is used to store measurement data and buffer telemetry packets. An Actel FPGA RT1280 implements the serial data transfer protocol to the IMA DPU with an interface identical to the corresponding hardware used on the Rosetta mission. All interface lines are buffered *via* special circuits to protect the instrument from external noise effects. The DPU board is controlled by a 24 MHz crystal, which is divided down to 12 MHz and buffered inside the FPGA before it is used for FPGA and processor operations, and on the housekeeping board. The main FPGA requires a 2.5 V operational voltage. This is generated by dedicated regulators directly on the DPU and on the HK board. The watchdog circuit inside the FPGA can be enabled by software. Then the watchdog circuit requires resetting by software access. Otherwise the watchdog circuit issues a hardware reset to the DPU board (after 16 s). Except for a special error message, the reset behavior is identical to a boot sequence initiated after power-up. The DPU board controls most detector voltages with direct access to the HVPS board (NPI, NPD) or the ELS power supply.

3.7.2. *Housekeeping Board (HK)*

A separate RT54SX32S FPGA, which maps all input, controls the HK board and output functions into standard bus address space. It also implements the needed counters for detector pulses from NPI and ELS. Four eight-channel analog multiplexers select one out of 32 analog voltages to be monitored. They are digitalized by one 14-bit ADC LTC1419. Another 14-bit ADC monitors the ELS deflection voltage. Two 8-bit DACs generate control voltages for NPD, NPI, ELS and the scanner *via* 8 latching buffers. The sun sensor electronics is implemented on the HK board.

3.7.3. *Software*

When the instrument is switched on, a boot loader copies the basic program from the bipolar PROM into RAM, switches the PROM off and starts monitoring the TC interface for possible boot instructions. This allows the configuration of the instrument to be flexible, but in a safe manner. If a start configuration is defined *via* TC, the software continues accordingly. Otherwise the default start configuration inside the EEPROM is used. If this is corrupted, the original default configuration from PROM will be used. The software is built around a real-time system with a scheduler and an interrupt handler. All executable routines are defined inside a routing table, which resides in EEPROM and can be modified during flight. In this way, new or modified software routines can be stored inside a free area of the EEPROM, verified and added to the operating software by including their start address into this routing table. A macro feature of the TC handler offers the possibility to generate sequences of standard TC automatically according to a pre-defined list, reducing the need for complex TC groups to be uplinked over and over again. Besides detector activation and parameter control, compression and averaging of measurement data allow the reduction of the amount of telemetry generated.

3.7.4. *DC/DC Electronics and High Voltage Power Supply (HVPS)*

The ASPERA-3 DC/DC electronics is build around four Interpoint DC/DC converters type HL which provide ±5 V, ±12 V, +5 V, and −5 V respectively. There are no redundant converters. Because of the extremely tight mass budget it was decided not to have separate switches for the individual sensors (NPI, NPD1, NPD2, ELS). Separate switching is only implemented for IMA (±5 V, ±12 V, +5 V, and −5 V). All sensors are powered when the instrument is switched on. There are, however, individually controlled 28 V switches for each sensor, which provide power for the sensor high voltage supplies.

The MU HVPS provides high voltages for NPI, NPD1, and NPD2. The general design is similar to the IMA high voltage power supply. A single high voltage supply provides a base voltage, which is regulated by AMPTEK HV601B optocouplers for the sensor use. The regulation accuracy is 256 steps for each range which is sufficient for this application. The base supply uses a common coil transformer followed by a custom-made doubler space qualified in a number of missions.

The NPI HVPS uses two base supplies generating two voltages, namely, an MCP bias in the range from 0 V down to −4300 V and a deflector bias from 0 to +5000 V. The fast (1 ms fall/rise time) alternative mode for the deflector voltage is provided by an AMPTEK HV601B optocoupler. Each NPD sensor has an individual HVPS which is built around two base supplies. The single polarity supply provides one base voltage from 0 to 3000 V which is regulated by two AMPTEK optocoules to bias individually START and STOP (all three at one) MCP assemblies. The second double polarity supply provides two voltages from 0 V to +5000 V and −5000 V for the NPD double polarity deflector.

3.8. Solar Sensors and MCP Protection

Direct solar light reaching NPI and NPD aperture would cause overload of the MCP based detectors. If the Sun illuminates the sensor only for a short period during scanning, the expected extracted charge over the nominal mission would be around 0.1–0.2 Coulomb/cm^2 for the NPD MCPs and 2 Coulomb/cm^2 for NPI. These levels of the extracted charge are a factor of 100 above those resulting in the MCP gain drop by a factor of 2 in chevron configuration (Malina and Coburn, 1984). Therefore, a system to protect the MCPs was implemented for the ENA sensors. The charged particle sensors, ELS and IMA, are basically immune to the direct solar light. Since the spacecraft attitude and the instrument accommodation were such that the Sun unavoidably reaches the NPI and NPD apertures when the instrument is in the scanning mode, a system to decrease the respective MCP bias voltages on approximately 30% (around 1 kV) for the required period was implemented. The system consists of a high voltage shutter (HVS) which operates when the instrument is in a scanning mode only and includes three different HVS based on different criteria for MCP voltage reduction control. The three HVS are NPD Count HVS, External HVS, and Solar Sensor HVS. Selection of the HVS type is made by a TC. Only one HVS can function at any time.

The NPD Counts HVS is based on continous monitoring of the NPD START count rate (non-correlated). When the count rate exceeds a certain (TC given) threshold, the MCP bias voltages are reduced for a certain (also TC given) period. The disadvantages of this method are a long response time, sensitivity to non-Sun related disturbances, and the difference between the NPI, NPD1, and NPD2 field of view. External HVS reduces the MCP bias voltages over a certain range of scanner positions given by a TC. The required range is defined on the ground from the analysis of the spacecraft altitude. This method was found not to be flexible enough and cannot be used when the spacecraft is in non-Sun related pointing modes, for example, nadir pointing.

The most advanced HVS uses autonomous detection of the Sun position during scanning with two solar sensors. The solar sensors are mounted inside the NPI sensor (Figure 8). Each sensor has $4.2^\circ \times 90^\circ$ field of view (see their apertures in Figure 7) and built around a photodiode sensitive to the solar UV. The bore-sight direction of each sensor makes the 75° angle to the instrument scanning axis. If the Sun is within the solar sensor field of view, the sensor generates a TTL signal. When the Solar Sensor HVS is enabled, the instrument first makes a scan with no HV switched-on and identifies the Sun position. The DPU calculates the respective scanner positions for the NPI, NPD1, and NPD2 sensors where the respective HV must be reduced.

3.9. Ground Support Equipment

Each sensor, ELS, NPI, NPD, and IMA have their individual EGSEs (Electrical Ground Support Equipments). The sensor EGSEs are used during calibrations and verification on the sensor level. The sensor EGSEs are built around Linux PCs with peripheral interface electronics to communicate with the sensors. External standard laboratory voltage supplies are used to power the sensor. The instrument level EGSE emulates the spacecraft OBDH (On Board Data Handling) and power system. It was also built around a Linux PC with the peripheral interface electronics and power system.

The instrument contains contamination sensitive detectors (MCPs) and surfaces (NPD STOP surfaces). To protect the instrument during ground activities, red-tagged covers enclose all sensor apertures and both MU and IMA were constantly purged by nitrogen. The purging inlets were installed directly in the covers, one for IMA, and three for MU (NPD1, NPD2, and ELS/NPI assembly). Purging Ground Support Equipment (PGSE) provided the constant flow (1 l/min) of nitrogen and the distribution between different units.

4. Instrument Accommodation and Operations

ASPERA-3 is externally mounted on the Mars Express bus (Figure 37). The accommodation was chosen (1) to minimize blocking of the instrument field of view by the spacecraft body, (2) to co-aligned the central plane of the IMA field of view with the ecliptic plane when the spacecraft is in the Earth pointing mode, (3) to locate the instrument away from the altitude thruster plumes, (4) to satisfy the bus mechanical requirements. The instrument scanning axis is co-aligned with the +Zb axis (Figure 37). During scanning the +Zb hemisphere is covered.

The main pointings used throughout the mission are the nadir pointing (the Zb axis points toward the local nadir) and the Earth (communication) pointing when −Xb points toward the Earth and Yb is perpendicular to the ecliptic plane. A number of special pointings are also available but most of the time the following profile is applied. The spacecraft is in the Earth pointing throughout the entire orbit except for 40 min around pericenter when it is in the nadir pointing for planetary surface and atmosphere observations. Slews to change from one pointing to another normally take 20–30 min. Therefore, the solar direction (close to the solar wind flux) is always within the IMA field of view when the spacecraft is in the Earth pointing. When the spacecraft is in the Nadir pointing, Mars is within NPI and NPD field of view. For electrons the spacecraft pointing is less critical. The scanner parking position is either the ELS/NPI central axis points in the direction of the −Yb axis, so called 90° position (as shown in Figure 37), or it points in the direction of +Xb, 0° parking position. If the instrument is parked in the 90° position, NPI cannot be operated in the Earth pointing because the Sun is in its field of view.

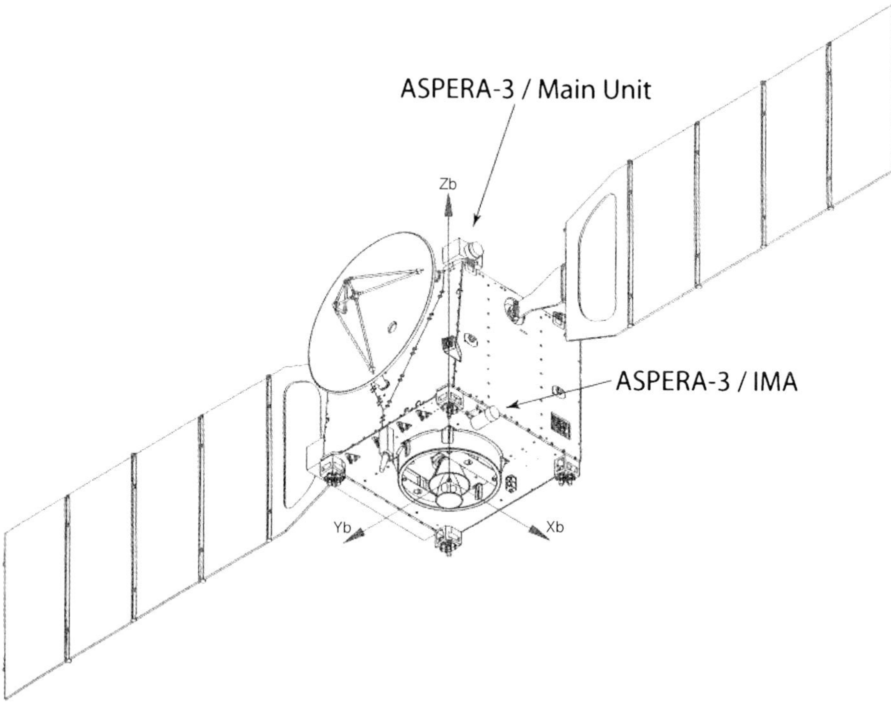

Figure 37. ASPERA-3 accommodation on the Mars Express bus (Courtesy of ASTRIUM, Toulouse).

The typical ASPERA-3 operational profile is as follows. IMA and ELS begin science data taking 20 min prior to the modeled inbound bow shock crossing and stop 20 min after the outbound crossing. ELS is run in the 128 energy step mode (full spectrum per 4 s). IMA is run in the full mode giving 32 masses $\times 16$ azimuths $\times 96$ energies $\times 16$ elevations matrix per 192 s. NPI and NPD are operational during nadir pointing only. NPI provides measurements of the signals from 32 directions with the sampling time 1 s. During different phases of the missions NPD is run in either the binned matrix mode or TOF mode. Because of very high TM demand, the RAW mode is used very seldomly.

5. Summary

The ASPERA-3 experiment is a comprehensive plasma package used to measure ions, electrons, and ENAs. It is for the first time such detailed particle measurements are conducted at Mars. However, because of the absence of field and wave experiments onboard Mars Express, the ASPERA-3 objectives concentrate on the studies of the solar wind impact with the Martian atmosphere. ASPERA-3 is instrumental in defining the current escape rates of the Martian atmosphere, and thus, defining the evolutionary impact of the solar wind interaction.

For the first time, ASPERA-3 performs ENA imaging of a non-magnetized atmospheric body. Due to severe mass constrains, the ENA sensor geometrical factors could not be made large by simply increasing the size of the instrument. Therefore, an entirely new technique, ENA surface reflection, was developed and implemented. The ENA results reported in this issue as well as previous publications (see references in this issue ENA papers) clearly indicate that the technique works.

A replica of the ASPERA-3 experiment, ASPERA-4, is currently operational on an orbit at Venus onboard the Venus Express mission launch 2003. This makes the ASPERA-3 experiment a unique tool to be used in comparative magnetospheric studies.

Appendix. List of Acronyms

ADC	Analog-To-Digit Converter
ASPERA	Analyzer Of Space Plasmas And Energetic Atoms
DAC	Digit-To-Analog Converter
DigTOF	Digital TOF Electronics
DPU	Digital Processing Unit
EEPROM	Electrically Erasable Programmable Read Only Memory
EGSE	Electrical Ground Support Equipment
ELS	Electron Spectrometer
ENA	Energetic Neutral Atoms
ESA	Electrostatic Analyzer
FEE	Front End Electronics
FPGA	Field Programmable Gate Array
FWHM	Full Width At Half Maximum
HK	House Keeping
HVS	High Voltage Shutter
HVPS	High Voltage Power Supply
ICA	Ion Composition Analyzer
IMA	Ion Mass Analyzer
IMAGE	Imager For Magnetopause-To-Aurora Global Exploration
IMI	Ion Mass Imager
IMIS	Iona Mass Imaging Sensor
LENA	Low Energy Neutral Atoms
MAG/ER	Magnetometer And Electron Reflectrometer
MCP	Microchannel Plate
MEDUSA	Miniaturized Electrostatic Dual-Top-Hat Spherical Analyzer
MGS	Mars Global Surveyor

MOCAD	Monolitic Octal Charge Amplifier/Pulse Discriminator
MU	Main Unit
NPI	Neutral Particle Imager
NPD	Neutral Particle Detector
OBDH	On Board Data Handling
RAM	Random Access Memory
PGSE	Purging Ground Support Equipment
PHD	Pulse Height Distribution
PIPPI	Prelude In Planetary Particle Imaging
PROM	Programmable Read Only Memory
SE	Secondary Electrons
SRAM	Static RAM
TC	Telecommand
TDC	Time-To-Digit Converter
TICS	Three-Dimensional Ion Composition Spectrometer
TM	Telemetry
TMU	TDC Management Unit
TOF	Time-Of-Flight
TTL	Transistor-Transistor Logic

Acknowledgements

The ASPERA-3 experiment on the European Space Agency Mars Express mission is a joint effort between 15 laboratories in 10 countries, all sponsored by their national agencies as well as the various departments/institutes hosting these efforts. We also wish to acknowledge the Swedish National Space Board for their support of the main Principle Investigator institute, Swedish Institute of Space Physics, Kiruna, and we are indebted to European Space Agency for its courage in embarking on the Mars Express program, the first European mission to the red planet. We acknowledge contributions from Imperial college, London, UK for providing the IEEE-1335 link chips used in the IMA sensor and NASA NASW-0003 for providing the ELS sensor.

References

Acuña, M. H., *et al.*: 1998, *Science* **279**, 1676.

Alsop, C., Free, L., and Scott, S.: 1996, UV rejection design and performance of the Cluster PEACE 'top-hat' electrostatic analyser, submitted to *Proc. AGU Chapman Conference on Measurement Techniques in Space Plasmas*.

Barabash, S.: 1995, *IRF Scientific Report* **228**.

Barabash, S., Fedorov, A., Lundin, R., and Sauvaud, J.-A.: 2007, *Science* (in press).

Barabash, S., Holmström, M., Lukyanov, A., and Kallio, E.: 2002, *J. Geophys. Res.* **107**(A10), 1280, doi:10.1029/2001JA000326.

Bertaux, J.-L., Leblanc, F., Witasse, O., Quemerais, E., Lilensten, J., Stern, S. A., *et al.*: 2005, *Nature* **435**, doi:10.1038/nature03603.

Bibring, J.-P., Langevin, Y., Gendrin, A., Gondet, B., Poulet, F., Berth, M., *et al*: 2005, *Science* **307**, 1576.

Brecht, S. H.: 1997, *J. Geophys. Res.* **102**, 11287.

Brinkfeldt, Klas, Instrumentation for Energteic Neutral Atom Measurements at Mars, Venus, and the Earth: 2005, *IRF Scientific Report 288*, Ph.D. thesis.

Brinkfeldt, K., Gunell, H., Brandt, P., Barabash, S., Frahm, R. A., Winningham, J. D., *et al.*: 2006, *Icarus* **182**, 439.

C:son Brandt, P., Barabash, S., Wilson, G. R., Roelof, E. C., and Chase, C. J.: 2000, *J. Atmos. Solar Terrestrial Phys.* **62**, 901.

Carlsson, E., Fedorov, A., Barabash, S., Budnik, E., Grigoriev, A., Gunell, H., *et al.*: 2006, *Icarus* **182**, 320.

Collier, M. R., Moore, T. E., Ogilvie, K. W., Chornay, D., Keller, J. W., Boardsen, S., *et al.*: 2001, *J. Geophys. Res.* **106**, 24893.

Crider, D., Acuña, M., Connerney, J., Vignes, D., Ness, N., Krymskii, A., *et al.*: 2002, *Geophys. Res. Lett.* **29**(8), 1170.

Early, D. S., and Long, D. G.: 2001, *IEEE Trans. Geosci. Remote Sensing* **39**, 291.

Fox, J. L.: 1997, *Geophys. Res. Lett.* **24**, 2901.

Futaana, Y., Barabasha, S., Grigoriev, A., Winningham, D., Frahm, R., and Lundin, R.: *Space Sci. Rev.*, this issue, doi: 10.1007/s11214-006-9026-9.

Galli, A., Wurz, P., Barabash, S., Grigoriev, A., Gunell, H., Lundin, R., *et al.*: *Space Sci. Rev.*, this issue, doi: 10.1007/s11214-006-9088-8.

Grigoriev, A., Futaana, Y., Barabash, S., and Fedorov, A.: *Space Sci. Rev.*, this issue, doi: 10.1007/s11214-006-9121-y.

Gunell, H., Holmström, M., Barabash, S., Kallio, E., Janhunen, P., Nagy, A. F., *et al.*: 2006, *Plane. Space Sci.* **54**, 117.

Holmström, M., Barabash, S., and Kallio, E.: 2002, *J. Geophys. Res.* **107**(A10), 1277, JA000325.

Jans, S.: 2000, Ionization of energetic neutral atoms for application in space instrumentation, Diplomarbeit der Philosophisch-naturwissenschaftlichen Fakultät det Univerität Bern.

Johnstone, A. D., Alsop, C., Burge, S., Carter, P. J., Coates, A. J., Coker, A. J., *et al.*: 1997, in C. P. Escoubet, C. T. Russell, and R. Schmidt (eds.), Kluwer Academic: Dordrecht, Netherlands, *Space Sci. Revs.* **79**, 351.

Kallio, E.: 1996, *J. Geophys. Res.* **101**, 111333.

Kallio, E., Luhmann, J. G., and Barabash, S.: 1997, *J. Geophys. Res.* **102**, 22183.

Kallio, E., and Janhunen, P.: 2001, *J. Geophys. Res.* **106**, 5617.

Kallio, E., and Barabash, S.: 2000, *J. Geophys. Res.* **105**, 24973.

Kallio, E., and Barabash, S.: 2001, *J. Geophys. Res.* **106**, 165.

Kallio, E., Barabash, S., Brinkfeldt, K., Gunell, H., Holmström, M., Futaana, Y., *et al.*: 2006, *Icarus* **182**, 448.

Krymskii, A. M., Breus, T. K., *et al.*: 2003, *J. Geophys. Res.* **108**(A12), 1431.

Lammer, H., Lichtenegger, H. I. M., Kolb, C., Ribas, I., Guinan, E. F., Abart, R., *et al.*: 2003, *Icarus* **165**, 9.

Lichtenegger, H., Lammer, H., and Stumptner, W.: 2002, *J. Geophys. Res.* **107**(A10), 1279, doi: 10.1029/2001JA000322.

Luhmann, J. G., and Kozyra, J. U.: 1991, *J. Geophys. Res.* **96**, 5457.

Luhmann, J. G., *et al.*: 1992, *Geophys. Res. Lett.* **19**, 2151.
Lundin, R., *et al.*: 1991, *Geophys. Res. Lett.* **18**, 1059.
Lundin, R., Barabash, H., Andersson, M., Holmström, A., Grigoriev, M., Yamauchi, J.-A., *et al.*: 2004, *Science* **305**, 1933.
Lundin, R., Winninghan, D., Barabash, S., Frahm, R., Holmström, M., Sauvaud, J.-A., *et al.*: 2006, *Science* **311**(5763), 980.
Malina, R. F., and Coburn, K. R.: 1984, *IEEE Trans. Nuclear Sci.* **NS-31**, 404.
Marshall, F. J., Hardy, D. A., Huber, A., Pantazis, J., McGarity, J., Holeman, E., *et al.*: 1986, *Rev. Sci. Instrum.* **57**(2), 229.
Mura, A., Milillo, A., and Orsini, S.: 2002, *J. Geophys. Res.* **107**(A10), 1278, doi: 10.1029/2001JA 000328.
Nilsson, H., Carlsson, E., Gunell, H., Futaana, Y., Barabash, S., Lundin, R., *et al.*: *Space Sci. Rev.*, this issue, doi: 10.1007/s11214-006-9030-0.
Nairn, C. M. C., Grard, R., Skalsky, A., and Trotignon, J. G.: 1991, *J. Geophys. Res.* **96**, 11227.
Norberg, O., Yamauchi, M., Lundin, R., Olsen, S., Borg, H., Barabash, S., *et al.*: 1998, *Earth, Planets, Space* **50**, 199.
Norberg, O., Winningham, J. D., Lauche, H., Keith, W., Puccio, W., Olsen, J., *et al.*: 2001, *Ann. Geophys.* **19**, 593.
Roelof, E. C., and Skinner, A. J.: 2000, *Space Sci. Rev.* **91**, 437.
Sablik, M. J., Scherrer, J. R., Winningham, J. D., Frahm, R. A., and Schrader, T.: 1990, *IEEE Trans. Geosci. Remote Sensing* **28**, 1034.
Terada, N., Machida, S., and Shinagawa, H.: 2002, *J. Geophys. Res.* **107**, doi:10.1029/2001JA009224.
Verigin, M., *et al.*: 1991, *J. Geophys. Res.* **96**, 19315.

PLASMA MOMENTS IN THE ENVIRONMENT OF MARS
Mars Express ASPERA-3 Observations

M. FRÄNZ[1,*], E. DUBININ[1], E. ROUSSOS[1], J. WOCH[1], J. D. WINNINGHAM[2], R. FRAHM[2], A. J. COATES[3], A. FEDOROV[4], S. BARABASH[5] and R. LUNDIN[5]
[1]*MPI für Sonnensystemforschung, 37191 Katlenburg-Lindau, Germany*
[2]*Southwest Research Institute, San Antonio, TX 78228-0510, USA*
[3]*Mullard Space Science Laboratory, University College London, Surrey RH5 6NT, UK*
[4]*Centre d'Etude Spatiale des Rayonnements, BP-4346, F-31028 Toulouse, France*
[5]*Swedish Institute of Space Physics, Box 812, S-98 128, Kiruna, Sweden*
(**Author for correspondence: E-mail: fraenz@mps.mpg.de*)

(Received 19 April 2006; Accepted in final form 13 November 2006)

Abstract. We present the first electron and ion moment maps (density, velocity and temperature) of the martian plasma environment, using data from the ELS and IMA sensors of the ASPERA-3 experiment onboard Mars Express. Moments are calculated by integration and by Gaussian fits to the phase space distribution. The methods of calculation and the calibration parameters relevant for the calculation are described in detail in the first part of the paper. The estimation of ionospheric electron densities assumes that the thermal electron temperature can be determined by the instrument – despite a cut-off by a negative spacecraft potential. The spacecraft potential is estimated by the location of photoelectron peaks in the energy spectrum. For the magnetosheath we separate the low energy part of the electron spectrum – presumably spacecraft photo electrons and the high energy part. For ions, we present maps for solar wind protons and alpha particles. Protons with energies below 500 eV which may play an important role in the ionosphere are not measured by the instrument. As well the low speed solar wind protons are not sampled very well. The maps reveal all the boundaries of the Mars-solar wind interaction and give a good qualitative description of the plasma behavior at the different interaction regions.

Keywords: Mars, magnetosphere, plasma moments

1. Introduction

The plasma environment of Mars has been keeping many secrets up to the present day (see the reviews by Nagy *et al.* (2004) and Luhmann and Brace (1991)). While the average location of the main plasma boundaries (bow shock and magnetic pile-up boundary, MPB) have been studied in depth using the Mars Global Surveyor (MGS) magnetometer instrument (Mazelle *et al.*, 2004; Vignes *et al.*, 2000) the fundamental question of how the pressure balance between ionosphere of Mars and the solar wind is achieved remains unsolved. The reason for this has been the insufficient instrumentation for plasma investigations on previous missions. In this paper we follow the terminology of Nagy *et al.* (2004) by calling the region between bow shock and MPB magnetosheath. With the spatial resolution used in

Space Science Reviews (2006) 126: 165–207
DOI: 10.1007/s11214-006-9115-9

this paper we cannot identify an ionopause (if it is different from the MPB) and call the region inside the MPB ionosphere. Only during the Viking lander missions altitude profiles of the ionospheric plasma densities and temperatures have been obtained. The Phobos-2 mission, which had the ASPERA-1 plasma instrument onboard, had too short of a lifetime to deliver enough statistics on the plasma parameters (Lundin *et al.*, 1993). The electron-reflectometer on MGS has been giving excellent results on the morphology of the ionosphere (Brain *et al.*, 2003) but it was so far not possible to extract plasma densities, velocities and temperatures from the data because of instrumental restrictions. With two years of operation of the ASPERA-3 instrument on board the Mars Express spacecraft it is for the first time possible to determine large scale statistics of plasma moments in the environment of Mars. In this paper we present and discuss data obtained by the ELS electron sensor and IMA ion sensor of the ASPERA-3 experiment between February 2004 and January 2006. Unfortunately there are again severe instrumental restrictions for the analysis: (1) Electron spectra are strongly influenced by the charging of the spacecraft with respect to the local plasma environment (*spacecraft potential*). Fortunately the energy resolution of the ELS sensor is good enough to resolve peaks in the spectrum caused by ionospheric photo electrons. These peaks allow an estimation of the spacecraft potential in the ionosphere. Outside of the ionosphere we can only calculate electron moments by assuming different levels of (positive) spacecraft potential. (2) the IMA sensor does not measure protons below a threshold of about 500 eV (depending on instrument mode), so that we cannot give an estimate of proton moments in the ionosphere. For heavier ions the separation of O^+,O_2^+ and CO_2^+ is rather difficult (Carlsson *et al.*, 2006), thus we will discuss moments of heavy ions in a later paper and present only proton and alpha particle moments in this paper.

2. Instrumentation

The ASPERA-3 instrument on board of Mars Express consists of 4 sensors: the ELS sensor for thermal and energetic electrons, the IMA sensor for protons, helium and heavy ions, the time-of-flight neutral particle sensor NPD and the neutral particle sensor NPI. A general description of the instrument is given in an accompanying paper (Barabash *et al.*, 2006). In the first part of this paper we describe how one can obtain plasma moments from the electron sensor ELS and the ion sensor IMA. For the IMA sensor we will only discuss the derivation of proton and alpha moments. Respective calculations for other ions can be made in a similar way. Specifically for heavier ions there is an additional problem of species separation. The purpose of the first part of the paper is to give a guideline and reference for the calculation of moments from the ASPERA-3 detectors. Actual calibration factors might change as the data analysis develops but we expect that the principal methods described here remain valid. Statistics of the derived moments for the environment of Mars are presented in the last section of this paper.

For the following calculations we have been using some technical documents describing calibration parameters of the ASPERA-3 sensors. These documents are not published in a journal but can be obtained from the authors. We will refer to these documents by title and main author with the remark *personal communication*.

2.1. Instrumental Coordinates

Instrumental coordinate systems for ASPERA-3 are described in the document ASPERA-3 sensor numbering, 3.1, (S. Barabash, pers. comm.). We use a coordinate system (ASP) which is defined as

$$X_{\rm ASP} = Z_u = Z_{\rm SR}, \quad Y_{\rm ASP} = Y_u = -X_{\rm SR}, \quad Z_{\rm ASP} = -X_u = -Y_{\rm SR} \tag{1}$$

in relation to the ASPERA-3 main unit system (X_u, Y_u, Z_u) and the MEX spacecraft reference system ($X_{\rm SR}, Y_{\rm SR}, Z_{\rm SR}$). We use the same coordinate system for all ASPERA-3 sensors with azimuthal angle φ and polar angle ϑ such that, $\vartheta = 0^\circ$ is the $Z_{\rm ASP}$ axis and $\varphi = 0^\circ$ is the positive $X_{\rm ASP}$ axis. The ASPERA-3 main unit is mounted on a rotating platform (scanner) but during the first two years of operation discussed in this paper the scanner was not operating. That means the ASP system as it is used here is fixed with respect to the spacecraft frame.

3. Plasma Moment Calculation from Particle Counters

For general introductions on moment calculations we refer the reader to textbooks, e.g. (Hutchinson, 2002; Kallenrode, 1998; Parks, 1991; Paschmann *et al.*, 2000). But since textbook usually lack applied examples we list in the following the principal equations used for this paper.

We assume that each particle species can be described by a distribution function $f(\mathbf{v})(\mathbf{v})$ in velocity space. Macroscopic properties of the particle distribution can be described by integrals of this function folded with powers of the velocity vector:

$$\mathbf{M}^k = \int f(\mathbf{v})(\mathbf{v})^k d^3\mathbf{v}, \tag{2}$$

where $\mathbf{M}$ is a tensor of order k.

For $k = 0$ we get the particle number density n, $k = 1$ gives the velocity vector, normalized by n, while $k = 2$ gives us the pressure tensor. The measurable quantity for particle counters is the differential flux $J(E, \Omega, \mathbf{r})$ for particles of energy E, at a position $\mathbf{r}$, within a solid angle dΩ. If m is particle mass, the relation between the distribution function and the differential flux is:

$$J(E, \Omega, \mathbf{r}) = \frac{v^2}{m} f(\mathbf{r}, \mathbf{v}) = \frac{2E}{m^2} f(\mathbf{r}, \mathbf{v}) \tag{3}$$

3.1. Moments by Integration

Using Equations (2) and (3) we will derive explicit forms of the moment equations. For $k = 0$ in Equation (2), we get the following expression for the density:

$$n = \int_{\mathbf{V}} f(\mathbf{v}) d^3\mathbf{v} = \int d\varphi \int d\vartheta \sin\vartheta \int dv v^2 f(v, \vartheta, \varphi) \tag{4}$$

In the case of an isotropic plasma we get:

$$n = 4\pi \int_{\mathbf{V}} f(\mathbf{v}) v^2 dv \tag{5}$$

Using Equation (3), $v(E) = \sqrt{\frac{2E}{m}}$, and because $dE = mvdv$:

$$n = \int d\varphi \int d\vartheta \sin\vartheta \int dv f(\mathbf{v}) v^2 = \int d\varphi \int d\vartheta \sin\vartheta \int dE \frac{J}{v} \tag{6}$$

If $c(E, \varphi, \vartheta)$ are the detector counts, $G(E, \varphi, \vartheta)$ the geometric factor of the detector, τ the acquisition time and $\Delta E = E_{n+1} - E_n$ the energy width of the n-th energy channel, then:

$$J = \frac{c(E, \varphi, \vartheta)}{G(E, \varphi, \vartheta)\tau \Delta E} \tag{7}$$

Note, that here we use the solid geometric factor $G = \Delta A \Delta\Omega$ (sensitive surface $\times$ solid angle) of the detector, which is usually multiplied by the detector energy resolution to define the energy geometric factor $G_E = G\frac{\Delta E}{E}$. IF $\sum_E \Delta E$ gives the full energy width of the sensor, we can substitute integrals with sums and $dE \equiv \Delta E$. Then:

$$n = \sum_{\varphi} \Delta\varphi \sum_{\vartheta} \Delta\vartheta \sin\vartheta \sum_{E} \frac{c(E, \varphi, \vartheta)}{G(E, \varphi, \vartheta)\tau v(E)}. \tag{8}$$

The general expression for the bulk velocity V ($k = 1$)is:

$$n\mathbf{V} = \int_{\mathbf{V}} \mathbf{v} f(\mathbf{v}) d^3\mathbf{v} \tag{9}$$

or explicitly:

$$\begin{aligned} nV_x &= \int d\varphi \cos\varphi \int d\vartheta \sin^2\vartheta \int dE J(E, \vartheta, \varphi) \\ nV_y &= \int d\varphi \sin\varphi \int d\vartheta \sin^2\vartheta \int dE J(E, \vartheta, \varphi) \\ nV_z &= \int d\varphi \int d\vartheta \sin\vartheta \cos\vartheta \int dE J(E, \vartheta, \varphi) \end{aligned} \tag{10}$$

For $k = 2$ we get the thermal pressure tensor:

$$\mathbf{P} = m \int_{\mathbf{V}} (v_i - V_k)(v_i - V_k) f(\mathbf{v}) d^3\mathbf{v} = m\mathbf{M}^2 - nmV_i V_k, \tag{11}$$

where:

$$\mathbf{M}^2 = \int_\Omega \int_\mathbf{V} v_i v_k f(\mathbf{v}) d\Omega d\mathbf{v}, \tag{12}$$

and $i, k = (x, y, z)$ respectively, and V_i is the bulk flow.

P is a symmetric tensor with 9 directional elements. However, due to spatial coverage limitations of the ASPERA-3 sensor (see Section 5), we will only estimate the three diagonal terms, in the ASPERA-3 coordinate system. For typical solar wind and magnetosheath plasma distributions the off-diagonal terms are negligible.

$$\begin{aligned}
P_{xx} &= m \int d\varphi \cos^2\varphi \int d\vartheta \sin^3\vartheta \int dE v J(v, \vartheta, \varphi) - mV_x^2 n \\
P_{yy} &= m \int d\varphi \sin^2\varphi \int d\vartheta \sin^3\vartheta \int dE v J(v, \vartheta, \varphi) - mV_y^2 n \\
P_{zz} &= m \int d\varphi \int d\vartheta \sin\vartheta \cos^2\vartheta \int dE v J(v, \vartheta, \varphi) - mV_z^2 n
\end{aligned} \tag{13}$$

and

$$P = \frac{P_{xx} + P_{yy} + P_{zz}}{3}, \tag{14}$$

$$T = \frac{P}{nK} \text{ or } T[\text{eV}] = 6241 \frac{P[\text{nPa}]}{n[\text{cm}^{-3}]}, \tag{15}$$

where $K = \frac{1\text{eV}}{11605\text{Kelvin}}$ is the Boltzmann constant and the factor 6241 comes from conversion of units. For comparison and later use, note also:

$$1\text{nPa} = 10^{-8}\text{dyn/cm}^2 = 5.403 \cdot 10^{-12} \sqrt{\text{eV} m_e},$$

where m_e is the electron mass.

3.2. Moments by Fitting

A different method to calculate moments of a plasma distribution is by assuming that the phase space density of particles has a Maxwellian distribution in velocity space:

$$\bar{f}(v) = C \cdot e^{-\frac{(v-\bar{v})^2}{v_t^2}}, \tag{16}$$

where $\bar{v}$ is the bulk velocity which may be determined by integration. The constant C is determined by Equation (4):

$$C = \frac{n}{(\sqrt{\pi} v_t)^3}. \tag{17}$$

Replacing the thermal velocity v_t by the thermal energy using $v_t = \sqrt{\frac{2E_t}{m}}$ allows to express the phase space density as

$$\bar{f}(E) = n \cdot \left(\frac{m}{\pi 2E_t}\right)^{3/2} \cdot e^{-\frac{E-2\sqrt{E\bar{E}}+\bar{E}}{E_t}}, \tag{18}$$

where for the mean energy we may use: $\bar{E} = \frac{\int_E E \cdot f(E)}{\int_E f(E)}$.

Expressing E in [eV], n in [1/cm^3] and m in electron masses m_e, and using $1\text{eV} = 1.7588 \cdot 10^{15} m_e \frac{\text{cm}^2}{\text{s}^2}$ we get:

$$\bar{f}(E)[\text{s}^3/\text{km}^6] = 0.8608 \cdot 10^6 n \cdot \left(\frac{m}{E_t}\right)^{3/2} \cdot e^{-\frac{E-2\sqrt{E\bar{E}}+\bar{E}}{E_t}}. \tag{19}$$

On the other hand the phase space density for each energy channel can be expressed by the omni-directional differential flux $J(E)$ as:

$$\bar{f}(E) = \frac{J(E)m^3}{p^2} = \frac{J(E)m^2}{2E}. \tag{20}$$

Using Equation (7) we get

$$\bar{f}(E) = \frac{m^2}{2G\tau} \frac{c(E)}{\Delta E \cdot E}. \tag{21}$$

Again expressing E in [eV] and m in electron masses m_e, this is:

$$\bar{f}(E)[\text{s}^3/\text{km}^6] = \frac{0.1616\, m^2}{G\tau} \frac{\bar{c}(E)}{\Delta E \cdot E}. \tag{22}$$

Demanding equality between Equations (19) and (22) allows to determine density n and thermal energy E_t by fitting to the measured spectrum of $f(E)$. When there is a positive spacecraft potential E_p the energy for each step has to be replaced by $E - E_p$. Since spacecraft potentials are typically less than 20 eV, this correction is only important for electrons. On the other hand if plasma bulk speeds are below 300 km/s, we have $\bar{E} < 1$ eV for electrons, such that we may use $\bar{E} = 0$ in this case.

4. Electron Moments from ASPERA-3 ELS

This section describes the implementation of the moment calculation for the ELS sensor of the ASPERA-3 instrument onboard Mars Express. The ELS sensor has an energy range of 0.4 eV to 26 keV which is split into 512 energy channels. We use the ELS Calibration Recon_5 of 13 Oct,2005 (R. Frahm, pers.comm.). In normal operation mode the energy steps are sampled into 128 channels. The energy allocated for each channel and the efficiency are calculated from the deflection voltages which are transmitted every 32 s with the engineering data set. The geometric factor is $G = 5.88 \cdot 10^{-4}\text{cm}^2\text{sr}$ for each anode, but is multiplied by an efficiency factor which is linearly dependent on energy. The acquisition time for each energy channel $\tau = 3.6/128$ s, at a sampling rate of 4 s.

For the first two years of operation the ELS sensor is measuring in the plane $\vartheta_{\mathrm{ASP}} = 90^\circ$ and we assume spherical symmetry. This assumption may be dropped when the ASPERA-3 scanner starts operating in 2006. In this paper we assume that the scanner is not operating. Also we neglect effects of shading of the instrument by the spacecraft since the bulk flow of electrons is negligible compared to their thermal speed. We effectively only loose about a quarter of the distribution by shading resulting in a relatively small underestimation of the density. This is different for higher energetic electron beams in the ionosphere which we do not discuss in this paper.

The quantity defined via the calibration procedure for each anode and energy step is the differential flux

$$J(E,\varphi)[\mathrm{c/(cm^2\ s\ sr\ eV)}] = \frac{c(E,\varphi)S_{\mathrm{adj}}(\varphi)}{G_{\mathrm{eff}}(E,\varphi)\tau} \tag{23}$$

where $c(E,\varphi)$ are the raw counts for each bin, $S_{\mathrm{adj}}(\varphi)$ is a time constant science adjustment for each anode and $G_{\mathrm{eff}}[\mathrm{cm^2\ sr\ eV}]$ contains energy resolution and efficiency of each anode and is further described in the calibration document.

Using this we get from Equation (8) the final expression for the density:

$$n[\mathrm{cm^{-3}}] = \frac{\pi}{4}\cdot 1.686\cdot 10^{-8}\sqrt{m[m_e]}\sum_{\varphi}\sum_{E}\frac{J(E,\varphi)\Delta E}{\sqrt{E[\mathrm{eV}]}} \tag{24}$$

The factor appears as a result of the conversion of Joule to eV (see Equation (19)) and $\Delta\varphi\int_0^\pi d\vartheta\,\sin\vartheta = \frac{\pi}{4}$, if we assume that the value observed at each anode is valid for all values of ϑ. Alternatively one can regard this factor as the anode average multiplied by 4π. ΔE is the energy width of each channel obtained by taking the difference between the center energies.

The calculation of the three velocity components is done with respect to the ASP coordinate system (see Section 2.1). Since ELS is scanning on a plane, it is not possible to estimate the velocity in the z-direction (this will change if the ASPERA-3 scanner operates.) Therefore the measurement is only dependent on the angle φ, between the x-direction as defined in the ASP coordinate system and the viewing direction of each of the 16 ELS anodes. For $\vartheta = 90^\circ$ by using Equations (3), (7), and (10) we get:

$$\begin{aligned} nV_x[\mathrm{km/s}] &= \frac{\pi^2 10^{-5}}{16}\sum_{\varphi}\cos(\varphi)\sum_{E}J(E,\varphi)\Delta E \\ nV_y[\mathrm{km/s}] &= \frac{\pi^2 10^{-5}}{16}\sum_{\varphi}\sin(\varphi)\sum_{E}J(E,\varphi)\Delta E \\ nV_z &\equiv 0 \end{aligned} \tag{25}$$

where $c(E, \varphi)$ are the counts recorded by each ELS anode at an angle φ with respect to the x-direction. The factor $\pi^2/16$ defines the solid angle of integration: $\Delta\varphi \int d\vartheta \sin^2 \vartheta$.

For the thermal pressure, as in the case of the velocity calculation, we can analyze only the dependency on the φ angle and therefore we set $P_{zz} \equiv 0$. In total:

$$\begin{aligned} P_{xx}[\mathrm{nPa}] &= \frac{5.403 \cdot 10^{-12}\pi}{12} \sum_{\varphi} \cos^2\varphi \sum_{E} \sqrt{E} J(E, \varphi)\Delta E - mV_x^2 n \\ P_{yy}[\mathrm{nPa}] &= \frac{5.403 \cdot 10^{-12}\pi}{12} \sum_{\varphi} \sin^2\varphi \sum_{E} \sqrt{E} J(E, \varphi)\Delta E - mV_y^2 n \\ P_{zz} &= 0 \end{aligned} \tag{26}$$

where $\pi/12 = \Delta\varphi \int d\vartheta \sin^3 \vartheta$ and a factor from conversion of units (see Equation (15)). The velocity and pressure formulas are provided only for completeness. Since measurement of the bulk flow with a planar sensor, which is partly shadowed by the spacecraft, produces large errors it is better to calculate the thermal pressure from the thermal temperature obtained by fitting the energy spectrum.

4.0.1. *Spacecraft Potential*

Figure 1 shows data obtained by the ELS-sensor for the period 2004-06-02 05:30 UT to 06:30 UT. The third panel from top shows the energy spectrum in raw counts obtained by the sensor. We generally work with spectra obtained by integrating all 16 anodes. The drop in counts below 5 eV is caused by a −5 V repeller voltage applied to protect the anode counters from saturation. In consequence the low energy part of the spectrum is hidden from observation.

Between 04:30 and 04:50 the sensor observes high count rates above 20 eV. Here the spacecraft crosses the magnetosheath. Between 04:50 and 05:55 the spacecraft crosses the magnetosphere with a wake crossing from 05:10 to 05:40. After 05:55 the spacecraft crosses the sheath again and enters the solar wind at 06:15. Experience with other missions shows that in the ionosphere the spacecraft is usually negatively charged and positively outside. To estimate the spacecraft potential for a specific time we first have to determine whether the distribution is ionospheric or not ionosperic. For this discrimination we use the ratio of counts obtained above and below 20 eV (panel 2 from top). If this ratio exceeds the value 3 we call the distribution ionospheric otherwise non-ionospheric. We observe that e.g. in the wake this criterion declares distributions non-ionospheric. For non-ionospheric distributions we assume a constant potential of either 0 V or +5 V. Unfortunately with this criterion sometimes spectra obtained in the solar wind are also classified ionospheric when SC photo electrons are present. To avoid this we apply an additional criterion demanding that a photo electron peak determination

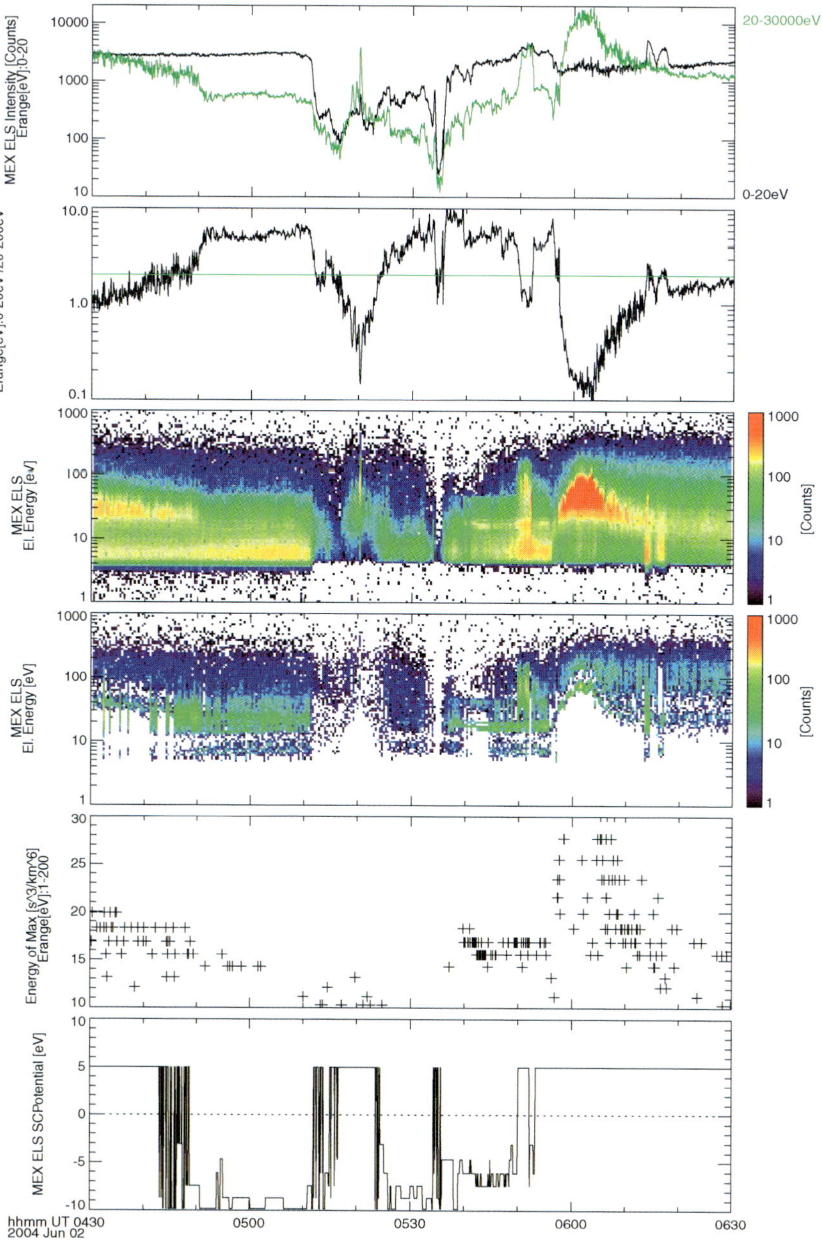

Figure 1. Data obtained by the ASPERA-3 ELS sensor in 2004-06-02 04:30 UT to 06:30 UT. Data are sampled over 4 s. From top to bottom: (a) counts sampled by all anodes in the energy range 0–20 eV (black) and in the energy range 20–30000 eV (green), (b) ratio of the two quantities plotted above, (c) energy spectra sampled by all anodes (counts/12 s), (d) same as above after subtracting an exponential fit to the phase space density for each record, (e) Energy of maximum flux in the subtracted spectrum in the range 10–30 eV, (f) resulting spacecraft potential assuming CO_2-peak energy at 23 eV.

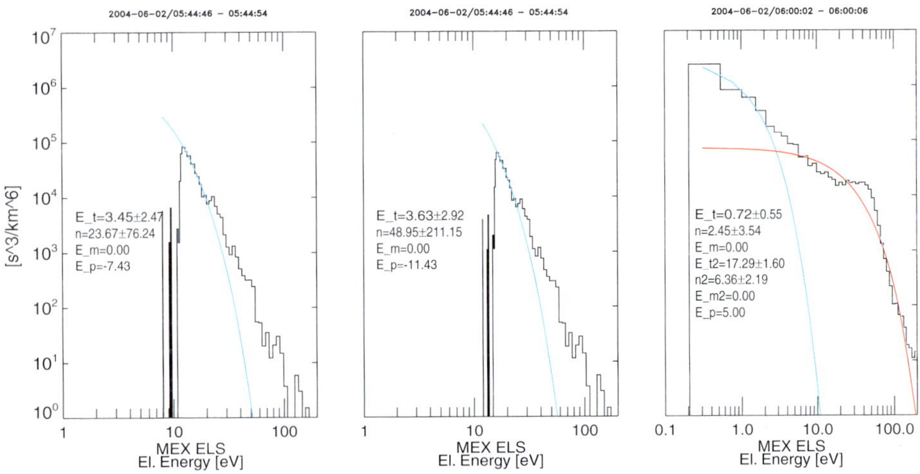

Figure 2. Phase space density as a function of energy, obtained by the ASPERA-3 ELS sensor in the ionosphere at 2004-06-02 05:44:46 UT to 05:44:58 UT. Data are sampled over 4 s. The black lines are the measured data, the blue lines are exponential fits assuming the spacecraft potential value E_p (CO_2 peak at 23 eV) and mean energy $E_m = 0$ eV. The red lines are fits to the high energy part of the distribution. Fit parameters n [1/cm^3] and E_t [eV] are given for each fit with respective standard deviations.

must be possible within a 20 min time interval around the timetag of the data sample.

As can be seen from Figure 2 for ionospheric spectra a local peak in the spectrum can sometimes be observed between 10 and 25 eV (here especially between 05:40 and 05:55). This peak corresponds to O_2 and CO_2 photo electrons and is expected to have an actual energy between 21 and 28 eV (Mantas and Hanson, 1979). Actually there are two peaks expected (and sometimes observed) at 23 and 27 eV energy. The identification of these photo electrons is discussed in detail by Frahm *et al.* (2006). We determine the observed energy of this peak by subtracting the low energy thermal part of the spectrum using an exponential fit to the phase space density (panel 4 from top of Figure 1). The energy of the maximum phase space density in the range 10 to 30 eV is shown in the bottom panel. We now subtract either 23 eV or 27 eV from this energy to estimate the SC potential for ionospheric distributions. If for a specific point in time a CO_2-peak cannot be determined we take the value from the spectrum closest in time for which a value can be determined. To correct the distribution for the spacecraft potential we subtract the potential from the instrumental energy for each channel. Note, that this method does not exclude SC photo electrons or secondary electrons. The geometric factor and energy resolution of the sensor are a function of the actually measured energy, thus they are not affected by the shift in energy applied to the data after applying the geometric factors.

4.1. Discussion of ELS Spectra and Moments

Figure 2 shows phase space densities as a function of energy calculated from the same data as in Figure 1 for some 4s-spectra obtained in the ionosphere (a,b) and magnetosheath (c). In Figure 2a we assume the CO_2-peak to be at 23 eV to determine the SC potential (E_p). We further assume that the mean electron energy (E_m) is 0 eV. Then we fit an exponential to the 10 energy bins with energies larger than the bin with maximum flux. The resulting fit is shown in blue and the fitted values for temperature E_t and density n are given with their standard deviations. We observe that the standard deviation for the temperature is usually lower than E_t, while the density has very large standard deviations. This reflects the uncertainty in the extrapolation to 0 eV energy which essentially determines n.

Figure 2b shows the same distribution assuming that the CO_2-peak is at 27 eV. We observe that the density is rather sensitive to the assumed spacecraft potential.

Figure 2c shows a distribution measured in the magnetosheath assuming a SC potential of +5 V. Here the high energy part of the distribution (from 10 channels above the energy of maximum flux) is fitted as well (red line) resulting in the partial density n_2 and temperature E_{t2}. Note, that the bump at 50–80 eV is better fitted assuming a high mean energy E_{m2} for the high energy part. But since this would correspond to a very high differential streaming velocity of the high energy part we assume $E_{m2} = 0$.

Figure 3 shows again data obtained by the ELS-sensor for the same time interval as Figure 1. The top panel shows an uncalibrated energy spectrum in counts/s taken from 12 s averages of the original data. The second panel shows the SC potential estimate, which here is set to +5 V for non-ionospheric distributions and calculated from a 23 eV CO_2 peak for ionospheric distributions (black line). When setting the potential to 0 V for non-ionospheric distributions and calculate it from a 27 eV CO_2 peak for ionospheric distributions we obtain the potential shown by the red line. The third panel shows electron density estimates by integration (black), by fitting the low energy peak (green) by fitting low and high energy parts with the first spacecraft potential estimate (blue) and with the second spacecraft potential estimate (red). One can see that for ionospheric distributions the integrated density is far off from the expected value, while outside the ionosphere they are off by a factor 2. Note also, that for ionospheric distributions there is no fit to the high energy part of the spectrum.

The bottom panel shows temperature estimates by the same methods. Here we observe that for ionospheric distributions fitted and integrated temperatures are comparable, while in the sheath and solar wind the high energy part determines the total temperature. Total temperature has been calculated by

$$T_{\text{total}} = \frac{T_{\text{low}} n_{\text{low}} + T_{\text{high}} n_{\text{high}}}{n_{\text{low}} + n_{\text{high}}}. \tag{27}$$

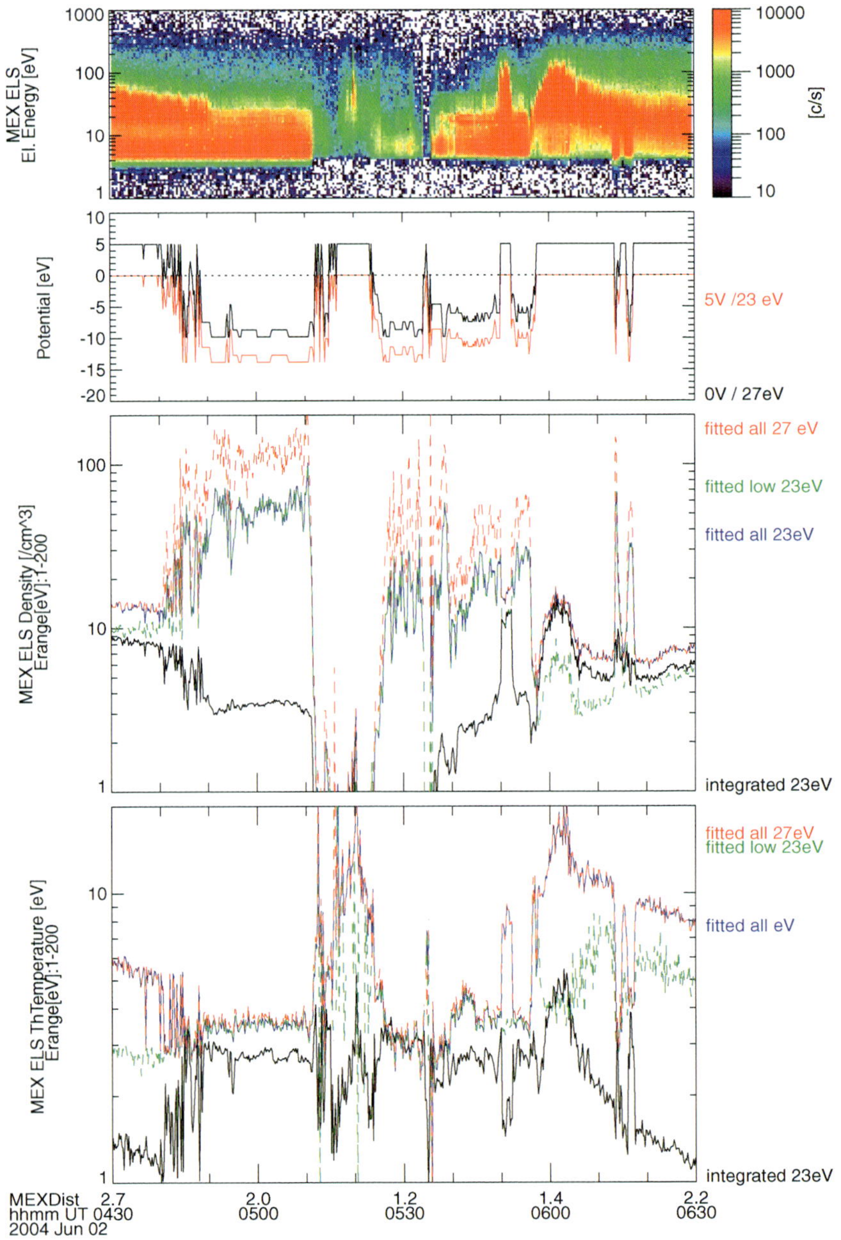

Figure 3. Data obtained by the ASPERA-3 ELS sensor in 2004-06-02 05:30 UT to 06:30 UT. Data are sampled over 4 s. From top to bottom: (a) energy colour spectrum of the integrated counts/12 s of all anodes, (b) spacecraft potential assuming +5 V outside the ionosphere and CO_2 peak energy at 23 eV, (b) densities [/cm^3] derived from calibrated data assuming the SC potential above by integration (black), by fitting over the low energy part of the spectrum (green) and the complete spectrum (blue), (c) total temperatures [eV] by the same methods.

We observe that for the alternative spacecraft potential estimate for non-ionospheric densities values do not change significantly, while for ionospheric densities values increase by a factor 2. Temperatures are unaffected for both cases.

For magnetosheath and solar wind the observed values of 1–10 e/cm^3 appear to be reasonable, but in the ionosphere at altitudes of about 300 km we might expect densities well above 1000 e/cm^3 – as reported by plasma frequency measurements (E. Nielsen, MARSIS – personal comm.). Though there is a lot of uncertainty in this value at higher altitudes. The reason for this discrepancy is probably that at lower altitudes the electrons have a core temperature of less than 1eV such that the energy resolution of the ELS sensor is insufficient when the spectrum is shifted by a negative SC-potential.

Electron velocity and temperature determinations are discussed at the end of this paper in context with IMA observations. On about one orbit per month the sensor is operated in *linear stepping mode*. In this mode the energy range of the sensor is restricted to 0–128 eV divided into linear steps of 1 eV. In this mode the −5 V repeller voltage mentioned above is switched off. Analysis of sample spectra shows that the problems related to the extrapolation of the spectrum at negative spacecraft potential is present as well and the quality of derived moments does not increase substantially in this mode.

5. Ion Moments from ASPERA-3 IMA

The IMA sensor of the ASPERA-3 instrument is a combined electrostatic energy- and magnetic mass-analyzer. It measures mass/charge and energy/charge of ions in the ranges 1–30 amu/e and 10–30000 eV/e. The instrument uses 16 anodes covering the ASP-XY plane. The ASP polar angle is measured by electrostatic deviation covering $45^\circ \leq \vartheta_{\mathrm{ASP}} \leq 135^\circ$ in 16 sectors during a sampling time of 192 s. For each anode and sector mass/charge is measured in 32 channels (massrings) by magnetic deviation and energy/charge in 96 energy channels.

The instrument operates at 3 different post-acceleration voltages (PAC) to allow increased energy and mass resolution depending on the plasma environment. These PAC levels are: PAC0: 90 V, PAC1:2433 V, PAC2:4216 V. The efficiencies and energy range depend on the PAC level. The PAC level for a specific data record can in principle be taken from the PAC high voltage monitor variable in the IMA house keeping (HK) dataset. In practice there is a timetag mismatch between HK timetags and data record timetags. We apply an algorithm which searches for the last valid HK record for each data record.

The ground calibration of the instrument is described in *MEX IMA Calibration. Final Report. V3.0,2005 (A. Fedorov, pers.comm, cited hereafter as IMACalRep).* This report essentially covers the determination of geometric factors as a function of energy for different ion species by lab measurements. Inflight calibrations are covered by the documents *Mars Express ASPERA-3.Flight Tables, 2006 (A. Fedorov, pers.comm)* and *IMA ASPERA-3 MEX. What happened with the low energy ions?,2006 (A. Fedorov, pers.comm)*. The essential result of these calibrations is that the effective number of energy channels is reduced to 54 – caused by an unexpected voltage on one of the deflector plates.

In the following sections we discuss observations made when analysing the actual inflight data. We try to determine noise reduction algorithms and efficiencies of the different massrings and anodes before we proceed to the moment calculations. There are many different ways to reduce noise in measured data. We here describe the methods which we regard as best suited for the IMA dataset.

Geometric factors GF_L for He^{++},O^+ and O_2^+ for the 3 PAC levels have been calculated in the IMACalRep from lab measurements. These factors contain the energy resolution $\Delta E/E$ and are integrated over the polar angle ϑ. That means the angular width $\Delta\vartheta$ must not be applied in the integrations and we define $G = GF_L E/\Delta E$ in the following equations. We use Tables V–VII of the IMACalRep and apply the He^{++} factors to H^+ and He+, the O^+ factors to O^+ and O^{++} and the $O2^+$ factors to heavier ions. As mentioned above we do not apply the massring dependency of GF_L discussed in Section 7 of the IMACalRep. We also do not take account of the φ and ϑ dependencies of GF_L within each anode and sector range. The principal dependence on φ and ϑ is covered by the corrections discussed above. Towards the borders of each anode the efficiency decreases by about 50%. A correction for this effect would alter the calculated density by less than 2. We expect that application of the minor efficiency corrections do not have a significant influence on the derived moments. Another problem is the sampling of a cold ion beam – as the solar wind, which usually has an angular spread of less than 5°. Since the IMA sensor has angular bins of $22.5^\circ \times 5.8^\circ$ the beam should usually be observed in only one bin and the geometric factor will be overestimated. In fact there seems to occur some scattering in the sensor which causes signals in the neighboring spatial bins as well. For this reason when calculating core densities we sum the counts of all spatial bins but take the geometric factor for one bin only as we do in this paper. But when calculating the spectrum for non-beam distributions (e.g. pick-up ions) it is better to take the maximum spatial bin value for each energy level. This is discussed in the context of pick-up ions in another paper (Dubinin *et al.*, 2006).

The integrated density can either be determined by integrating over a complete 192 s scan or by just using one φ-scan for a fixed ϑ. For a complete scan we get:

$$n[\mathrm{cm}^{-3}] = \frac{\pi \cdot 7.1987 \cdot 10^{-7}\sqrt{m[\mathrm{amu}]}}{8G[\mathrm{cm}^2\mathrm{rad}]\tau[\mathrm{s}]} \sum_{\varphi}\sum_{\vartheta} \sin\vartheta \sum_{E} \frac{c(E,\varphi,\vartheta)}{\sqrt{E[\mathrm{eV}]}} \tag{28}$$

where we use $\Delta\varphi = 2\pi/16$ and unit conversion as in Equation (19).

For the velocity we have to use the complete scan:

$$nv_x[\text{km/s}] = \frac{\pi 10^{-5}}{8G[\text{cm}^2\text{rad}]\tau[\text{s}]} \sum_{\varphi} \cos\varphi \sum_{\vartheta} \sin^2\vartheta \sum_{E} c(E, \vartheta, \varphi)$$

$$nv_y[\text{km/s}] = \frac{\pi 10^{-5}}{8G[\text{cm}^2\text{rad}]\tau[\text{s}]} \sum_{\varphi} \sin\varphi \sum_{\vartheta} \sin^2\vartheta \sum_{E} c(E, \vartheta, \varphi) \tag{29}$$

$$nv_z[\text{km/s}] = \frac{\pi 10^{-5}}{8G[\text{cm}^2\text{rad}]\tau[\text{s}]} \sum_{\varphi} \sum_{\vartheta} \sin\vartheta \cos\vartheta \sum_{E} c(E, \vartheta, \varphi)$$

But velocity and pressure can only be determined when the bulk flow is in the instruments ϑ-range. Fortunately this is the case for most orbits when the spacecraft is in the solar wind and magnetosheath. For regions with low bulk flow speed as for example heavy ions in the ionosphere one has to apply a correction for the partial field of view of IMA.

For the integrated kinetic pressure we might again assume spherical symmetry, such that we get from Equation (13) and (15):

$$P_{xx}[\text{nPa}] = \sqrt{m[m_e]} \frac{5.403 \cdot 10^{-12}\pi}{8G[\text{cm}^2\text{rad}]\tau[\text{s}]} \sum_{\varphi} \cos^2\varphi \sum_{\vartheta} \sin^3\vartheta \sum_{E} \sqrt{E[\text{eV}]}c(E, \vartheta, \varphi) \tag{30}$$

$$P_{yy}[\text{nPa}] = \sqrt{m[m_e]} \frac{5.403 \cdot 10^{-12}\pi}{8G[\text{cm}^2\text{rad}]\tau[\text{s}]} \sum_{\varphi} \sin^2\varphi \sum_{\vartheta} \sin^3\vartheta \sum_{E} \sqrt{E[\text{eV}]}c(E, \vartheta, \varphi) \tag{31}$$

$$P_{zz}[\text{nPa}] = \sqrt{m[m_e]} \frac{5.403 \cdot 10^{-12}\pi}{8G[\text{cm}^2\text{rad}]\tau[\text{s}]} \sum_{\varphi} \sum_{\vartheta} \sin\vartheta \cos^2\vartheta \sum_{E} \sqrt{E[\text{eV}]}c(E, \vartheta, \varphi) \tag{32}$$

The thermal pressure is calculated from this by

$$\mathbf{P}_{\text{th}}[\text{nPa}] = \mathbf{P}_{\text{kin}} - m[m_e]9.1 \cdot 10^{-10}\mathbf{V}^2 n \tag{33}$$

and total temperature by

$$T_{\text{th}}[\text{eV}] = 6241 \frac{\text{Trace}(\mathbf{P}_{\text{th}})}{3n}. \tag{34}$$

As for the ELS sensor the better estimate of the total pressure can be obtained by using a fit to the energy spectrum since the integrated pressure is rather sensitive to noise in the data.

5.1. Noise Reduction

Observation of the measured spectra shows that the IMA sensor is sensitive to different sources of noise: electronic noise of the amplifiers, noise caused by penetrating electrons and high energy protons depending on solar activity, noise caused by UV light depending on spacecraft orientation.

Figure 4 shows energy spectra obtained by IMA for the same period of time as discussed in the previous section for the ELS sensor. The top panel shows the unreduced counts integrated over all 16 anodes and 32 massrings for each ϑ-sector sampled over 12 s. The polar scan covering 192 s shows up in the repeatable pattern of the data. In the ionosphere before 05:40 data seem to contain just noise, between 05:40 and 06:10 the noise level increases – presumably caused by UV-light, after 05:55 solar wind ions appear at 1–4 keV energy.

We observe that the noise affects all energy channels and massrings in comparable levels. We apply two different techniques to reduce the noise level: 1. (reduce by maximum method) Since valid ion data are usually observed in the energy range 50 eV–8 keV we determine for each anode and sector the maximum count rate above 8 keV and below 40 eV (top and bottom 15 energy channels) and subtract this rate from all bins. 2. (reduce by average method) We subtract from all bins of the $96 \times 32 \times 16 \times 16$ matrix the mean counts of all non-zero bins.

Results of the two noise reduction methods are shown in the lower panels of Figure 4. We observed qualitatively that the first method is sufficient for most data records in reducing the noise level such that the reduced background is negligible compared to the valid signal. Only for records with severe UV contamination the second method is more recommendable. We also tested a third method which neglects all bins of the $96 \times 32 \times 16 \times 16$ matrix which contain just a single count, but we observed that this method severely affects the valid signal. Nevertheless this method has been applied onboard to IMA data after Oct 11, 2005 because high solar activity in summer 2005 increased the IMA background noise level in such a way that the allowed data transmission rate was exceeded.

5.2. Mass Ring Efficiencies

Figure 5a shows an IMA energy/mass matrix without noise reduction for 2004-06-02 06:17-06:19UT (PAC1). Overplotted are expected ion species ranges for PAC1 (see Section 5.4). The track marked by black lines is for protons, the one by green lines for He^{++}-ions. We observe: 1. The efficiency of different massrings is different, specifically massrings 1,5 and 17 have higher count rates than the

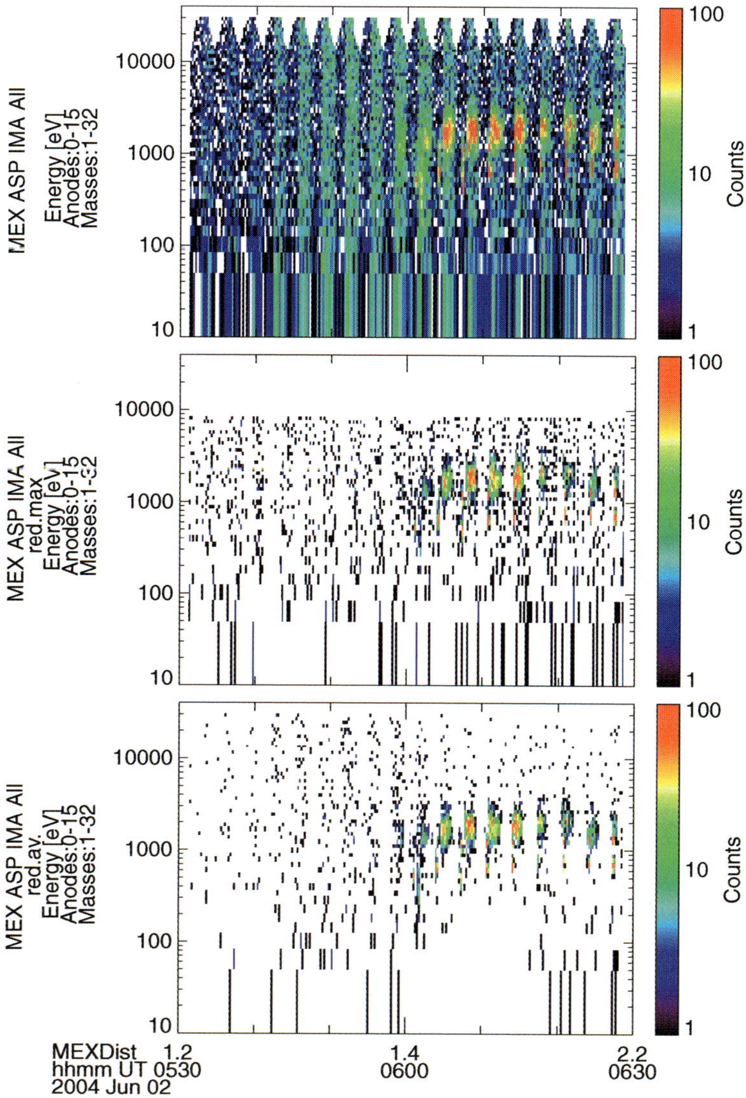

Figure 4. Noise reduction in absolute counts obtained by the ASPERA-3 IMA sensor in 2004-06-02 05:30 UT to 06:30 UT as a function of energy/charge. Shown are sums over all 16 anodes and 32 massrings. Data are sampled over 12 s. Panels show from top to bottom: (a) Unreduced counts, (b) counts after subtracting the maximum count observed in the 15 top and bottom energy channels, (c) counts after subtracting the average counts for each 192 s dataset from each bin of the 96 $\times$ 32 $\times$ 16 matrix.

neighboring rings. 2. There is a trace of higher count rates between 800 and 1500 eV which is presumably caused by protons for which the massring allocation did not work. We call these spill-over protons since their signal is appearing in mass rings were no protons are expected.

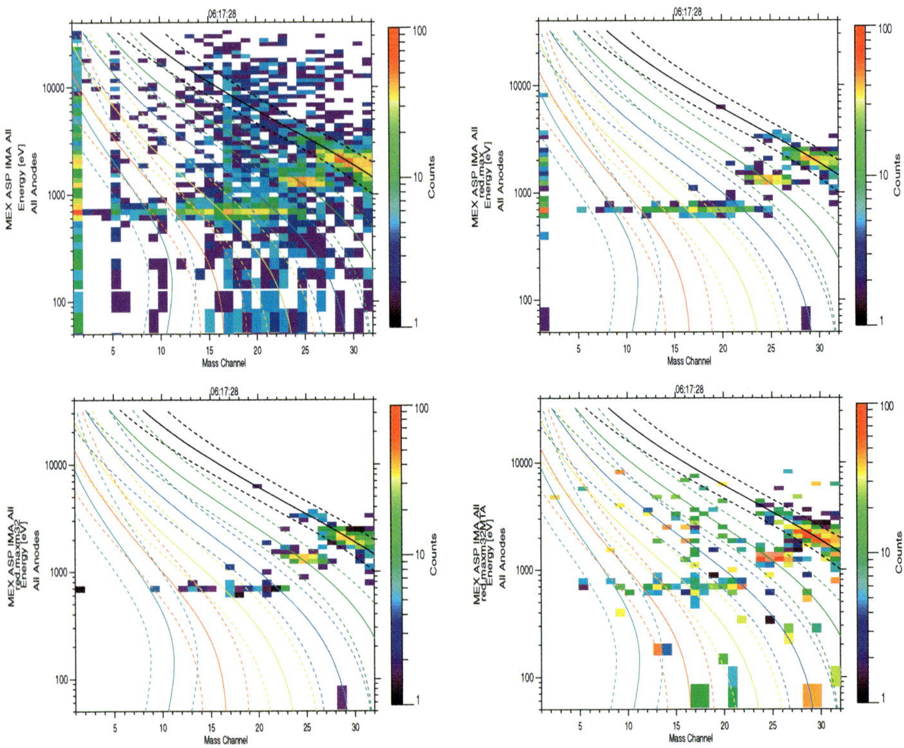

Figure 5. ASPERA-3 IMA energy/mass matrix obtained on 2004-06-02 06:17-06:19 UT. Overplotted are expected ion species ranges for PAC1. Absolute count rates: (a) without reduction, (b) after maximum reduction, (c) after proton subtraction, (d) after application of efficiencies.

To take account of both effects we calculated integrated count rates for each massring accumulated between 2004-02-01 and 2005-10-10 (Figure 6). We observe that massring 1 contains most counts by far. The other massrings behave similarly for the different PAC levels. Massrings 11 and 23 are virtually empty, massrings 5 and 17 are unusually high, above massring 15 even massrings have lower efficiency. Inspection of individual matrices also revealed that the high counts in massring 1 coincide with spill-over protons. We now assume that the massring 1 counts for each energy level can be taken as a measure of spill-over protons in that energy channel. To obtain the relative massring contamination by spill-over protons we integrated the massring totals only for those periods where massring 32 contains more than 2000 counts. We assume that these periods are representative for high proton contamination. To subtract spill-over protons from a data record we subtract from each bin in the $96 \times 32 \times 16 \times 16$ matrix the massring 1 counts multiplied by the massring efficiency. The result is shown in Figure 5c. We observe that the reduction is not perfect but by far the best method we have obtained so far.

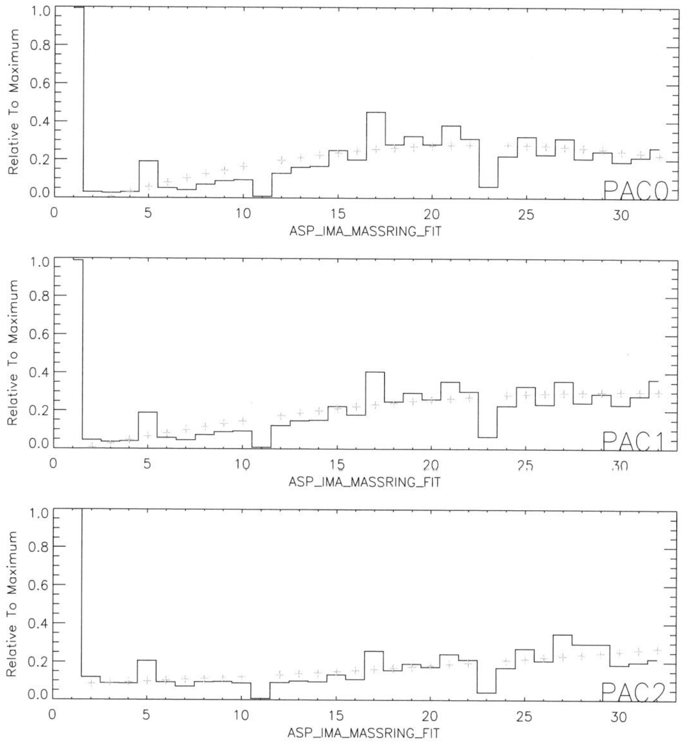

Figure 6. ASPERA-3 IMA total counts for the 32 different massrings for the 3 PAC levels for data obtained between 2004-02-01 and 2005-10-10. The blue crosses show a 3rd order polynomial fit to the data – excluding bins without cross.

To determine the relative efficiencies of neighboring massrings we apply a 3rd order polynomial fit to the massring totals summed over energy and spatial bins – relative to the counts observed in massring 1 observed for periods with massring 32 counts of less than 2000. This gives us a measure of the background noise in the different massrings. The result is shown in Figure 6. Each massring gets an efficiency correction defined by the ratio between the measured counts and the fit.

5.3. Anode and Theta Efficiencies

Figure 7 shows the total counts (integrated over energy and mass) relative to the maximum total counts observed for each 192 s-record for the 16 anodes and 16 ϑ-sectors of the IMA sensor. We observe that anodes 1–3 have a count level about 4 times higher than the other anodes. One reason for this is that the sensor orientation is on most orbits such that these anodes obtain the bulk solar wind flux. On the other hand we would then expect a comparable flux level in the neighboring anode 16. Lab measurements did not show significant differences in anode efficiency

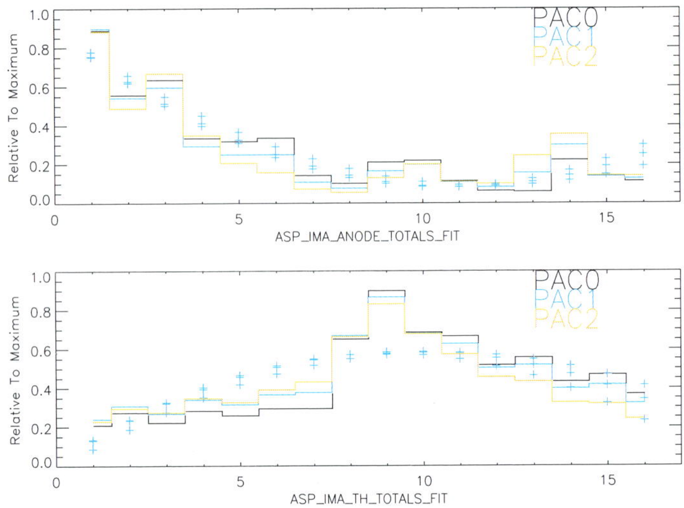

Figure 7. ASPERA-3 IMA total counts (relative to maximum) for the 16 different anodes (top) and the 16 different sectors (bottom) for the 3 PAC levels (colored) for data obtained between 2004-02-01 and 2005-10-10.

(A. Fedorov, pers.comm.). We assume that the large differences in anode count rates are partially caused by shading of the sensor by the spacecraft and generally apply the correction factors corresponding to Figure 7a.

The later is also true for the sector efficiency (Figure 7b). Sectors 1–8 are looking towards the spacecraft and show much lower count rates. In principle plasma moments can only be determined when the bulk flow of plasma is in sectors 9–14. In Figure 5d massring, anode and sector efficiencies have been applied.

5.4. Species Separation

For each PAC level the IMA sensor has a different measurement range for the ion species in massring and energy. These ranges are taken from a formula in *Mars Express ASPERA-3.Flight Tables, 2006 (A. Fedorov, pers.comm)*. The formula delivers an upper and lower massring number for a given mass/charge for each energy and PAC level.

Figure 8a shows an uncalibrated spectrum for orbit 539, which was discussed in Carlsson *et al.* (2006). We take this orbit as an example to discuss species separation because it contains solar wind light ion and ionospheric heavy ion observations.

Figure 9 (top left) shows an IMA energy/mass matrix obtained at 15:02 on this orbit. Overplotted are expected ion species ranges for PAC1 for mass/charge ratios 1(H^+), 2(He^{++}), 4(He^+), 8(O^{++}), 16(O^+) and 32(>O^+). From the species ranges

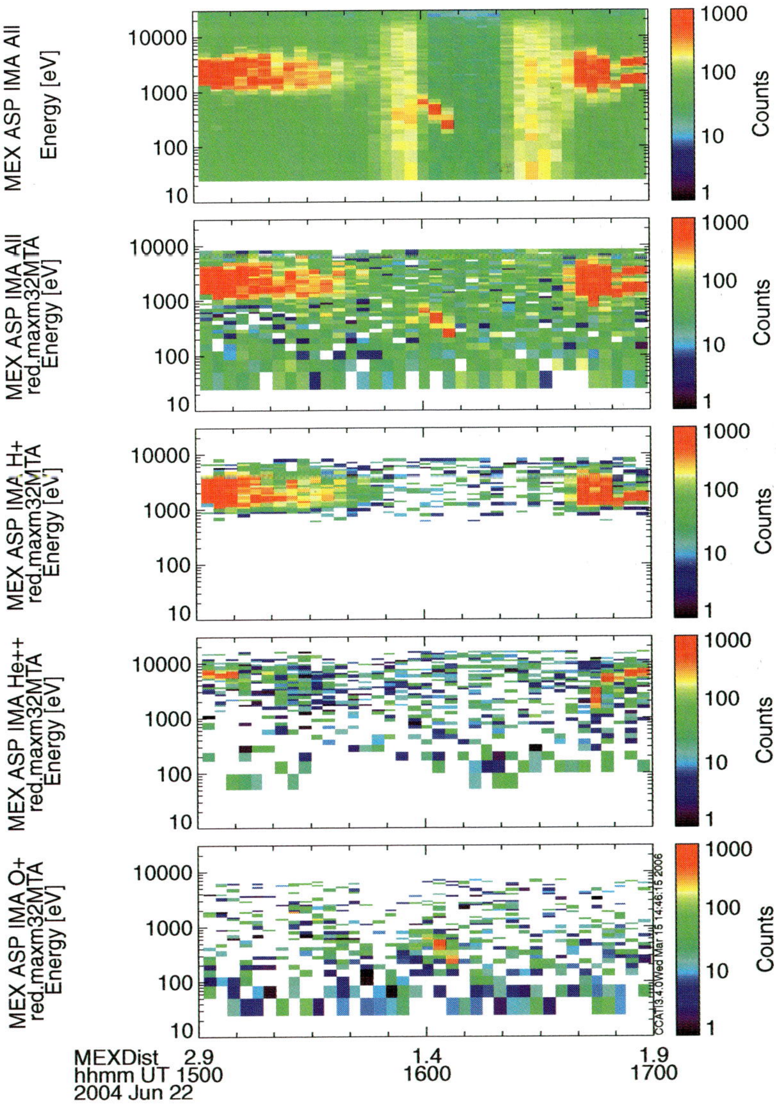

Figure 8. ASPERA-3 IMA energy spectra obtained between 2004-06-22 15:00 and 17:00UT. From top to bottom: (a) non-reduced, uncalibrated counts integrated over anodes,massrings and sectors, (b) after noise reduction and calibration, (c) after selection of H^+ trace , (d) after selection of He^{++} trace, (e) after selection of O^+ trace.

we calculate a probability matrix for each species where for each energy channel E and massring m the probability to contain species s is:

$$P_s(E, m) = e^{\left(\frac{m - c_s(E)}{b_s(E)}\right)^6}, \tag{35}$$

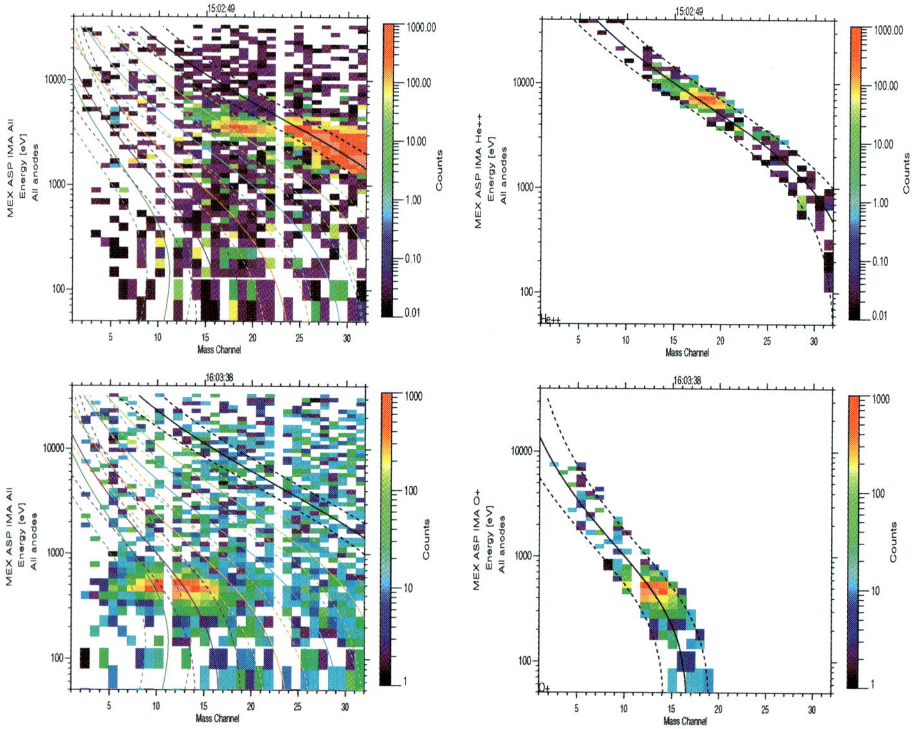

Figure 9. ASPERA-3 IMA energy/mass matrix obtained on 2004-06-22 15:02 (top) and 16:03 (bottom). Top panels are noise reduced. All panels are calibrated. The left panels contain all counts. Right panels just He^{++} (top) and O^{+} (bottom) tracks. Overplotted are expected ion species ranges for PAC1.

where $c_s(E)$ is the center massring number and $b_s(E)$ the massring range for each species and energy step. While the probability distribution underlying the tracks is Gaussian (Carlsson *et al.*, 2006, IMACalRep), we empirically determined the exponent 6 to avoid the loss of counts within the species range. On the other hand we loose valid counts outside of the range. In principle one has to apply an efficiency correction taking account of a decrease in efficiency with distance from the track center as described in Section 7 of IMACalRep, but since this is a second order effect we here do not apply this correction.

Figure 9 (top right) shows the matrix after multiplication with $P_{\mathrm{He}}^{++}(E, m)$ and multiplying energies by a factor 2 to account for the ion charge. Figure 9(bottom left) shows the matrix obtained at 16:03 containing heavy ions. This matrix is also shown in Figure 3 of Carlsson *et al.* (2006). Note, that in Carlsson *et al.* (2006) massring 11 – which is virtually empty – is replaced by an average of the neighboring massring counts to achieve a smooth dataset for species fitting. Note, also that for the heavy ion matrix we did not apply a noise reduction to get a comparable result to Carlsson *et al.* (2006).

Figure 9 (bottom right) shows the matrix after multiplication with $P_{\rm O}^{+}(E, m)$. The lower panels of Figure 8 show the energy spectra as a function of time after applying $P_s(E, m)$ for different ion species.

5.5. Resulting IMA Moments

5.5.1. *Proton Density*

Figure 10 shows data obtained by ASPERA-3 ELS and IMA obtained between 2004-08-01 03:00 and 04:00UT. The top panel shows the energy spectrum obtained by the ELS sensor. The interval contains a period in undisturbed solar wind (until 03:15), magnetosheath (until 03:40), ionosphere (until 03:55) and wake (from 03:55). We calculate high energy electron densities (second panel) of 2–3/cm^3 in the solar wind, and of 4–10/cm^3 in magnetosheath. In the ionosphere fitted electron densities are calculated at 20–40/cm^3.

The third panel from top of Figure 10 shows the sectorized uncalibrated IMA energy spectrum. The fourth panel shows proton densities calculated by integration without application of noise reduction and calibration factors (but application of geometric factors). If we calculate densities for each sector separately (black line) we assume symmetry of the distribution when rotating around the sector ring. This only makes sense when the peak flux is contained in the respective sector ring. The 4π-integrated density (green line) is essentially the integral over the sectorized densities.

In the fifth panel of Figure 10 we apply noise reduction and calibration factors before calculating proton densities. Since here we take account of the sector efficiency we regard the integrated average density (blue line) as the best estimate for the proton density. If we remember that application of additional corrections for massring and anode efficiency might increase the intensity by a factor 2, agreement with the fitted high energy electron densities in solar wind and magnetosheath is rather good.

Figure 11 shows fits to the IMA proton spectra for the first 30min of the time interval shown in Figure 10. The fit range is restricted to the 10 energy bins around the peak flux (blue fit). The energy bins above that range are fitted by a second Gaussian (red fit). The first four spectra are obtained in the solar wind. One can see that the fitted density is around 1.0/cm^3 indicating that the cold solar wind beam is probably slightly underestimated. One can also observe a high energy component which increases when approaching the magnetosheath. The bimodal distribution observed after 03:13 is not well fitted by one gaussian, only the heated distributions after 03:26 are fitted well and give densities in agreement with calculation by integration. After 03:42 in the ionosphere no fit is possible.

This figure also shows the effect of the energy cut-off of the instrument which is between 500 and 700 eV depending on PAC level. This corresponds to bulk speeds of 310 and 370 km/s respectively but distributions with higher bulk speeds

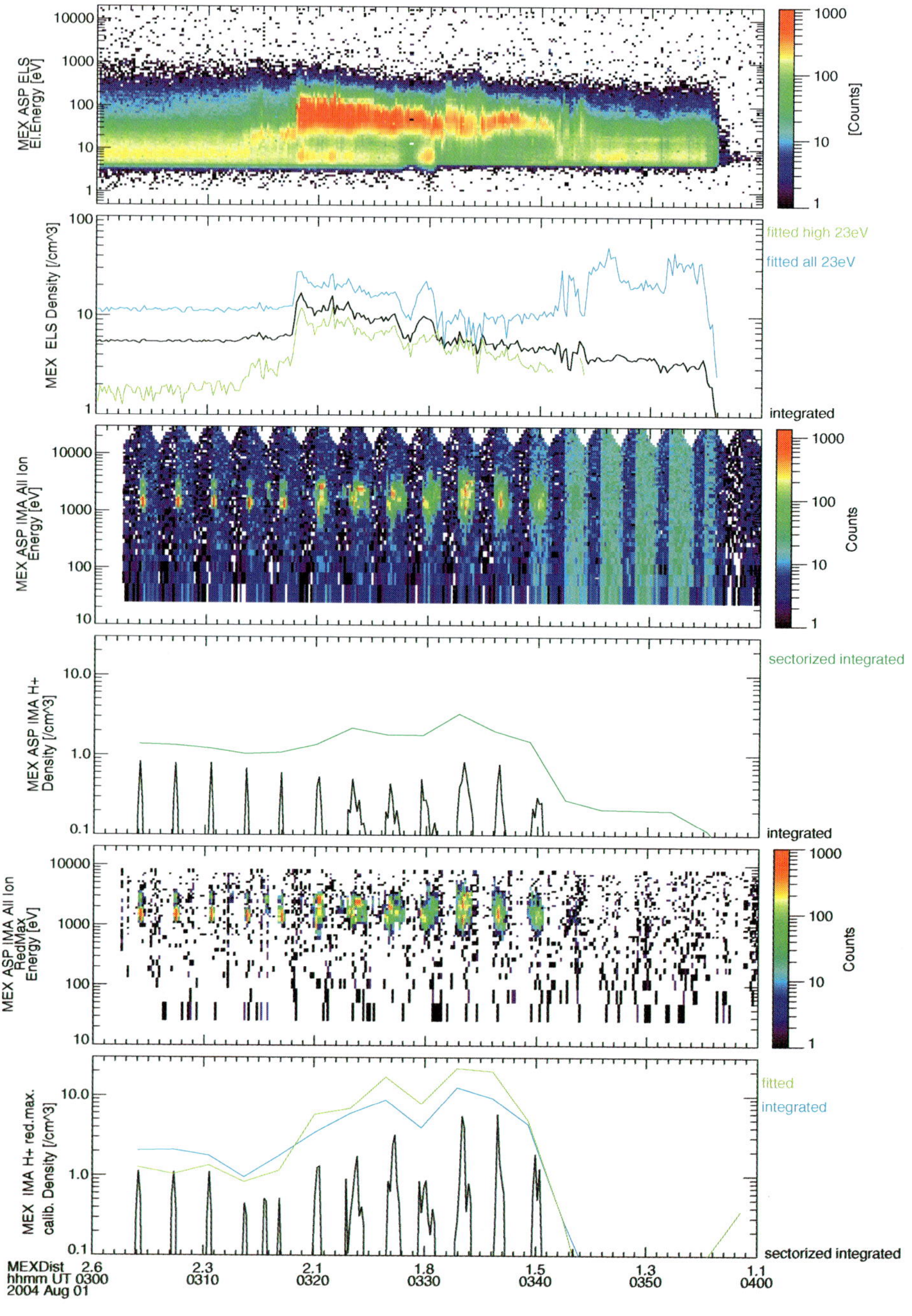

Figure 10. ASPERA-3 ELS and IMA data obtained between 2004-08-01 03:00 and 04:00 UT. From top to bottom: (a) uncalibrated ELS energy spectra, (b) integrated and fitted ELS electron densities, (c) non-reduced, uncalibrated IMA energy spectra, (d) non-reduced integrated IMA H^+ densities for each sector (black) and averaged over all sectors (green), (e) reduced IMA energy spectrum (f) reduced, calibrated IMA H^+ densities for each sector (black), after calibration (blue) and fitted (green).

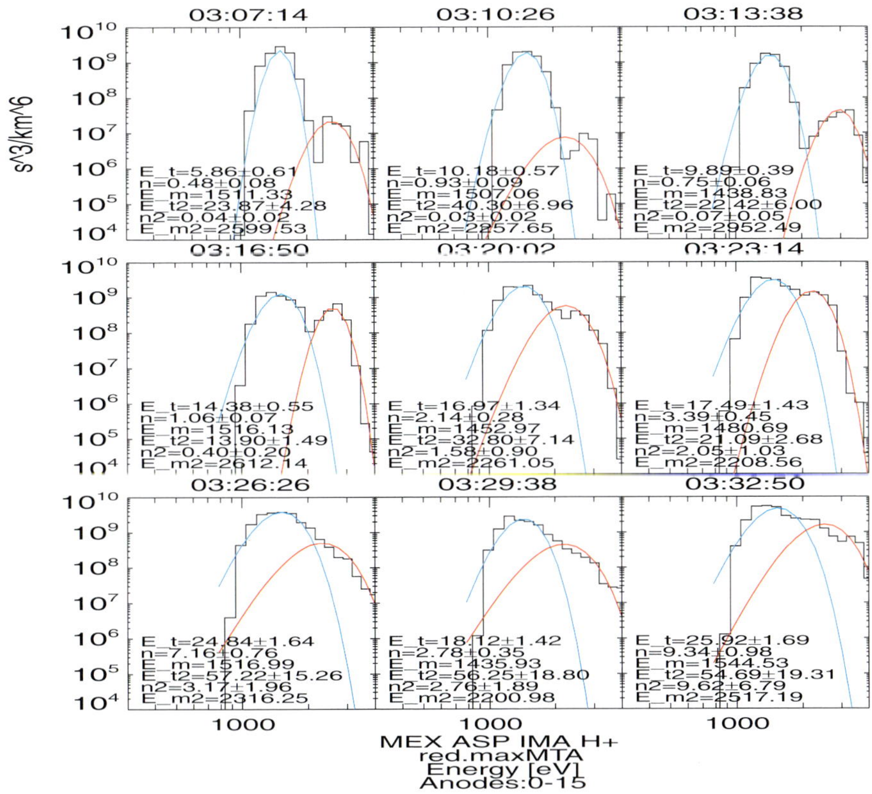

Figure 11. Phase space density of H^+ as a function of energy, obtained by the ASPERA-3 IMA sensor in 2004-08-01 03:00 UT to 03:30 UT. Data are sampled over 192 s. The black lines are the measured data, the blue lines are fits assuming a Gaussian distribution around the maximum of the spectrum. The red lines are a Gaussian fit to the high energy part of the spectrum. Fit parameters n [1/cm^3], E_t [eV] and E_m [eV] are given for each fit.

are also affected such that we can say that only distributions with bulk speeds above 400 km/s can be properly measured.

Figure 12 shows fits to the IMA He^{++} spectra for the same time interval. These spectra have been obtained after proton spill-over subtraction. We expect the peak-energy to be four times the proton peak energy (the instrument measures E/Q, here the instrumental E/Q has been multiplied by 2). For the solar wind spectra that should be at about 5–7 keV where a high energy peak is actually observed. We interpret the peak at low energies as a residual of proton spill-over. Also the temperature of the He^{++}-peak in the cold solar wind seems to be over-estimated by the fit.

5.5.2. *Ion Velocities*

Figure 13 shows plasma velocity and temperature determinations by ELS and IMA for the same time interval as Figure 10. The two top panels show the energy spectra

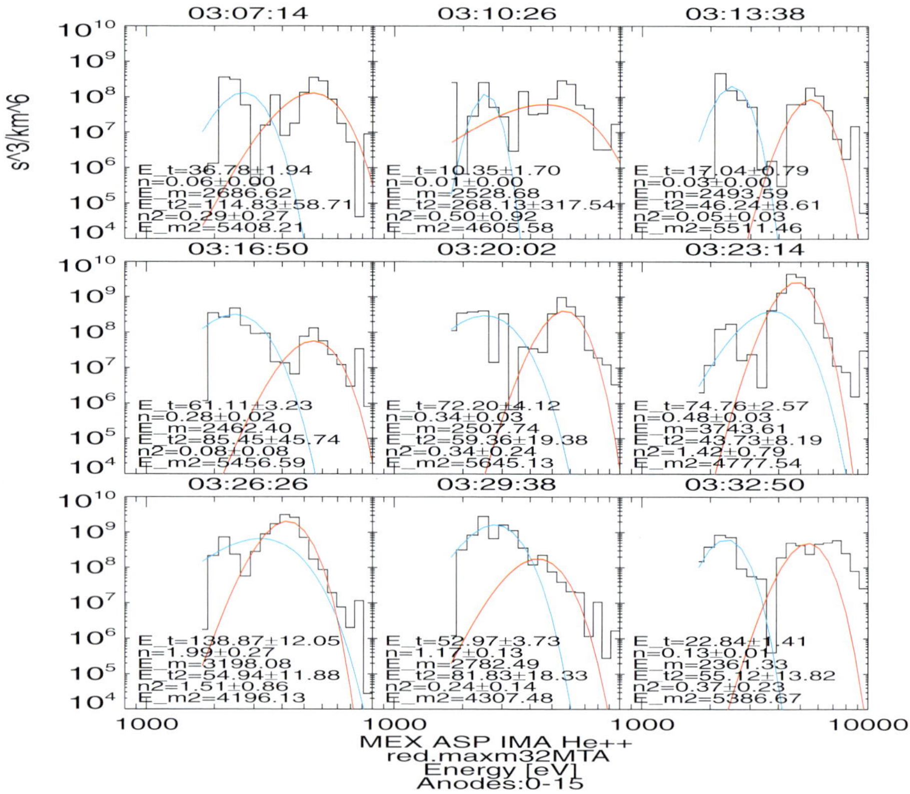

Figure 12. Phase space density of nominal He^{++} as a function of energy, obtained by the ASPERA-3 IMA sensor in 2004-08-01 03:00 UT to 03:30 UT. Data are sampled over 192 s. The black lines are the measured data, the blue lines are fits assuming a Gaussian distribution around low energy part of the spectrum. The red lines are a Gaussian fit to the high energy part of the spectrum. Fit parameters n [1/cm^3], E_t [eV] and E_m [eV] are given for each fit.

for reference. The next two panels show IMA proton and He^{++} velocity components determined by integration after calibration. The fifth panel from top shows the ϑ and φ components of the proton velocity in the instrumental (ASP) system: whenever ABS(φ) $> 90^\circ$ and $\vartheta > 90^\circ$ the ion distribution might be partly shadowed by the spacecraft (the 90° line is shown in red for reference). We observe that in the solar wind (before 03:18 UT) both ions show $V_x \sim -500$, $V_z = 0$, $V_y \sim -100$ km/s. In the magnetosheath the V_z components agree, while V_y and V_x do not. Specifically the He^{++} total velocity decreases while the proton velocity stays constant. We interpret this as an effect of the low energy cut-off of the proton distribution. That means, the He^{++}-velocity is usually a better measure of the ion speed but with the draw back of lower statistics. Note also, that after 03:27 UT the velocity components in the ASP frame indicate that the distribution might be affected by shadowing. We do not show ELS electron velocities in this paper since the ELS measurements are planar and can only qualitatively represent the plasma velocity.

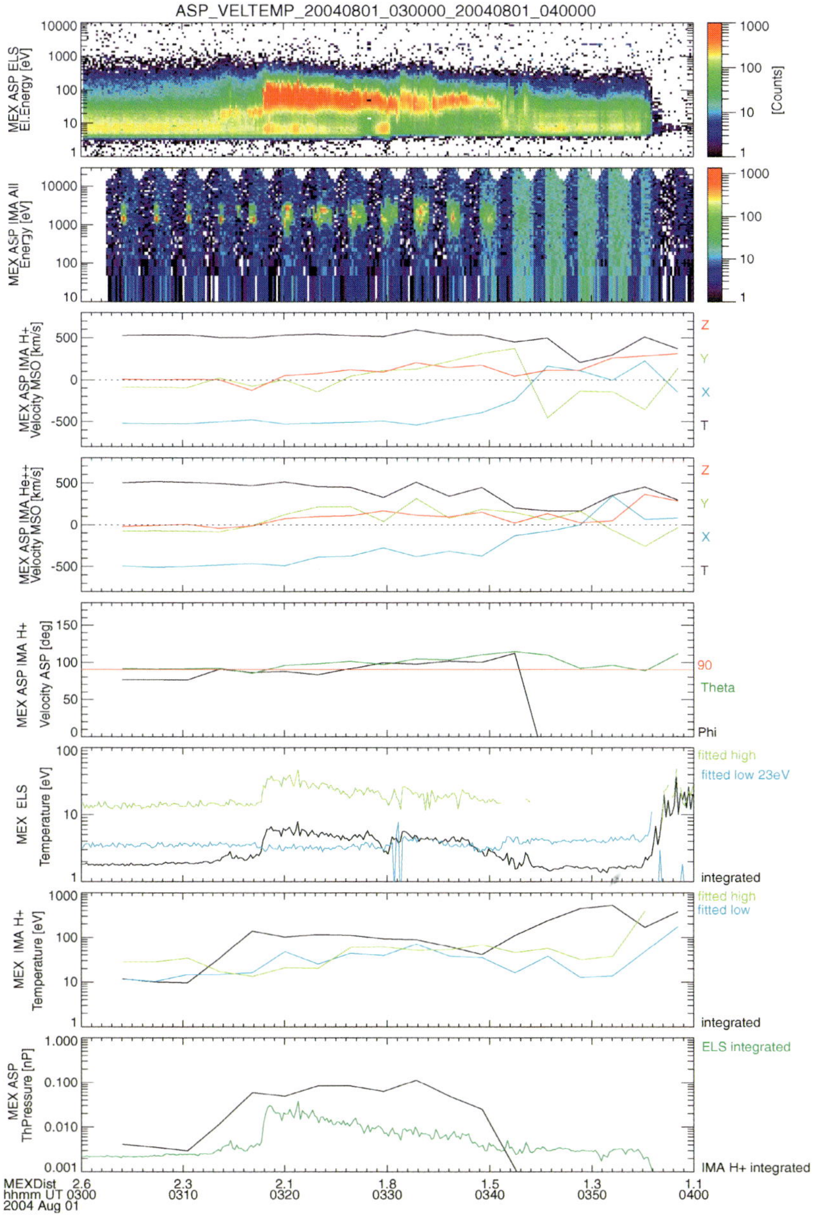

Figure 13. ASPERA-3 ELS and IMA data obtained between 2004-08-01 03:00 and 04:00 UT. From top to bottom: (a) ELS electron energy spectrum, (b) IMA all ion energy spectrum, (c) IMA proton velocity with anode and sector calibration *xyz* and total components (MSO system), (d) IMA He^{++} velocity with anode and sector calibration *xyz* and total components (MSO system), (e) IMA proton velocity ASP ϑ and φ components, (f) ELS electron temperatures by integration (black), fitting low energies (blue) and high energies (green). (g) IMA proton temperatures by integration (black), fitting low energies (blue) and high energies (green). (h) IMA total proton thermal pressure (black) and ELS total electron thermal pressure.

5.5.3. *Temperature and Pressure*

The lowest three panels of Figure 13 show electron and proton thermal pressures and temperatures. The third panel from bottom shows the electron temperature by integration (black), low-energy fitting (blue) and high-energy fitting (green). We assume that for the solar wind and magnetosheath integration and low-energy fitting deliver most of the times the temperature of spacecraft photo electrons only. Only in the ionosphere (after 03:42 UT) – when the spacecraft potential is negative – these measures might be reasonable. But, it should be noted that also here distributions with temperatures of less than 1 eV cannot be measured due to the shift in the spectrum caused by the negative spacecraft potential and the repellent voltage of ELS. The high-energy fitted temperature of ~15 eV in the solar wind and 20–40 eV in the sheath are the better measure in these regions and in agreement with the fitted proton temperature. The bottom panel of Figure 13 shows solar wind electron and proton pressures in the range 1–5 pPa, magnetosheath electron pressures of 10–30 pPa, and proton pressures of 80–200 pPa.

6. Plasma Moment Statistics in the Environment of Mars

In the following we apply the moment calculations discussed in the previous sections to the complete ASPERA-3 ELS and IMA data sets obtained between 1 Feb 2004 and 1 Feb 2006 at full time resolution (4 s for ELS and 192 s for IMA). For ELS we exclude periods of linear stepping mode, for IMA we only use PAC level 1 and 2 data, exclude periods of spacecraft shading of the sensor by taking records with a bulk speed inside the core field-of-view of the sensor. Also we only use spectra where the integrated proton density is larger than $0.1/cm^3$. The last condition excludes most spectra obtained with IMA inside of the MPB where light ions are rarely observed.

As discussed in the beginning of this paper plasma moments can be obtained either by integration of the energy spectra or by fitting a Gaussian to the phase space distribution function. While the integration usually covers the complete energy spectrum, we fit low and high energy parts of the spectrum separately for distributions outside the ionosphere. The parameters discussed in the following are: 1. the low and high energy electron density by fitting, 2. the low and high energy electron temperature by fitting, 3. the proton density, velocity and temperature by fitting and integration, and 4. the alpha density by fitting and integration.

6.1. Spatial Binning

In this paper we only discuss mean and maximum values of plasma moments sampled over the first two years of operation of Mars Express in orbit. We use the MEX orbital data in the Mars-Solar-Orbital system (MSO) where the positive

X-axis is defined by the instantaneous Mars-Sun line and the Y-axis points against the Mars orbital velocity vector. The Z-axis then points approximately in ecliptic north direction. We calculate mean values by binning the data on a spatial grid where the X-axis is defined by the MSO X-axis and the Y-axis by $R_{cyl} = \sqrt{Y_{MSO}^2 + Z_{MSO}^2}$, that is we assume cylindrical symmetry with respect to the Mars-Sun line. We do not take account of aberration effects by the Mars orbital speed (24.1 km/s), since it is low compared to the errors of measurement. Dawn-dusk or North-South asymmetries are usually related to the orientation of the interplanetary magnetic field. Since Mars Express does not have a magnetometer on board, the IMF orientation can only be estimated by using a proxy from MGS data. We are planning to bin the data according to IMF orientation in a later paper. The bin size we are using is 0.05 Martian radii or 170 km for electrons and 0.1 Martian radii or 340 km for ions. Since electrons are sampled at 4 s per spectrum we get more than 100 samples/bin for most regions covered by the orbits (Figure 14 top). For ions the acquisition time is only 192s per full 3D spectrum such that we use a coarser grid and get between 10 and 100 samples per bin (Figure 14 bottom). In all figures black shaded bins denote a value which is equal or less than the minimum value of the color bar, red shaded bins denote values which are higher than the maximum value of the color bar. White space means that no valid samples were taken here. We use three different statistical measures to determine moment levels for each spatial bin: the median (value for which same number of samples have value above and below), the mean (sum over all samples divided by number of samples) and the maximum value observed during the measurement interval 1 Feb 2004 to 1 Feb 2006. Since data of particle counters are typically influenced by disturbances which may only show up sporadically – like solar UV-light on the sensors, there are outliers in the data which can falsify the maximum values observed but also the mean values. Thus the median is usually the most robust measure of the average of the data.

6.2. Electron Densities

As discussed in the first part of this paper it is a general problem of electron counters flying in space that electrons of energies with less than about 10 eV are not registered by the instrument when the spacecraft potential is negative or overenhanced when the spacecraft potential is positive. In addition electrons with less than 5 eV are reflected by the an additional grid to avoid counter saturation. For ionospheric spectra the energy resolution of the ELS sensor of the ASPERA-3 experiment is good enough to observe the non-thermalized photo electron peaks expected in the energy range 20–30 eV. The location of these peaks allows one to estimate the spacecraft potential and subsequently we extrapolate the spectrum to energies below 10 eV. For non-ionospheric spectra we do not have an indication of the spacecraft potential and can only extrapolate the low energies assuming fixed values of the

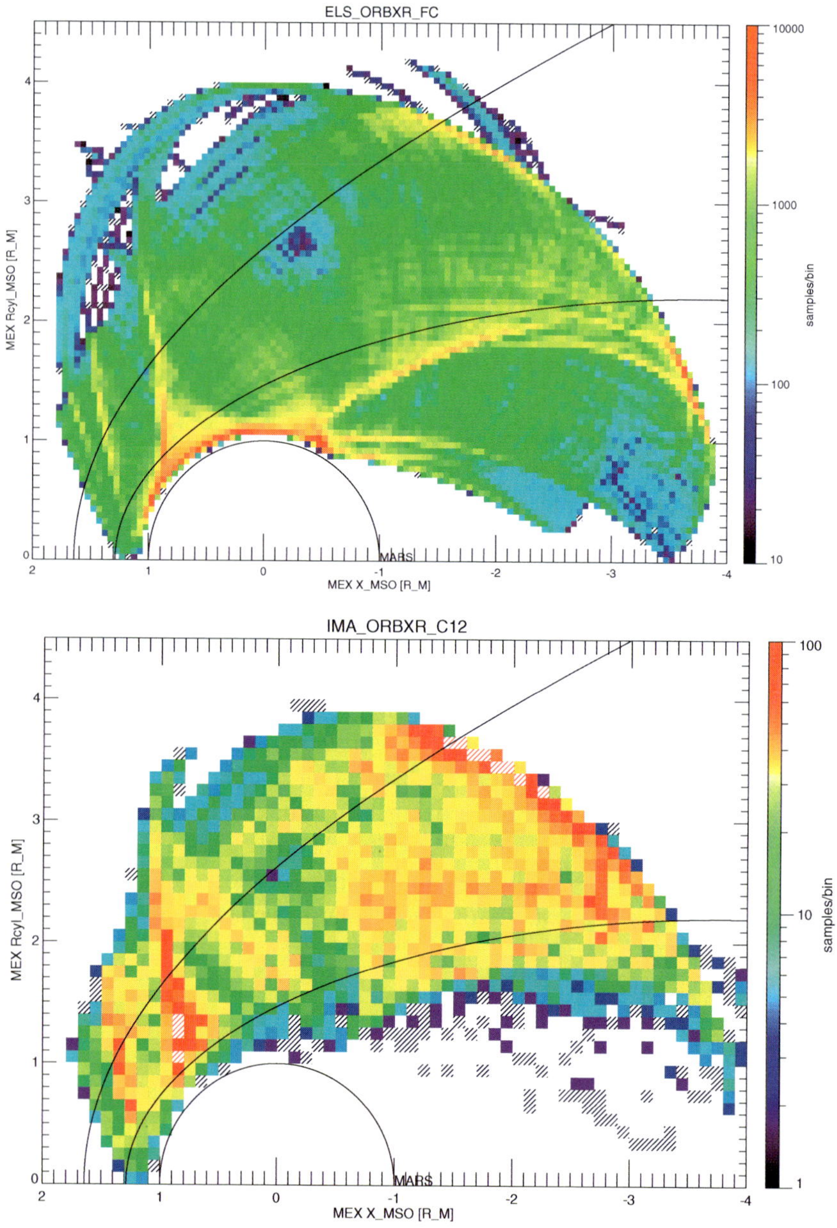

Figure 14. Number of data samples per bin obtained by the ELS sensor in logarithmic stepping mode (top) and the IMA sensor for PAC level 1 and 2 with $n_p > 0.1/\text{cm}^3$ between 01 Feb 2004 and 01 Feb 2006 binned on an MSO X-Rcyl grid with a gridsize of 0.05 and 0.1 R_M respectively.

potential. Note, that the energy determining geometric factors and efficiency of each energy channel is the actually measured energy while the energy shift by spacecraft potential is applied after applying these factors. Thus the effect of the shift on the efficiency is covered.

Figure 15 top shows the median electron density calculated by integration with spacecraft potential correction. Here the spacecraft potential has been estimated as +5 V for non-ionospheric spectra and calculated assuming photo electron peak energy at 23 eV. The separation method for ionospheric and non-ionospheric spectra has been discussed in Section 4.1. A respective map of the resulting ionospheric spacecraft potentials is shown as Figure 15 bottom. Note, that potential values are determined for each 4 s spectral value separately depending on the closest observation of a photoelectron peak. Since the spacecraft potential is being determined whenever a 4s spectrum is classified ionospheric by the ratio of low energy to high energy counts, this map also shows that this criterion is fullfilled by some spectra in the magnetosheath and solar wind. To reduce this effect we impose the additional criterion for ionospheric spectra that the point of measurement must be not more than 0.5 Martian radii away from the MGS MPB and that the closest observed CO_2 peak must not be more than 10min away in measurement time.

In solar wind, sheath and ionosphere the integrated density gives a wrong measure of the actual density – only in the magnetotail with sparse distributions it can be assumed a better measure than the fitted values. We show the map here mainly as a guide to the actual measured counts. Overplotted as black lines on all figures are the bowshock and MPB location as observed by the MGS magnetometer (Vignes *et al.*, 2000).

In Figure 16 top we show median fitted densities calculated from the high energy part of the spectrum only. Here we observe densities of 1–3/cm^3 for the solar wind, which agrees with the proton observations (see below). For the magnetosheath the same influence of spacecraft photoelectrons prevails such that here also Figure 16 top gives the best density estimate very much in agreement with proton density observations. We also observe that the presumed positive spacecraft potential value does only have a minor influence on solar wind and magnetosheath densities. Figure 16 bottom shows the median fitted low-energy electron density for ionospheric spectra. We regard the high density values observed for zenith angles larger than 100 degree as artefacts of an erroneous spacecraft potential estimation or bad fitting by low counting statistics.

For the ionosphere the determination of electron densities is much more problematic. The minimum altitudes reached by Mars Express is about 260 km. Electron densities for altitudes below 300 km have been determined on previous missions by radar sounding and radio occultation (Kliore, 1992). While maximum densities of 10^5/cm^3 are reported below 200 km altitude, for solar zenith angles below 45 degree densities fall to 10^3/cm^3 at 300 km altitude. The median densities we observe for the lowest MEX altitudes are only 20–40/cm^3 when assuming a 23 eV photoelectron

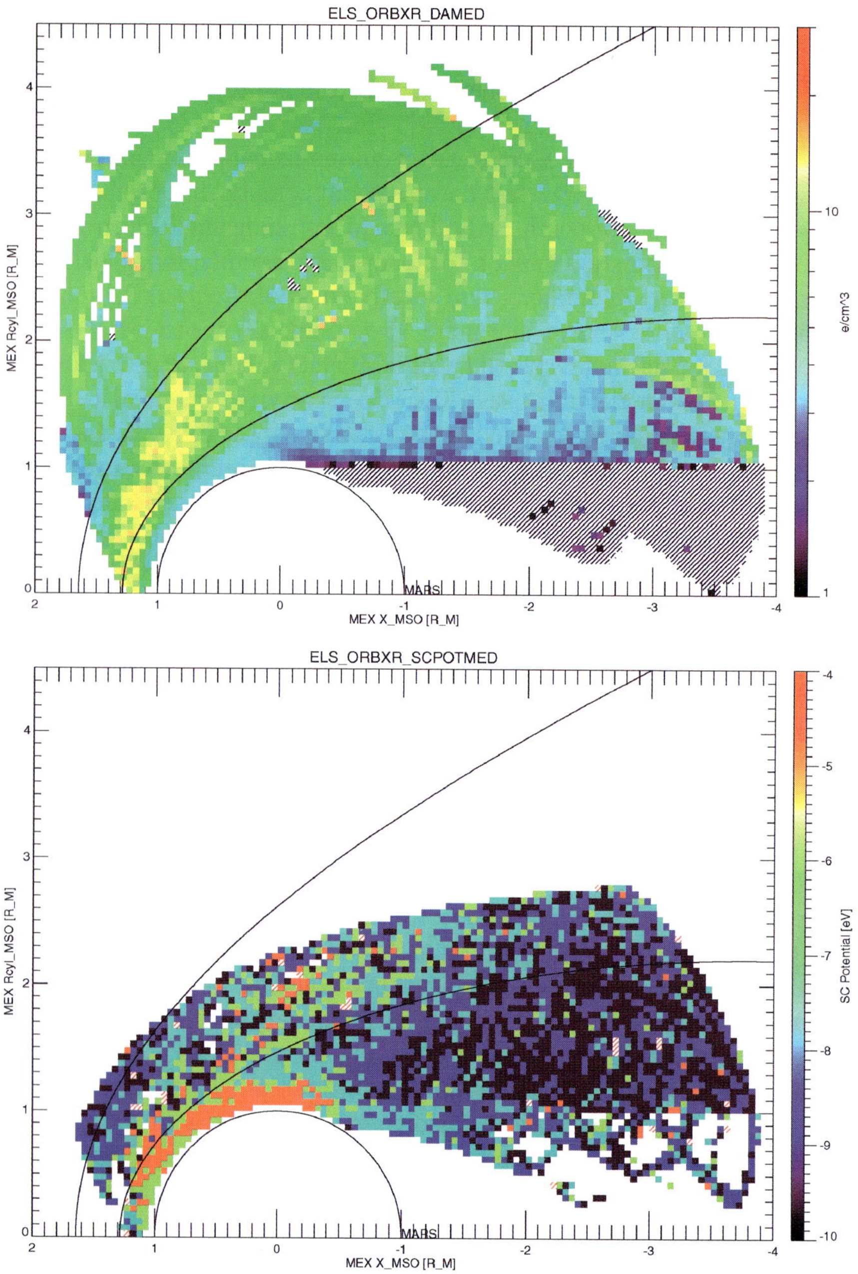

Figure 15. Median integrated electron density over all energies (top) and median ionospheric spacecraft potential estimate (bottom) assuming CO_2 peaks at 23 eV. Observations by the ELS sensor between 01 Feb 2004 and 01 Feb 2006 binned on an MSO X-Rcyl grid with a gridsize of 0.05 R_M.

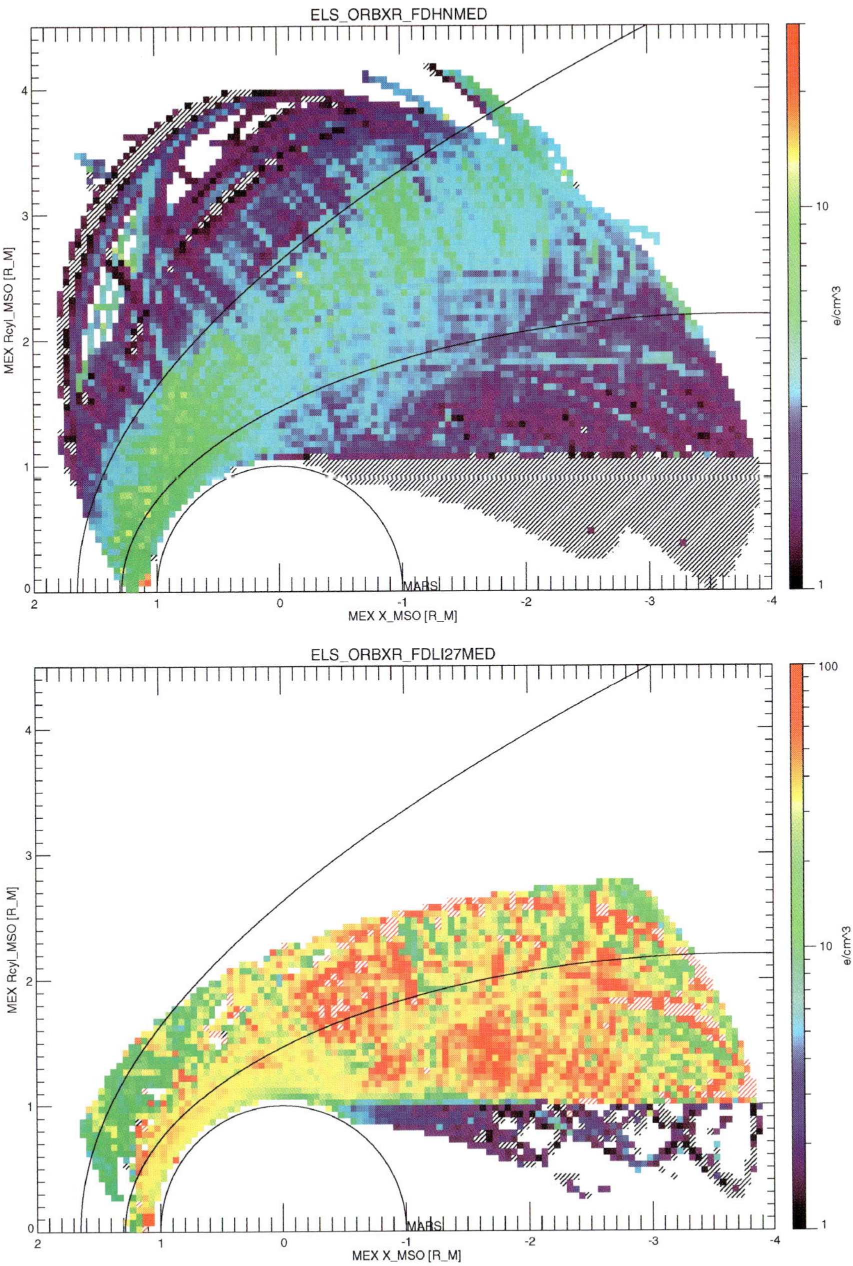

Figure 16. Median of fitted high-energy electron density for non-ionospheric spectra (top) and low-energy electron density for ionospheric spectra (bottom) observed by the ELS sensor between 01 Feb 2004 and 01 Feb 2006 binned on an MSO X-Rcyl grid with a gridsize of 0.05 R_M.

peak (not shown here) or 40–70/cm^3 when assuming a 27 eV photoelectron peak (Figure 16 bottom). A more detailed comparison of ionospheric densities with radio sounding results of the MARSIS experiment on Mars Express has to be done to resolve this issue.

6.3. Electron Temperatures

Figure 17 top shows the median electron temperature for the high-energy part of the spectrum and non-ionospheric spectra. In the solar wind we observe temperatures of 10–20 eV which is higher than the expected 1–5 eV (Schwenn, 1991). This might be an artefact of a bad separation of the high-energy tail of the spectrum in the solar wind. Towards the bow shock the temperature seems to increase which might be an effect of upstreaming electrons or just the fluctuation of the bow-shock position. In the magnetosheath we think that as for the densities the high-energy part of the spectrum (Figure 17 bottom) will give the better estimate. Here we observe temperatures of 20–40 eV for zenith angles smaller than 90 degree and slightly lower for larger angles. For the ionosphere only the low-energy part (Figure 17 bottom) is relevant and the map seems to indicate that temperatures decrease with altitude at solar zenith angles smaller than 90 degree. But minimum temperatures in the ionosphere are about 4 eV. Hanson and Mantas (1988) give temperatures of only 0.5 eV for 300 km altitude. While the ELS sensor has an energy resolution sufficient to measure such low energies, we think that in a region of negative spacecraft potential P the minimum temperature which can be measured is given by eP, which is about 4 eV.

6.4. Proton Densities

Figure 18 shows median proton densities obtained by integrating and fitting the spectra of the IMA sensor. We observe typical proton densities of about 1/cm^3 outside of the bow-shock and 1–3/cm^3 by integration and 3–5/cm^3 by fitting in the magnetosheath. This difference is probably caused by the instrumental cut-off below 1 keV which is better extrapolated by the fitting. The fitted values agree with observations by the ASPERA-1 experiment on Phobos-2 (Lundin *et al.*, 1993). At the MPB densities drop well below 1/cm^3. The median fall-off location of the proton density seems also to agree with the MGS MPB.

6.5. Ion Velocities

Figure 19 top shows the total proton velocity observed by the IMA sensor at typical median values of 500 km/s in solar wind and magnetosheath. Because of the low-energy cut-off of the sensor there is a strong bias towards high velocities in the proton data. As estimated above (Section 5.5.1) only distributions with bulk speeds

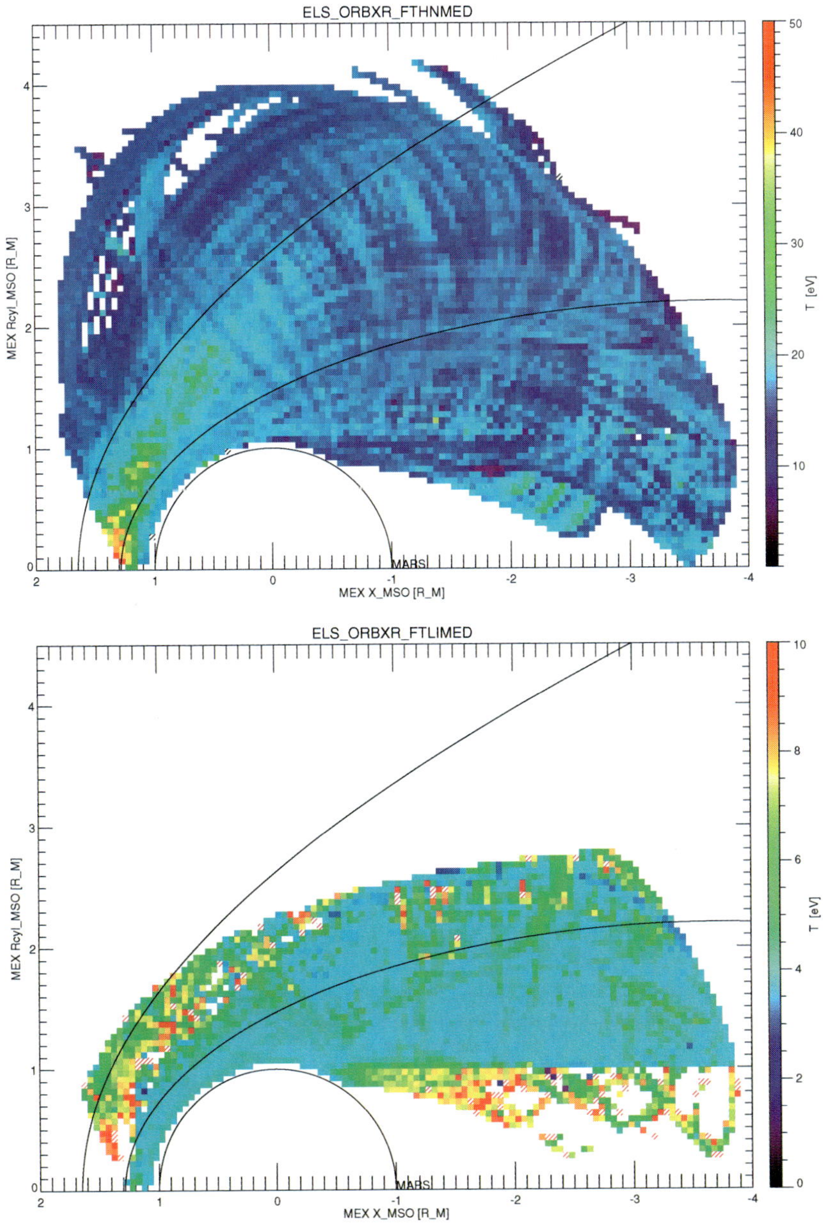

Figure 17. Median of fitted high-energy electron temperature for non-ionospheric spectra (top) and of fitted low-energy electron temperature for ionospheric spectra (bottom) observed by the ELS sensor between 01 Feb 2004 and 01 Feb 2006 binned on an MSO X-Rcyl grid with a gridsize of 0.05 R_M.

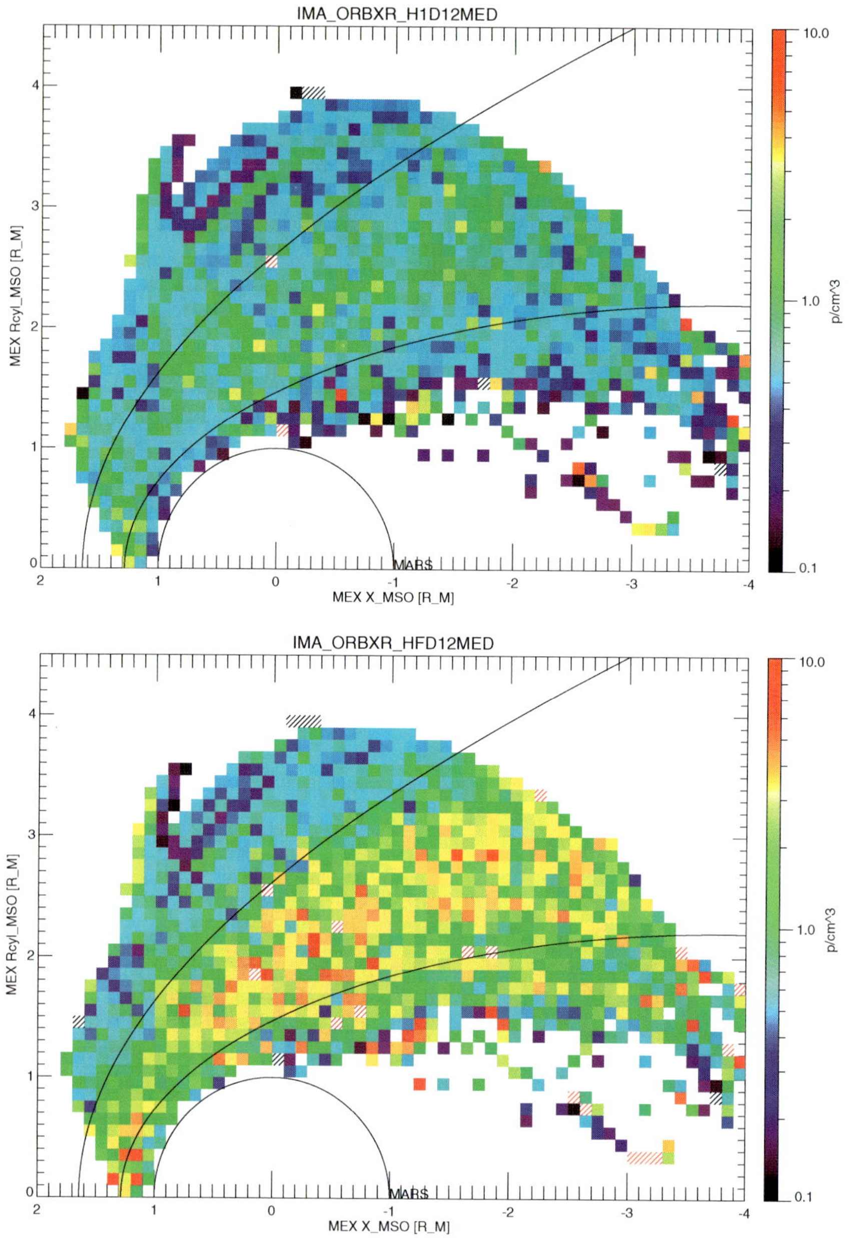

Figure 18. Median of integrated (top) and fitted (bottom) proton density observed by the IMA sensor in PAC 1 and 2 between 01 Feb 2004 and 01 Feb 2006 binned on an MSO X-Rcyl grid with a gridsize of 0.1 R_M.

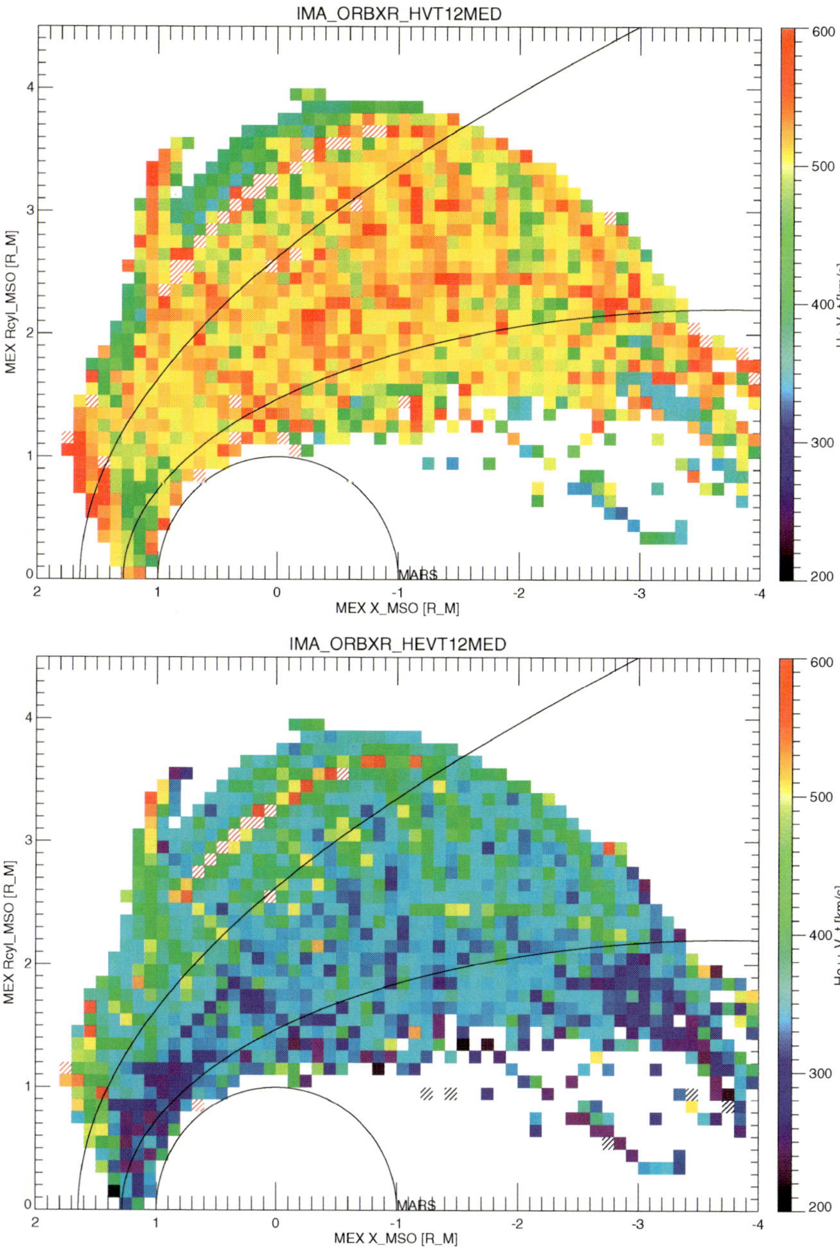

Figure 19. Median of integrated total proton velocity (top) and of integrated total He^{++} velocity (bottom) observed by the IMA sensor in PAC 1 and 2 between 01 Feb 2004 and 01 Feb 2006 binned on an MSO X-Rcyl grid with a gridsize of 0.1 R_M.

over 400 km/s are correctly sampled. A better estimate is the He^{++}-velocity shown in Figure 19 bottom. Here we observe a more reasonable median solar wind speed of 300–400 km/s and braking of the speed by about 50 km/s at the bow shock.

Figure 20 shows the median velocity vector orientation in MSO cylindrical coordinates for protons (top) and He^{++} (bottom). Here we can observe nicely the deviation of the solar wind by the obstacle: in the inner magnetosheath at solar zenith angles of 45 to 90 degrees vectors are almost parallel to the MGS MPB. Since the velocity vector orientation does not depend strongly on the energy range, there are only small differences between proton and He^{++} observations.

6.6. Proton Temperatures

Figure 21 shows the median proton temperature calculated by integration (top) and by fitting (bottom). The fit here includes the low and high energy part of the spectrum. For the temperature the two methods show very different values: integrated temperatures are at around 100 eV and above in the magnetosheath, while fitted values are at about 10–30 eV in the solar wind and 30–50 eV in the sheath. The fitted values agree with expected values for the solar wind Schwenn (1991) which indicates that the high temperatures calculated by integration might be caused by high energy noise or by the bad spatial resolution of the IMA sensor. For the magnetosheath the bad spatial resolution is much less important. The fitting of proton spectra is done only for limited energy ranges (see Section 5.5). This limits the maximum temperatures which can be fitted. Thus we may assume that for the magnetosheath the integrated values may be more representative since they also show maximum temperatures at the nose as one would expect. On the other hand the integrated temperature is more influenced by the low-energy cut-off than the fitted values. Here the ASPERA-1 experiment also reported very high values of around 600 eV (Lundin *et al.*, 1993), which is much higher than our median values.

6.7. Alpha Densities

Figure 22 shows the integrated and fitted alpha particle densities observed by the IMA sensor. Here we calculated the densities over the complete energy range. As discussed in Section 5.5 this might overestimate the densities since at lower energies the alpha track is contaminated by spill-over protons. Still the observed values are on the order of 0.2–0.3/cm^3 which is about 10% of the proton densities. Fitted values show much better the density increase at the bow shock.

7. Summary and Conclusions

In the first part of this paper we have presented methods to derive plasma moments from the ion sensors of the ASPERA-3 experiment onboard Mars Express. This

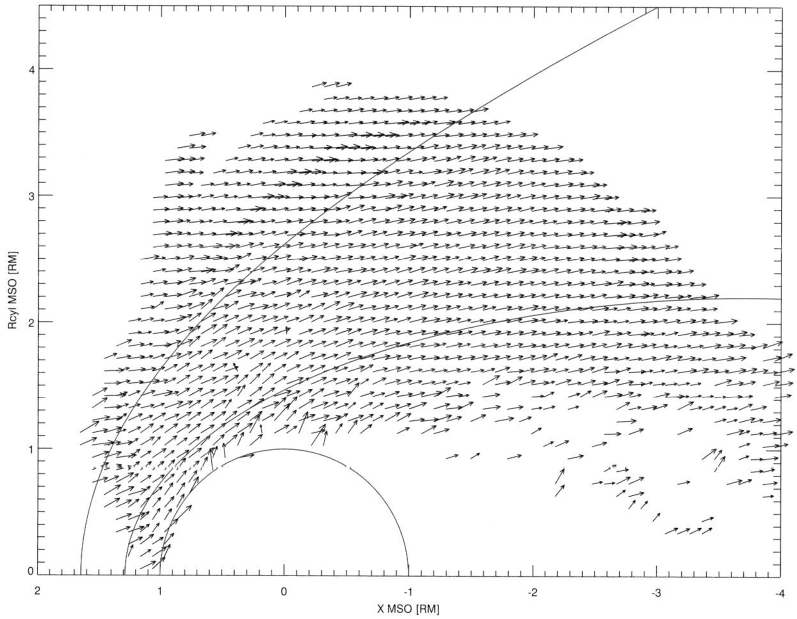

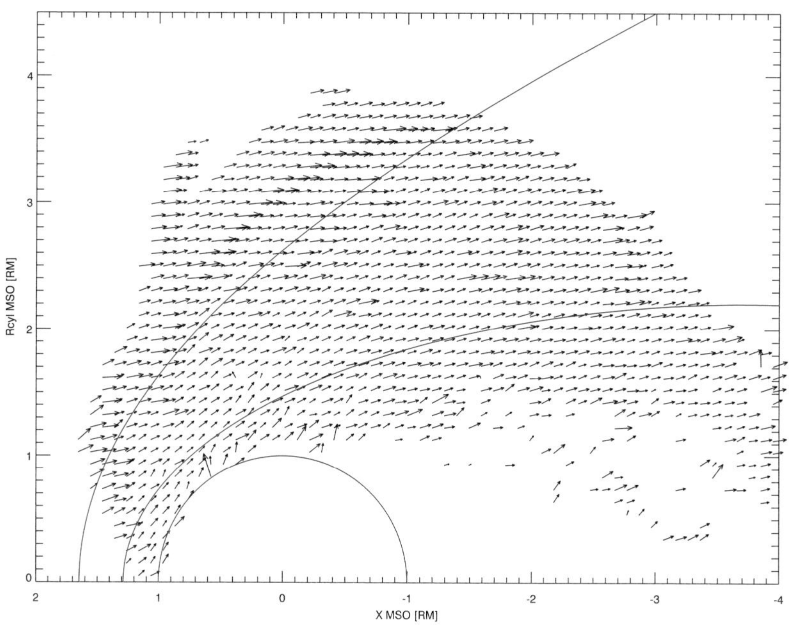

Figure 20. Median of integrated proton velocity (top) and He^{++} velocity (bottom) vector orientation in the MSO system observed by the IMA sensor in PAC 1 and 2 between 01 Feb 2004 and 01 Feb 2006 binned on an MSO X-Rcyl grid with a gridsize of 0.1 R_M.

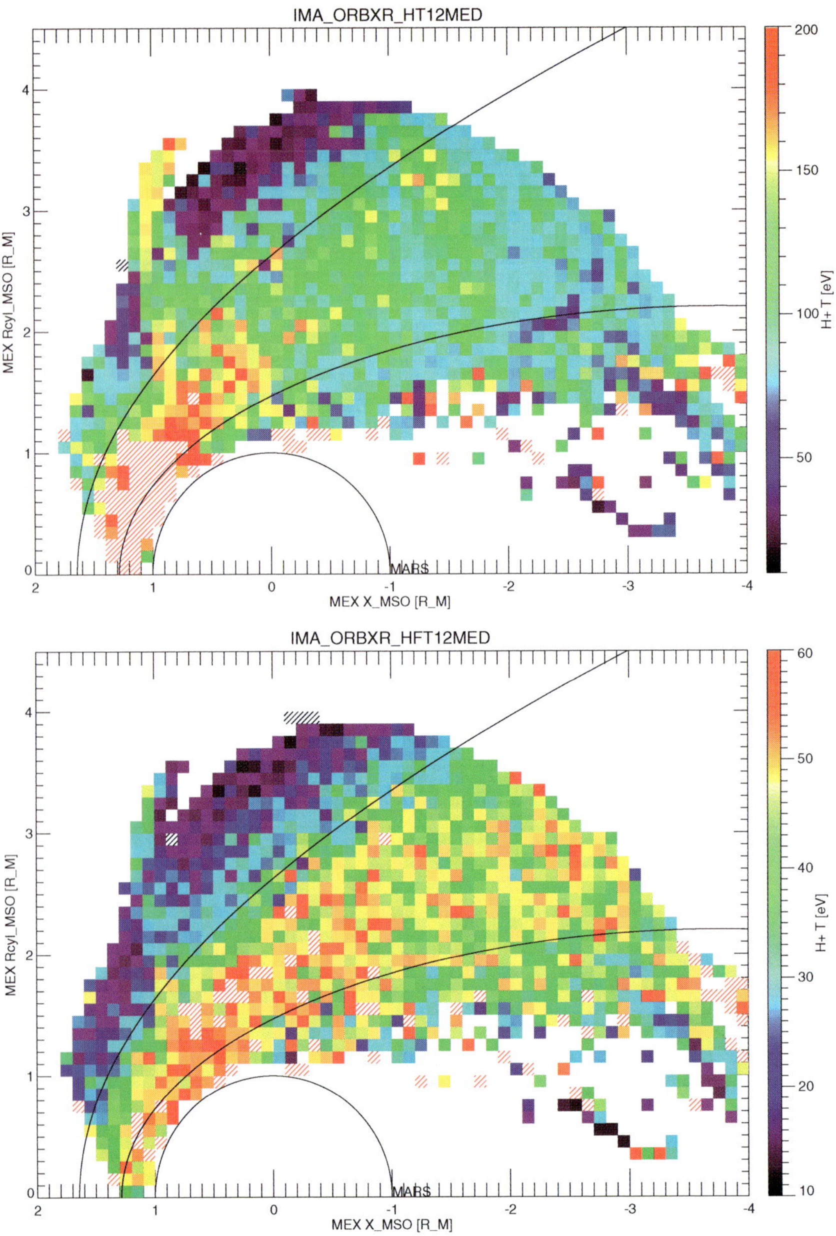

Figure 21. Median of integrated (top) and fitted (bottom) proton temperature observed by the IMA sensor in PAC 1 and 2 between 01 Feb 2004 and 01 Feb 2006 binned on an MSO X-Rcyl grid with a gridsize of 0.1 R_M.

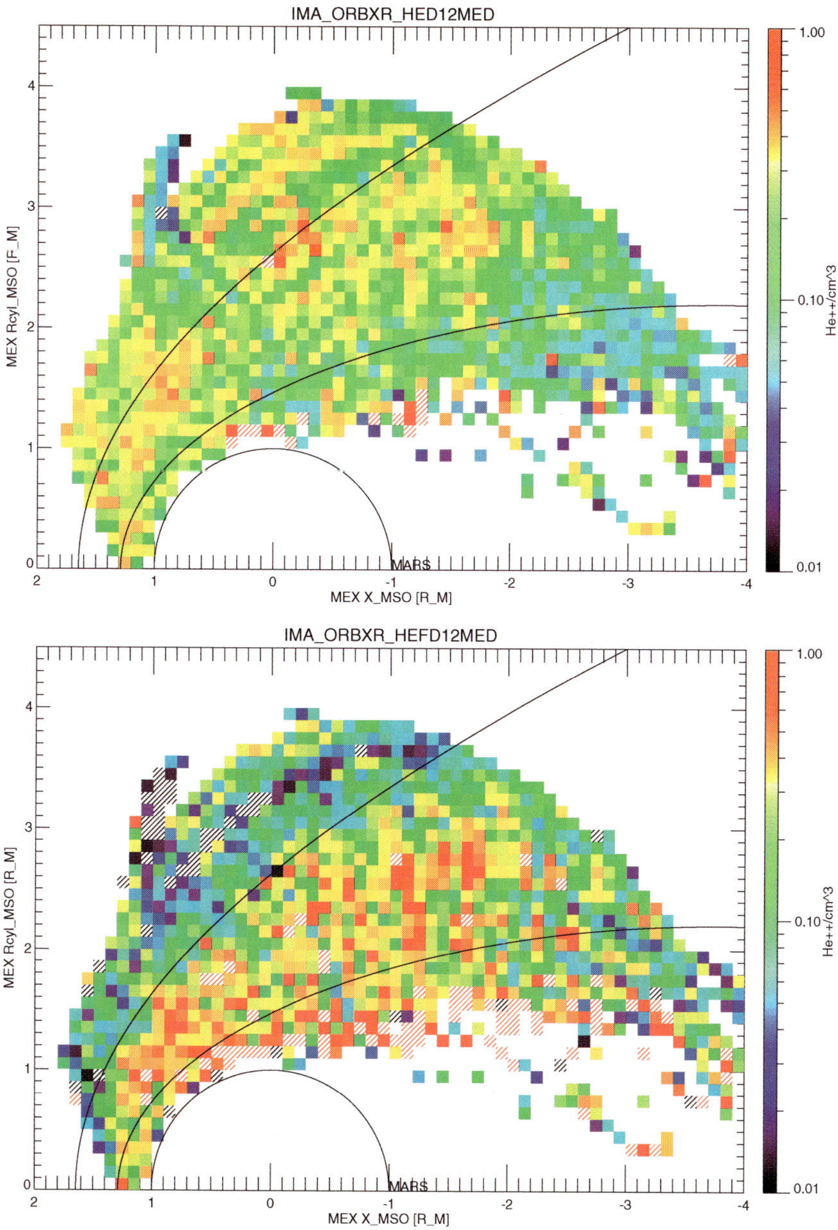

Figure 22. Median of integrated (top) and fitted (bottom) alpha particle density observed by the IMA sensor in PAC 1 and 2 between 01 Feb 2004 and 01 Feb 2006 binned on an MSO X-Rcyl grid with a gridsize of 0.1 R_M.

is the first time that plasma moments have been determined in a wide range of conditions in the environment of Mars. After two years of operation of the ASPERA-3 instrument we are still learning about the instruments behavior under changing conditions. Thus we expect that certain calibration parameters and interpretations will change in the coming years of ASPERA data analysis. But we expect that the principles described in this paper remain valid and that the derived moments will not change dramatically. The values we derive for density, temperature and velocity of electrons and protons are very reasonable when the spacecraft crosses solar wind and magnetosheath regions. For the ionosphere we have the specific problem of very low energy electron and ion distributions. Only further comparisons with other instruments (MARSIS) will show how good our determinations are for this region.

Each ion species has its specific problems for the moment calculation: for electrons we have the influence of the spacecraft potential and local photo electrons. In addition the planar measurement without a magnetometer onboard prohibits the determination of the electron velocity vector. For protons we have the problem of an energy cut-off at about 500 eV in addition to the high noise level of the instrument. For helium and heavier ions we have the problem of spill-over protons – that is protons which are erroneously registered with a high mass/charge value. Nevertheless we think that we have shown in this paper that we can derive moments which are consistent in very different plasma conditions.

In the second part of this paper we have been presenting the first maps of plasma moments for the space environment of Mars obtained by the ASPERA-3 experiment on board Mars Express. These moments include densities and temperatures for electrons and protons, densities for alpha particles and velocities for protons. Proton and alpha moments are strictly valid for solar wind and the magnetosheath only because of the low energy cut-off of the instrument. Moments of heavier ions will be treated in a later paper. We observe median density values of 2–3/cm^3 and proton temperatures of 20–30 eV in the solar wind as expected for solar distances of 1.5 AU. In the magnetosheath densities increase by a factor 2–3 and ion temperatures by a factor 2. Recently these values have been compared with a 3D hybrid simulation of the Martian plasma environment (Bößwetter *et al.*, 2006). The results showed qualitative agreement for most parameters and quantitative agreement for electron and proton densities and temperatures.

We think that using the correction by spacecraft potential we are even able to estimate ionospheric electron moments, which is difficult when using particle counters with a low energy cut-off. Also the spatial binning used for this paper is too coarse to determine densities at lowest altitudes. Still the maximum values observed at 300 km altitude indicate that densities reach up to 10^3/cm^3 in agreement with radio occultation observations. To resolve the long-standing question whether the ionospheric particle pressure is sufficient to balance the solar wind pressure, we need additional investigations using the heavy ion data of ASPERA-3.

Acknowledgements

We thank all members of the ASPERA-3 team for the big effort which led to the successfull operation of the instrument and the calibration of the data. For this paper we are especially gratefull to Emmanuel Penou, CESR, for data preparation. We wish to acknowledge support from DLR grant 50QM99035 and Deutsche Forschungsgemeinschaft grant WO 910/1-1, support through the National Aeronautics and Space Administration (NASA) contract NASW-00003 in the United States, and the Particle Physics and Astronomy Research Council (PPARC) in the United Kingdom.

References

Barabash, S., Lundin, R., Andersson, H., Brinkfeldt, K., Grigoriev, A., Gunell, H., *et al.*: *Space Sci. Rev.*, this issue, doi: 10.1007/s11214-006-9124-8.

Bößwetter, A., Simon, S., Bagdonat, T., Motschmann, U., Fränz, M., Roussos, E., *et al.*: 2006, *Ann. Geophys.*, submitted.

Brain, D., Bagenal, F., Acuña, M., and Connerney, J.: 2003, *J. Geophys. Res.* **108**(A12), 482, doi:10.1029/2002JA009.

Carlsson, E., Fedorov, A., *et al.*: 2006, *Icarus* **182**(2), 320.

Dubinin, E., Fraenz, M., Woch, J., Barabash, S., Lundin, R., and Yamauchi, M.: 2006, *Geophys. Res. Lett.*, submitted.

Frahm, R., Winningham, J., *et al.*: 2006, *Icarus* **182**(2), 371.

Hanson, W., and Mantas, G.: 1988, *J. Geophys. Res.* **93**, 7538.

Hutchinson, I.: 2002, *Principles of Plasma Diagnostics*, 2nd edn, Cambridge University Press, Cambridge, UK.

Kallenrode, M. B.: 1998, *Space Physics: An Introduction to Plasmas and Particles in the Heliosphere and Magnetospheres*, Springer, Berlin, Germany.

Kliore, A.: 1992, *AGU Geophys. Monograph* **66**, 265.

Luhmann, J., and Brace, L.: 1991, *Rev. Geophys.* **29**(2), 121.

Lundin, R., *et al.*: 1993, in Gombosi, T. (eds.), *Plasma Environment of Non-Magnetic Planets*, Pergamon p. 311.

Mantas, G., and Hanson, W.: 1979, *J. Geophys. Res.* **84**, 369.

Mazelle, C., Winterhalter, D., Sauer, K., Trotignon, J., Acuña, M., Baumgartel, K., *et al.*: 2004, *Space Sci. Rev.* **111**(1–2), 115.

Nagy, A., Winterhalter, D., Sauer, K., Cravens, T., Brecht, S., Mazelle, C., *et al.*: 2004, *Space Sci. Rev.* **111**(1–2), 33.

Parks, G.: 1991, *Physics of Space Plasmas: An Introduction*, Perseus, Cambridge, Mass.

Paschmann, G., Fazakerley, A., and Schwartz, S.: 2000, *Analysis Methods for Multi-Spacecraft Data*, ISSI/ESA, Bern, CH.

Schwenn, R.: 1991, *Large-Scale Structure of the Interplanetary Medium* in *Physics of the Inner Heliosphere*, vol. 1, chap. 3. Springer.

Vignes, D., Mazelle, C., Rème, H., Acuña, M., Connerney, J., Lin, R., *et al.*: 2000, *Geophys. Res. Lett.* **27**(1), 49.

PLASMA MORPHOLOGY AT MARS. ASPERA-3 OBSERVATIONS

E. DUBININ[1,*], M. FRÄNZ[1], J. WOCH[1], E. ROUSSOS[1], S. BARABASH[2], R. LUNDIN[2], J. D. WINNINGHAM[3], R. A. FRAHM[3] and M. ACUÑA[4]
[1]*MPI für Sonnensystemforschung, 37191 Katlenburg-Lindau, Germany*
[2]*Swedish Institute of Space Physics, Kiruna, Sweden*
[3]*Southwest Research Institute, San-Antonio, USA*
[4]*NASA Goddard Space Flight Center, Greenbelt, USA*
(**Author for correspondence: E-mail: dubinin@mps.mpg.de*)

(Received 15 March 2006; Accepted in final form 26 September 2006)

Abstract. A total of about of 400 orbits during the first year of the ASPERA-3 operation onboard the Mars Express spacecraft were analyzed to obtain a statistical pattern of the main plasma domains in the Martian space environment. The environment is controlled by the direct interaction between the solar wind and the planetary exosphere/ionosphere which results in the formation of the magnetospheric cavity. Ionospheric plasma was traced by the characteristic "spectral lines" of photoelectrons that make it possible to detect an ionospheric component even far from the planet. Plasma of solar wind and planetary origin was distinguished by the ion mass spectrometry. Several different regions, namely, boundary layer/mantle, plasma sheet, region with ionospheric photoelectrons, ray-like structures near the wake boundary were identified. Upstream parameters like solar wind ram pressure and the direction of the interplanetary electric field were inferred as proxy from the Mars Global Surveyor magnetic field data at a reference point of the magnetic pile up region in the northern dayside hemisphere. It is shown that morphology and dynamics of the main plasma domains and their boundaries are governed by these factors as well as by local crustal magnetizations which add complexity and variability to the plasma and magnetic field environment.

Keywords: Mars: magnetosphere, Mars: ionosphere, sun: solar wind

1. Introduction

Previous missions to Mars have established the existence of the main plasma regions near Mars. Mariner 4 which passed within $3.9R_M$ of Mars in 1965 has detected a bow shock. At the bow shock, solar wind is deflected around the Martian obstacle. However, as the previous spacecraft (except the Viking landers which have not carried an onboard magnetometer) have not approached Mars closer than ~850 km, the nature of the obstacle to the solar wind was not finally resolved before the Mars Global Surveyor (MGS) mission. The MGS measurements have shown that at present Mars does not possess a global intrinsic magnetic field which could be an obstacle for the solar wind as for most of other planets in our solar system (Acuña *et al.*, 1998). Instead, MGS has detected localized, rather strong magnetic anomalies of a crustal origin. Due to the absence of a magnetic obstacle at Mars the solar wind directly interacts with its upper atmosphere and ionosphere and induces a magnetosphere by the pile up of the interplanetary magnetic field. Such an induced

Space Science Reviews (2006) 126: 209–238
DOI: 10.1007/s11214-006-9039-4

magnetosphere can screen the ionosphere from the direct exposure to the solar wind. The formed magnetic barrier separates the solar wind from the ionosphere and acts as an effective obstacle deflecting the magnetosheath plasma.

A similar type of interaction occurs around another nonmagnetized planet, Venus, and was extensively explored by the Pioneer-Venus-Orbiter in over 14 years of operation (see, for example, Russell, 1992). Although the PVO mission has provided a wealth of excellent in-situ data about the solar wind/ionosphere interaction for a wide range of solar wind conditions, the plasma ion component in the energy range ~10 eV–10 keV was studied rather poorly because of instrument and telemetry constraints. The MGS science payload does not include a plasma instrument for the measurement of ion components at Mars, and therefore only the MEX mission and the ASPERA-3 in-situ measurements fill this gap (curiously, there is no magnetometer on MEX).

It is also worth noting that active comets interacting with solar wind develop similar plasma field and magnetic structures as Mars or Venus (Slavin *et al.*, 1986; Neubauer, 1987; Raeder *et al.*, 1987; Mazelle *et al.*, 1989).

The most convincing evidence of the formation of the magnetic barrier at Mars was the observations of the magnetic pile up boundary (MPB), a sharp boundary with a strong jump in the magnetic field strength, a drop in the magnetic field fluctuations and a strong decrease in the superthermal electron fluxes (Acuña *et al.*, 1998). According to Bertucci *et al.* (2003) the MPB is also characterized by an increase in the magnetic field line draping. Downstream from the MPB, a region called the magnetic pile up region (MPR) is characterized by a sustained high magnetic field. It was believed, despite of a lack of ion measurements on MGS, that the MPB separates the region of shocked solar wind (magnetosheath) from the induced magnetosphere. Such an assumption was supported by the Phobos-2 observations (Breus *et al.*, 1991; Pedersen *et al.*, 1991; Dubinin *et al.*, 1996). It will be shown subsequently that, indeed, a magnetospheric cavity almost void of the solar wind plasma is formed at Mars.

There is also a somewhat different view. Mitchell *et al.* (2001) have suggested that another boundary, "ionopause," observed at lower altitudes separates ionospheric and solar wind plasmas. This boundary was detected by the comparison of electron spectra, with magnetosheath-like solar wind electrons above the boundary and ionospheric photoelectrons below the boundary. Its median altitude at solar zenith angles (SZAs) of about 80° was estimated as 380 km. In between, Mitchell *et al.* (2001) identified a "transition region" which the authors compare with the Venusian plasma mantle whose the lower boundary is the ionopause (Spenner *et al.*, 1980). Recall that the term ionopause was introduced to describe the direct interaction between the solar wind plasma and ionosphere at Venus. The currents flowing in the thin layer (ionopause), where the external hot solar wind magnetized plasma and cold ionospheric plasma balance each other, screen the magnetic field from the ionosphere. They cause a pileup of magnetic field lines in front of the ionopause. A magnetic field barrier of piled up field lines almost balances the solar wind pressure.

On the other hand, the magnetic field pressure balances the thermal ionospheric pressure at lower altitudes. As a result, the real obstacle to the solar wind is observed at the magnetic barrier whose position is further from the planet than the ionopause (see, for example, Zhang *et al.*, 1991). If the ionosphere is resistive the ionopause is broadened and the magnetic field penetrates deeper into the ionosphere. This happens, for example, when the solar wind pressure increases and the ionopause moves to lower altitudes where there are more collisions between particles.

However, as it will be shown in this paper, a stoppage of the solar wind at Mars occurs at higher altitudes, at the boundary, identified earlier as MPB.

The Martian ionosphere, formed by the photoionization of the major neutral constituents CO_2 and O with subsequent molecular reactions giving rise to O_2^+ as the major ionospheric ion species and O^+ becoming comparable at altitudes $\geq$300 km is poorly explored as compared to Venus. The measurements of the main ionospheric characteristics at Mars were made in-situ by the two Viking landers (Hanson *et al.*, 1977; Hanson and Mantas, 1988), that provided us with two ionospheric height profiles, and by radio occultation experiments (Kliore, 1992). Recently new radio occultation and sounding measurements were carried out onboard the MEX spacecraft (Pätzold *et al.*, 2005; Gurnett *et al.*, 2005). Most of the radio occultation profiles show a relatively extended ionosphere without clear ionopause structure. On the other hand, a decrease in the magnetic field value within the ionosphere observed by MGS (Acuña *et al.*, 1998) is a typical feature of the ionopause.

In the ASPERA-3 data, ionospheric plasma is well traced by the characteristic "spectral lines" of photoelectrons which are resolved due to a high energy resolution of the electron spectrometer (Lundin *et al.*, 2004; Frahm *et al.*, 2006a,b). It will be shown here that ionospheric electrons are observed in a wide range of altitudes and the boundary of the photoelectrons (PEB) is often located at higher altitudes than it was reported by Mitchell *et al.* (2001) (see also Frahm *et al.*, 2006b). It is not clear yet whether PEB and ionopause are collocated since the lowest energy part of the plasma distribution which primary contributes to the thermal pressure has not been measured yet.

It is worth noting that the region below the MPB remains a mysterious one. It will be subsequently shown that the main fluxes of escaping planetary ions are clustered in this region. Energy characteristics of ion beams yield an estimate of electric fields responsible for ion energization. The values of electric field are close to the typical values of the interplanetary motional electric field that implies an effective penetration of solar wind electric field deep into the magnetosphere and effective scavenging of planetary ions (Dubinin *et al.*, 2006a).

The induced magnetosphere contains several different subregions. The boundary layer/mantle dominated by planetary plasma was identified in the previous missions (Vaisberg, 1992; Lundin *et al.*, 1990a; Dubinin *et al.*, 1996). This boundary layer can be considered as a site where the momentum of the solar wind is transferred to the planetary plasma (Lundin *et al.*, 1991; Lundin and Dubinin, 1992). Ray-like structures stretched in the tailward direction were measured on Phobos-2 as well as on the MEX spacecraft (Dubinin *et al.*, 2001, 2006b). It is shown in this paper that both these regions, namely, the boundary layer and plasma rays are important channels for transportation of planetary ions to the tail.

The magnetotail of Mars consists of two lobes of opposite polarity separated by plasma sheet (Yeroshenko *et al.*, 1990). The plasma sheet consists primarily of planetary ions which are accelerated up to keV energies by the magnetic field tensions (Dubinin *et al.*, 1993). The Phobos-2 observations in the tail at distances of $\sim 2.8 R_M$ from the planet have revealed signatures of field lines of crustal origin (Dubinin *et al.*, 1994) that implies a complicated magnetic structure of the tail due to reconnection of the IMF and crustal field lines. Large-scale modification of the plasma flow in the tail due to the crustal field contribution was observed in 3D-MHD simulations (Harnett and Winglee, 2005).

Crustal fields add complexity and variability to the Martian magnetic environment (Brain *et al.*, 2003, 2006). The strongest crustal source was detectable up to altitudes of 1300–1400 km and, as it will be shown subsequently, it shifts the magnetospheric boundary upwards (see also Crider, 2004; Fraenz *et al.*, 2006a). To date it is not clear whether the local crustal fields are able to balance the thermal pressure of the magnetosheath plasma or the upward motion of the magnetospheric boundary occurs due to local ionospheric inflations caused by a lift of the ionospheric electrons. The crustal field also shields the localized regions from intrusion of the magnetosheath plasma (minimagnetospheres) (Brain *et al.*, 2005; Fraenz *et al.*, 2006a).

In this paper we have analyzed about 400 orbits during the first year (Feb.–Dec. 2004) of the ASPERA-3 operation onboard the Mars Express spacecraft. In some cases, when we did not use an information about the upstream solar wind and IMF parameters, we have analyzed the observations of two years (2004–2005). MEX ASPERA-3 data provide information about the main plasma domains of the Martian space environment. We present an analysis of the morphology of these regions and their boundaries. We analyze the MGS data to infer the upstream parameters, namely, ram pressure of the solar wind and the direction of cross flow component of the IMF. We then explore the influence of these parameters on the plasma distribution within the magnetosphere and the position of boundaries. The influence of crustal sources is also studied.

2. Observations

The Mars Express spacecraft was inserted into an elliptical orbit around Mars in January 2004. This eccentric elliptical orbit has a periapsis altitude of about 275 km, an apoapsis of about 10000 km, an orbital inclination of 86° and a period of 6.75 h. The scientific payload includes the ASPERA-3 instrument with several sensors to measure electrons, ions and energetic neutral atoms (ENAs). The ASPERA-3 (Analyzer of Space Plasma and Energetic Atoms) experiment is a combination of *in-situ* and remote diagnostics of atmospheric escape induced by the solar wind. It comprises the Ion Mass Analyzer (IMA), ELectron Spectrometer (ELS), Neutral Particle Imager (NPI) and Neutral Particle Detector (NPD) (Barabash *et al.*, 2004). In this paper we discuss the results obtained from the IMA and ELS sensors. The IMA sensor measures 3D-fluxes of different ion species with m/q resolution (m and q are respectively mass and electric charge) in the energy range 10 eV/q–30 keV/q with a time resolution of ~3 min and a field of view of $90^\circ \times 360^\circ$ (electrostatic sweeping provides elevation coverage $\pm 45^\circ$). Mass (m/q) resolution is provided by a combination of the electrostatic analyzer with deflection of ions in a cylindrical magnetic field set up by permanent magnets. The ELS instrument measures 2D distributions of the electron fluxes in the energy range 0.4 eV–20 keV ($\delta E/E = 8\%$) with a field of view of $4^\circ \times 360^\circ$ and a time resolution of ~4 s. In many cases the grid biased at -5 V cuts the low energy ionospheric electrons. A spacecraft potential which is usually positive in solar wind and magnetosheath and negative in a dense ionosphere also strongly influences the measurements in the low energy part of the distribution function. The bulk parameters of plasma were obtained by using algorithms discussed in (Fraenz *et al.*, 2006b).

Figure 1 shows spectrograms of the electron fluxes measured by ASPERA-3 and describing the different domains of the Martian plasma environment. The dotted curves depict the altitude of the spacecraft over the Mars surface. The respective scale in km is given on the right vertical axes. The corresponding MEX orbits in cylindrical coordinates (with the X-axis directed from the Mars center towards the Sun and the radial distance R taken from the X-axis) are shown in Figure 2. In all these cases the spacecraft subsequently crossed the bow shock, magnetosheath, entered the magnetosphere and moving further along the outbound leg of the orbits recorded all these characteristic regions in the opposite order. The nominal positions of the bow shock (BS) and the magnetic pile-up boundary (MPB) (which can also be referred to as the boundary of the induced magnetosphere, MB), determined from Mars Global Surveyor (MGS) measurements (Vignes *et al.*, 2000) are also given. Pile up of the IMF accompanied by a drop of the solar wind electrons was observed at the MPB (Acuña *et al.*, 1998).

The magnetosheath region bounded by the BS and MPB is well displayed in Figure 1 by the appearance of solar wind electrons heated at the bow shock. The cavity void of magnetosheath electrons (the top panel) tells us about the existence of a magnetospheric obstacle to the solar wind. Since Mars has no global intrinsic field the magnetosphere is formed by the pile up of the interplanetary magnetic field (IMF), carried by the solar wind, and a draping of the field lines around the ionospheric obstacle. Indeed, the electron spectra within the Martian magnetosphere contain clear signatures of the ionosphere. The peaks in the electron fluxes near ~20–30 eV appear due to the absorption of the strong solar He II line at 304 Å in the carbon dioxide dominated atmosphere of Mars (Mantas and Hanson, 1979; Frahm *et al.*, 2006a). These peaks can be used for tracing of ionospheric photoelectrons. The interesting feature is that photoelectrons are often observed not only near the

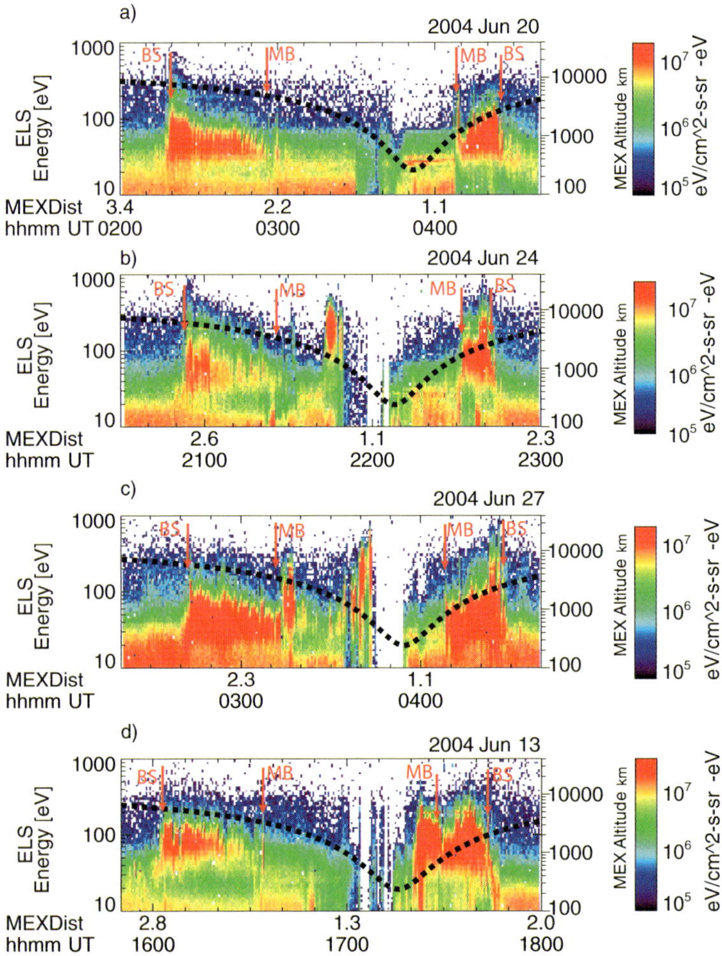

Figure 1. Spectrograms of electron fluxes along the several similar MEX orbits. Dotted curves show the MEX altitude (scale in km is given on the right vertical axis). Positions of the bow shock (BS) and magnetospheric boundary (MB) are marked by arrows.

periapsis, but also near the magnetospheric boundary. For example, ionospheric signatures are seen at an altitude of about ∼900 km, close to the MB on the outbound leg (∼0408 UT June 20, 2004). Moreover, traces of CO_2 photoelectrons are detected at much higher altitudes, up to ∼5000 km (∼0300 UT) close to the inbound MB. The thick blue segments along the MEX orbit in Figure 2 depict the region where the photoelectrons were observed. In most cases a gap (small or large) exists between the MB identified by a drop of the sheath electrons and the photoelectron boundary (PEB). The presence of this gap clearly shows that MB (or MPB) and PEB are indeed two distinct boundaries.

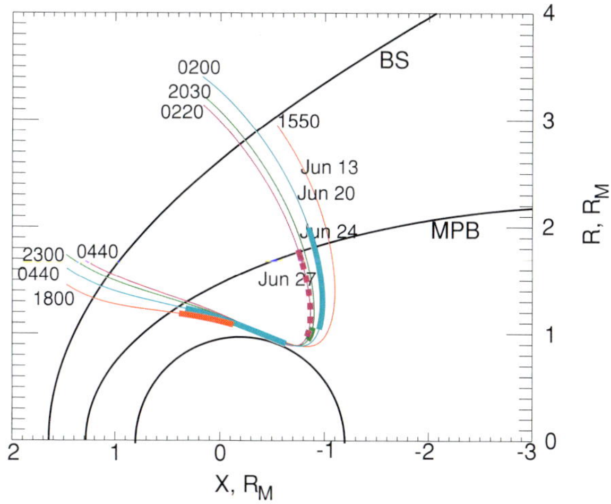

Figure 2. Orbits of MEX in cylindrical coordinates. The spacecraft enters the magnetosphere at $X \sim -1 R_M$ and exits at the dayside. The thick color segments show the location of different plasma domains discussed in the text.

New features in the electron fluxes appear within the magnetosphere on the panel (b) of Figure 1. A spatially narrow plasma structure composed of magnetosheath-like electrons is observed near the wake boundary i.e. the boundary of the geometrical shadow (~2150 UT and the thick green orbital segment in Figure 2). The peak energy of the electrons exceeds their peak-energy at the BS. Plasma in such structures is primarily of planetary origin (O^+ and O_2^+ ions). Different mechanisms were discussed (Dubinin *et al.*, 2006b) to explain the appearance of such structures. One scenario assumes the existence of efficient plasma transport channels into the magnetosphere in magnetic polar regions. In this description the position of the equatorial plane is controlled by the IMF direction, the equatorial plane contains the solar wind velocity and the IMF vector in the undisturbed solar wind. The magnetic field tensions of the draped field lines which become dominant near the MPB (Bertucci *et al.*, 2003) accelerate plasma in the polar regions and push it into the magnetosphere. Such a mechanism suggests a gradual formation of a plasma sheet which separates the two magnetic tail lobes. According to another possible mechanism, reconnection between the crustal and draped IMF field lines can open the inner magnetospheric regions up to solar wind electrons. As a result, magnetic field configurations with "auroral field lines" similar as at Earth, may appear (Lundin *et al.*, 2006).

The narrow structures near the wake boundary stretching in the tailward direction are similar to rays, composed of escaping suprathermal ionospheric O^+ ions, observed at Venus (Brace *et al.*, 1987). Luhmann (1993) suggested that these structures appeared from a thin source region around the terminator where the solar wind convection electric field penetrates into the oxygen-dominated high altitude terminator ionosphere. Dubinin *et al.* (1991) have also observed such structures in the Martian tail. Most of the events were centered near the wake boundary.

On some orbits, an additional appreciable heating of the sheath electrons is observed in the region adjacent to the MB (the panel (c) in Figure 1, ∼0315 UT). The location of this narrow region is marked in Figure 2 by the thick violet segment. Ion composition measurements show that the plasma in such structures consists of planetary O^+ and O_2^+ ions. The top panel in Figure 3 presents the spectrogram of He^{++} and O^+ ions. Alpha-particles are used as tracers of the solar wind plasma while oxygen ions have a planetary origin. Planetary ions occupy a broad boundary layer marked in Figure 2 by the dotted violet segment. A similar, although not so appreciable structure is seen on the panel (b) at ∼2130 UT. The bottom panels in Figure 3 depict the normalized to the solar wind conditions number densities of electrons, protons, atomic (O^+) and molecular (O_2^+) oxygen ions, and electron temperature. Electron heating and a density increase associated with the appearance of planetary ions near the magnetospheric boundary (MB) at 0312 UT is observed. Another feature observed at ∼0340, near the wake boundary, is a ray structure similar to one seen on the panel (b). Note that the cutoff of the low energy ionospheric electrons strongly reduces the measured electron number densities at low altitudes.

A change of the ion composition in the boundary layer/mantle is the characteristic feature of the transition. Similar observations by the Phobos-2 spacecraft have suggested that the magnetospheric boundary at Mars is also the ion composition boundary to emphasize a sharp transition from the solar wind to planetary plasma. As a matter of fact, all these boundaries at a macroscopic scale are collocated (Dubinin *et al.*, 1996; Nagy *et al.*, 2004).

Pioneer-Venus-Orbiter observations made at another nonmagnetized planet, Venus, have shown the existence of a boundary layer with enhanced wave activity (Perez-de-Tejada *et al.*, 1993). Its appearance was attributed to a "friction" action between the shocked solar wind and planetary plasma (Perez-de-Tejada, 1979). According to Perez-de-Tejada (1993) this so-called "intermediate transition" is characterized by a decrease in the magnetic field which is not the case of the MPB/MB. Although the terms "viscosity" and "friction" are not well determined in a collisionless plasma, dissipative processes associated with the transport of the solar wind momentum to the planetary plasma could be responsible for the observed electron heating.

The panel (d) in Figure 1 demonstrates the existence of a boundary layer with an additional heating of magnetosheath electrons on the outbound leg of the orbit (∼1725 UT) when the spacecraft crossed the near terminator MB. The location of the layer is marked in Figure 2 by the red orbital segment. Figure 4 presents the normalized number densities of electrons, protons, atomic (O^+) and molecular (O_2^+) oxygen ions, and the electron temperature. Note here, that the boundary layer (mantle) composed of planetary ions is not always accompanied by appreciable electron heating as for the outbound crossing (1832 UT) (see, for example, the inbound crossing at 1632 UT). The inconsistency between the electron and ion

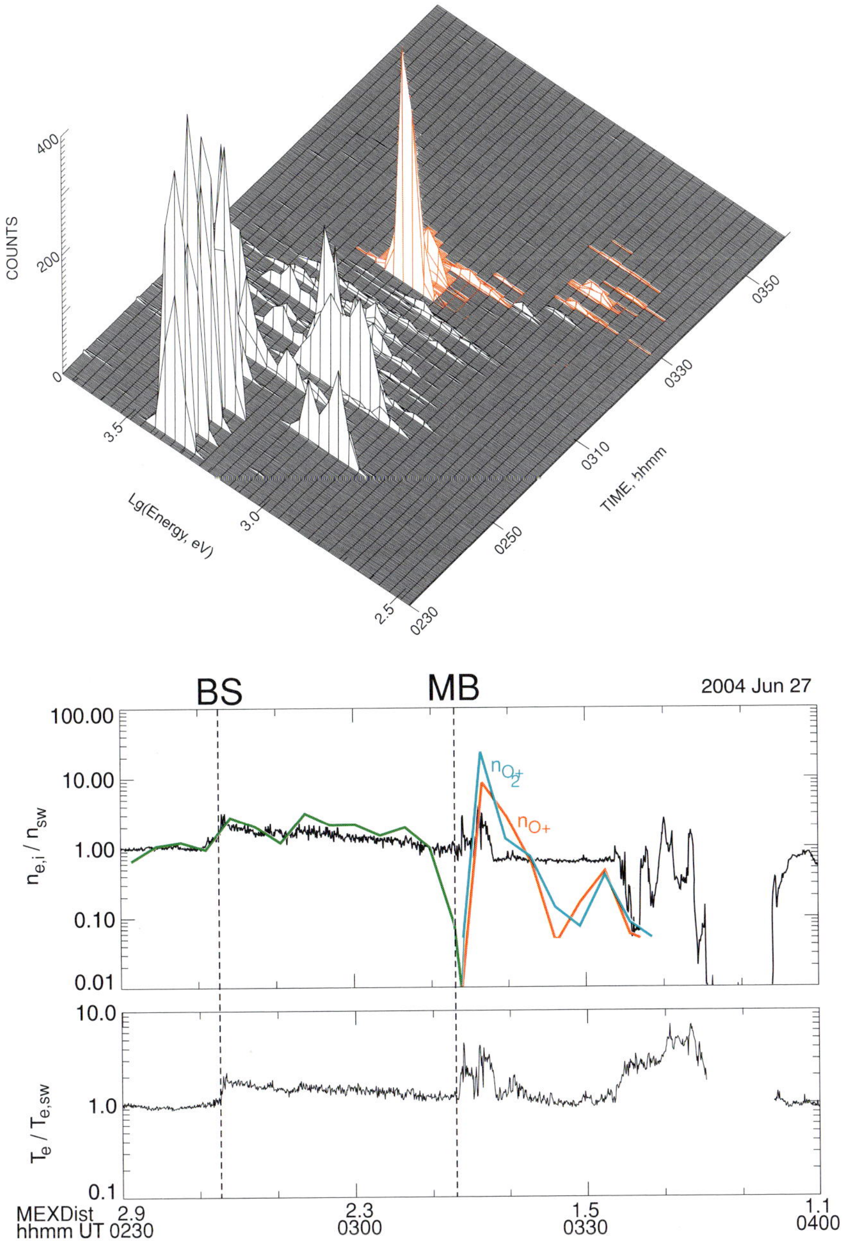

Figure 3. From top to bottom: Spectrograms of He^{++} (the black curves) and O^+ (the red curves) ions along the MEX trajectory on June 27, 2004. The oxygen ions dominate in the boundary layer/mantle adjacent to the MB crossed at ~0312 UT. Variations of the densities of the electrons (the black solid curve), protons (green curve), O^+-ions (red curve), O_2^+ ions (blue curve) and the electron temperature. The parameters for the electrons and proton are normalized to the their upstream solar wind values. Note that the electron measurements are carried out at $E_e > 5$ eV (repelling grid). The spacecraft potential also shifts the measured electron distribution.

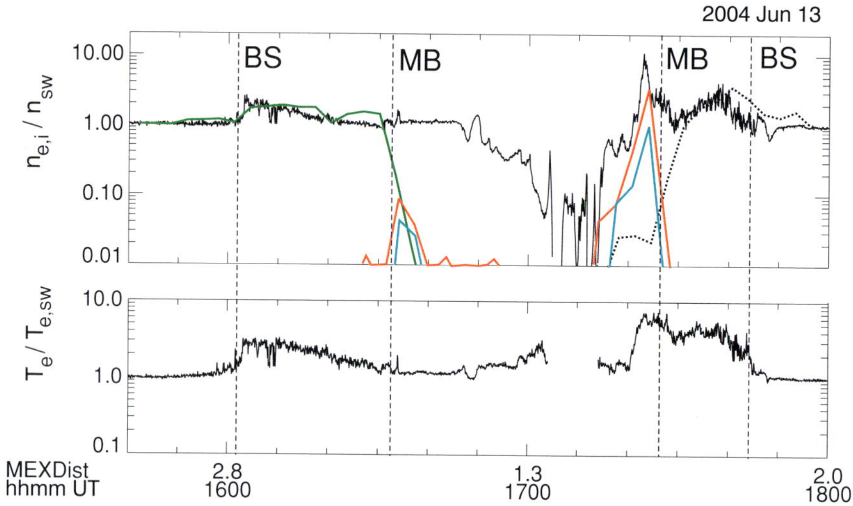

Figure 4. The normalized bulk parameters of the electrons and ions along the orbit on June 13, 2004.

number densities in the inbound magnetosphere (after 1632 UT) is due to the instrumental "gaps" in the measurements of the low-energy parts of the electron and ion distributions.

The above examples display the different characteristic features of the main plasma regions which were used to trace and explore their morphology.

2.1. Magnetospheric Boundary

We have analyzed the position of the magnetospheric boundary characterized by a drop of the magnetosheath electrons using MEX-ASPERA-3 data from February 2004–December 2004. Figure 5 presents the position of the boundary crossings plotted in cylindrical coordinates. Superposed on the data points red and blue curves depict the position of the bow shock and magnetic pile up boundary from Vignes *et al.* (2000) (MGS data) and the bow shock and planetopause (PP) from Trotignon *et al.* (1996) (Phobos-2 data), respectively. Different names of boundaries introduced from single instrument observations, as a matter of fact, correspond to the same and one magnetospheric boundary (Dubinin *et al.*, 1996; Nagy *et al.*, 2004). It is observed that at small solar zenith angles (SZAs) the position of the boundary is closer to the planet and in a better agreement with the PP position derived from the Phobos-2 measurements although the solar activity during these missions was very different (Figure 6). In contrast, at larger SZAs, the positions of the MB and MPB are in a reasonable agreement. The difference between two model curves is not statistically significant since the vast majority of the Phobos-2 crossings of the MB was at the night side, and a lack of the MGS measurements at low SZAs.

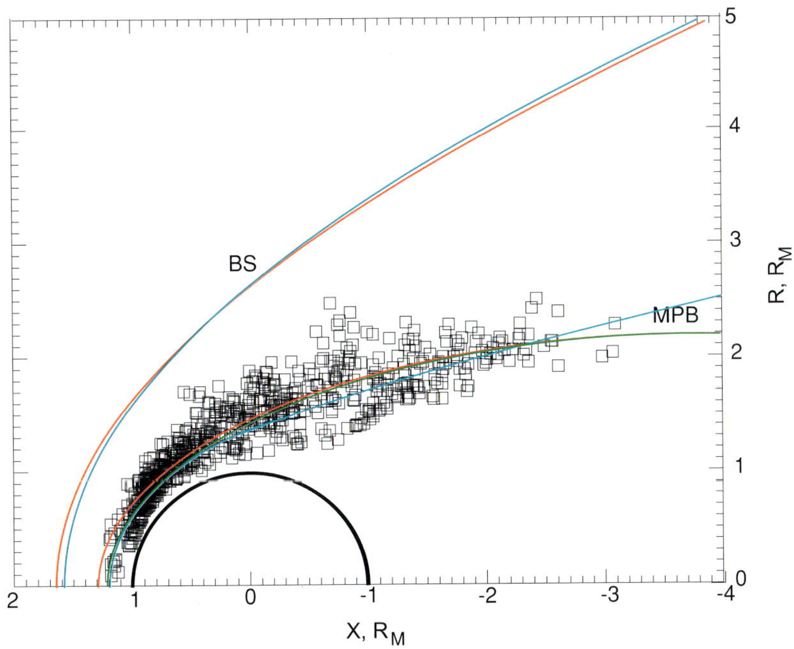

Figure 5. Positions of the magnetospheric boundary in cylindrical coordinates. Red and blue curves depict the nominal positions of the bow shock (BS) and the magnetospheric boundary inferred from the MGS and Phobos-2 observations, respectively. The green curve presents fits to the ASPERA-3 observations on the MEX orbits.

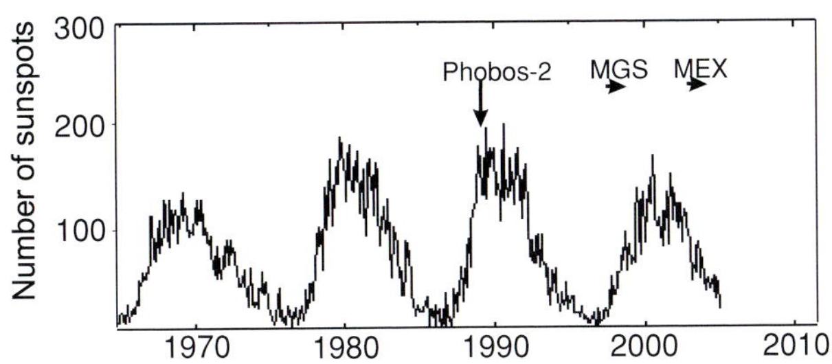

Figure 6. Solar cycle variations during the periods when the PHOBOS-2, MGS and MEX observations discussed in this paper were made.

The used equation of the MPB surface (in assumption of a cylindrical symmetry along the X-axis) in polar coordinates was (Vignes *et al.*, 2000)

$$r = \frac{L}{1 + \epsilon \cos \theta}. \tag{1}$$

Here $L = 0.96 R_M$ and $\epsilon = 0.9$ are the semi-latus tectum and the eccentricity, respectively. Polar coordinates (r, θ) are measured about the focus located at the point

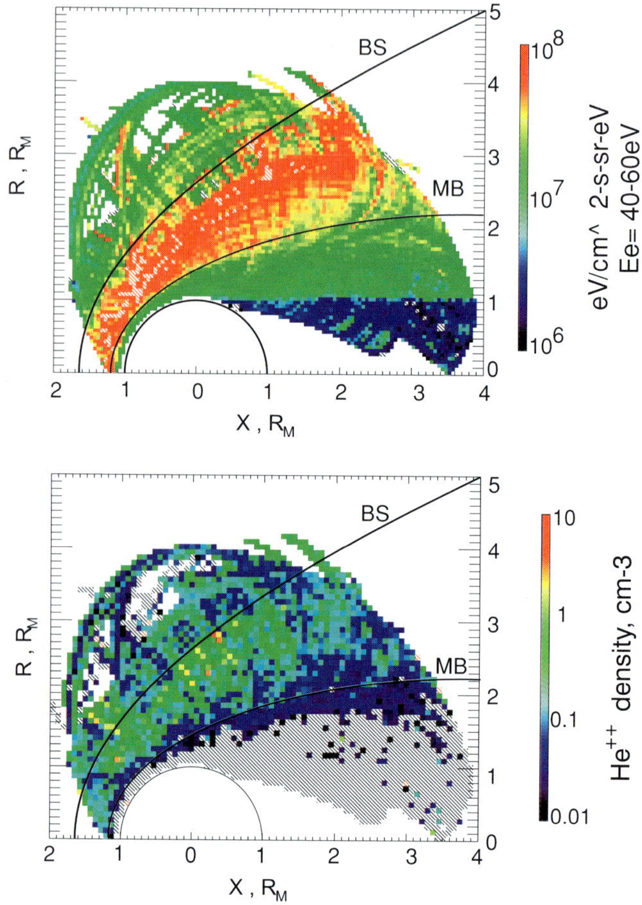

Figure 7. Images of fluxes of electrons ($E_e = 40$–$60\,\mathrm{eV}$) and density of He^{++} ions in cylindrical coordinates plotted on the total of the MEX orbits during 2004–2005. Positions of the MB (the MEX data) and bow shock (BS) are also given.

($x_0 = 0.78, 0, 0$). A better agreement with the MEX-ASPERA-3 observations, in particularly, at small solar zenith angles can be obtained by using the same values for L and ϵ, but moving the focus to $x_0 = 0.7$ (the green curve in Figure 5).

Figure 5 also shows that a scatter of the data points with respect to the nominal boundary position, increases with the solar zenith angle.

The boundary determined from a drop of the magnetosheath electrons coincides with a boundary of a "stoppage" of the solar wind. Figure 7 compares the median distributions of fluxes of the $E_e = 40$–$60\,\mathrm{eV}$ electrons and the number densities of He^{++} ions. The data set contains the measurements carried out by ASPERA-3 over two years (2004–2005). The magnetosphere almost void of solar wind particles can well be seen. Since the magnetic pile up boundary is also characterized by a drop of the magnetosheath electrons, MPB is the magnetospheric, obstacle boundary which determines the position of bow shock and plasma flow around Mars. We used here a term MB for definition of the magnetospheric boundary , because of the lack of the magnetic field measurements on MEX.

The existence of an extended magnetospheric cavity for median conditions does not imply that solar wind can not penetrate to closer altitudes above the planet. Magnetospheric "images" plotted for maximum values of fluxes and densities in each bin reveal a significant contraction of the magnetosphere (not shown here) for extreme conditions in the solar wind. Among the main factors which are expected to account for the observed variations of the boundary position are the solar wind dynamic pressure, local crustal magnetic field sources and orientation of the interplanetary electric field $-\boldsymbol{V}_{\mathrm{sw}} \times \boldsymbol{B}_{\mathrm{IMF}}$.

2.1.1. *Solar Wind Dynamic Pressure Dependence*

In this paper we use a MGS proxy for the solar wind RAM pressure monitoring. It is assumed that the solar wind dynamic pressure is balanced at the induced magnetospheric boundary (MPB) by the magnetic field pressure of the draped IMF tubes. The pileup of the magnetic field and formation of the induced magnetic barrier occurs over a short distance, that accounts for a sudden drop of the solar wind electron and proton fluxes. The magnetic field value remains approximately constant for several hundred km in the magnetic pile up region (MPR) (Crider *et al.*, 2003). On mapping orbits, the MGS spacecraft moves along a circular 0200-LT/1400-LT polar trajectory at the altitude of ~400 km, crossing the MPR in the northern hemisphere. Since the magnetic field at middle latitudes of the northern hemisphere is primarily of induced origin, we can use its value as a proxy for the magnetic field pressure which stops the solar wind, and readily infer a proxy value for the solar wind dynamic pressure (Spreiter and Stahara, 1992)

$$kP_{\mathrm{dyn}} \cos^2\theta = \frac{B^2}{2\mu_o}, \qquad (2)$$

where $k \sim 0.88$ and θ is the solar zenith angle and the magnetic field B is measured on each MGS orbit on the dayside at the reference point $\theta \sim 45^\circ$. This proxy solar wind dynamic pressure P_{dyn} is adjusted to the times of the magnetospheric boundary crossings. It is worth noting that Vennerstrom *et al.* (2003) and Crider *et al.* (2003) have also successfully used the MGS data as a proxy for solar wind pressure. Brain *et al.* (2005) have shown that MGS was outside of the MPR, in the magnetosheath as much as 20–25% of the time during mapping orbits. For these orbits, the inferred RAM pressure is likely even higher than predicted by this method.

Figure 8 compares variations of the inferred solar wind dynamic pressure and the ratio $r_{\mathrm{obs}}/r_{\mathrm{ave}}$ which characterizes the difference in the measured and averaged boundary positions. Here r_{obs} is the length of the radius-vector between the focus point $(x_o, 0, 0)$ and the observation point of the MB, and r_{ave} is the distance from

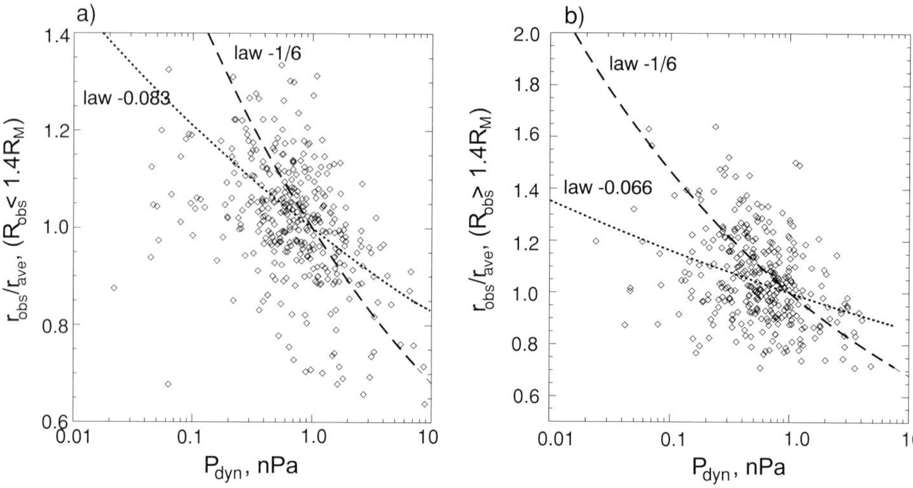

Figure 8. Variations in the MB positions as a function of the solar wind RAM pressure. The data are separated on two groups, $R < 1.4R_M$ (a) and $R > 1.4R_M$ (b), where R is a radial distance from the X-axis to the MB crossings. Dashed curves are the power law $P_{sw}^{-1/6}$ dependence. Dotted curves are the power law fits.

the focus to the crossing point of the average boundary surface and the vector $\boldsymbol{r}_{obs}$. The MEX data are separated on two groups of $R_{obs} > 1.4R_M$ and $R_{obs} < 1.4R_M$, where R_{obs} is the radial distance from the X-axis to the observation point. The small $R_{obs} < 1.4R_M$ group corresponds to solar zenith angles less than 60–70°. It is observed that the response of the boundary position to the RAM pressure is better visible at smaller zenith angles. If the MB is asymmetrically shaped as suggested by Crider *et al.* (2004) and Brain *et al.* (2005), then at high SZA there should be larger scatter about the mean position of the boundary – making difficult to see the effects of pressure.

The dashed curves in Figure 8 show a power law ($P_{dyn}^{-1/6}$) dependence. Verigin *et al.* (1993) have shown that the diameter of the Martian tail D is proportional to $P_{dyn}^{-1/6}$ what is expected if Mars would have an intrinsic magnetosphere. A similar dependence was noted by Dubinin *et al.* (1996) although the authors have argued in favor of an induced magnetosphere. For the small R_{obs} group a power law fit is given by $r_{obs}/r_{ave} \sim P_{dyn}^{-0.053}$ that is in a good agreement with the MGS data, $k = -0.0546$ (Crider *et al.*, 2003). If we exclude the data points for small values of the RAM pressure ($P_{dyn} > 0.133$ nPa) then the power law index $k \sim -0.083$ (the dotted curve in Figure 8a). For the large R_{obs} group, the index $k = -0.065$ (the dotted curve in Figure 8b).

Thus the MEX data as well as the MGS observations show a weaker dependence between the RAM pressure and variations in the MB location than it is expected for a magnetic dipole obstacle. Nevertheless a power law dependence is still revealed.

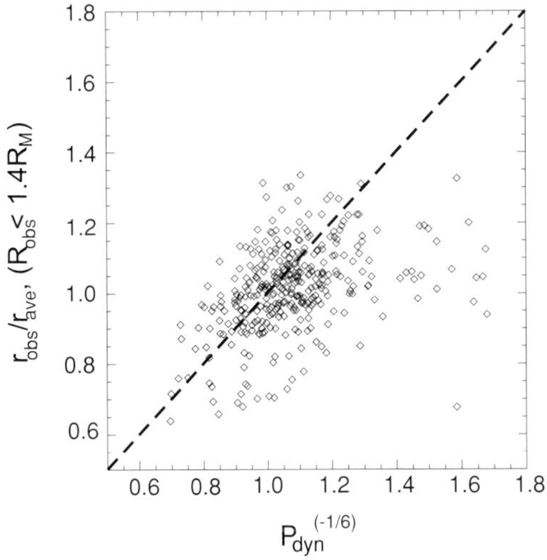

Figure 9. Variations in the MB position as a function of $P_{sw}^{-1/6}$.

Such dependence becomes weaker and ceases for small P_{dyn} that is better seen in Figure 9 which depicts the r_{obs}/r_{ave} as a function of $P_{dyn}^{-1/6}$.

It is worth noting that although an induced origin of the obstacle to solar wind at Mars is well established now, a question whether or not a power law dependence exists, remains important and is closely related to a question what makes an induced magnetosphere. An induced magnetosphere can be created by induction currents flowing in a conductive ionosphere or within the bodies (e.g. in molten cores) (see, for example, Luhmann *et al.*, 2004). Two types of induction mechanisms are usually considered, an unipolar induction where the current is driven by the $-\boldsymbol{V}_{sw} \times \boldsymbol{B}_{IMF}$ electric field or a classical electromagnetic induction associated with temporal variations (in direction or value) of the external magnetic field. Here, in a case of a unipolar induction, we do not separate unipolar currents flowing in a conducting body from currents flowing in a mass-loaded plasma (in both cases, currents are driven by the motional electric field). It has been shown that both types of induction may contribute to induced magnetic fields (Podgorny *et al.*, 1982). Temporal variations of the IMF induce a dipole magnetic field due to the currents in a conducting ionosphere (or/and interior), and a power-law dependence with index $k = -1/6$ of the boundary position as a function of solar wind dynamic pressure seems not to be unreasonable. Indeed, Brecht (1995) have observed a such dependence of the magnetotail width on the RAM pressure in hybrid simulations of the solar wind interaction with a "conducting" body. On the other hand, unipolar currents which bound the draped IMF induce a weakly dependent on a distance magnetic field (similar as the magnetic field within a solenoid). The observations of a weak power-law dependence show that both mechanisms probably contribute to the induced magnetic field at Mars.

While comparing the Phobos-2 and MGS, MEX observations it is also necessary to recall that solar wind pressure in the Phobos-2 data has been measured in-situ. On the other hand, the sampling was poorer.

2.1.2. *Interplanetary Electric Field Dependence*

For the study of the solar wind interaction with planets like Mars or Venus having draped magnetospheric configurations, the IMF reference frame is the most natural one. This coordinate system has the X^*-axis antiparallel with the upstream solar wind flow and Y^*-axis along the cross-flow magnetic field component of the IMF. Then the motional electric field $-\boldsymbol{V}_{\text{sw}} \times \boldsymbol{B}_{\text{IMF}}$ is always along the Z^*-axis. Since there is no magnetometer on the MEX spacecraft the only way to infer an information about the IMF is the MGS observations in the MPR. IMF directions have been previously derived from MGS data by Crider *et al.* (2001) for aerobraking data and by Brain *et al.* (2006) for mapping orbits. Assuming that the clock-angle of the IMF is not changed while the field lines are draped around Mars we can infer a proxy direction of the cross-flow magnetic field component and construct the IMF coordinate system. We used the same reference point in the dayside northern hemisphere as for the determination of a proxy RAM pressure. As a matter of fact, the IMF system is inadequate to observe simultaneously in two dimensions a possible "north-south" asymmetry due to the motional electric field and a "dawn-dusk" draping asymmetry, if different B_x polarities of the IMF for the same sector polarities are analyzed. Moore *et al.* (1990) have used a combination of rotations and foldings (see also Dubinin *et al.*, 1996). However, in our case, the lack of information about the X-component of the IMF does not allow to apply such foldings.

Normalizing a boundary position to average solar wind conditions ($P_{\text{dyn}} =$ 1 nPa) by using the power law fit dependence we can test a possible asymmetry of the magnetosphere in the IMF coordinate plane. Figure 10 shows $r_{\text{obs}}/r_{\text{ave}}$ in the plane Y^*Z^*. We observe only a certain elongation of the magnetospheric shape in the "north-dawn" direction for $R_{\text{obs}} > 1.4R_M$ probably caused by two factors: (i) a preferential pile up of the IMF in the "northern" hemisphere and (ii) a "dawn-dusk" asymmetry of the draping due to X-component of the IMF. It will be shown subsequently that a similar trend is observed in the distribution of CO_2 photoelectrons. It is worth noting that draping directions in the subsolar region and in the reference point at the middle latitudes of the northern hemisphere which was used to infer the IMF direction may be somewhat different due to 'weathervaning' effects (see e.g. Brain *et al.*, 2006). Then the overall pattern must be rotated clockwise at $\sim$30–40° and a 'north-south' asymmetry related to the motional electric field will be better noticeable.

Observations near Venus have shown that the piled up magnetic field is stronger in the Z^*-hemisphere into which the motional electric field is pointing (Luhmann *et al.*, 1985). A similar effect is found at Mars (Vennerstrom *et al.*, 2003) as well as in 3-D hybrid simulations of the solar wind interaction with Mars (Bößwetter *et al.*, 2004; Modolo *et al.*, 2005). Therefore it is might be expected that the position of the

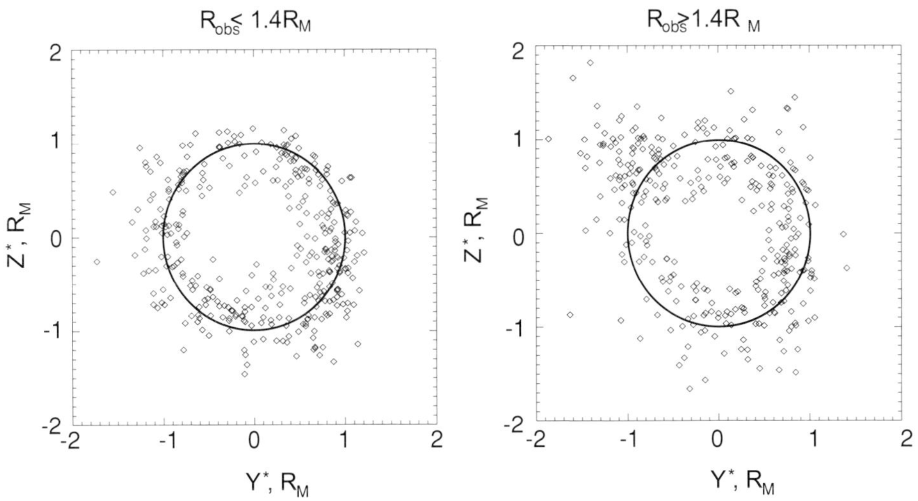

Figure 10. Variations in the MB position in the Y^*Z^*-plane, where the Y^*-axis is along the cross-flow component of the IMF, and Z^*-axis is along the motional electric field in the solar wind.

magnetospheric boundary is further from the planet in the $+Z^*$-hemisphere where mass-loading effects could be more essential (Dubinin *et al.*, 1998). On the other hand, effects of a finite proton Larmour radius can lead to an opposite asymmetry (Brecht, 1997; Bößwetter *et al.*, 2004). Further observations are necessary for better understanding of different controlling factors which interfere the general pattern of the Martian magnetosphere.

2.1.3. *Crustal Field Dependence*

The crustal magnetic fields can also influence the position of the magnetospheric boundary as the magnetic pressure in some localized regions may be high enough to balance the solar wind dynamic pressure. Crider *et al.* (2002) have found that the MPB distance increases with increasing southern latitude. Using the electron measurements by ASPERA-3-ELS, Fraenz *et al.* (2006a) have shown that the altitude of the intruded magnetosheath electrons ($E_e = 80$–100 eV) increases with the strength of the crustal field. Figure 11a shows a relative shift of the boundary in the dayside southern hemisphere with respect to its averaged position (r_{obs}/r_{ave}) as a function of the strength of the crustal magnetic field. We used the crustal field strength interpolated on a regular grid for an altitude of 400 km from the MGS MAG/ER observations as presented by Connerney *et al.* (2001). Although the sampling of measurements above the strong crustal sources is small an upward motion of the boundary with increasing magnetic field strength is clearly observed. There is a reasonable agreement with the picture of the intrusion of magnetosheath electrons as a function of crustal field strength (Figure 11b).

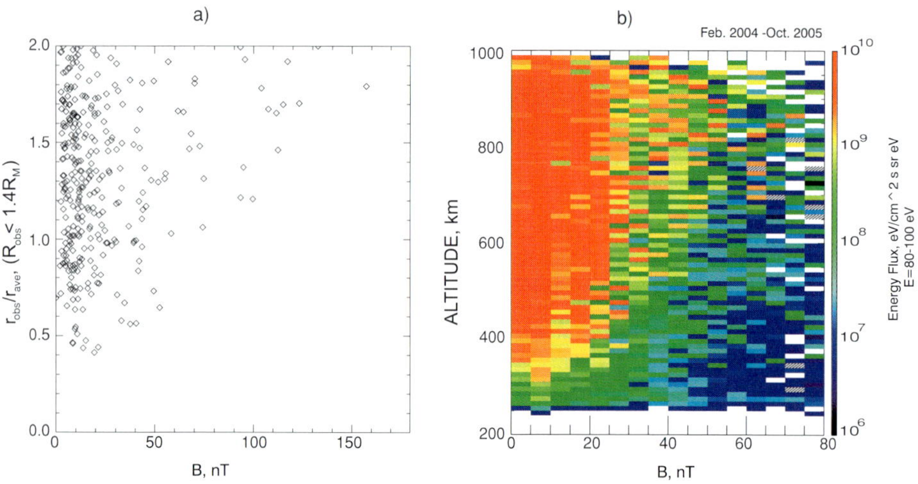

Figure 11. (a) Variations in the MB position in the southern dayside hemisphere as a function of the strength of the crustal magnetic field at 400 km. (b) Maximum fluxes of the electrons with $E_e = 80$–100 eV observed at different altitudes during the MEX observations (Feb 2004–Oct 2005) on the dayside as a function of the magnetic field strength of the crustal sources at altitude of 400 km.

2.2. Ionospheric Photoelectrons

The ionospheric electrons are well traced by the peaks in the energy spectra of the electrons in the range of 20–30 eV. Observations of such electrons can be used to probe the Martian ionosphere. Figure 12 shows the distribution of the energy flux of CO_2 – photoelectrons in the energy range ($\delta E = 4$ eV) centered near its characteristic "spectral lines" (20–30 eV) in cylindrical coordinates. Floating of these spectral peaks due to spacecraft potential variations was taken into account. The ionospheric electrons are observed at altitudes up to ~7000 km. Statistics of their occurrence at different altitudes is discussed in more detail by Frahm *et al.* (2006b). Another interesting feature is that the photoelectrons are often detected close to the nominal magnetospheric boundary almost filling the whole dayside magnetosphere. In many cases the photoelectrons are also observed close to the distant positions of the magnetospheric boundary. These features probably imply an important role of the ionospheric plasma as an obstacle to solar wind.

Figure 13 shows the radial distance of the MB crossings versus the highest radial distance at which ASPERA-3 records the photoelectrons. It is observed that a gap between the MB and PEB can be rather small even at large distances from Mars. Unsolved yet is the question, does a drop of photoelectrons (PEB) near the magnetospheric boundary (MB/MPB) correspond to the ionopause (if we speak in terms of pressure balance)?

According to the MGS aerobraking observations (Mitchell *et al.*, 2000), at solar zenith angles (SZAs) ~80°, the transition from the region occupied by the shocked

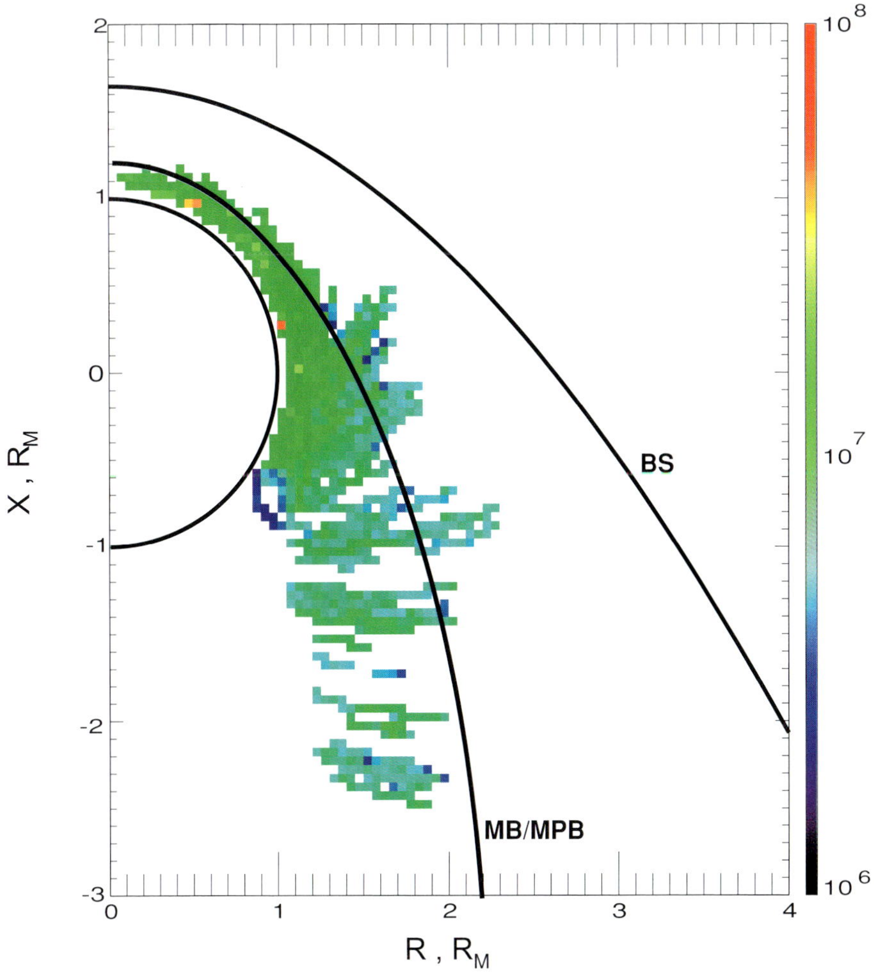

Figure 12. Regions in cylindrical coordinates $R - X$ where CO_2-photoelectrons were observed. The color shows the energy flux of the photoelectrons.

solar wind electrons to the ionosphere characterized by the appearance of Auger electrons (~500) eV and poorly resolved photoionization peaks at 20–50 eV occurs in the altitude range 180–800 km with a median value of 380 km. The electron spectrometer (ELS) of the ASPERA-3 experiment due to a higher energy resolution was able to identify the boundary of photoelectrons with a better accuracy as a position where fluxes of CO_2-photoelectrons cease. It is shown that a drop of the magnetosheath electrons ($E_e = 100$ eV) on the dayside approximately coincides with MB and there is a clear gap between MB and PEB.

To date reliable ionospheric profiles near the MB/MPB are absent. Recent MARSIS ionospheric soundings performed on MEX have shown that the

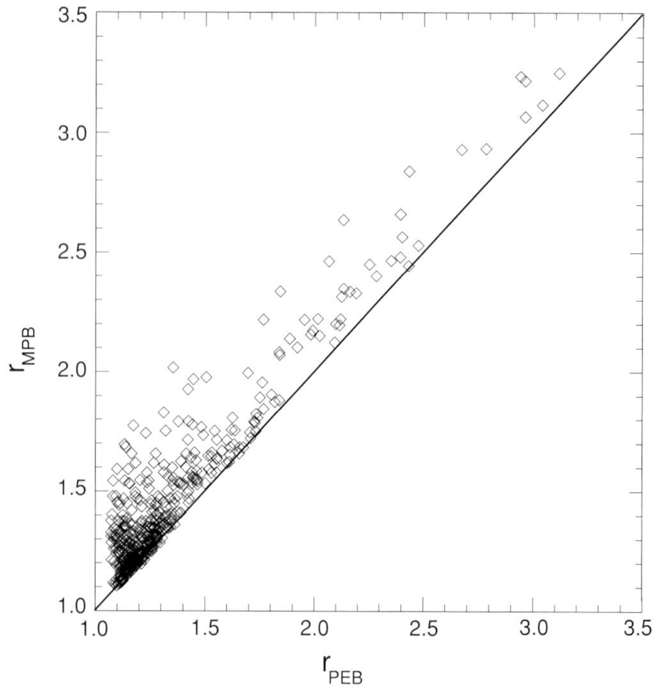

Figure 13. Positions of the MB and PEB (the radial distances from Mars) for the orbits at which both boundaries were observed.

ionospheric number density at altitude of ~400 km near the terminator is about of $3 \times 10^3\ \mathrm{cm}^{-3}$ (Gurnett *et al.*, 2005). This implies a possible essential ionospheric contribution to the pressure balance at altitudes of the magnetospheric boundary. However, it is unlikely that the ionospheric pressure at PEB altitudes is able to stop the solar wind. We may assume that some part of the momentum of the solar wind can be transferred to the ionosphere via the magnetic field stresses driving the ionospheric plasma into the bulk motion. This motion can explain the observations of ionospheric photoelectrons far in the tail. The photoelectrons can also lift up along the magnetic field lines and, particularly, along the reconnected crustal field lines which are stretched into the tail ("polar wind" at Mars). The existence of field-aligned fluxes of photoelectrons in the $+Z^*$ hemisphere at the negative Y^* values can be tentatively observed in Figure 14 which presents the fluxes of CO_2-photoelectrons in the IMF coordinate system. However, since photoelectron fluxes are mainly contained within the magnetospheric cavity a bulk transport is likely a dominant process. A small bulge in the $(-Y^* + Z^*)$ – hemisphere is similar to a bulge in the position of the magnetospheric boundary (Figure 10) implying a contribution of the ionospheric plasma to the formation of the obstacle. The observed "dawn-dusk" asymmetry can be caused by different tension forces of the draped field lines due to the presence of the X-component of the IMF. Since the motion of

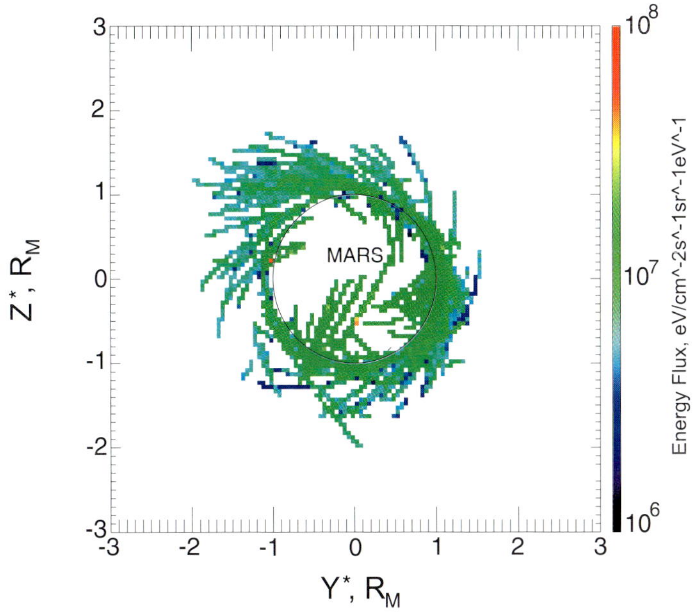

Figure 14. Regions in the IMF Y^*Z^* plane where CO_2-photoelectrons were observed. The same dataset as in Figure 13 is presented.

low-energy ionospheric plasma is not quantified yet it is difficult to estimate escape fluxes of oxygen from the topside ionosphere.

2.3. Ray Structure Near the Wake Boundary

The ASPERA-3 experiment has often observed a spatially narrow structure composed of hot sheath-like electrons and planetary ions near the wake boundary (see the second and third panels in Figure 1 and Dubinin *et al.*, 2006b). The structure appears near the terminator plane and is stretched, like a ray into the tail. Figure 15a shows in $R - X$ coordinates locations of the events observed in 2004. Figure 15b gives the image of electron fluxes in the energy range of 80–100 eV along the orbits on which ray-electron structures were observed. Such rays are important erosion channels through which planetary ions are transported to the tail. That can be readily inferred from Figure 15c which shows density fluxes of oxygen ions along the same set of MEX orbits. It was suggested (Dubinin *et al.*, 2006b) that draped field lines slipping along the magnetospheric surface near the MPB, around the "magnetic poles" can push planetary ions into the magnetosphere. This mechanism also explains the formation of the plasma sheet which separates two magnetic field lobes in the induced tail. Recent hybrid simulations (Bößwetter *et al.*, 2004; Modolo *et al.*, 2005) have shown a distinct asymmetry in the strength of the field

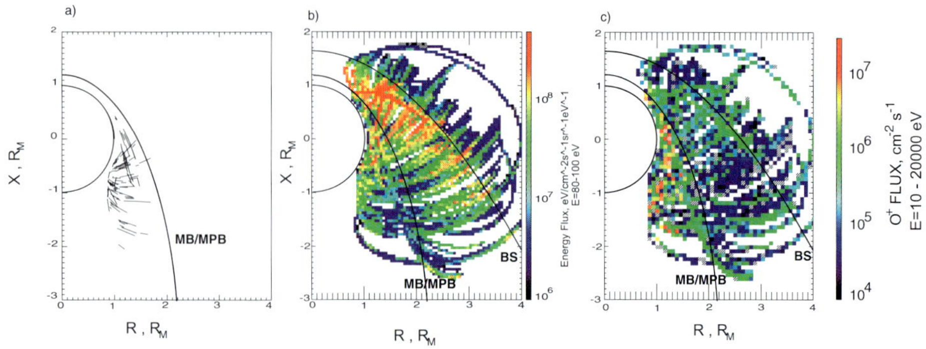

Figure 15. (a) Orbital segments in cylindrical coordinates at which electron signatures of the ray-like structures near the wake boundary were observed. (b) The fluxes of electrons with $E_e = 80$–$100\,\text{eV}$ measured on the same set of orbits on which ray structures were detected. (c) Fluxes of oxygen ions on these orbits. The positions of the bow shock (BS) and magnetospheric boundary (MB) respectively inferred from the MGS and MEX observations are also shown.

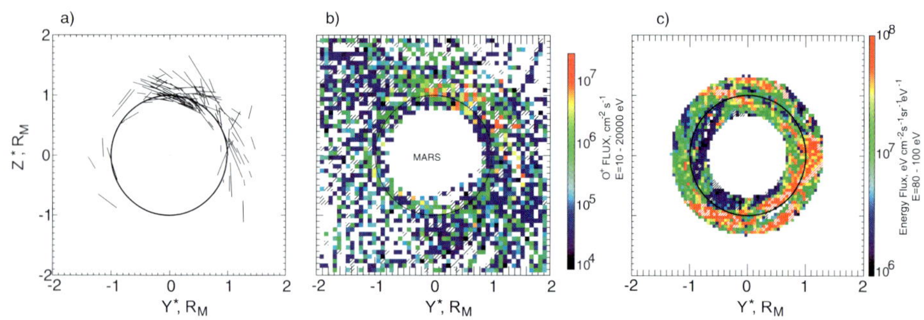

Figure 16. (a) Orbital segments in the Y^*Z^*-plane of the IMF coordinate system at which the electron signatures of ray-like structures near the wake boundary were observed. (b) the fluxes of oxygen ions along these MEX trajectories transformed into the IMF reference frame. (c) Maximum fluxes of 80–100 eV electrons in the bins of the ring-area 0.7–$1.3R_M$ around Mars for two year observations.

at the MPB. The maximum intensity of the draped magnetic field is observed in the hemisphere into which the motional electric field is pointing (the "northern" hemisphere in the IMF coordinate system). Therefore, if this mechanism works, one would expect a preferential observation of ray structures in the $+Z^*$ hemisphere near the pole. Figure 16a depicts the locations of the orbital segments along which ray-events were observed in the IMF Y^*Z^*-plane. It is seen that most of the events are clustered near the "northern magnetic pole." There are also events near the "magnetic equator" which could be the counterparts of stretched ray-like structures in the "magnetic equatorial plane" observed in 3D-hybrid simulations (Bößwetter *et al.*, 2004; Modolo *et al.*, 2005). A force which pushes planetary ions along the field lines is probably a day-night thermal pressure gradient. The asymmetry of ray structures is also revealed on the right panel in Figure 16b which shows the fluxes of oxygen ions along the orbits in which the ray features were observed in the electron data.

Another mechanism which associates the events with auroral inverted "V" structures suggests their appearance in the southern hemisphere where the shear flows at the boundary of open, draped IMF field lines and closed field lines from crustal sources can generate field-aligned currents and the parallel electric fields (Lundin *et al.*, 2006). Figure 16c depicts the maximum fluxes of the 80–100 eV electrons in the ring-area within 0.7–1.3R_M of thc Mars-Sun line at $X < 0$ during two years. The fluxes near wake boundary dominate in the southern hemisphere. Thus both mechanisms probably contribute to the occurrence of ray-like structures.

2.4. Boundary Layer and Plasma Sheet

Another important reservoir of planetary ions is the boundary layer. The existence of the boundary layer/mantle in the Martian magnetosphere has been shown during the first Soviet space missions to Mars (Vaisberg, 1992) as well as in the Phobos-2 observations (Lundin *et al.*, 1990a; Breus *et al.*, 1991; Dubinin *et al.*, 1996). Moreover, it was assumed that the boundary layer is a main channel for the escape of planetary ions (Lundin *et al.*, 1990b). Figure 17 (left panel) shows in the $R - X$ plane the orbital segments near the MB along which planetary ions were detected. The right panel depicts the values of oxygen ion fluxes measured during these intervals. The main fluxes are observed within the magnetosphere although on some orbits remarkable fluxes of planetary ions were also recorded in the

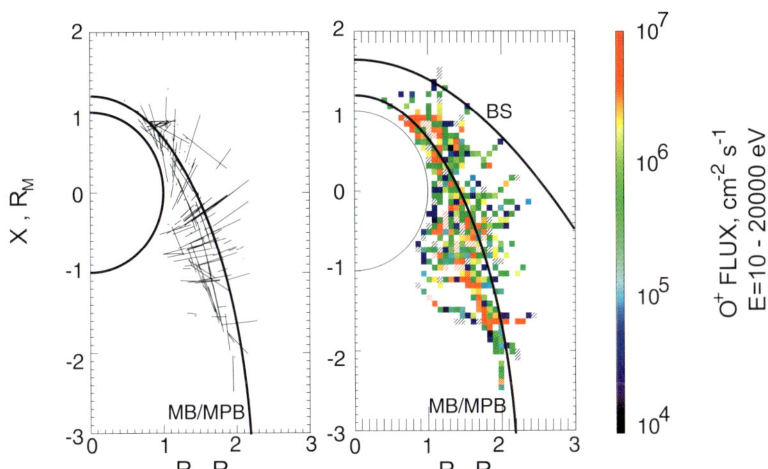

Figure 17. (a) Orbital segments in the cylindrical coordinates at which the fluxes of planetary ions were detected in the boundary layer/mantle. The right panel shows the values of oxygen fluxes measured during these intervals. The positions of the bow shock (BS) and magnetospheric boundary (MB) respectively inferred from the MGS and MEX observations are also shown.

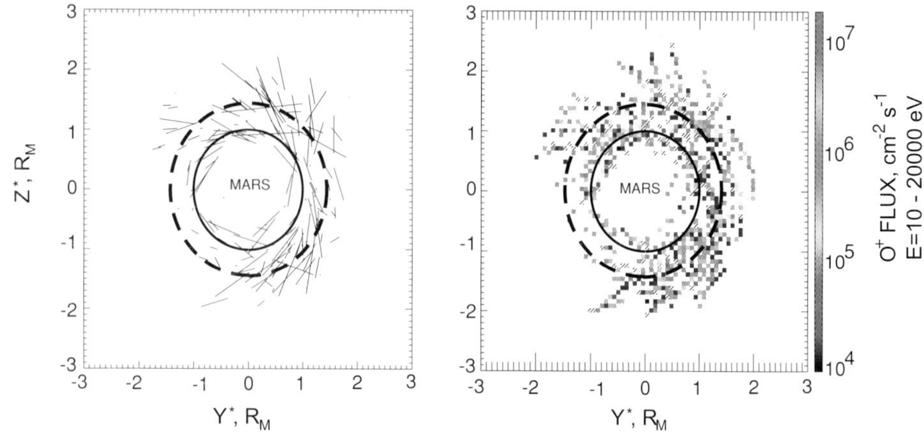

Figure 18. Orbital segments in the Y^*Z^* -plane of the IMF coordinate system at which the signatures of the boundary layer were found. The right panel depicts the fluxes of oxygen ions along these orbital intervals. Dashed circle depicts the nominal location of the MB in the terminator plane.

adjacent magnetosheath. The values of fluxes in the boundary layer often exceed 10^7 cm^{-2} s^{-1}.

The geometry of the outflowing plasma is very important for calculations of the total escape rate of planetary matter. Analyzing the ASPERA data on Phobos-2 Lundin *et al.* (1989, 1990b) have suggested that a primary solar wind induced escape with a total rate of about 2.5×10^{25} s^{-1} occurs through a cylindrically symmetric boundary layer. Verigin *et al.* (1991) have made the assumption that the main channel for the loss of planetary ions is the plasma sheet. Correspondingly, the estimated total outflow rate in this case is significantly less ($\sim 5 \times 10^{24}$ s^{-1}). Figure 18 presents the data set of the observations made in the boundary layer with ASPERA-3 on MEX in the IMF coordinate system. A strong "dawn-dusk" asymmetry is probably related with the different draping features due the X-component of the IMF. If we assume that planetary oxygen ions emanate from an asymmetric ring-shaped area $0.8 R_M$ in thickness around the terminator and typical fluxes of ions are of the order of $\sim 10^6$–10^7 cm^{-2} s^{-1}, the total escape rate would be about 6×10^{23}–6×10^{24} s^{-1}. These estimates rather correspond to the maximum escape fluxes since the boundary layer was observed only in ~20–25% of the orbits. The absence of the boundary layer in ~80% of cases implies that there are probably other, unknown yet factors, than the geometry of the IMF, which control the escape processes. Recall here, that the MEX measurements were carried out close to solar minimum conditions while the Phobos-2 spacecraft has operated near Mars at solar maximum when the oxygen exosphere was expected to be denser.

It was observed (see Section 1) that on some orbits the boundary layer is characterized by a sudden additional heating of magnetosheath electrons. Spectra of electrons in these cases become similar to the spectra observed in ray-structures

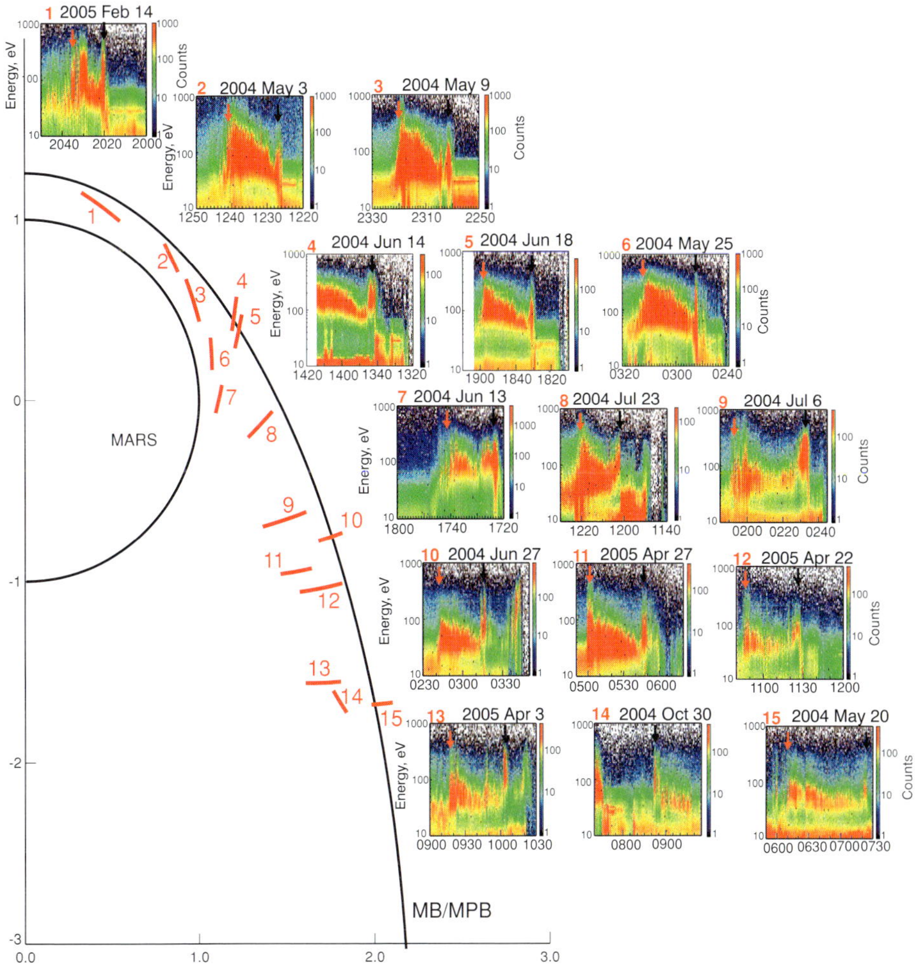

Figure 19. Sites near the inner boundary of the magnetosheath where the sheath electrons inhibit an additional heating. The spectrograms of electron fluxes which display these events are shown on the small right panels. Red and black arrows show the positions of the bow shock and boundary events (BE), respectively.

or in the plasma sheet. The ion composition is dominated by O^+ and O_2^+ ions. A change of ion composition of the plasma within these structures implies that the observed spikes of heated electrons at the inner edge of the sheath are not related to temporal variations in the magnetosheath caused by the passage of different types of inhomogeneities and discontinuities in the solar wind, but that they are an inherent boundary layer feature. Figure 19 shows the position of sample events in cylindrical coordinates. The corresponding spectrograms of electron fluxes with clear spikes of electron heating near the MB are also shown. The inner part of the magnetosphere is readily recognized by the absence of magnetosheath-like electrons. The positions of the bow shock (BS) and the boundary events (BE) are also marked by red and black arrows, respectively. In IMF coordinates the BEs appear in the $+Z^*$-hemisphere. More analysis is required to understand the origin of these events.

The magnetosphere structure within the optical shadow of Mars ($R < 1R_M$) is still poorly covered by the ASPERA-3 measurements. The observations of the plasma sheet carried out in 2004 yield a similar morphological pattern as for the ray-structures (see Figures 15 and 16) which may imply that they have a common root. The values of oxygen fluxes in the plasma sheet are somewhat higher than in the boundary layer and often exceed 10^7 cm^{-2} s^{-1}. Fedorov *et al.* (2006) have also distinguish two different escape channels for planetary ions, a layer adjacent to the MB/MPB and the planetary shadow. Authors showed that mechanisms of ion acceleration in the boundary layer and wake can be different and controlled by the IMF direction.

3. Summary

We explored the morphology of the main plasma regions and their boundaries by analyzing MEX ASPERA-3 data collected in 2004.

1. It is shown that a magnetospheric cavity strongly depleted in solar wind particles is formed. The position of its boundary determined by a drop of fluxes of ~50 eV magnetosheath electrons coincides with a boundary determined by a drop of solar wind ions. This implies that the magnetospheric boundary is collocated with the MPB which is also characterized by a drop of the magnetosheath electrons.
2. We have analyzed the position of the magnetospheric boundary and compared it with Phobos-2 and MGS observations. Good agreement with Phobos-2 observations at small solar zenith angles and with MGS data for larger angles is observed. A general reasonable agreement in the MB position observed at different phases of solar activity implies that it is not sensitive to this parameter. A similar conclusion was made by Vignes *et al.* (2000) while comparing the Phobos-2 and MGS data.
3. Variations in the MB location increase with increasing SZA.
4. We have analyzed the dependence of MB locations on solar wind dynamic pressure. We used a MGS proxy for solar wind RAM pressure assuming that the RAM pressure is balanced at the MPB by the magnetic field pressure. It is generally observed that variations of the MB position are in a reasonable agreement with a magnetic origin of the obstacle to the solar wind (an obstacle formed by a barrier of the piled up IMF field lines). It is shown that a response of the MB to the RAM pressure is revealed more clearly at SZA $\leq 60^\circ$–70°. The K–H instability of shear flows near the MB may result in large inward-outward

motions of the MB at larger zenith angles providing a significant "scattering" in the MB locations.

5. The ASPERA-3 data show a weaker power law dependence between the RAM pressure and variations in the MB location than can be expected for the magnetosphere created only by currents of the electromagnetic induction.
6. In the IMF coordinate system, determined by the cross-flow component of the IMF, a "north-south" asymmetry in the MB location caused by mass loading effect in the electric field pointing hemisphere is only revealed if a weathervaning of the draped field lines is taken into account while inferring the IMF direction.
7. Although the sampling of MB measurements above strong crustal source is poor, an upward lift of the MB is observed. This trend is also confirmed by an altitude-crustal field dependence of protrusion of magnetosheath electrons.
8. Ionospheric photoelectrons traced by their characteristic peaks in energy spectra are used to identify the photoelectron boundary PEB and explore their distribution within the Martian magnetosphere. Photoelectrons can be observed close to the MB locations implying an important role of the ionospheric component in dynamic processes responsible for the formation of the magnetospheric obstacle at Mars. It is unlikely that PEB and ionopause (as a pressure balance boundary) are collocated. It is assumed that some part of the momentum from solar wind is transferred to the ionosphere driving it into a convective motion. This motion together with a mechanism of "polar wind" along "open" field lines can explain the observation of ionospheric photoelectrons at distances more than $3R_M$ far in the tail.
9. In the IMF reference frame the distribution of photoelectrons reveals a similar asymmetry as the magnetospheric boundary.
10. It is shown that the position of ray-like structures centered close to the wake boundary are governed by the IMF direction. The events are clustered in the hemisphere of locally upward convective electric field. This supports the suggestion that these structures are formed in a process of scavenging of planetary plasma by draped magnetic field lines near the "magnetic poles." However their dominance in the southern hemisphere also implies a possible important role of auroral-like acceleration processes at Mars. A "dawn-dusk" asymmetry due to draping features is also revealed.
11. It is shown that the boundary layer/mantle is an important channel for planetary ions escaping from the Martian space. A strong "dawn-dusk" asymmetry in IMF coordinates appeared due to a draping asymmetry. Estimates of outflowing fluxes of oxygen ions yield $6 \times 10^{23} - 6 \times 10^{24}$ s^{-1}. However, these values may be somewhat revised after the final instrumental calibration.
12. If PEB is not a boundary at which the solar wind pressure is balanced by the thermal pressure of the cold ionospheric plasma then plasmas of ionospheric and atmospheric origin which fill the region between MB and ionopause must be driven into a convective motion.

13. An interesting class of events is observed close to the inner boundary of the magnetosheath. These boundary events are characterized by an abrupt additional heating of magnetosheath electrons and remarkable fluxes of planetary ions. It is not clear yet whether such events are the manifestation of a transition, "viscous-like" layer as observed near Venus or crossings of a plasma sheet near the MB.

Acknowledgements

Authors wish to acknowledge very useful comments of the referees. The ASPERA experiment on the European Space Agency (ESA) Mars Express mission is a joint effort between 15 laboratories in 10 countries, all sponsored by their national agencies as well as the various departments/institutes hosting these efforts. We wish to acknowledge support from Deutsche Forschungsgemeinschaft for supporting this work by grant WO 910/1-1 and DLR grant 50QM99035. We also wish to acknowledge the Swedish National Space Board for their support of the main PI-institute and we are indebted to ESA for their courage in embarking on the Mars Express program, the first ESA mission to the red planet. We wish to acknowledge support of NASA contract NASW00003 for the support of the design, construction,operation for the Electron Spectrometer through the Discovery Program Mission of Opportunity.

References

Acuña, M. H., Connerney, J., and Wasilewski, P., *et al.*: 1998, *Science* **279**, 5357, 1676.

Barabash, S., Lundin, R., Andersson H., *et al.*: 2004, *ESA Publication* **SP-1240**, 121.

Bertucci, C., Mazelle, C., Crider D., *et al.*: 2003, *Geophys. Res. Lett.* **30**, 1876, doi:10.1029/2002GL015713.

Bößwetter, A., Bagdonat, T., Motschmann, U., and Sauer, K.: 2004, *Annal. Geophys.* **22**, 4363.

Brace, L. H., Kasprzak, W. T., Taylor, H. A., Theis, T. F., Russell, C. T., Barnes, A. *et al.*: 1987, *J. Geophys. Res.* **92**, 15.

Brain, D. A., Bagenal, F., Acuña, M., and Connerney, J. E.: 2003, *J. Geophys. Res.* **108**, 1424, doi:10.1029/2002JA009482.

Brain, D. A., Halekas, J. S., Lillis, R. J., Mitchell, D. L., Lin, R. P., and Crider, D. H.: 2005, *Geophys. Res. Lett.* **32**, 18, doi: 10.1029/2005GL023126. L18203.

Brain, D. A., Mitchell, D. L., and Halekas, J. S.: 2006, *Icarus* **182**, 464.

Brecht, S. H.: 1995, *Geophys. Res. Lett.* **22**, 1181.

Brecht, S. H.: 1997, *J. Geophys. Res.* **102**, 4743.

Breus, T., Krymskii, A., Lundin, R. , Dubinin E., *et al.*: 1991, *J. Geophys. Res.* **96**, 11165.

Connerney, L. E., Acuña, M. ,Wasilewski, P., Kletetschka, G., Ness, N. F., Reme, H. *et al.*: 2001, *Geophys. Res. Lett.* **28**, 4015.

Crider, D. H.: 2001, *Adv. Space Res.* **27**, 1831.

Crider, D. H.: 2004, *Adv. Space Res.* **33**, 152.

Crider, D. H., Vignes, D., Krymskii, A., Breus, T., Ness, N., Mitchell, D. *et al.*: 2003, *J. Geophys. Res.* **108**, 1461, doi:10.1029/2003JA009875.

Dubinin, E., Lundin, R., Riedler, W., Schwingenschuh, K., Luhmann, J., Russell, C. T. *et al.*: 1991, *J. Geophys. Res.* **96**, 11189.
Dubinin, E., Lundin, R., Koskinen, H., and Pissarenko, N.: 1993, *J. Geophys. Res.* **98**, 3991.
Dubinin, E., Lundin, R., and Schwingenschuh, K.: 1994, *J. Geophys. Res.* **99**, 21233.
Dubinin, E., Sauer, K., Lundin, R., Norberg, O., Trotignon, J.-G., Schwingenschuh, K. *et al.*: 1996, *J. Geophys. Res.* **101**, 27061.
Dubinin, E., Sauer, K., Delva, M., and Tanaka, T.: 1998, *Earth Planets Space* **50**, 873.
Dubinin, E., Winningham, J. D., Fraenz, M., Woch, J., *et al.*: 2006a, *Icarus* **182**, 343.
Dubinin, E., Lundin, R., Fraenz, M., Woch, J., *et al.*: 2006a, *Icarus* **182**, 337.
Fraenz, M., Winningham, J. D., Dubinin, E., Roussos, E., *et al.*: 2006a, *Icarus* **182**, 406.
Fraenz, M., Dubinin, E., Roussos, E., Woch, J.: 2006b, *Space Sci. Rev.*, this volume, doi: 10.1007/s11214-006-9115-9.
Frahm, R., Winningham, J. D., Sharber, J. R., *et al.*: 2006a, *Icarus* **182**, 371.
Frahm, R., Winningham, J. D., Sharber J. R., *et al.*: 2006b, *Space Sci. Rev.*, this volume, doi: 10.1007/s11214-006-9119-5.
Gurnett, D. A., Kirchner, D. L., Huff, R. L., Morgan, D., *et al.*:2005, *Science* **310**, 1929.
Hanson, W. B., Sanatani, S., and Zuccaro, D. R.: 1977, *J. Geophys. Res.* **82**, 4351.
Hanson, W. B., and Mantas, G. P.: 1988, *J. Geophys. Res.* **93**, 7538.
Harnett, E. M., and Winglee, R. M.: 2005, *J. Geophys. Res.* **110**, A07226, doi: 10.1029/2003JA010315.
Kliore, A. J.: 1992, in J. G. Luhmann, M. Tatrallyay, and R. O. Pepin, (eds.), *Venus and Mars: Atmospheres, Ionospheres and Solar Wind Interactions*, AGU monograph, **66**, Washington, DC, p. 265.
Luhmann, J. G., Russell, C. T., Spreiter, J. R., and Stahara, S. S.: 1985, *J. Geophys. Res.* **5**(4), 307.
Luhmann, J. G.: 1993, *Adv. Space Res.* **98**, 17615.
Lundin, R., and Dubinin, E.: 1992, *Adv. Space Res.* **12**(9), 255.
Lundin, R., Zakharov, A., Pellinen, R., *et al.*: 1989, *Nature* **341**, 609.
Lundin, R., Zakharov, A., Pellinen, R., *et al.*: 1990a, *Geophys. Res. Lett.* **17**, 873.
Lundin, R., Zakharov, A., Pellinen, R., *et al.*: 1990b, *Geophys. Res. Lett.* **17**, 877.
Lundin, R., Dubinin, E., Koskinen, H., Norberg, O., Pissarenko, N., and Barabash, S.: 1991, *Geophys. Res. Lett.* **18**, 1059.
Lundin, R., Barabash, S. , Andersson, H., *et al.*: 2004, *Science* **305**, 1933.
Lundin, R., Winningham, J. D., Barabash, S., *et al.*: 2006, *Science* **311**, 980.
Mantas, G. P., and Hanson, W. B.: 1979, *J. Geophys. Res.* **84**, 369.
Mazelle, C., Reme, H., Sauvaud, J.-A., D'Uston, C., and Carlson, C. W.: 1989, *Geophys. Res. Lett.* **16**, 1035.
Mitchell, D. L., Lin, R. P., Mazelle, C., *et al.*: 2001, *J. Geophys. Res.* **106**, 23419.
Modolo R., Chanteur, G., Dubinin, E., Matthews, A.: 2005, *Annal. Geophys.* **23**, 433.
Moore, K. R., McComas, D., Russell, C. T., and Mihalov, J. D.: 1990, *J. Geophys. Res.* **95**, 12005.
Nagy, A. F., Winterhalter, D., Sauer, K., *et al.*: 2004, *Space Sci. Rev.* **111**(1), 33.
Neubauer, F. M.: 1987, *Astron. Astrophys.* **187**(1–2), 73.
Pätzold, M., Tellmann, S., Häusler, B., Hinson, D., Schaa, R., and Tyler, G. L.: 2005, *Science* **310**, 837.
Pedersen, A., Nairn, C., Grard, R., and Schwingenschuh, K.: 1991, *J. Geophys. Res.* **96**, 11243.
Perez-de-Tejada, H.: 1979, *J. Geophys. Res.* **84**, 1555.
Perez-de-Tejada, H., Intriligator, D. S., and Strangeway, R. J.: 1993, *Geophys. Res. Lett.* **20**, 991.
Podgorny, I., Dubinin E., and Israelevich, P.: 1982, *Moon and Planets* **27**, 397.
Raeder, J., Neubauer, F., Ness, N. F., Burlaga, L. F.: 1987, *Astron. Astrophys.* **187**(1–2), 61.
Russell, C. T.: 1992, in J. G. Luhmann, M. Tatrallyay, and R. O. Pepin (eds.), *Venus and Mars: Atmospheres, Ionospheres and Solar Wind Interactions*, AGU monograph, **66**, Washington, DC, p. 225.

Slavin, J., Smith, E. J., Tsurutani, B. T., Siscoe, G. L., Jones, D. E., and Mendis, D. A.: 1986, *Geophys. Res. Lett.* **13**, 283.

Smith, E. J., Davis, L., Coleman, P. L., and Jones, D. E.: 1965, *Science* **149**, 1241.

Spenner, K., Knudsen, W. C., Miller, K. L., Novak, V., Russell C. T., and Elphic, R. C.: 1980, *J. Geophys. Res.* **85**, 7655.

Spreiter, J. R., and Stahara, S. S.: 1992, in J. G. Luhmann, M. Tatrallyay, R. O. Pepin, (eds.), *Venus and Mars: Atmospheres, Ionospheres and Solar Wind Interactions*, AGU Monograph, **66**, Washington, DC, p. 345.

Trotignon, J.-G., Dubinin, E., Grard, R., Barabash, S., and Lundin, *J. Geophys. Res.* **101**, 24965.

Vaisberg, O.: 1992, *Venus and Mars: Atmospheres, Ionospheres and Solar Wind Interactions*, in: J. G. Luhmann, M. Tatrallyay, R. O. Pepin, (eds.), AGU monograph, **66**, Washington, DC, p. 311.

Verigin, M., Shutte, N., Galeev, A., Gringauz, K. I., *et al.*: 1991, *Planet. Space Sci.* **39**, 131.

Verigin, M., *et al.*: 1993, *J. Geophys. Res.* **98**, 1303.

Vennerstrom, S. Olsen, N., Purucker, M., and Acuña, M.: 2003, *Geophys. Res. Lett.* **30**, 1369, doi:10.1029/2003GL016883.

Vignes. D., Mazelle, C., Reme, H., Acuña, M., *et al.*: 2000, *Geophys. Res. Lett.* **27**, 49.

Yeroshenko, Ye. Riedler, W. R., Schwingenschuh, K., Luhmann, J. G., Ong, M., and Russell, C. T.,: 1990, *Geophys. Res. Lett.* **17**, 885.

Zhang, T. L., Luhmann, J. G., and Russell, C. T.: 1991, *J. Geophys. Res.* **96**, 11145.

IMF DIRECTION DERIVED FROM CYCLOID-LIKE ION DISTRIBUTIONS OBSERVED BY MARS EXPRESS

M. YAMAUCHI[1,*], Y. FUTAANA[1,7], A. FEDOROV[2], E. DUBININ[3], R. LUNDIN[1], J.-A. SAUVAUD[2], D. WINNINGHAM[4], R. FRAHM[4], S. BARABASH[1], M. HOLMSTROM[1], J. WOCH[3], M. FRAENZ[3], E. BUDNIK[2], H. BORG[1], J. R. SHARBER[4], A. J. COATES[5], Y. SOOBIAH[5], H. KOSKINEN[6,17], E. KALLIO[6], K. ASAMURA[7], H. HAYAKAWA[7], C. CURTIS[8], K. C. HSIEH[8], B. R. SANDEL[9], M. GRANDE[10], A. GRIGORIEV[1], P. WURZ[11], S. ORSINI[12], P. BRANDT[13], S. MCKENNA-LAWLER[14], J. KOZYRA[15] and J. LUHMANN[16]

[1]*Swedish Institute of Space Physics, Box 812, SE-98128, Kiruna, Sweden*
[2]*Centre d'Etude Spatiale des Rayonnements, BP-4346, F-31028 Toulouse, France*
[3]*Max-Planck-Institut für Sonnensystemforschung, D-37191 Katlenburg-Lindau, Germany*
[4]*Southwest Research Institute, San Antonio, TX 7228-0510, USA*
[5]*Mullard Space Science Laboratory, University College London, Surrey RH5 6NT, UK*
[6]*Finnish Meteorological Institute, Box 503 FIN-00101 Helsinki, Finland*
[7]*Institute of Space and Astronautical Science, 3-1-1 Yoshinodai, Sagamichara, Japan*
[8]*Department of Physics, University of Arizona, Tucson, AZ 85721, USA*
[9]*Lunar and Planetary Lab, University of Arizona, Tucson, AZ 85721, USA*
[10]*Rutherford Appleton Laboratory, Chilton, Didcot, Oxfordshire OX11 0QX, UK*
[11]*University of Bern, Physikalisches Institut, CH-3012 Bern, Switzerland*
[12]*Instituto di Fisica dello Spazio Interplanetari, I-00133 Rome, Italy*
[13]*Applied Physics Laboratory, Johns Hopkins University, Laurel, MD 20723-6099, USA*
[14]*Space Technology Ltd., National University of Ireland, Maynooth, Co. Kildare, Ireland*
[15]*Space Physics Research Lab., University of Michigan, Ann Arbor, MI 48109-2143, USA*
[16]*Space Science Lab., University of California in Berkeley, Berkeley, CA 94720-7450, USA*
[17]*University of Helsinki, Department of Physical Sciences, Box 64, FIN-00014, Helsinki, Finland*
*(*Author for correspondence: E-mail: m.yamauchi@irf.se)*

(Received 6 April 2006; Accepted in final form 31 October 2006)

Abstract. Although the Mars Express (MEX) does not carry a magnetometer, it is in principle possible to derive the interplanetary magnetic field (IMF) orientation from the three dimensional velocity distribution of pick-up ions measured by the Ion Mass Analyser (IMA) on board MEX because pick-up ions' orbits, in velocity phase space, are expected to gyrate around the IMF when the IMF is relatively uniform on a scale larger than the proton gyroradius. Upstream of bow shock, MEX often observed cycloid distributions (two dimensional partial ring distributions in velocity phase space) of protons in a narrow channel of the IMA detector (only one azimuth for many polar angles). We show two such examples. Three different methods are used to derive the IMF orientation from the observed cycloid distributions. One method is intuitive (intuitive method), while the others derive the minimum variance direction of the velocity vectors for the observed ring ions. These velocity vectors are selected either manually (manual method) or automatically using simple filters (automatic method). While the intuitive method and the manual method provide similar IMF orientations by which the observed cycloid distribution is well arranged into a partial circle (representing gyration) and constant parallel velocity, the automatic method failed to arrange the data to the degree of the manual method, yielding about a 30° offset in the estimated IMF direction. The uncertainty of the derived IMF orientation is strongly affected by the instrument resolution. The source population for

Space Science Reviews (2006) 126: 239–266
DOI: 10.1007/s11214-006-9090-1

these ring distributions is most likely newly ionized hydrogen atoms, which are picked up by the solar wind.

Keywords: IMF, Mars, ion gyration, pick-up process

1. Introduction

The European Space Agency Mars Express (MEX) carries the Analyzer of Space Plasma and EneRgetic Atoms (ASPERA-3) experiment (Barabash *et al.*, 2004), which measures hot plasma and energetic neutral atoms (ENA), but MEX does not carry a magnetometer. Without the magnetic field data, it is difficult to interpret plasma processes and ENA formation processes. Therefore, any method by which the magnetic field direction can be obtained aids in interpreting the ASPERA-3 data. In this regard, Fedorov *et al.* (2006) used the 400 km circular orbit Mars Global Surveyor (MGS) magnetometer data (Acuna *et al.*, 1998) to derive the interplanetary magnetic field (IMF) direction. Although MGS is not always on the dayside of Mars where the IMF direction can be estimated, MGS produced estimated IMF direction data that is sufficient for large-scale statistics.

Here, we propose an alternative method to utilize the three-dimensional (3-D) ion distribution, based on the work of Mukai and coworkers (Mukai *et al.*, 1986a,b; Terasawa *et al.*, 1986). They derived the direction of the magnetic field in comet Halley's sheath region from ion data obtained by the Suisei spacecraft. The principle takes advantage of the gyration of ions of cometary origin around the IMF. In the solar wind frame, no electric field is imposed on the ions, causing the ions to perform simple spiral motions. In velocity phase space, a spiral motion forms a two-dimensional (2-D) ring trajectory with a constant velocity along the magnetic field. The ring's plane is perpendicular to the local magnetic field. In the actual Suisei data, a 3-D shell-like distribution is observed instead of a 2-D ring, and the orientation of the symmetry axis is considered as parallel to the IMF direction. This principle was also applied to both electron and ion data near the Moon during the flyby of the Nozomi spacecraft (Futaana *et al.*, 2003). In order to obtain the symmetry axis in 3-D velocity space, a measurement of the 3-D proton distribution is required.

The ring distribution is also found in the upstream foot region of the Earth's bow shock, but the source ions are not the newly born ions but the reflected solar wind. Using ISEE-1 and -2 data, Paschmann *et al.* (1981) and Sckopke *et al.* (1983) showed that the ion distribution in the upstream foot region of a quasi-perpendicular bow shock is consistent with a partial ring distribution that originates from the reflected solar wind at the bow shock. Later, AMPTE and Cluster observations further demonstrated that the ring distribution exists only within a gyroradius (few hundred km) upstream of the bow shock (Sckopke *et al.*, 1990; Möbius *et al.*, 2001).

For the Martian case, both the newly ionized neutrals and the reflected solar wind can be substantial sources, and therefore, we can expect the ring or shell-like distribution to be detected upstream and/or downstream of the bow shock. To obtain the magnetic field direction from the actual MEX data, it is important that the magnetic field direction is nearly uniform over a distance greater than an ion gyroradius and a duration of the observation cycle of 3-D ion distribution measurement. Obviously, the best place for such an attempt is the upstream region of the front-side bow shock, where we actually have observed many cycloid-like ring distributions. It might also be possible to derive the magnetic field direction from the data within the magnetosheath, but this is beyond the scope of this paper.

During 2004 and 2005, the Ion Mass Analyser (IMA) of the ASPERA-3 experiment on board MEX measured the 3-D ion distribution with a nearly 3-min cycle (the operation mode is different in 2006). One 3-D measurement cycle of IMA corresponds to a distance of about 550 km or one gyroradius for a 1 keV proton in an 8 nT magnetic field. Therefore, IMA is capable of providing data to derive the IMF orientation if the cycloid distribution is observed and if the IMF is constant and uniform during the observation. However, IMA operation is optimized to separate heavy ions (e.g., atomic ions and molecular ions) of Martian origin and the majority of observations were performed in operational modes in which IMA hardly detects ring-distributed protons. Because of this, only one bow-shock crossing in 2004 was observed when IMA was in the appropriate operational mode. In this paper, we use this observation together with one of best observations from 2005 to illustrate the technique of determining the IMF direction from the ring distribution observed by IMA.

2. Instrument

The IMA and ELectron Spectrometer (ELS) on board MEX are parts of ASPERA-3 experiment (Barabash *et al.*, 2004). ELS has a $4^\circ \times 360^\circ$ field of view that is divided into 16 azimuthal sectors, each 22.5° wide. The sensor consists of a top hat electrostatic analyzer in a very compact design. ELS measures electrons in the energy range from 1 eV to 20 keV in logarithmically scaled energy steps every 4 sec. For the detail of the ELS instrument, refer to Barabash *et al.* (2004) and Winningham *et al.* (2006).

IMA is a top hat instrument that combines an electrostatic energy analyzer with a magnetic mass analyzer. IMA has a $4.6^\circ \times 360^\circ$ field of view that is divided into 16 azimuthal sectors, each 22.5° wide. IMA measures ions in the energy range from 10 eV/q to 30 keV/q in logarithmically scaled energy steps every 12 sec. In order to produce a 3-D particle measurement on the 3-axis stabilized MEX spacecraft, IMA has an electrostatic deflection system (or elevation analyzer) at its entrance, which scans from -45° to $+45^\circ$ (16 elevations) in approximately 3 min. The actual entrance angle of the ions is slightly energy dependent. The overall field-of-view is approximately 360° (16 sectors) $\times$ 90° (16 elevations).

IMA is primarily designed to examine ions of Martian origin, with an option to sample the solar wind. The mass analyzer (magnets) is designed to deflect the incident solar wind protons away from its position-sensitive detector (microchannel plate or MCP) so that observations do not suffer from contamination by solar wind protons. In order to sample the solar wind with this design, IMA contains an adjustable electrostatic post-acceleration (PA) system between its electrostatic analyzer and magnetic analyzer. With the highest PA voltage, incident solar wind protons are accelerated to fast enough to reach the MCP detector before being deflected significantly by the magnetic mass analyzer. However, this mode was rare during 2004–2005.

IMA has three PA settings: PA = 0 (nearly no acceleration, about 0.3 kV), PA = 1 (about 2.4 kV), and PA = 2 (highest acceleration, about 4.2 kV). The PA = 0 mode is optimized to separate heavy ions, and solar wind protons are not detected unless the solar wind is extremely fast. The PA = 2 mode is optimized to detect the solar wind. The PA = 1 mode is a marginal mode which detects only a small part of the solar wind (alpha particles and superthermal protons) in most cases. For details of the IMA instrument, refer to Barabash *et al.* (2004), Lundin *et al.* (2004), and Fedorov *et al.* (2006).

3. Observations

IMA data from January 2004 through June 2005 were examined for operation in the PA = 2 mode (including a mode which has an alternating PA for every other full scan). Only one bow shock crossing was identified in 2004, while 38 bow shock crossings were identified during the first half of 2005. The observation from 27 April 2005 near 1337 UT is presented below as one of the best examples observed, followed by the observation from 22 March 2004 near 1230 UT, the first IMA measurement of a bow shock crossing with PA = 2.

3.1. 27 April 2005, 1330 UT

Figure 1 shows the MEX orbit and energy-time spectrograms of the electron (ELS) and ion (IMA) data during 1331–1357 UT on 27 April 2005. All axes references are made in the Mars-Sun Orbit (MSO) Cartesian coordinate system, with the $+X$ direction pointing sunward, the $+Y$ direction duskward, and the $+Z$ direction toward the north ecliptic pole, and $R^2 = X^2 + Y^2$. The nearly 3-min (192 sec) cycle seen in the IMA data is due to the scanning cycle of the IMA entrance direction from about -45° (elevation = 0) to about $+45^\circ$ (elevation = 15). Figure 1 contains 7 full scans of IMA data. In the present case, elevation = 0 corresponds to the northward viewing sector (detecting 45° southward traveling ions) and elevation = 15 corresponds to the southward viewing sector (detecting 45° northward traveling

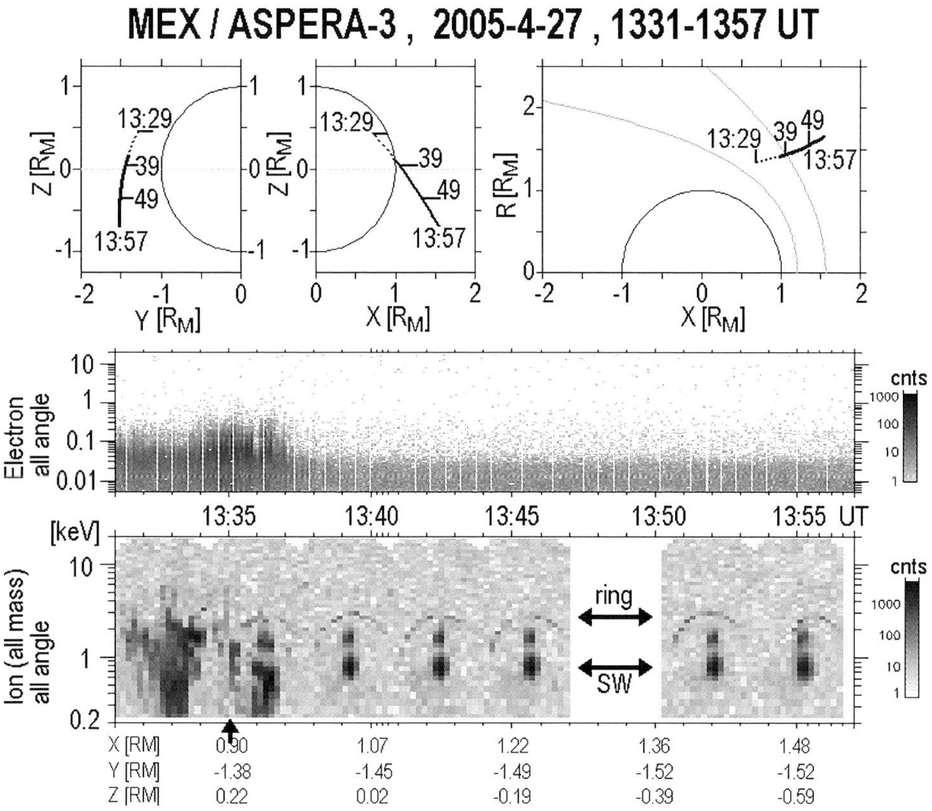

Figure 1. Overview of the MEX orbit and hot plasma data during 1331–1357 UT on 27 April 2005. The upper part shows the MEX orbit in the Mars-Sun Orbit (MSO) Cartesian coordinate system, with the $+X$ direction pointing sunward, the $+Y$ direction duskward, and the $+Z$ direction toward the north ecliptic pole, and $R^2 = X^2 + Y^2$. The unit "R_M" is the Martian radius. The average boundary positions (bow shock and induced magnetosphere boundary) are drawn with grey lines in the upper right panel. The MEX traversal (IMA operational) is drawn by a thick line: the solid line corresponds to MEX outside the bow shock where the partial ring distributions are observed and the dashed line corresponds to MEX inside the bow shock where the magnetosheath-like distributions are observed. The lower panels show the energy-time spectrograms of electrons (from ELS, 5 eV–20 keV) and ions (from IMA, 0.2–20 keV). All mass and azimuthal angles are integrated. The nearly 3-min cycle seen in the IMA data is due to the electric scan of the entrance direction from nearly $-45°$ (elevation $=$ 0) to nearly $+45°$ (elevation $=$ 15). From both ELS and IMA, the bow shock outbound is identified at around 1337 UT. Horizontal arrows in the IMA data indicate the cycloidal ions (see text) and the solar wind protons, and vertical arrow at the bottom (at around 1335 UT) indicates a high count rate discussed at the end of Section 3.1.

ions), while azimuthal sector 2 is pointing toward $+X$ (detecting tailward traveling ions), azimuth $= 14$ is pointing toward $+Y$ (detecting ions traveling toward the $-Y$ direction), and azimuth $= 6$ is pointing toward $-Y$ (detecting ions traveling toward the $+Y$ direction) in the MSO coordinates.

The spacecraft traversed the bow shock on the dawn side (outbound) at around 1337 UT, as identified by the sudden change of the energization/thermalization level of the solar wind as observed by both IMA and ELS. During the next 20 min (5 full elevation scans of IMA), IMA detected a partial ring-like distribution of ions at around 2–3 keV (indicated by the upper horizontal arrow shown in Figure 1), well above the solar wind alpha particles ($\sim$1.6 keV) and the solar wind protons ($\sim$0.8 keV, indicated by the lower horizontal arrow shown in Figure 1). Since the time axis is the same as the elevation scan within each 3-min scanning cycle, this ring-like structure actually means that the velocity depends on the direction as one would expect with a partial ring-like distribution in velocity space.

Figure 2 shows the energy-time spectrograms of IMA organized by the mass-charge ratio (protons in the lower half and alpha particles in the upper half) and by azimuthal sectors (in individual panels) during 1334–1344 UT, i.e., covering the 2nd, 3rd, and 4th full scans of Figure 1. The mass channel selection can be confirmed by Figure 3, which shows the energy-mass matrix during two full scans (1337–1344 UT). At 1339:20 UT and 1342:30 UT in Figure 2, the solar wind is clearly separated into protons at around 0.8 keV and alpha particles at around 1.6 keV, and they are registered at azimuthal scan 2 and elevation scan 9 (closer to scan 8 than scan 10). The separation between alpha particles and protons is clearer in Figure 3 (lower middle panel). Note that solar wind protons strongly contaminate all mass channels (contamination is observed at all mass channels at around 0.8 keV in Figure 3).

The partial ring distribution is recognized during all 3 full scans in Figure 2, and is detected at a single azimuthal sector (azimuth = 3) for a wide range of elevation angles (from elevation = 2 at 1341:00 UT to elevation = 15 at 1343:40 UT for the third full scan, and from elevation = 4 at 1338:10 UT to elevation = 15 at 1340:30 UT for the second full scan). The counts at elevation = 3/azimuth = 2 at 1338:00 UT during the second full scan are also connected to the ring, but no other counts are found at azimuth = 2 or azimuth = 4 in conjunction with the ring at azimuth = 3. Thus, these ions are distributed in a 2-D plane rather than in a 3-D shell. Its direction (azimuth = 3) is one sector (22.5°) off from the solar wind direction. In Figure 2, the ring distribution is recognized in the proton channel only, and its composition is confirmed from the energy mass matrix shown in Figure 3 (upper three panels).

For the purpose of further analysis, we list the energy and direction of this ring distribution during the full scan of 1344–1347 UT in Table I. The first three columns are elevation scans (El), azimuthal sectors (Az), and the corresponding viewing directions (unit vector components) in the MSO coordinates (see Figure 1 for the X, Y, and Z directions). The center energy (keV) of the ring distribution is listed in the next column, and this energy is converted into the velocity components (km/s) using MSO coordinates (the last three columns). The information in Table I is basically enough to derive the IMF orientation.

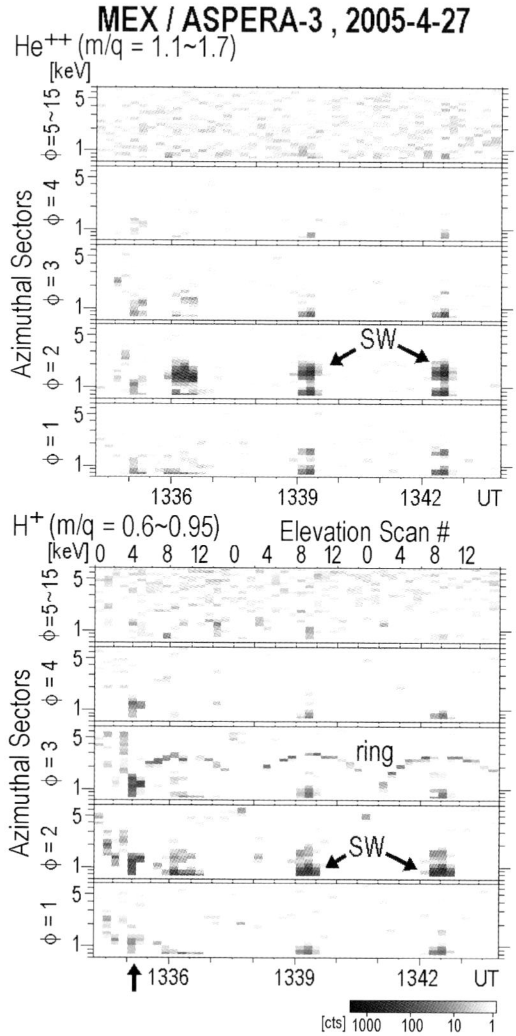

Figure 2. Energy-time spectrograms of light ions between 0.7 and 7 keV during 1334:00–1343:40 UT on 27 April 2005. Five different azimuthal sectors (ϕ) are presented for both the proton mass-channel (lower 5 panels) and the alpha particle mass-channel (upper 5 panels). These 5 azimuthal sectors are: 1, 2, 3, 4, and the integration of 5 through 15. Sector 0 is not presented because it gathers contamination from all of the other sectors. Note that IMA scans 16 elevation directions in nearly 3 min, and the presented period corresponds to 3 complete scans, with the first pixel on the time axis corresponding to elevation = 0 (about -45° at 2 keV) and the last pixel on the time axis corresponding to elevation = 15 (about $+40^\circ$ at 2 keV) as indicated at the top of proton panel. The solar wind is seen at elevation = 9 (weak at elevation = 8). The counts that are seen above the solar wind in the proton channel are due to contamination from alpha particles and the counts that are seen below the solar wind in the alpha particle channel are due to contamination from protons. The vertical arrow at the bottom indicates the intense count that is shown in Figure 1.

TABLE I

Direction and energy of the registered ring during 1344–1347 UT, 27 April 2005.

El Az	Sensor direction (X, Y, Z)	E (keV)	V_X (km/s)	V_Y (km/s)	V_Z (km/s)
02 03	(0.81, −0.29, 0.51)	1.37	−414	148	−260
03 03	(0.85, −0.30, 0.42)	1.86	−508	181	−251
04 03	(0.89, −0.32, 0.33)	2.27	−584	208	−218
05 03	(0.91, −0.33, 0.24)	2.49	−630	225	−164
08 03	(0.94, −0.34, −0.05)	2.74	−680	243	39
09 03	(0.93, −0.33, −0.15)	2.74	−673	241	110
10 03	(0.91, −0.33, −0.25)	2.74	−659	236	180
12 03	(0.85, −0.30, −0.43)	2.49	−585	210	298
13 03	(0.80, −0.29, −0.52)	2.27	−529	190	341
14 03	(0.75, −0.27, −0.60)	2.00	−464	167	370
15 03	(0.69, −0.25, −0.68)	1.68	−392	141	382

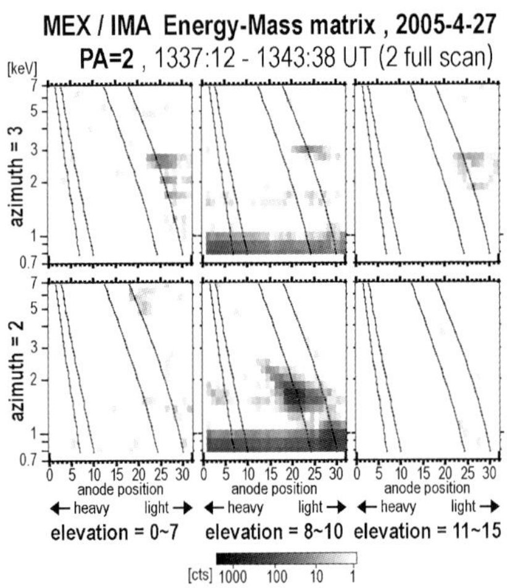

Figure 3. Energy-Mass matrices during 2 full scans (1337:10–1343:40 UT) of Figure 2. The horizontal axis of each panel is the detector position that corresponds to a different mass value for each given energy. The curved lines in each panel correspond to (from right to left) mass per charge $m/q = 1$, $m/q = 2$, $m/q = 16$, and $m/q = 32$. The upper 3 panels show data from azimuthal sector 3 where the partial ring distribution is found in Figure 2 and the lower 3 panels show data from the azimuthal sector 2 where the solar wind is detected for elevations = 8–10. In the left, middle, and the right panels, the counts are integrated over elevation = 0–7 (-45° to 0°), elevation 8–10 (0° to $+15^\circ$), and elevation 11–15 ($+15^\circ$ to $+45^\circ$), respectively. Note that solar wind protons with an energy of about 0.7–1 keV strongly contaminate all mass channels due to a mode-dependent instrumental effect (marginally deflected protons by the magnetic mass analyzer to hit the outer boundary of the instrument near the MCP, and scatter randomly).

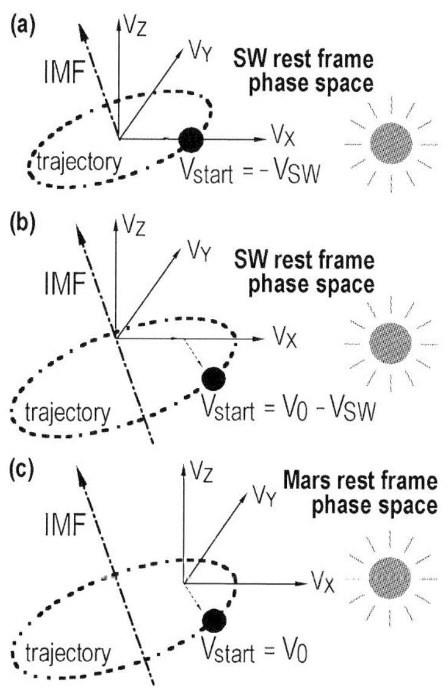

Figure 4. Illustration of ion motion in the solar wind ($\mathbf{V}_{SW}$) with respect to the IMF in a velocity space diagram (V_X, V_Y, X_Z) where X, Y, Z directions are given in Figure 1. (a) An ion with zero initial speed (i.e., negligible compared to the solar wind velocity) in the Martian rest frame has an initial velocity of $-\mathbf{V}_{SW}$ in the solar wind frame. Since there is no electric field in the solar wind frame, the ion performs a simple spiral motion around the IMF (the magnetic field direction is about the same in both the Martian rest frame and the solar wind frame). Any spiral motion is represented by a circle about the magnetic field in velocity space. (b) For the finite initial velocity ($\mathbf{V}_0$) in the Martian rest frame, the start position is shifted by $\mathbf{V}_0$ from (a), and hence the radius of the ring changes. However, the ring is still found around the IMF and the ion speed (distance from the origin of coordinates) is constant. (c) The Martian rest frame trajectory of what is shown in (b). The difference between (b) and (c) is only a constant velocity $+\mathbf{V}_{SW}$.

To understand the relation between the ring-like distribution and the IMF orientation, we illustrate velocity space ion motion (V_X, V_Y, V_Z) in the solar wind in Figure 4. Note that the magnetic field is almost the same (non-relativistic limit) between the Martian rest frame and the solar wind frame. We first consider a newly ionized neutral atom with nearly zero initial velocity in the Martian rest frame. Such an ion has an initial velocity of $-\mathbf{V}_{SW}$ (where $\mathbf{V}_{SW}$ is the solar wind velocity) in the solar wind frame as illustrated in Figure 4a.

Since there is no electric field in the solar wind frame, any ion with a velocity different from the solar wind performs a simple spiral motion (circular motion around the magnetic field plus constant motion along the magnetic field) with a constant speed in the solar wind frame. The spiral motion of an ion is represented by a ring trajectory which is symmetric about the magnetic field in velocity space. The constant speed in velocity space means that the symmetry axis of the trajectory (i.e., magnetic field) passes through the origin in the solar wind frame. Figure 4b illustrates the general case of a finite initial velocity $\mathbf{V}_0$ in the Martian rest frame (initial velocity $\mathbf{V}_0 - \mathbf{V}_{SW}$ in the solar wind frame). While the start position and radius of the ring in velocity space are different from the previous case (Figure 4a), the orientation of the ring is still the same as that in Figure 4a; i.e., the ring plane is again perpendicular to the magnetic field, and the symmetry axis (magnetic field) again passes through the origin of coordinates in the solar wind frame.

Strictly speaking, feedback of ring ions to the magnetic field (e.g., diamagnetic effect) may deforms the IMF orientation if the mass flux of these ring ions is significant. Such deformation is indeed important near the ionopause or induced magnetosphere boundary. However, the mass flux of the ring component is much lower than that of the solar wind (Figures 1 and 2), and the orientations of the rings are nearly the same for many full elevation scans. Thus, the feedback from the ring ions to the IMF orientation is ignorable inside the solar wind, and the above test particle approximation is appropreate in deriving the IMF orientation.

Figure 4c illustrates the same ion motion as Figure 4b in the Martian rest frame, i.e., with initial velocity $\mathbf{V}_0$. Although what the particle instrument detects is the energy seen in the spacecraft rest frame, the spacecraft velocity relative to Mars is negligible compared to the solar wind velocity for the MEX observations. The Lorenz transform in velocity space from the solar wind frame to the Martian rest frame means an addition of the constant velocity $\mathbf{V}_{SW}$ to the ring trajectory illustrated in Figure 4b. Therefore, the radius and orientation of the ring is the same between Figure 4b and Figure 4c; i.e., the ring plane in velocity space is always perpendicular to the IMF in both rest frames. The resulting ring trajectory in velocity space does not necessarily have its symmetry axis (IMF axis) aligned to any of V_X, V_Y, or, V_Z axes, but this symmetry axis always intersects the $V_Y - V_Z$ plane in velocity space at $(V_X, V_Y, V_Z) = (-V_{SW}, 0, 0)$.

The motion depicted in Figure 4c can also be seen as the motion of an arbitrary ion under a constant IMF and a constant solar wind electric field $\mathbf{E}_{SW} = -\mathbf{V}_{SW} \times \mathbf{B}$ in the Martian rest frame (where $\mathbf{B}$ is the IMF vector). Since $\mathbf{E}_{SW}$ does not have any parallel component to $\mathbf{B}$, the acceleration direction of the ion is always perpendicular to the IMF, making the velocity space trajectory stay within a plane (a ring under constant fields) perpendicular to the IMF. This E $\times$ B drift has an average velocity in the Martian rest frame of $\mathbf{E}_{SW} \times \mathbf{B}/B^2 = (\mathbf{V}_{SW})_\perp$ for the velocity component perpendicular to the IMF and $(\mathbf{V}_0)_{//}$ for the velocity component parallel to the IMF.

Since the location of the symmetry axis depends only on the solar wind velocity and the IMF orientation, the velocity space trajectory is the same for different IMF strengths as long as the IMF orientation is the same. For example, if $\mathbf{V}_0 = 0$ (e.g., pick up of newly ionized hydrogen corona by the solar wind), the trajectory in velocity space is a circle passing through the origin in Figure 4c. The ion trajectory in velocity space thus reflects the IMF orientation and the initial velocity of the ion, but not the strength of the IMF.

In observations, we deal with an ensemble of ions with different initial velocities, and in this case, the velocity space trajectories of these ions are not necessarily the same. Yet, if the observed distribution is symmetric around an axis, this axis is most likely aligned to the magnetic field direction. Mukai *et al.* (1986b) used this information to derive the magnetic field direction from a shell-like distribution observed by the Suisei spacecraft.

Unlike in the comet observations of Mukai *et al.* (1986b), IMA detected a 2-D ring distribution. It is very difficult for all the trajectories to lie within a plane unless the spread of the initial velocity (thermal velocity) is much smaller than the solar wind velocity (about 400 km/s in the present case). A small spread in the initial velocity means that the original ion distribution in the Martian rest frame must be either beam-like or of nearly zero velocity (e.g., newly ionized hydrogen corona) upstream of the satellite for about a proton gyroradius. In other words, the responsible source ions must be either a stable beam (duration more than the observation time, i.e., more than 20 min) or at nearly zero velocity, both for a wide region (more than one proton gyroradius) upstream of the satellite location.

In both cases, the IMF direction must be perpendicular to the plane occupied by the ring distribution (we hereafter call it a "ring's plane"). In the present case, the IMF is perpendicular to the azimuth = 3 meridian. The attitude of the spacecraft is such that azimuth = 3 corresponds to a meridian plane (perpendicular to the X–Y plane) lying about $\pi/8$ westward of the X–Z plane. Therefore, the IMF is lying nearly in the X–Y plane ($B_Z \ll B$) with $B_X/B_Y \sim \tan(\pi/8)$.

There is an ambiguity with the sign; i.e., we do not know if $B_Y > 0$ or $B_Y < 0$. Determining the sign of IMF orientation is not easy. Although the ions perform right-handed gyro orbits in both real space and velocity space, this knowledge helps very little in finding out which way the velocity vectors of the observed gyrating ions have evolved along the ring trajectory (e.g., clockwise or counter-clockwise in Figure 4 when looking from $+Z$). The Table I data does not give information whether the ring ions evolve from elevation = 2 to elevation = 15 (corresponding to $B_Y < 0$) or elevation = 15 to elevation = 2 (corresponding to $B_Y > 0$).

In Table I, the ring distribution reaches the maximum speed at azimuth = 3/elevation = 9. This direction is only one azimuthal sector away (the same elevation scan) from the solar wind direction (or $-V_X$ axis). This tilt angle (about $\pi/8$) is the same as the angle between the Y–Z plane and the IMF direction derived above. This is not a coincidence. In Figure 4c, for small $\mathbf{V}_0$, the angle between the direction of the maximum speed and the $-V_X$ direction is the same as the angle between the IMF direction and the Y–Z plane. Therefore, if this angle is small, B_X is most likely small compared to B_Y or B_Z.

Note that the observed ring distribution covers only a part of the expected circle because IMA can only detect protons with energies higher than few hundred eV, even in the operational mode designed to observe protons (PA = 2 mode). As shown in Figure 4c, the ring distribution in velocity space is shifted by the solar wind velocity, making its energy in the spacecraft coordinate high in the $-V_X$ direction and low in the V_X direction. Therefore, an instrument with a limited energy coverage can only detect a part of the total ring distribution (more than a certain distance away from the origin in velocity space (see Figure 4c)). For example, we can detect the ring distribution in only one of two oppositely looking sectors. For the present case, we did not observe the ring distribution at azimuthal sector 11 that is 180° away from sector 3. No outstanding signature is recognized in the top and 6th panels of Figure 2, which show the integrated ion counts for azimuthal sectors 5–15.

The ring's plane (or its normal direction) can also be obtained from the minimum variance method (Sonnerup and Cahill, 1967; see also Sonnerup and Scheible, 1998) by applying it to the registered velocities of the ring distribution if the data set is clean (i.e., composed of the gyrating component only). This method determines the direction in which the data scatter the least. This minimum variance direction is represented by an eigenvector for the minimum eigenvalue of a matrix calculated from the data.

One major problem with using the minimum variance method is the cleaning of the data (to choose only the ring data) because the ring distribution is a minor population compared to the solar wind. The solar wind counts severely deform the minimum variance direction. Furthermore, there are other counts that do not belong to either the ring or the solar wind. Extracting only the ring population (which is just one of the secondary populations) from the data is not a simple task. The most reliable way is to visually-identify the ring population and construct a reliable set of data. Table I is an example of such a manually constructed data. The minimum variance direction obtained from the velocity data of Table I is $(bx, by, bz) = \pm(0.34, 0.94, -0.003)$ in MSO coordinates. The intuitive method agrees well with the minimum variance result on this manually selected data.

It would be ideal if one could automatically filter the raw data to construct a set composed only of the ring component. We have tried various filters, such as a count filter (optimized filter is $7 \leq \text{count} \leq 40$ for the present case) and an energy filter (optimized filter is $K_{SW}/4 < K_{REF} < 1.5^* K_{SW}$ for the present case, where K_{REF} is the kinetic energy of the ion in the solar wind frame and K_{SW} is the solar wind energy).

Some of the results using various filter thresholds together with the results from the manual method are listed in Table II. The first column describes the method used to obtain the ring data. The square root of the maximum, medium, and minimum eigenvalues (in km/s) are listed in the next three columns, followed by the minimum variance directions (unit vector components) in MSO coordinates. The angle between the derived minimum variance direction and that obtained from Table I (they are given in first two rows in Table II) are listed in the last two columns.

The automatic filter removes the majority of the solar wind; however, this is not enough. The minimum variance direction for the automatically filtered 1337:00–1357:00 UT data with the optimum threshold is $(bx, by, bz) = \pm(-0.20, 0.95, -0.23)$ in MSO coordinates, which is 34° off from the manual method result. To illustrate the difference, we plot the observed velocities in the three different coordinates in Figure 5.

Figure 5 shows the velocity scatter plots in (a) MSO (*XYZ*) coordinates, (b) local Cartesian (LMN) coordinates determined by the minimum variance method which

TABLE II

Minimum variance analyses for various data sets.

Data set	L_L*	L_M*	L_N*	**B** direction (bx, by, bz)	offset**	offset**
Table I	248.5	102.9	0	$\pm(0.337, 0.942, -0.003)$	–	11.8°
Table I***	255.8	113.2	37.5	$\pm(0.516, 0.854, -0.062)$	11.8°	–
7≤count≤40	203.3	153.7	77.7	$\pm(-0.195, 0.953, -0.232)$	33.7°	43.3°
7≤count≤40****	236.5	188.8	130.2	$\pm(0.733, 0.661, -0.161)$	29.5°	17.7°
5≤count≤25	202.3	149.6	80.6	$\pm(-0.259, 0.958, -0.120)$	35.4°	46.2°
10≤count≤50	205.5	162.7	62.5	$\pm(-0.272, 0.909, -0.315)$	40.1°	49.0°

*Square root of the eigenvalues (in km/s).
**Angle from minimum variance direction obtained from Table I.
***Add one data point (elevation = 2/azimuth = 2 at 1371 eV at 1334 UT) to Table I (see Section 4.2).
****From 1334–1357 UT data. Otherwise, from 1337–1357 UT data (see Section 4.2).

is applied to Table I data (manual method), and (c) LMN coordinates determined from automatically filtered data (automatic method). The LMN coordinates are defined by the maximum variance direction (**L**) and the minimum variance direction (**N**) (Sonnerup and Cahill, 1967).

One can recognize that the variance in the $\mathbf{V}_N$ direction (horizontal alignment of the data in the L-N projection and M-N projection) is zero in Figure 5b but not in Figure 5c. The zero variance in the $\mathbf{V}_N$ direction means that the velocity is constant in the $\mathbf{V}_N$ direction, i.e., ions receive no electromagnetic force in this direction. The L-M projection plot is closer to a circle in Figure 5b than in Figure 5c, and the center of the circle is located where many non-ring data (representing the solar wind) are clustered. The good fit to a circle means that the observed ring distribution is composed of gyrating ions around the IMF with the same initial velocity, and that the center of the circle represents the effective solar wind velocity by which the ring ion is exposed to the convection ($-\mathbf{V}_{\mathrm{SW}} \times \mathbf{B}$) electric field in the Martian rest frame.

These facts mean that the LMN coordinates in Figure 5b (with **N** direction = (0.34, 0.94, −0.003)) determine the ring's plane much better than the LMN coordinates in Figure 5c (with **N** direction = (−0.22, 0.96, −0.18)). Thus, the IMF orientation should be estimated from the **N** direction of Figure 5b, but not of Figure 5c. Furthermore, all ring data during 1334:00–1357:00 UT nearly lie on the same circle. This means that the ring's orientation and diameter (i.e., the IMF orientation and the solar wind velocity component perpendicular to the IMF) did not change very much during these 6 full scans. The estimated IMF, which is nearly uniform and constant during the observation, is pointing about 20° off from the $+Y$ direction toward the $+X$ direction for the 27 April 2005 event. The automatic method gives the IMF direction about 30° off from this best estimate.

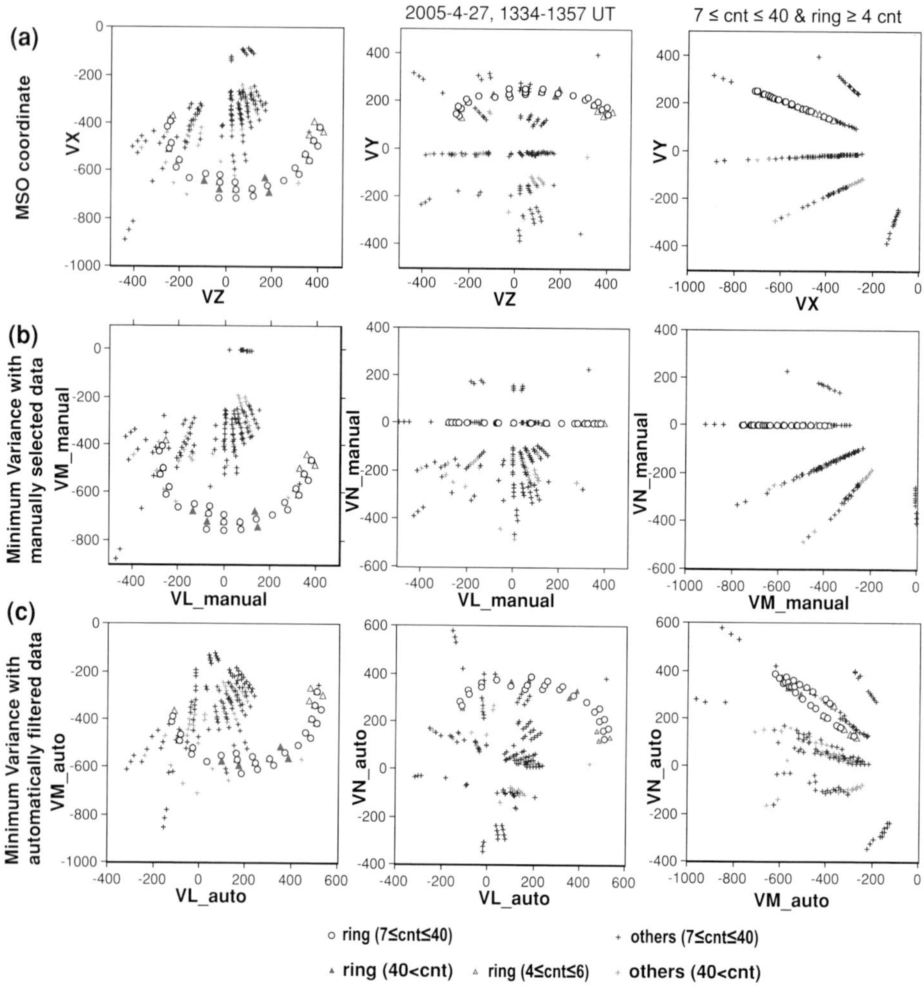

Figure 5. Velocity scatter plots of the ion data (all axes are in km/s) during 1334–1357 UT, 27 April 2005, in (a) MSO coordinates, (b) local LMN (minimum variance) coordinates determined from Table I (manual method) and (c) local LMN coordinates determined from automatically filtered data during 1337–1357 UT. The selected data are ion mass range 0.5–1.1 (proton channel: it is slightly contaminated by the solar wind alpha particles as seen in Figure 2). Plotted data consist of points that have more than 3 counts and azimuth = 0 data are not shown (total 450 points). The circles and triangles denote to the point belong to the ring distribution by a manual inspection, with circle corresponding to 7–40 counts, filled triangle corresponding to more than 40 counts, and empty triangle corresponding to 4–6 counts. Non-ring data are shown using plus marks (black: count is 7–40, grey: count more than 40).

For the second and the third full scans in Figure 2, the start direction of the ring distribution is close to the direction in which a high count rate is observed at 1335:00 UT (azimuth = 2–3 and elevation = 4–5 as indicated by a vertical arrow at the bottom of Figures 1 and 2) with exactly the same energy as the ring distribution (around 1.3 keV at azimuth = 3/elevation = 4–5). Although we cannot distinguish whether this high count rate at 1335:00 UT represents a spatial structure (i.e., beam) or a temporal structure (i.e., narrow boundary) from a single spacecraft, the perfect match in directions between this high count rate and the ring distribution suggests that they could be related. The peak count rate is found at the velocity (−440 km/s, −20 km/s, 260 km/s) in MSO coordinates, i.e., at exactly the same velocity as the solar wind but with a substantial Z component.

Since the ring distribution covers nearly half a circle in velocity space, the registered counts of the ring distribution provides nearly half the ring distribution flux. The total number of counts registered for the ring distribution is about 250–300 counts/scan (nearly constant during 1337–1357 UT, i.e., nearly the same for all full scans) in the energy range of 2–3 keV. Meanwhile, solar wind alpha particles are registered with a count rate of about 2000 counts/scan (again, nearly constant) at an energy of about 1.6 keV. Since the count rate is roughly proportional to the energy flux, the total mass flux of the ring distribution (entire ring circle) is about 20% of the solar wind alpha particle flux. As mentioned in Section 2, we cannot get the total proton flux because of the instrumental limit at low energies (Fedorov *et al.*, 2006). The registered solar wind proton count rate (we took the mass range to be $m/q = 0.5$–6.0 at energy 0.6–1.0 keV) is about 5000 counts/scan (again, nearly constant), setting an upper limit to the alpha/proton mass flux ratio of about 10%. Combined, the flux of the ring distribution is less than 2% of the solar wind flux.

3.2. 22 March 2004, 1230 UT

The previous example is an ideal case because the ring distribution is detected in a single azimuthal sector over many elevation scans. However, the ring distribution is generally registered over different azimuthal sectors. Here, we present one such case. Figure 6 shows the MEX orbit and energy-time spectrograms of electrons and ions during 1223–1240 UT on 22 March 2004 (5 full scans). This is the only bow-shock crossing with PA = 2 mode (optimized for proton detection) during 2004.

The spacecraft traversed the bow shock (outbound) at around 1231 UT (end of the ramp) to 1240 UT (end of the foot): the ion data shows nearly undisturbed solar wind after 1231 UT while the electron data shows an extended foot (slightly heated from the solar wind) until about 1240 UT. The partial ring-like distribution is seen during the last full elevation scan at 1237–1240 UT (indicated by the upper horizontal arrow), well above the solar wind (1 keV for protons, indicated by the lower horizontal arrow). The highest energy portion of the ring is also recognized in the previous two full scans at around 4 keV.

Figure 7 shows the energy-time spectrograms from IMA, separated into protons and alpha particles at different azimuthal sectors at 1230–1240 UT, i.e., during the last 3 full scans of Figure 6. The solar wind is detected at elevation = 8/azimuth =

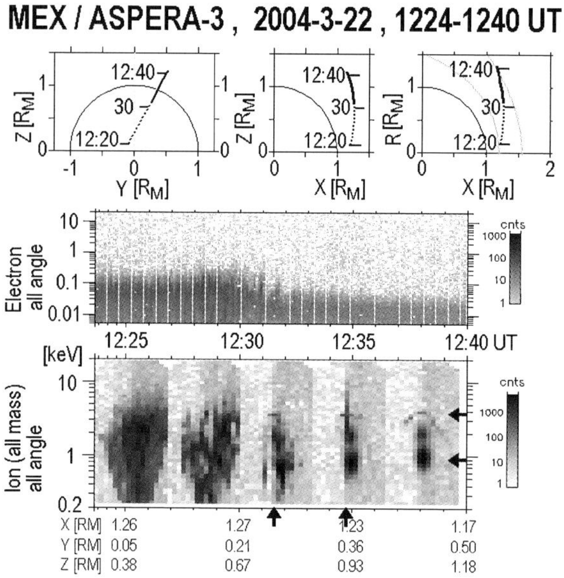

Figure 6. Overview of the MEX orbit and hot plasma data during 1223–1240 UT on 22 March 2004. The format is the same as Figure 1. ELS data indicates the bow shock outbound at around 1240 UT (end of the foot), while IMA data show nearly undisturbed solar wind already after 1231 UT (end of the ramp).

3 with a proton energy of about 1 keV and an alpha particle energy of about 2 keV. The ring distribution during the last full scan at 1237–1240 UT consists of protons in the energy range between 2 to 4 keV. The energy and direction of the identified ring distribution during the last full scan is summarized in Table III (same format as Table I).

The ring is again distributed in a plane rather than on a 3-D shell, passing exactly through the solar wind direction (elevation = 8/azimuth = 3) at its highest energy of 4 keV (or about 900 km/s) that corresponds to twice the solar wind velocity. In the first 2 full scans, only the highest energy portion is visible at around 1231:30 UT (elevation = 6–7/azimuth = 3) and 1234:50 UT (elevation = 8/azimuth = 3), although the rings' directions of these 2 full scans are not clear due to the bow shock crossing.

As mentioned in Section 3.1, the matching of directions between the solar wind and the ring's highest energy suggests that the IMF B_X is small compared to the other components. Therefore, the IMF direction inferred from this ring plane lies on a plane that azimuth = 7 (or 15) covers, and is tilted more than 45° from the center elevation direction (elevation = 8) of azimuth = 7 (or 15). Since this center elevation direction is nearly in the Z direction, the IMF direction is estimated as lying almost within the Y–Z plane and with $B_Z/B_Y \leq 1$. Note that for such an intuitive derivation, we have to compare three directions in velocity space, one at

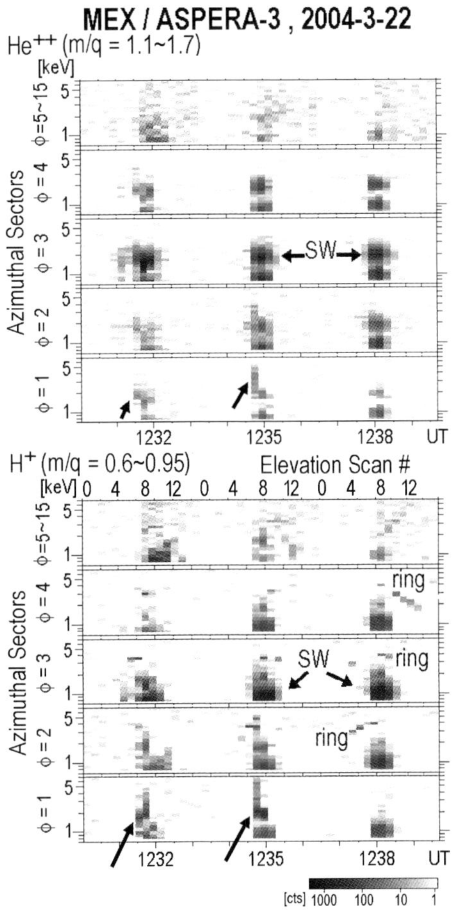

Figure 7. Energy-time spectrograms of light ions between the energies of 0.7 and 7 keV during 1230–1240 UT on 22 March 2004. The format is the same as Figure 2, except that there is no data between elevations 2 through 6 of azimuth = 1. Also, the quadrant of azimuth 9 through 15 and elevation 0 through 7 is blocked for this particular event.

the highest energy, and the others at the same (lowest possible) energy at both sides relative to the viewing direction that registers the highest energy of the ring.

To confirm this intuitive result, we again employ the minimum variance method. The minimum variance direction obtained from the velocity data shown in Table III is $\pm(-0.01, 0.88, 0.47)$ in MSO coordinates; i.e., the IMF orientation for the 22 March 2004 event is estimated as $B_X \ll |B|$ and $B_Z/B_Y \sim +0.5$ (lying almost within the Y–Z plane and tilted nearly 30° from the Y axis). The intuitive method again agrees with the minimum variance result from the manually selected data. The velocity scatter plots in both the MSO coordinates and the local Cartesian LMN coordinates determined by the minimum variance method are shown in Figure 8 (the same format as Figure 5). The scatter in the $\mathbf{V}_N$ direction is due to the $\pi/8$

TABLE III
Direction and energy of the registered ring during 1237–1239 UT, 22 March 2004.

El Az	Sensor direction (X, Y, Z)	E (keV)	V_X (km/s)	V_Y (km/s)	V_Z (km/s)
04 02	(0.91, 0.28, −0.29)	3.02	−693	−215	219
05 02	(0.94, 0.19, −0.28)	3.31	−747	−151	221
06 02	(0.96, 0.09, −0.27)	3.98	−836	− 82	231
07 02	(0.97, −0.00, −0.25)	3.98	−844	2	217
08 03	(0.99, −0.02, 0.15)	3.98	−861	22	−132
09 03	(0.98, −0.12, 0.17)	3.63	−813	101	−140
10 04	(0.83, −0.15, 0.53)	2.90	−618	110	−397
11 04	(0.81, −0.24, 0.54)	2.40	−546	164	−363
12 04	(0.77, −0.34, 0.54)	2.05	−484	210	−335
13 04	(0.73, −0.43, 0.53)	1.86	−437	253	−315

resolution of the instrument, and it cannot be reduced by tilting the N axis toward the L axis or M axis. The alignment to a circle (not ellipse) as shown in the $L - M$ plot of Figure 8b guarantees that the magnetic field orientation is approximately in either the $+\mathbf{N}$ or $-\mathbf{N}$ direction within this coordinate system.

It is worth noting that we need the energy information (i.e., the velocity vectors) to derive the correct result. If we use only the unit vectors instead of the velocity vectors, the minimum variance direction becomes $\pm(-0.99, 0.11, -0.11)$, which is nearly 90° off from the correct direction. Similarly, the intuitive method requires the energy information if the ring distribution is registered over different azimuthal sectors (as in the present case).

The minimum variance direction determined by the automatically filtered data (the same criterion as the previous event) is $\pm(-0.06, 0.99, -0.09)$, which is again about 30° off from the result obtained using the manual method. The observed ring distribution in the L-M projection plots is not as well arranged to a circle in Figure 8c as in Figure 8b. Thus, the automatically filtered data again failed to determine the magnetic field direction to the same degree as the manual method.

Similar to the previous event, one can recognize high count rates during a short period corresponding to only one elevation scan (elevation = 7) during the first 2 full scans in Figures 6 and 7 (1231:20 UT and 1234:30 UT, indicated by vertical arrows). Figure 9 shows the corresponding energy-mass matrix. Unlike the previous example, these short-time high count rates are repeated in the same direction (azimuth and elevation) at two consecutive full scans, which are more than 3 min apart. Therefore, they are most likely a spatial structure (i.e., keV ions flowing in a narrow direction (beam)) rather than a temporal structure (i.e., a narrow region of intense keV ion flux (boundary)), although the energies of these two beam-like ions are slightly different. The beam-like count is most intense at azimuth = 0/elevation = 7 (direction is (0.51, −0.12, −0.85) in MSO coordinates), and it is registered in

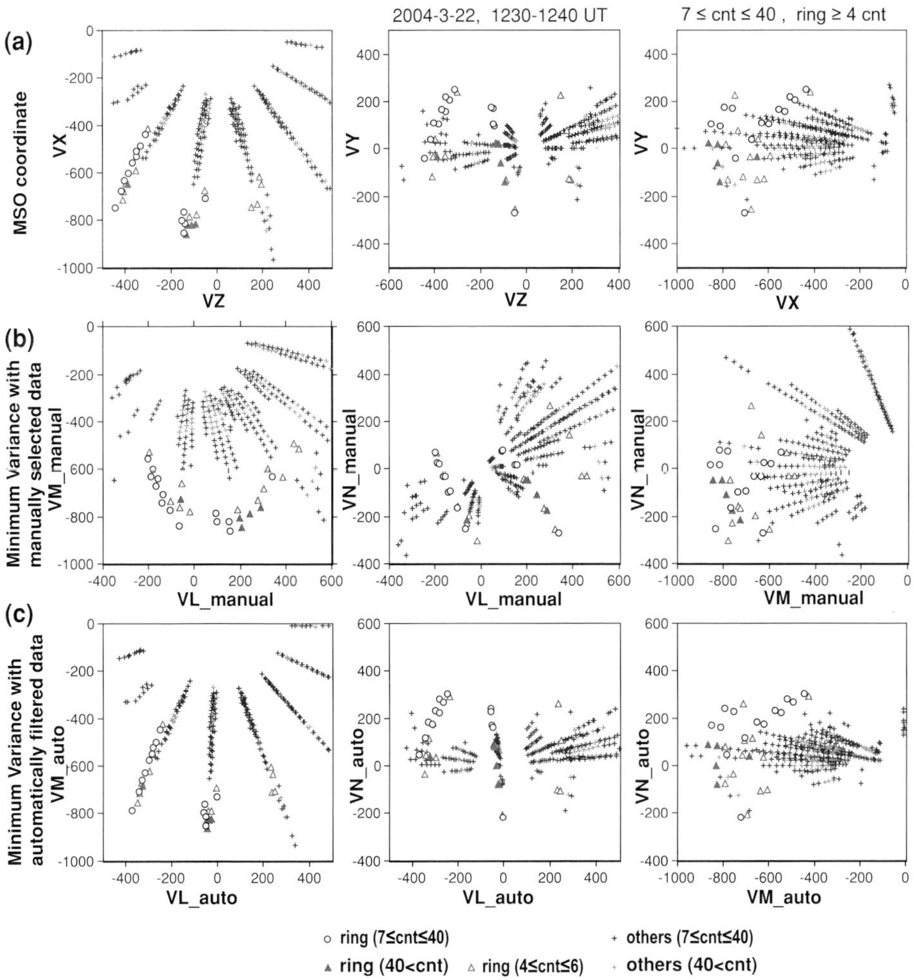

Figure 8. Velocity scatter plots of the ion data (in km/s) during 1230–1240 UT, 22 March 2004. The format is the same as Figure 5.

both the proton channel and the alpha particle channel. From its composition, these beam-like ions are of solar wind origin with energies of 2.5 keV for protons and 4.5 keV/q (9 keV) for alpha particles, i.e., about 2.5 times the solar wind energy. The energy ratio indicates that both protons and alpha particles are accelerated to nearly same velocity (650–700 km/s).

Since count rate registered at azimuth = 0 is a summation of actual count rate reaching azimuth = 0 and the contamination of all the other azimuths (from 1 to 15), it is generally not easy to extract the actual count only from azimuth = 0. However, for this particular event at azimuth = 0/elevation = 7, the count rate is largest among all azimuth (from 0 to 15), and hence this count rate cannot be the contaminated one from the other azimuth. The only other azimuth detecting

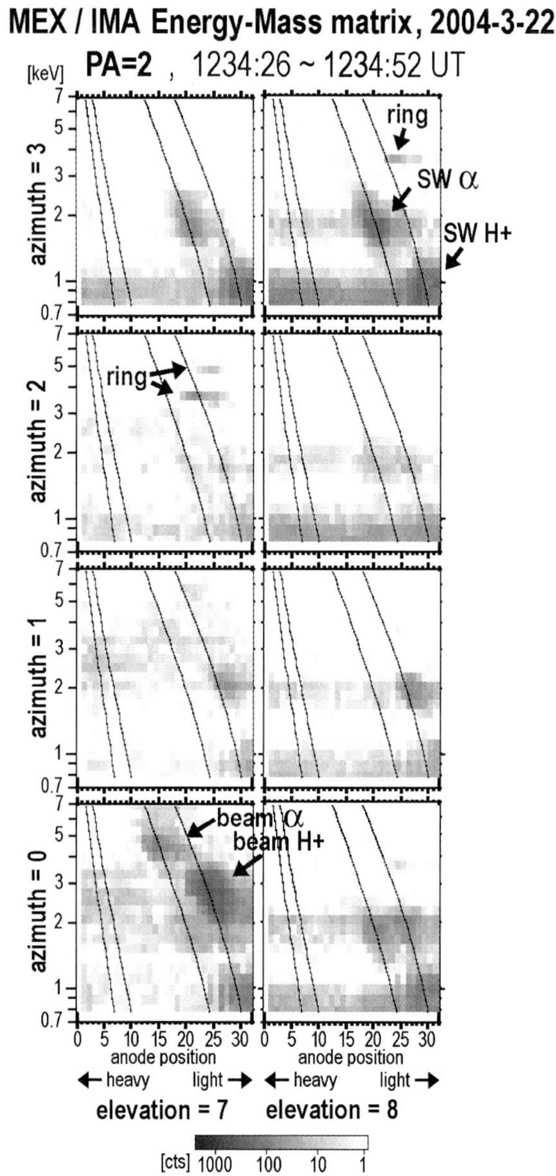

Figure 9. Energy-Mass matrices for elevation = 7 (left: 1234:27–134:39 UT) and elevation = 8 (right: 1234:39–134:51 UT) during the middle of Figure 6. The format of each panel is the same as Figure 3. From top to bottom are the azimuthal sectors 3, 2, 1, and 0. Azimuth = 3/elevation = 8 corresponds to the solar wind direction. We show azimuth = 0 because the beam-like ions are registered much stronger in azimuth = 0 than in azimuth = 1. The beam-like ions are not registered at other azimuth directions (cf. Figure 6). Therefore, the intense count at azimuth = 0 is not due to contamination from the other sectors, but is a real feature.

this beam-like ions is azimuth = 1, but it is less intense than that registered at azimuth = 0. Therefore, the count rate at azimuth = 1 is most likely due to the effect of finite spread of direction. The center of the beam-like ions might even be shifted to azimuth = 15 because, for this event, the quadrant of azimuth = 9–15 and elevation = 0–7 is blocked. However, the similar (but less intense) count at elevation = 8 is registered only at azimuth = 0–1 and not found at azimuth = 13–14 (azimuth = 15 is blocked at elevation = 8). Therefore, the center of the beam-like count is more likely at azimuth = 0. No matter where is the center direction, the registered directions of this beam-like ions are outside of the ring's plane.

The likely direction and energy of this beam-like ions (at azimuth = 0/elevation = 7) correspond to a velocity of (−360 km/s, 80 km/s, 600 km/s) in MSO coordinates, i.e., flowing nearly northward with a 30° tilt toward the solar wind direction, and about 60° in pitch angle. With such a large pitch angle, we should be able to detect a substantial number of counts from other directions at different energies which correspond to the gyration in velocity space. In the solar wind frame, the above velocity (−360 km/s, 80 km/s, 600 km/s) corresponds to about 600 km/s northward with the other velocity component about 100 km/s. Since the IMF lies within the Y–Z plane about 30° from the Y axis, the expected velocity modulation is about ±500 km/s in the Z direction (0.5–4 keV for a proton in the Martian rest frame). However, we did not identify the expected population in the expected directions.

The observed ring distribution becomes clearer as the spacecraft travels farther away from the bow shock. Only the highest energy part of the ring distribution is detected (i.e., the lower energy part disappeared in the both directions) during the first 2 scans after IMA entered the solar wind from the magnetosheath. Since the count of this highest energy part is near constant with more than 100 counts for all three scans, sensitivity of the instrument is high enough to observe the lower energy part in the first 2 scans. We cannot simply attribute this to the termination of the ion cycloid motion at the bow shock because the ring ions are flowing anti-sunward at all gyration phases. No answer to this issue has been forthcoming.

3.3. Other Events

Ring distributions are often observed. Among 38 bow-shock crossings with PA = 2 mode during the first half of 2005, 30 cases show partial ring distributions from which the orientation of the ring plane can be estimated, and among these 30 cases, 7 cases show ring distributions that are extended in elevation as many sectors as the present case. For these 7 cases, the IMF direction can be derived with similar accuracy as in the examples presented here (see Section 4.2 below). For the rest of the 23 cases, the uncertainty in deriving the IMF direction might be larger, and this topic needs further investigation.

4. Discussion

4.1. Ring Shape

The magnetic field direction can be estimated not only from the minimum variance direction but also from the direction that makes the ring-like distribution lie on a circle (not an ellipse) in the L-M projection. If the **N** direction has an angle dN from the magnetic field direction, the circular distribution in the L-M projection becomes elliptic with a diameter ratio of $\cos(dN)$. Since we can observe only part of the distribution due to the low-energy instrumental limit, this method is less quantifiable than using the minimum variance direction. Yet the combination of these two methods improves the accuracy of determination of the orientation of the ring plane. Both Figures 5b and 8b show that the velocity scatter plot in the L-M projection fits well into a (non-elliptic) circle. This guarantees that the minimum variance direction **N** is oriented approximately along the magnetic field.

Once the magnetic field direction is estimated, one can estimate the radius of the ring distribution compared to the solar wind velocity, and the offset of the fitted circle from zero velocity in the Martian rest frame in the L-M plot. Radius and offset provide information on the initial velocity $\mathbf{V}_0$, and hence, the source population. Let us examine the 27 April 2005 event, in which the data lie well on the L-M plane ($V_N = 0$ plane) as shown in Figure 5b. In this case, we just have to apply a least square fitting technique assuming a circle. Applying a linear weighting proportional to the count rate for each point, we obtain a center position $(V_L, V_M) = (40$ km/s, -430 km/s$)$ and a radius of 320 km/s. This best-fit circle does not cross the origin (offset by about 110 km/s). However, an alternate fit circle with a radius of 420 km/s (about solar wind velocity) does pass through the origin. Furthermore, the uncertainty in the normal direction dN allows us to draw an ellipse (instead of a circle) passing through the origin. Thus, the least square fitting to a circle is not necessarily the appropriate solution.

The uncertainty in fitting to a circle/ellipse is more obvious for the 22 March 2004 event because V_N is no longer zero for many points. Furthermore, the best fit in the $L - M$ projection could be an ellipse rather than a circle, with a shorter diameter in the L direction than in the M direction. In this case, the N direction could be tilted toward the $\pm L$ direction. With such an uncertainty, we cannot derive $\mathbf{V}_0$ accurately enough to distinguish whether the source ions are newly ionized neutrals (zero velocity in Martian rest frame) or reflected solar wind.

No matter how we fit the data into a circle or an ellipse in the $L - M$ projection, the nearly entire circle/ellipse is located within $V_M < 0$. This means that the ring population is always flowing in the $+M$ direction, which is mostly anti-sunward. This is also seen in the MSO coordinate velocity plot of Figure 5a: the fitted ellipse in the X–Z plane stays entirely within $V_X < 0$; i.e., the ions that form the ring distribution flow anti-sunward all of the time.

4.2. Uncertainty

The uncertainty of the derived IMF direction is determined by a combination of limitations in the spatial resolution of instrument and the data statistics. Contrary to high-angular resolution magnetic field data, the instrumental limit is very essential here because the azimuthal resolution of the IMA design is only $\pi/8$ (22.5°). Such a low angular resolution is a common limitation for almost all space plasma instruments. In the present case, small spread of the gyrating ions in the flowing directions slightly improves the resolution. If the direction of the gyrating ions is near the edge of one azimuthal sector, the ions are normally registered in the neighboring azimuthal sector as well. Therefore, the effective instrumental resolution is probably 15°–20°. This is the basic accuracy in deriving the ring's normal direction, and the final accuracy (normally worse than this accuracy as shown below) is determined by the angular coverage of the detected ring ions in velocity space.

With IMA, one can only detect a partial ring (never complete) because of the low-energy limit of IMA (cf. Section 3.1). In such a case, separation between the minimum variance direction (**N**) and the other variance directions (**M** and **L**) becomes an important factor in determining the uncertainty. For example, in both the M-N plot and L-N plot of Figure 8b, the data do not form a horizontal line, and their vertical scatter can stay within the same magnitude if we tilt the M-N axis or L-N axis. This is the degree of the uncertainty.

Three different rough estimates of the uncertainty for high-angular resolution data (like the magnetic field) are given in Sonnerup and Scheible (1998) by their equations (8.23)–(8.30) using the eigenvalues of the matrix that is used for the minimum variance analyses (cf. Table II). However, these formulas give zero uncertainty as long as the ring is observed in only one azimuthal sector, like the 27 April 2005 event (Figure 5b), because the minimum eigenvalue becomes identically zero for any 2-D limited data (i.e., limited in one azimuthal sector). Furthermore, for the 22 March 2004 event (Figures 8b and 8c), the automatic method yields a much smaller estimated error than the manual method, according to these formulas. This is because the error analyses assume that the majority of the data is relevant to the phenomena, while the automatic method cannot provide such clean data.

We are not aware of any formula that determines the uncertainty of minimum variance direction from low (and non-uniform) angular resolution data like in the present case. Therefore, we cannot tell how much the instrumental uncertainty (about 15–20 degrees) is enlarged or reduced through the analysis. Yet the instrumental uncertainty is most likely the limiting factor for at least the 27 April 2005 event (Figure 5b), because Figure 5a is only 20° off from the anticipated direction (Figure 5b) and the fitting to a circle is apparently worse. The uncertainty for the 22 March 2004 event (Figure 8b) is not clear because the scatter of data in the **N** direction is not small as shown by both the M-N plot and L-N plot. Such a large scatter in the **N** direction is mainly caused by the $\pi/8$ angular resolution in the sampling direction.

One pedagogic method to evaluate the uncertainty is to modify the input data and see the result. Here we added one possible point (elevation = 2/azimuth = 2 at 1371 eV at 1334 UT) to the data in Table I and calculated the minimum variance direction. This direction is in the next azimuthal sector to the monochromatic sector (azimuth = 3) found in Table I. The result is given in the third row of Table II (marked as "Table I***"). The minimum variance direction from this extended data is 12° away from the minimum variance direction using only Table I data.

A similar method can be used to evaluate the uncertainty in the automatic method. If we include the data from 1334:00–1337:00 UT (this part contains many non-ring signatures) into the 1337:00–1357:00 UT data that was used in Section 3.1, the resultant minimum variance direction (fourth row in Table II marked as "7≤count≤40*****") is about 60° away from the direction obtained without adding the data. Thus, the automatic method is unstable.

Another pedagogic method for the uncertainty evaluation is to move the N direction and see the change in circle fitting. For example, the $L - M$ plots of Figure 8b shows a rather good circle fit, but it is also possible to fit the data to an ellipse with the diameter ratio $\cos(30^\circ) = 0.87$. In other words, tilting the data 30° in the L-N direction still fits a circle (not a ellipse) that passes through the origin. Such a tilt does not affect the spread of V_N data in the L-N plot. In other words, the $\pi/8$ angular resolution of the instrument can easily be worsened to 30° through the spread of V_N if the ring is registered in more than one azimuthal sector.

4.3. Source Population

That the ring distribution is restricted to a plane indicates that the source population is either beam-like (just for the initial velocity) or has zero velocity in the Martian rest frame. The ring distribution is often observed many minutes after the bow shock crossing, with the flow direction mainly in the anti-sunward (e.g., see V_X values in Tables I and III). For the case of the 27 April 2005 event, a ring distribution with nearly zero parallel velocity was observed more than 1000 km upstream of where IMA crossed the bow shock (upper right panel of Figure 1). Since both the solar wind and IMF were stable during the observation period (cf. lower panel of Figure 1, and the last paragraph in Section 2.1), it is impossible that the bow shock moved outward significantly during this event. With the observed solar wind velocity (>400 km/s), the bow shock is expected to be closer to Mars than the model location shown in Figure 1. Therefore, the ring distribution existed more than 1000 km upstream the bow shock. In this sense, the formation mechanism of the ring distribution upstream of Mars is quite different from the Earth's case where the ring distribution is observed only inside the foot region (e.g., Sckopke *et al.*, 1990).

As mentioned in Section 4.1, all ions that belong to the ring distribution flow anti-sunward even for those that are not detected by IMA. Furthermore, a positive

V_Y in Table I (bow shock position is $Y < 0$) means that the ions are coming from position with a larger $|Y|$ than the MEX location. The estimated IMF direction is such that the spacecraft is not magnetically connected to any part of the bow shock during the last scan at 1354–1357 UT. From these facts, the ring distribution at 1354–1357 UT, with nearly zero parallel velocity, cannot be traced back to the bow shock even considering a finite gyroradius of the ring protons. In other words, the source population must have been transported to the upstream region from the bow shock beyond the proton gyroradius, and such transport is possible only if the source population is not ionized, i.e., in the form of the neutral hydrogen atoms. The atoms escape outward from the bow shock and start the cycloid motion after they are ionized, like the comet case. This is the so called ion pick-up by the solar wind.

According to our present knowledge, there are two types of sources for massive amounts of neutral hydrogen atoms escaping beyond the bow shock into the upstream region of the solar wind. The most probable source is the hydrogen corona that hydrostatically extends from the exobase (e.g., Lammer *et al.*, 2005). Since the distance between the Martian exobase and its bow shock is closer than that between the Earth's exobase and its bow shock, it is possible that the Martian hydrogen corona extends beyond the bow shock. The part of the hydrogen corona that is exposed to the solar wind is ultimately picked-up by the solar wind.

No solid observation exists on how far the hydrogen corona extends. If the source population of the ring distribution is the hydrogen corona, the distribution of the pick-up ions can provide the extent of the hydrogen corona (Barabash *et al.*, 1991; Barabash and Lundin, 1993; Dubinin *et al.*, 1994, 1995). This information determines the efficiency of the entire pick-up loss of the hydrogen from Mars. Since the pick-up loss is believed to contribute a large part to the hydrogen loss from Mars, deriving the coronal extent is important in understanding the evolution of the Martian atmosphere (and hence, Mars itself).

The other possible source is the newly found energetic neutral atom (ENA) jet emitting only from the subsolar bow shock region into a narrow angle mainly in the Y and Z direction (Futaana *et al.*, 2006; Gunell *et al.*, 2006). This ENA source is believed to be of solar wind origin from the energy information. The question is if the flux of the ENA jet and its ionization rate is high enough to produce the necessary newly born ions for the detected ring distribution. Unfortunately, the ENA instruments (NPD and NPI) on board MEX were turned off during both events, and we have to employ the preliminary statistics with the ENA flux of (4–7) $\times\ 10^5\ \text{cm}^{-2}\text{str}^{-1}\text{s}^{-1}$ (Futaana *et al.*, 2006). This corresponds to about two orders of magnitude less than the solar wind flux if we assume the angular extent of the ENA beam as about 1 str, and cannot be at moment dismissed because the flux of the ring distribution is also two orders of magnitude less than the solar wind flux (last paragraph in Section 3.1). Further examinations of this scenario require detailed modeling, which is beyond the scope of this paper. If we can eliminate this possibility, the frequent observation of the ring distribution provides a new way to study the Martian exosphere.

One may wonder whether the short-duration high count populations of at around 1335:10 UT (azimuth = 3/elevation = 4) in Figure 2 and at around 1231:20 UT and 1234:30 UT (azimuth = 0/elevation = 7) in Figure 6 are related to the ring distribution. The beam-like population in Figure 6 is directed in a direction away from the fitted circle or ellipse of the ring in velocity space, and it is difficult to consider a direct relation between this beam-like population and the ring population. On the other hand, the single high count event in Figure 2 falls into the fitted circle of the ring in velocity space, with the exception that this high count population has a non-zero parallel velocity (non-zero V_N), which may be attributed to instrumental uncertainty. Since this high count population have the same V_X component as the solar wind velocity pointing the anti-sunward direction with a minor (positive) V_Z component, this population cannot be the source of the ring distribution, but it could be a result of the ring distribution.

5. Conclusions

Using two examples (27 April 2005 event and 22 March 2004 event) of MEX/IMA observations, we have shown that the approximate IMF orientation can be derived from the 3-D ion velocity distribution as measured by the IMA instrument when the instrument observed ring-like distributed protons in a plane in velocity space. Since the ring distribution represents a gyration trajectory (cycloid motion) of an ensemble of ions with nearly the same initial velocity (beam-like or zero velocity), the ring's plane must be perpendicular to the IMF. Such ring distributions are observed (in the spectrogram like Figures 2 and 7) in 31 cases among 39 bow-shock crossings with the operation mode that can detect the solar wind protons during 2004 and first half of 2005.

On 27 April 2005 (1337–1357 UT) IMA detected the ring distribution in an ideal orientation, i.e., only in azimuthal sector 3 and for elevation scans from 2 through 15 (Table I). On 22 March 2004 (1230–1340 UT) the ring distribution was detected with a more general orientation, i.e., azimuthal sector 2 for elevation sector 2 to 7, azimuthal sector 3 for elevation sector 8 and 9, and azimuthal sector 4 from elevation sector 10 to 13 (Table III).

Three different methods are demonstrated to derive the IMF orientations for both events (Figures 5 and 8). The intuitive method and the minimum variance method using manually selected data agrees with each other, constructing a local Cartesian (LMN) coordinate that arranges the ring-like distribution along a circle in the $L - M$ projection with a constant V_N. The result guarantees that the magnetic field is pointing either in the $+\mathbf{N}$ or $-\mathbf{N}$ direction with the angular accuracy mainly determined by the instrumental limit and the statistics.

The minimum variance direction obtained from the automatically filtered data is more than 30° away from the direction derived from the manually selected data even for an ideal case presented here. The ring distribution projected into the derived

LMN coordinates is not as well arranged by the automatic method as by the manual method. This is because the simple automatic filters cannot provide an appropriately cleaned dataset that consist only of the ring distribution. Since the ring distribution is just one of the secondary populations seen in the solar wind, the solar wind and other components (e.g., beam-like population) severely deform the minimum variance direction.

The derived IMF direction (unit vector with arbitrary sign) in MSO coordinates is $\pm(0.34, 0.94, -0.003)$ for the 27 April 2005 event (20° from the Y axis within the X-Y plane), and $\pm(-0.01, 0.88, 0.47)$ for the 22 March 2004 event (30° from the Y axis within the Y-Z plane). The uncertainty in the derived IMF is determined by the instrumental angular resolution (15–20°) for the 27 April 2005 event, but is worsened to about 30° for the 22 March 2004 event.

The source of the ring distribution is most likely newly ionized hydrogen atoms (and picked-up by the solar wind) because the ring distribution is detected beyond the finite gyro-radius distance from the bow shock. The hydrogen corona is a strong candidate while we cannot eliminate at moment the newly found ENA jet as another possible candidate. The high count protons observed near the bow shock (at around 1335:10 UT), which shares the same velocity space as the ring distribution, cannot be the ultimate source of the ring distribution, but they could be the result of the ring distribution.

Acknowledgements

The Mars Express mission is the first European mission to the red planet managed by European Space Agency (ESA). The ASPERA-3 experiment on the Mars Express mission is a joint effort between 15 laboratories in 10 countries, all sponsored by their national agencies. We thank all these agencies as well as the various departments/institutes hosting these efforts. Work at the main PI-institute for the ASPERA-3 experiment is supported by the Swedish National Space Board, while the work at the main PI for the ELS instrument is supported by National Aeronautics and Space Administration (NASA) contract NASW-00003 in the United States. Y. Futaana is supported by Postdoctoral Fellowships for Research Abroad of the Japan Society for the Promotion Science.

References

Acuña, M. H., Connerney, J. E. P., Wasilewski, P., Lin, R. P., Anderson, K. A., Carlson, C. W., *et al.*: 1998, *Science* **279**(5357), 1676.

Barabash, S., and Lundin, R.: 1993, *Geophys. Res. Lett.* **20**, 787.

Barabash, S., Dubinin, E., Pissarenko, N., Lundin, R., and Russell, C. T.: 1991, *Geophys. Res. Lett.* **18**, 1805.

Barabash, S., Lundin, R., Andersson, H., Gimholt, J., Holmstrom, M., Norberg, O., *et al*: 2004, in *Mars Express: The Scientific Payload*, ESA SP-1240, pp. 121–139.

Dubinin, E., Lundin, R., Koskinen, H., and Norberg, O.: 1993, *Geophys. Res. Lett.* **98**(A4), 5617.

Dubinin E., Obod, D., Pedersen, A., and Grard, R.: 1994, *Geophys. Res. Lett.* **21**, 2769.

Dubinin E., Obod, D., Lundin, R., and Grard, R.: 1995, *Adv. Space Res.* **15**, 8(9), 423.

Fedorov, A., Budnik, E., Sauvaud, J.-A., Mazelle, C., Barabash, S., Lundin, R., *et al.*: 2006, *Icarus* **182**(2), 329, doi: 10.1016/j.icarus.2005.09.021.

Futaana, Y., Barabash, S., Grigoriev, A., Holmström, M., Kallio, E., C:son Brandt, P., *et al.*: 2006, *Icarus* **182**(2), 413–423, doi: 10.1016/j.icarus.2005.08.024.

Futaana, Y., Machida, S., Saito, Y., Matsuoka, A., and Hayakawa, H.: 2003, *J. Geophys. Res.* **108**(A10), doi: 10.1029/2002JA009366.

Gunell, H., Brinkfeldt, K., Holmström, M., Brandt, P., Barabash, S., Kallio, E., *et al.*: 2006, *Icarus* **182**(2), 431–438, doi: 10.1016/j.icarus.2005.10.027.

Möbius, E., Kucharek, H., Mouikis, C., Geogescu, E., Kistler, L. M., *et al.*: 2001, *Ann. Geophys.* **19**, 1411.

Mukai, T., Miyake, W., Terasawa, T., Kitayama, M., and Hirao, K.: 1986a, *Nature* **321**, 299.

Mukai, T., Miyake, W., Terasawa, T., Kitayama, M., and Hirao, K.: 1986b, *Geophys. Res. Lett.* **13**, 829.

Lammer, H., Lichtenegger, H. I. M., Penz, T., Amerstorfer, U. V., Kolb, C., and Biernat H. K.: 2005, in H. K. Biernat, H. Lammer, D. F. Vogl, and S. Mühlbachler (eds.), *Research Signpost*, Trivandrum, India, p. 209.

Lundin, R., Barabash, S., Andersson, H., Holmström, M., Grigoriev, A., Yamauchi, M., *et al.*: 2004, *Science* **305**, 1933.

Paschmann, G., Sckopke, N., Papamastorakis, I., Asbridge, J. R., Bame, S. J., and Gosling, J. T.: 1981, *J. Geophys. Res.* **86**, 4355.

Sckopke, N., Paschmann, G., Bame, S. J., Gosling, J. T., and Russell, C. T.: 1983, *J. Geophys. Res.* **88**, 6121.

Sckopke, N., Paschmann, G., Brinca, A. L., Carlson, C. W., and Luhr, H.: 1990, *J. Geophys. Res.* **95**, 6337.

Sonnerup, B. U. O., and Cahill, Jr., L. J.: 1967, *J. Geophys. Res.* **72**, 171.

Sonnerup, B. U. O., and Scheible, M.: 1998, in G. Paschmann, and P. W. Daly (eds.), *ISSI Scientific Report, ESA Publications Division*, Noordwijk, The Netherlands.

Terasawa, T., Mukai, T., Miyake, W., Kitayama, M., and Hirao, K.: 1986, *Geophys. Res. Lett.* **13**, 837.

Winningham, J. D., Frahm, R. A., Sharber, J. R., Coates, A. J., Linder, D. R., Soobiah, *et al.*: 2006, *Icarus* **182**(2), 360, doi: 10.1016/j.icarus.2005.10.033.

ENERGETIC HYDROGEN AND OXYGEN ATOMS OBSERVED ON THE NIGHTSIDE OF MARS

A. GALLI[1,*], P. WURZ[1], S. BARABASH[2], A. GRIGORIEV[2], H. GUNELL[2], R. LUNDIN[2], M. HOLMSTRÖM[2] and A. FEDOROV[3]
[1]*Physikalisches Institut, Sidlerstrasse 5, CH-3012, Bern, Switzerland*
[2]*Swedish Institute of Space Physics, Box 812,SE-981 28 Kiruna, Sweden*
[3]*Centre d'Etude Spatiale des Rayonnements, BP-4346, F-31028 Toulouse, France*
*(*Author for correspondence: E-mail: galli@space.unibe.ch)*

(Received 23 June 2006; Accepted in final form 25 October 2006)

Abstract. We present measurements of energetic hydrogen and oxygen atoms (ENAs) on the nightside of Mars detected by the neutral particle detector (NPD) of ASPERA-3 on Mars Express. We focus on the observations for which the field-of-view of NPD was directed at the nightside of Mars or at the region around the limb, thus monitoring the flow of ENAs towards the nightside of the planet. We derive energy spectra and total fluxes, and have compiled maps of hydrogen ENA outflow. The hydrogen ENA intensities reach 10^5 cm^{-2} sr^{-1} s^{-1}, but no oxygen ENA signals above the detection threshold of 10^4 cm^{-2} sr^{-1} s^{-1} are observed. These intensities are considerably lower than most theoretical predictions. We explain the discrepancy as due to an overestimation of the charge-exchange processes in the models for which too high an exospheric density was assumed. Recent UV limb emission measurements (Galli *et al.*, this issue) point to a hydrogen exobase density of 10^{10} m^{-3} and a very hot hydrogen component, whereas the models were based on a hydrogen exobase density of 10^{12} m^{-3} and a temperature of 200 K predicted by Krasnopolsky and Gladstone (1996). Finally, we estimate the global atmospheric loss rate of hydrogen and oxygen due to the production of ENAs.

Keywords: Martian atmosphere, atmospheric loss, energetic neutral atoms

1. Introduction

The emission of energetic neutral atoms (ENAs) from Mars has been addressed by several models (Kallio *et al.*, 1997; Holmström *et al.*, 2002; Barabash *et al.*, 2002; Lichtenegger *et al.*, 2002; Gunell *et al.*, 2006), but direct ENA measurements have become available only since the orbit insertion of ESA's Mars Express (MEX) spacecraft in December 2003. ENA images can give us a global picture of the interaction processes between the solar wind and the Mars atmosphere. If interpreted correctly, they allow us to deduce the physical properties of the exosphere and to quantify atmospheric loss processes. The production of ENAs in itself is not a major channel of the atmospheric escape for hydrogen or oxygen, but it scales with the ion escape rate (Barabash *et al.*, 2002). The first publications of ENA data from the ASPERA-3 experiment on MEX were concerned with the hydrogen ENAs seen on the dayside of Mars that are most probably neutralized solar wind protons; Futaana

Space Science Reviews (2006) 126: 267–297
DOI: 10.1007/s11214-006-9088-8

et al. (2006a) reported on the backscattered ENA albedo, and Futaana *et al.* (2006b) presented a study of the subsolar ENA jet, an intense and highly directional stream of hydrogen ENAs emitted from the subsolar region of Mars. The work we present here is dedicated to those ENA signals that were measured when Mars Express was on the nightside of Mars. Most observations were made for ENAs flowing away from the Sun towards the tail of the martian magnetosphere when the aperture plane of the neutral particle detector (NPD) was in the ecliptic plane, directed towards the sunward hemisphere. After the introduction we briefly characterize the NPD with which the data were obtained (Section 2) and we show the observation conditions (Section 3). Section 4 is dedicated to the hydrogen ENA signals: we present energy spectra of tailward flowing hydrogen ENAs, and we compare the global image of integral H-ENA intensities to model predictions. In Section 5 we show that no oxygen ENA signal above the detection limit has been found, and in Section 6 we deduce the global production rates of hydrogen and oxygen ENAs from our measurements. We conclude with Section 7 where we try to answer the following three questions: Can we distinguish planetary from solar wind hydrogen ENAs at the martian nightside? How does the global image of ENA fluxes compare to theoretical models? What is the loss rate of planetary hydrogen and oxygen due to the production of ENAs?

Because Mars lacks a significant intrinsic magnetic field, the solar wind directly interacts with the upper parts of the neutral atmosphere. The induced magnetosphere boundary (IMB) forces the solar wind plasma to flow around the denser part of the neutral atmosphere, but the upper atmosphere and ionosphere extend far beyond the IMB. The IMB is defined as the stopping boundary for the solar wind, the interior of the IMB is dominated by plasma of planetary origin. The location of the IMB is variable, for high solar wind pressure it is shifted to lower altitudes: on the dayside ASPERA-3 detected solar wind ions down to 300 km above the surface (Lundin *et al.*, 2004). The region outside the IMB, the magnetosheath, is dominated by the shocked solar wind plasma. Still further away from the planet (0.5 Mars radii (R_M) above the subsolar point according to the model of Kallio *et al.* (1997)) the bow shock separates the magnetosheath from the undisturbed solar wind. In regions directly exposed to the solar wind, the planetary atoms ionized by solar UV radiation, by charge-exchange or electron-impact processes, are picked up and are accelerated away from the planet in the solar wind electric field. This results in a strong erosion of the martian atmosphere. To quantify this loss process one has to measure local ion fluxes at many places inside the IMB and in the magnetosheath (Lundin *et al.*, 1989; Dubinin *et al.*, 2006a,b) or one can measure the flux of ENAs escaping the martian atmosphere to obtain an image from which a global estimate of the neutral escape can be made. Through forward modeling it is also possible to put constraints on the escape through pick-up ions.

An ENA is the product of a charge-exchange collision between an accelerated ion and an ambient neutral atom (Wurz, 2000). The ENA intensity J_{ENA} can be described by the line-of-sight (LOS) integral

$$J_{\mathrm{ENA}} = \sigma \int_{\mathrm{LOS}} ds\, n_{\mathrm{H}}(r) J_{\mathrm{p}}(s) \tag{1}$$

through the martian exosphere. Equation (1) holds for the simple case of a stream of hydrogen ions $J_{\mathrm{p}}(s)$ that are neutralized in a pure hydrogen exosphere with density $n_{\mathrm{H}}(r)$. In general, the charge-exchange cross section σ varies with ion energy and the ions may charge-exchange with other neutral species as well. Interpreting the data is therefore a complicated task as it requires models of the various ion populations. The exospheric density profiles of neutrals must also be known to interpret ENA data. A further complication is that MEX is not equipped with a magnetometer. In all studies of plasma data from the ASPERA-3 experiment the direction of the magnetic fields has to be inferred from Mars Global Surveyor (MGS) measurements or from indirect methods (Yamauchi *et al.*, this issue).

For the NPD measurements we present in this study, 'energetic neutral atom' refers to velocities between 100 and 1000 km s^{-1}, i.e., 0.1 to 10 keV (see Section 2). We only consider hydrogen and oxygen ENAs; they are the most abundant ENA species (data from the ion mass analyzer of ASPERA-3 show that O^+ ions are more abundant than O_2^+ and CO_2^+ in the martian exosphere (Carlsson *et al.*, 2006)), and other species, such as CO, CO_2, and O_2, result in ENAs out of the NPD measurement range. He-ENAs are in principle detectable by NPD, and the He^+ ion flux has been found to reach several $10^6\,\mathrm{cm}^{-2}\,\mathrm{s}^{-1}$ at some locations in the martian exosphere (Barabash *et al.*, 1995). However, these ion fluxes do not result in He-ENA intensities above the NPD detection threshold of $10^4\ \mathrm{cm}^{-2}\ \mathrm{sr}^{-1}\ \mathrm{s}^{-1}$ because the cross section for the charge-exchange between a 1 keV He^+ and a neutral hydrogen atom is two orders of magnitude lower (Macias *et al.*, 1983) than the cross section for the reaction between a proton and a neutral hydrogen atom (see Figure 1). In the NPD energy range the neutral gas species to be taken into account for ENA generation are H, H_2, and O. The neutral hydrogen population is the most important species because its density in the exosphere is higher than those of any other species due to its large scale height.

Energetic neutral hydrogen atoms observed at Mars either are solar wind protons that have been neutralized in the atmosphere (Futaana *et al.*, 2006a, b) or they are planetary hydrogen atoms that have been ionized and accelerated in the ambient electro-magnetic fields before charge-exchanging with the neutral gas again. The most important reactions that produce a hydrogen ENA are (Lichtenegger *et al.*, 2002):

$$\mathrm{H}_{sw}^{+} + \mathrm{O} \rightarrow \mathrm{H}_{sw}^{\mathrm{ENA}} + \mathrm{O}^{+}$$
$$\mathrm{H}_{sw}^{+} + \mathrm{H} \rightarrow \mathrm{H}_{sw}^{\mathrm{ENA}} + \mathrm{H}^{+}$$
$$\mathrm{H}_{sw}^{+} + \mathrm{H}_2 \rightarrow \mathrm{H}_{sw}^{\mathrm{ENA}} + \mathrm{H}_2^{+}$$

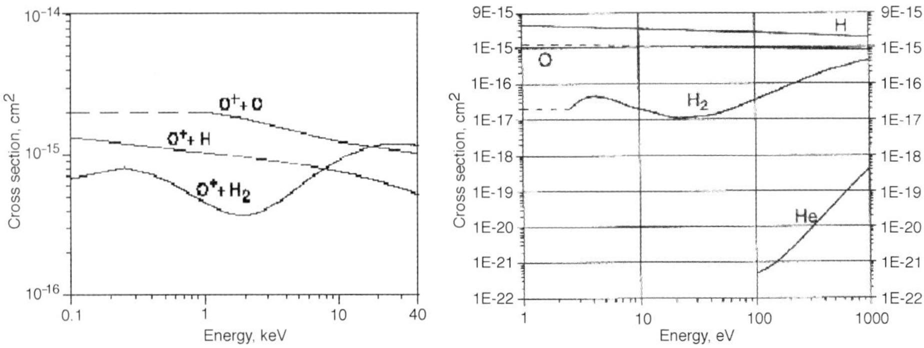

Figure 1. Cross sections of the most important charge-exchange reactions that produce oxygen (left panel) and hydrogen ENAs (right panel) in the martian exosphere. The diagrams are taken from Barabash *et al.* (2002) (left panel) and from Kallio *et al.* (1997) (right panel).

$$H_{pl}^{+} + O \rightarrow H_{pl}^{ENA} + O^{+}$$
$$H_{pl}^{+} + H \rightarrow H_{pl}^{ENA} + H^{+}$$
$$H_{pl}^{+} + H_2 \rightarrow H_{pl}^{ENA} + H_2^{+}$$

where H_{sw} stands for solar wind hydrogen and H_{pl} for planetary hydrogen. The right panel in Figure 1 shows the cross sections of the four charge-exchange reactions:

$$H^{+} + O \rightarrow H^{ENA} + O^{+}$$
$$H^{+} + H \rightarrow H^{ENA} + H^{+}$$
$$H^{+} + H_2 \rightarrow H^{ENA} + H_2^{+}$$

and

$$H^{+} + He \rightarrow H^{ENA} + He^{+}$$

The neutral helium will not be taken into account in the subsequent discussions because of its low charge-exchange cross section. The left panel in Figure 1 shows the charge-exchange cross sections for the reactions

$$O^{+}+O \rightarrow O^{ENA} + O^{+}$$
$$O^{+}+H \rightarrow O^{ENA} + H^{+}$$
$$O^{+}+H_2 \rightarrow O^{ENA} + H_2^{+}$$

In all cases the oxygen ENAs originate from Mars.

In analyzing the measurements of nightside ENAs we try to find answers to the three following questions:

1. Can we distinguish planetary from solar wind hydrogen ENAs?
   If we want to quantify the erosion processes in the martian atmosphere we need to image the planetary hydrogen ENAs. On the dayside, however, the measured H-ENA signals are dominated by neutralized solar wind protons (Futaana *et al.*, 2006a, b). Lichtenegger *et al.* (2002) suggest that on the nightside planetary

hydrogen ENAs constitute up to 15% of the tailward flow of H-ENAs. Inside the IMB planetary H-ENAs should be separable from solar wind H-ENAs because of their lower energies, whereas planetary ENAs originating in the outer magnetosheath and upstream of the bow shock are expected to have higher energies than that of solar wind ENAs.

2. Are the global ENA images consistent with theoretical predictions or do we need to revise the model input parameters?
   The temperature and density of the neutral atomic hydrogen in the exosphere, the strength of the solar wind, and the location of the bow shock and of the induced magnetosphere boundary (see Figure 2), are the most important parameters (Holmström *et al.*, 2002) for the empirical model of Kallio *et al.* (1997) of the solar wind-Mars interaction. This model in general provides a similar global flux distribution as the only parameterized model available (Gunell *et al.*, 2006).
3. What is the global production rate of planetary hydrogen and oxygen ENAs? How does it compare to measured ion escape rates (Lundin *et al.*, 1989) and to models of atmospheric escape processes (Kim *et al.*, 1998; Lammer *et al.*, 2005)?
   The tailward flux of planetary ENAs, integrated over the planet, gives a minimum estimate of the present loss rate of the martian atmosphere because the speed of the detected ENAs by far exceeds the escape velocity. Moreover, since the global production rate of ENAs depends on the available flux of ions, it can be used as a proxy for the global escape rate of ions as proposed by Barabash *et al.* (2002).

2. Instrumentation and Data Analysis

The ASPERA-3 instrument on board the MEX spacecraft comprises four different sensors. The ion mass analyzer (IMA) and the electron spectrometer (ELS) are used to measure local ion and electron densities, the NPD and the neutral particle imager (NPI) are used to detect energetic neutral hydrogen and oxygen atoms (for a more detailed description of all four sensors see for instance Barabash and Lundin (2006) or Barabash *et al.* (2004)). In the current report we restrict ourselves to ENA data that were measured by NPD. The NPD consists of two identical sensors NPD1 and NPD2 that are sensitive to ENAs in the energy range of 0.1 to 10 keV. The velocity of an incident particle can be reconstructed from the time-of-flight (TOF) between start and stop surface; each coincidence with one start and one stop pulse within the TOF range of 2048 ns is sampled in the corresponding TOF bin of 8 ns width. Each NPD sensor has one start and three stop surfaces, which provide an angular resolution of roughly 30° in azimuthal direction and 4° in elevation direction. Together, these six azimuth channels give an instantaneous field of view of NPD of $180° \times 4°$. The upper panel in Figure 3 shows a spacecraft-centered view from Mars orbit of the six NPD channels, numbered from NPD1_0 to NPD2_2.

Integration times of typically ten minutes are required to obtain a reliable TOF spectrum of ENAs in units of counts s^{-1}. First, we estimate the background of accidentals caused by UV photons and subtract it from the measured TOF spectrum.

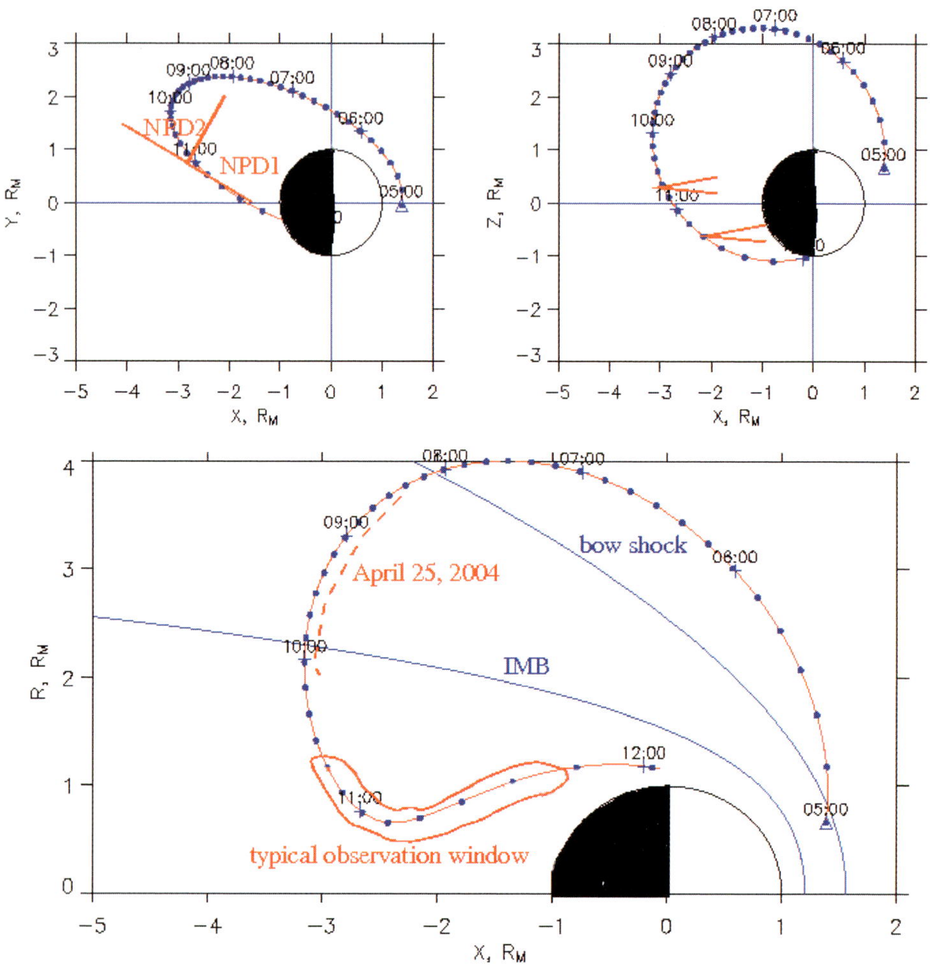

Figure 2. Typical NPD observation configuration. The time tags apply only to the orbit of April 9, 2004 (blue dots), but the shape of the orbit is representative for all orbits in April and May 2004. The upper panels show projections of the typical orbit onto the XY and the XZ plane, whereas in the lower panel the orbit is plotted in cylindrical coordinates, $R = \sqrt{Y^2 + Z^2}$ being the distance to the Mars-Sun-line X. The red wedges in the upper panels denote the fields-of-view of the two NPD sensors. In the lower panel the inner blue curve denotes the IMB, the outer curve denotes the bow shock between the undisturbed solar wind and the magnetosheath (Kallio *et al.*, 1997). Red encircled is the typical observation time of about half an hour (see Figure 3, upper panel, for a spacecraft-centered view).

The accidental count rates are spread equally across each TOF bin as they result from two uncorrelated UV photons that trigger a start and a stop pulse (see the dotted line in Figure 3). The height of the background is identified at the TOF bins from 200 to 256 that correspond to energies below 0.1 keV for which NPD is insensitive to ENAs. If an ENA signal beyond the background is recognizable, we apply a low-pass filter to eliminate any signal with periods shorter than 10 TOF

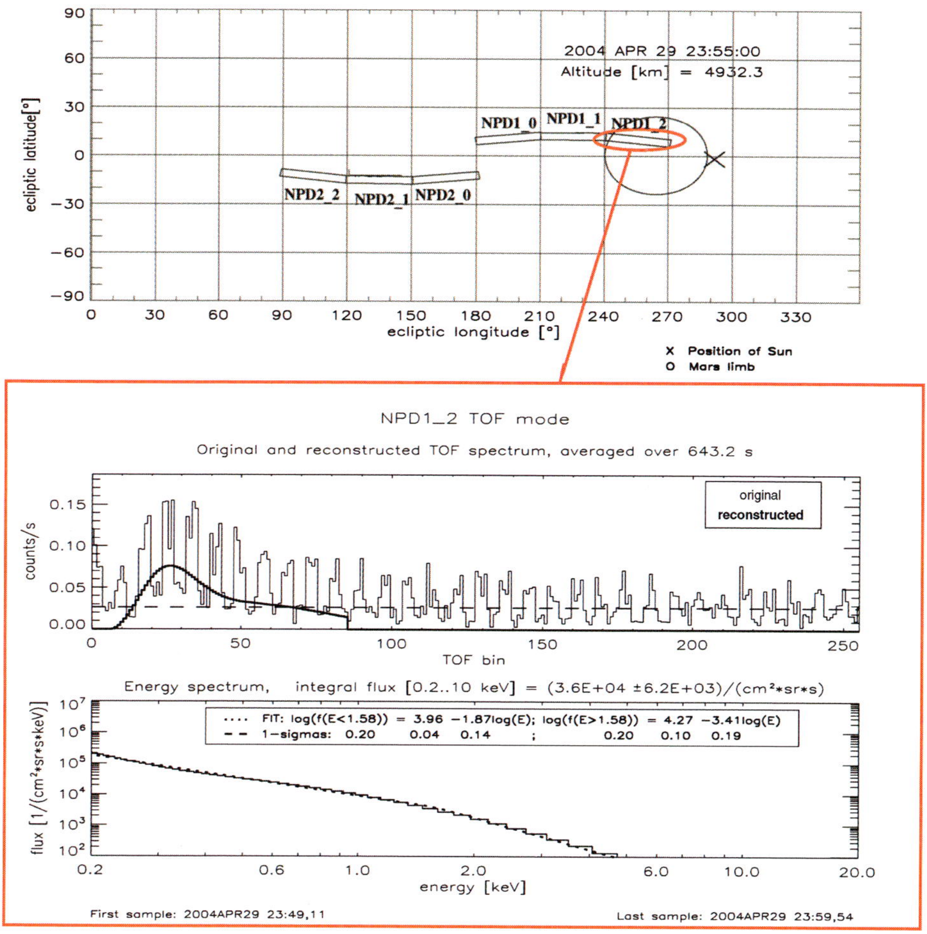

Figure 3. Hydrogen ENAs measured by ASPERA-3/NPD during a 10 min observation period on April 29, 2004. The upper panel gives the observation direction in the ecliptic reference frame, the spacecraft position is the same as in Figure 2 (red encircled area). The middle panel shows the measured (thin line) and the reconstructed (bold line) TOF spectrum. The peak between TOF bins 20 and 100 is due to hydrogen ENAs, the flat noise level of 0.03 counts per second is due to coincident UV photons (dashed line). The lower panel shows the reconstructed differential intensity that corresponds to the reconstructed TOF-signal (see Section 2).

bins, including the harmonic noise caused by the sensor electronics (see, e.g., the measured TOF spectrum in Figure 3). We then invert the instrument response by searching for an optimal fit function (the reconstructed TOF spectrum shown as bold curve in the middle panel of Figure 3), which, applied to the instrument response function, comes closest to the measured TOF spectrum. Finally, the reconstructed TOF spectrum is converted to a differential intensity energy spectrum in units of $cm^{-2}\,sr^{-1}\,s^{-1}\,keV^{-1}$ (lower panel of Figure 3) by dividing the count rate of each bin by the product of geometrical factor, energy-dependent efficiency and bin width in keV^{-1}, assuming either hydrogen or oxygen particles (see Galli *et al.*, 2006 for more details). Throughout this work, integral ENA intensities in units of cm^{-2} sr^{-1} s^{-1} are to be understood as differential intensity energy spectra integrated from 0.2 to 10 keV.

3. Observation Geometry

For the present analysis we have included all available NPD measurements from 2004 for which the NPD sensor was measuring ENAs at the nightside of Mars with a sufficiently high TOF resolution. These requirements limit our data base to the time period from February 10, 2004, to May 9, 2004. Within this period we have 21 different observation occasions, amounting to a total of ten hours of observation time. Figure 2 shows the observation configuration that is typical for the entire set of data, with the exception of the three measurements in February. The martian nightside is the black hemisphere, the Sun direction is in all images the positive x-axis. The blue dotted line indicates the spacecraft orbit, the red wedges (upper panels) indicate the NPD field-of-view directions projected onto the XY and the XZ plane. Usually, NPD1 was directed at the martian nightside while NPD2 was directed away from Mars towards the tail of the magnetosheath (see also Figure 3 for a spacecraft-centered view). For Figure 2 we use the Mars Sun Orbit reference frame, for which the + X axis is the direction from Mars to the Sun and the + Z axis is perpendicular to the orbit plane of Mars. The orbit parameters are similar for most observation occasions, thus, they can easily be compared to each other. Unfortunately, NPD was switched on only after entering the region inside the IMB, with the one exception of April 25 (red dashed curve in Figure 2). This is the only example where we see the tailward flow of ENAs in the region between the bow shock and the IMB. On all occasions the instrument had to be switched off before crossing the terminator to protect the NPD from direct sunlight.

4. Hydrogen ENAs

Hydrogen ENAs have been detected in the majority of NPD measurements on the nightside of Mars. With the exception of the three dates in February 2004 and the one measurement on April 25, 2004, all measurements were made inside the IMB (see Figure 2) at altitudes of at most 2 R_M above Mars. In the following presentation of results, we first define the typical ENA differential intensity energy spectrum measured inside the IMB and at the IMB itself (Section 4.1), and we compare this spectrum to the neutralized solar wind protons measured on the dayside of Mars (Section 4.2). Then we present the only spectrum measured in the magnetosheath, which clearly differs from the typical IMB spectrum (Section 4.3).

Finally, we construct a global picture of integral H-ENA intensities and interpret our measurements by comparing these values to theoretical predictions (Section 4.4).

4.1. The Typical Spectrum of Hydrogen ENAs

Figure 3 shows the hydrogen ENA spectrum detected in channel NPD1_2 on April 29, 2004, when the spacecraft was inside the IMB, 5000 km above the martian nightside surface. As usual, NPD1 was directed towards the sunward hemisphere, whereas NPD2 was pointed away from Mars towards the tail of the magnetosheath. As we shall see (Equation (3)), the shape of this ENA spectrum with a weak roll-over at 1.5 keV is typical for all tailward flowing ENAs within the IMB. The integral intensity of $(3.6 \pm 0.6) \times 10^4$ cm^{-2} sr^{-1} s^{-1} is rather low compared to other ENA signals because the field-of-view is directed at the planetary disk itself. In Section 4.4 we show that the intensities of tailward flowing H-ENAs vary between the detection limit and several 10^5 cm^{-2} sr^{-1} s^{-1}, the highest intensities are detected around the Mars limb towards the Sun.

As expected, the ENA streams from the nightside towards the sunward hemisphere, detected with NPD2, are even weaker, bordering to the detection threshold of 10^4 cm^{-2} sr^{-1} s^{-1}. The roll-over of the energy spectrum lies at lower energies between 0.5 and 1 keV, but there are less than 10 useful spectra to define a typical spectrum of sunward flowing ENAs.

The energy spectrum of the tailward ENA signals, on the other hand, can be well described by a two-component power law with two different slopes a_1, b_1, and a roll-over c:

$$j(E) = \begin{cases} a_0 E^{a_1} & \text{for } E < c \\ b_0 E^{b_1} & \text{for } E \geq c \end{cases} \tag{2}$$

The two-component power law in Equation (2) has been chosen because it reproduces the measurements well; it is inappropriate for only 6 of the 59 well-constrained spectra measured inside the IMB. There is no particular physical background to it, contrary to the Maxwell-Boltzmann parameterization shown in Figures 7 and 8. If one averages over the remaining 53 energy spectra of tailward flowing hydrogen ENAs, excluding only the measurements in February 2004 and on April 25 (see Section 4.3) that were obtained when MEX was at the surface of the IMB or in the magnetosheath, one finds the following median values:

$$j(E) = \begin{cases} a_0 E^{-1.7} & \text{for } E < 1.2\,\text{keV} \\ b_0 E^{-2.7} & \text{for } E \geq 1.2\,\text{keV} \end{cases} \tag{3}$$

To obtain these values for the typical spectrum in Equation (3) we averaged over all spectra of tailward flowing H-ENAs, whether they were observed from the planet itself or from the surrounding space. This is because we cannot define a typical eclipse or Mars limb spectrum. Most signals coming from the planet itself are

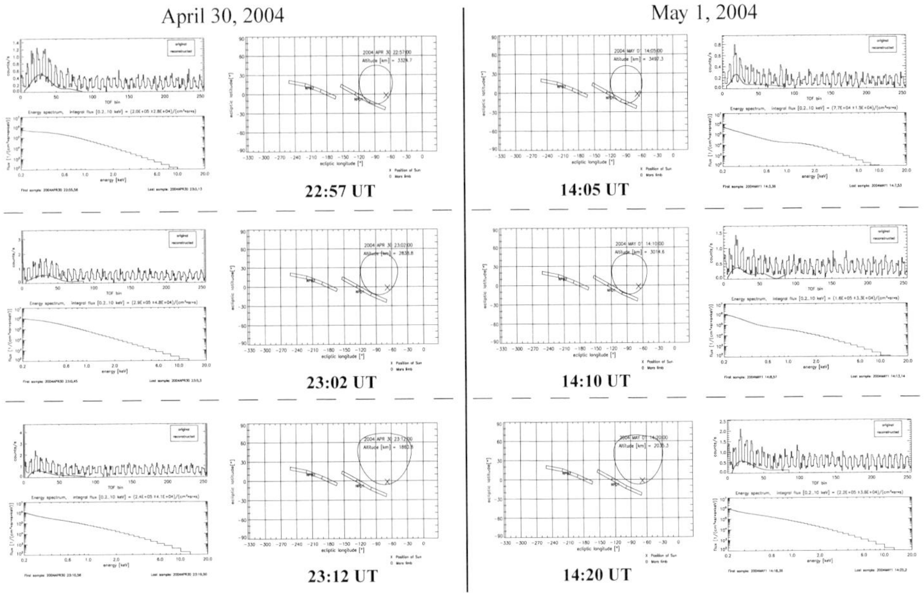

Figure 4. Temporal evolution of the H-ENA spectrum on the nightside of Mars for two consecutive measurements. The outer columns show the TOF and energy spectra measured by the NPD channel 1_2, the inner columns show the corresponding observation configuration. The format of the spectra and of the position plots is identical to Figure 3, the spacecraft position is the same as in Figure 2 (red encircled area). The ENA spectra in the two upper rows for observations close to the Mars limb show a high variability between April 30 and May 1. Once the field-of-view is out of the nightside and of the limb the spectra (bottom row) are similar to the typical nightside ENA spectrum as defined in Equation (3).

too weak (see the contour plots of integral intensities in Figure 9) to allow for a well-constrained energy spectrum, and the few well defined spectra show a high variability. This is illustrated by the time series in Figure 4 for two observations on April 30 and May 1, 2004 with almost identical orbit and viewing directions.

The energy spectra thus do not allow us to distinguish between two different ENA components inside the IMB. The only statistically significant pattern of the energy spectra inside the IMB is the correlation of the roll-over energy with distance to the Mars limb. Measurements made with the field-of-view covering the limb show a roll-over at higher energies than the signals observed from the planet itself or those far away from the limb. This is illustrated in Figure 5.

This trend reflects the variation in energy of the protons that give rise to the observed ENAs. Dubinin *et al.* (2006a, b) find, based on ion data obtained with IMA (see Section 2) in 2004, that inside the IMB the energy of H^+, H_2^+, O^+, and O_2^+ ions increases linearly with altitude from the planet because of the ambient electric field that accelerates the ions away from the planet. This increase in energy then stops in the magnetosheath (see Figure 9 in Dubinin *et al.* (2006b)). Outside the

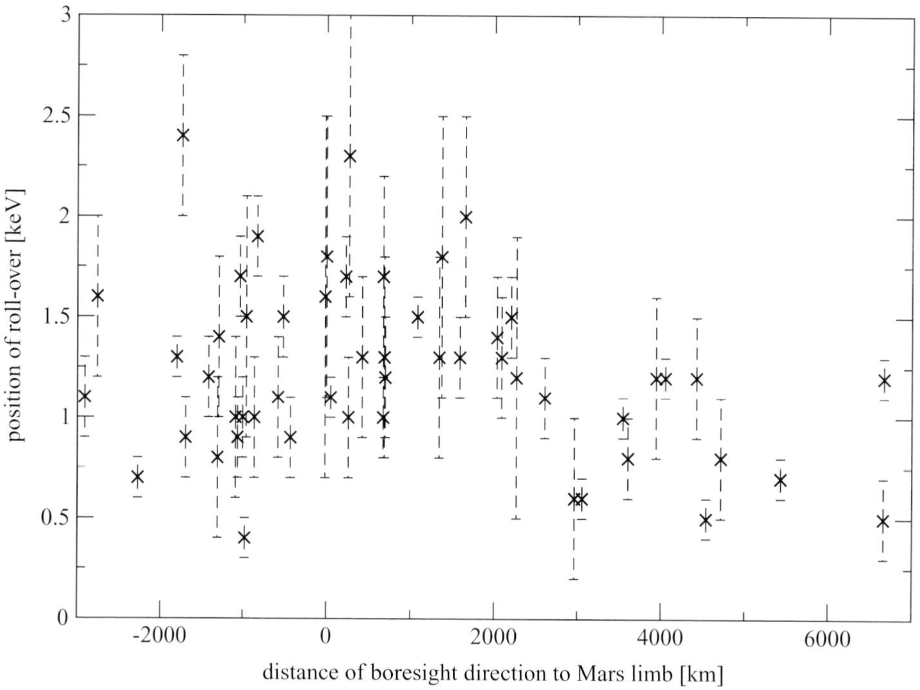

Figure 5. Place of the roll-over of observed H-ENA energy spectra plotted against the distance of the NPD LOS to the Mars limb. The roll-over is shifted to somewhat higher energies when the field-of-view is close to the Mars limb. All these measurements were obtained when MEX was inside the IMB above the martian nightside (red encircled area in Figure 2).

bow shock only solar wind protons with roughly 1 keV are available as parent ions for the production of ENAs. Based on these ion measurements one would predict the roll-over of ENA spectra in Figure 5 to be shifted to higher energies as the NPD LOS intersects ion populations at greater distances from the Mars exobase. Keep in mind, however, that unlike the local ion measurements the NPD measurements are to be interpreted as LOS integrals over regions of different ion populations. From Figure 5 it seems that the H-ENA signals observed from directions further than 2000 km away from the Mars limb are dominated by protons with solar wind energies that charge-exchange already on the dayside.

We conclude that the hydrogen ENA spectra measured inside the IMB are consistent with the ion measurements reported by Dubinin *et al.* (2006a, b). Unfortunately, they are no help to decide to what extent the observed H-ENA signals are due to planetary protons because the increase in energy with planetary distance is observed for planetary pick-up protons as well as for solar wind protons.

Contrary to the measurements in April and May 2004, the three observations in February 2004 were made when the spacecraft was at the boundary to the magnetosheath (see upper right panel in Figure 6). Again, it is not *a priori* clear whether the measured H-ENAs are neutralized solar wind protons or planetary ENAs flowing tailward along the IMB. The measured integral intensities are the highest in the entire data base as the NPD LOS is tangential to the IMB at some point close to the planet. The intensity reaches on all three occasions several 10^5 cm^{-2} sr^{-1} s^{-1}. The spectrum of the most intense ENA signal is shown in Figure 6. For this observation configuration the ENA signal in the NPD1 channels was probably more intense as they were directed to the Sun along the proton streamlines. Unfortunately, the NPD1 data of this observation are contaminated by Sun light. The median values of the 8 useful energy spectra cannot be discerned from the typical spectrum inside the IMB (Equation (3)). We find for the spectra at the boundary to the magnetosheath:

$$j(E) = \begin{cases} a_0 E^{-1.8} & \text{for } E < 1.1\,\text{keV} \\ b_0 E^{-2.9} & \text{for } E \geq 1.1\,\text{keV} \end{cases} \tag{4}$$

4.2. Comparison with Solar Wind ENAs

In order to compare to the typical night side ENA signal, Figure 7 shows an example of neutralized solar wind protons measured on the dayside of Mars on March 22, 2004. If an ENA spectrum reflects undisturbed solar wind protons that have been neutralized in the Mars exosphere before reaching the bow shock, we expect the ENAs to follow a maxwellian distribution as well. This also holds true for shocked solar wind protons in the magnetosheath as long as the protons are maxwellian with constant values for the thermal spread kT and for the bulk flow velocity $\mathbf{v}_{sw}$. Contrary to $j(E)$ (Equation (3)) for the typical spectrum of ENAs on the nightside, the neutralized solar wind spectrum in Figure 7 therefore can be compared to a fit function that has a theoretical motivation. The resulting ENA energy spectrum $j(v)$ (with $\mathbf{v}_{\text{ENA}} = \mathbf{v} \approx (v, 0, 0)$) follows (Holmström *et al.*, 2002)

$$j(v) = \int_{\text{LOS}} ds\, n_{\text{H}}(r) \frac{v}{m} n_p \sigma v_{sw,x} \left(\frac{m}{2\pi kT} \right)^{3/2} \exp\left(-\frac{m(\mathbf{v} - \mathbf{v}_{sw})^2}{2kT} \right), \tag{5}$$

where σ denotes the charge-exchange cross section, chosen to 2×10^{-15} cm^2 for the entire energy range of NPD (see Figure 1). $V_{sw,x}$ is the bulk velocity of the solar wind projected to the LOS of the detector, and $\Sigma_{\text{H}} = \int ds n_{\text{H}}(r)$ is the column density of neutral hydrogen along the LOS. Note that we have neglected the charge-exchange reactions with the neutral O and H_2, which is only legitimate if the LOS does not intersect the atmosphere below the exobase. Moreover, Equation (5) is strictly correct only for an infinitesimally small spatial aperture angle of the instrument as we have approximated $\mathbf{v}_{\text{ENA}} \approx (v, 0, 0)$ for the velocity distribution of ENAs. $T(s)$, $\mathbf{v}_{sw}(s)$, and $n_p(s)$ vary over the LOS integral for observations deep inside the bow shock, (for a model of these spatial variations see e.g. Kallio *et al.*,

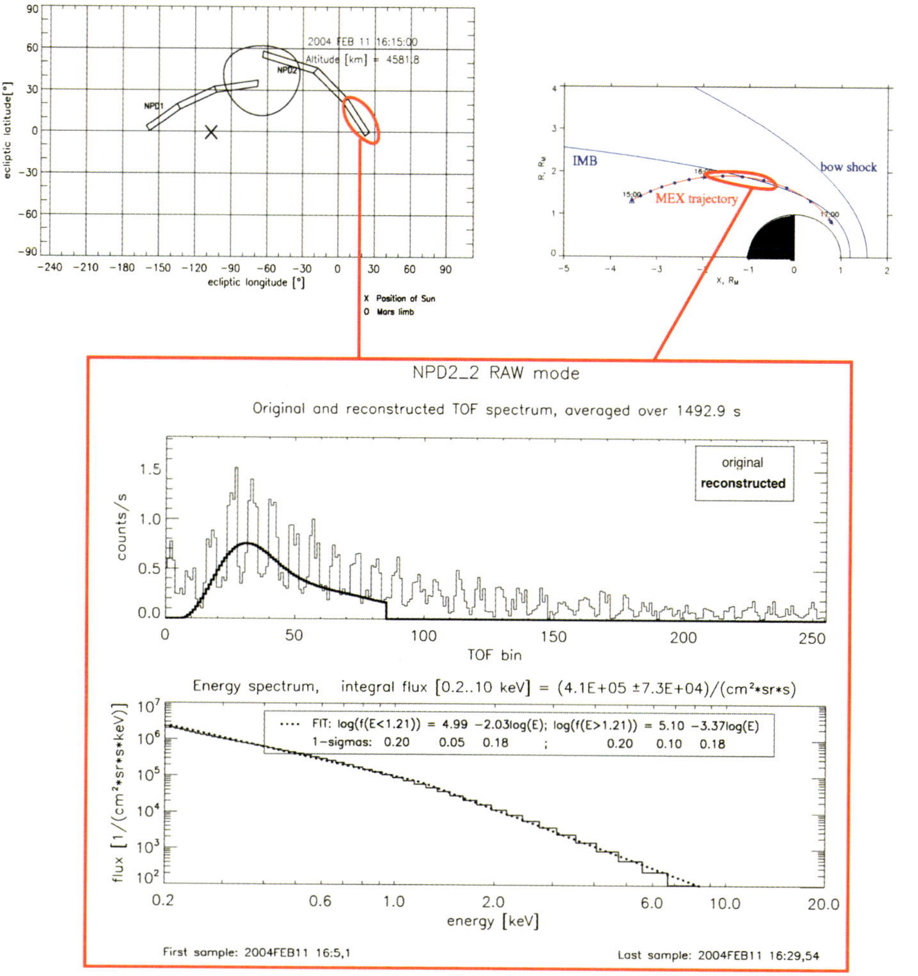

Figure 6. Hydrogen ENAs on the nightside of Mars, close to the tangential direction to the IMB surface. The data format is the same as in Figures 3 and 7. The energy spectrum (panel at the bottom) is very similar to the tailward flow of H-ENAs measured deep inside the IMB (see Figure 3).

1997), but in the case of solar wind protons with constant $T(s)$, $\mathbf{v}_{sw}(s)$, and $\mathrm{n}_p(s)$ Equation (5) can be written in units of $\mathrm{cm}^{-2}\ \mathrm{sr}^{-1}\ \mathrm{s}^{-1}\ \mathrm{keV}^{-1}$ as

$$j(E) = c_0\sqrt{E}\exp(-c_1 E + c_2\sqrt{E}). \tag{6}$$

Here, c_0, c_1, c_2 are three constants that depend on the thermal spread, the bulk flow velocity, the density of the solar wind, and on the hydrogen column density Σ_H along the LOS.

If we optimize the parameters c_0, c_1, c_2 in Equation (6) for the ENA spectrum of March 22, 2004, we obtain the fit that is plotted as dashed line in Figure 7. It implies a thermal spread of $kT = 93$ eV and a bulk velocity of $v_{sw,x} = 420$ km

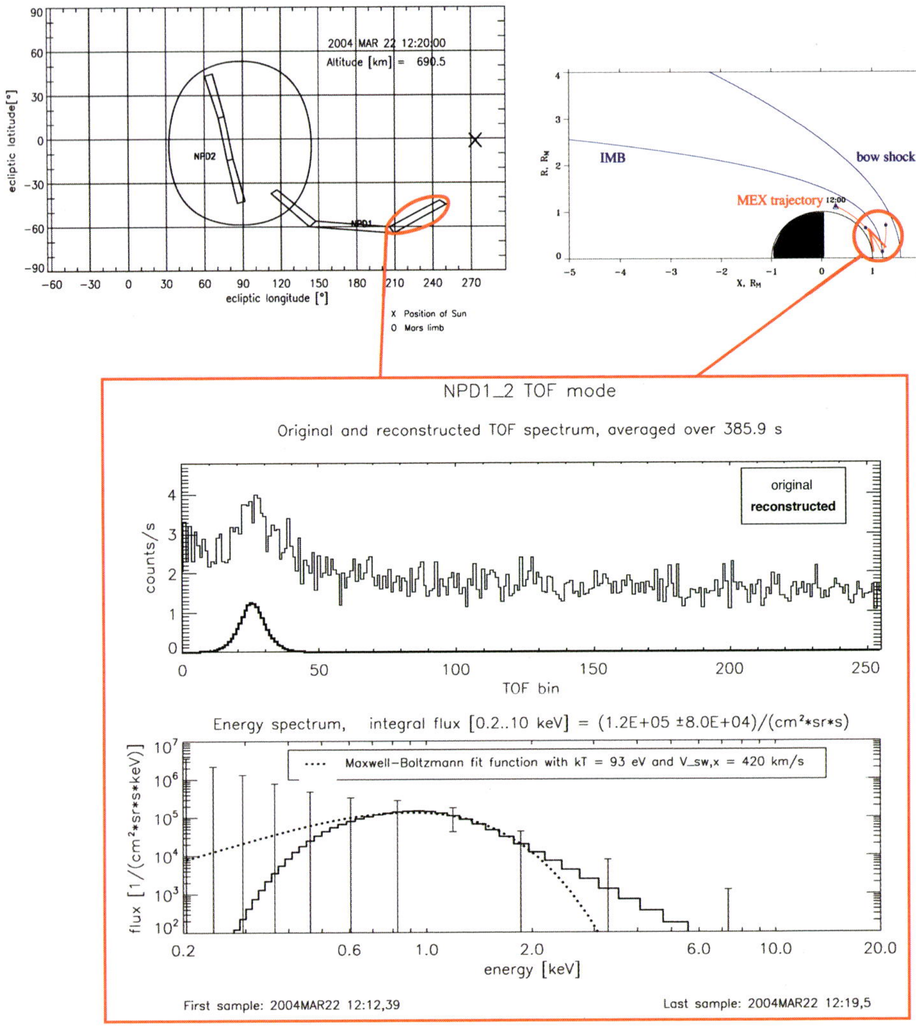

Figure 7. Solar wind ENAs on the dayside of Mars above the subsolar point. The data format is the same as in Figure 3, except for the error bars given for some of the energy bins. This measurement was made when the spacecraft was very close to the Mars surface, near the subsolar point, and the NPD channel 1_2 was pointed to the vicinity of the Sun. The energy spectrum (panel on the bottom) may easily be interpreted as maxwellian distributed solar wind protons that have been neutralized in the Mars hydrogen exosphere. Shortly after the indicated observation period the ENA intensity increased to several 10^6 cm^{-2} sr^{-1} s^{-1} and the NPD detector reached saturation.

s^{-1} for the parent proton distribution. Because of the short integration time and the high UV background level compared to the TOF signal the error bars (plotted in Figure 7 are the 1-σ error bars) of the single energy bins are huge. The optimized parameters for kT and $v_{sw,x}$ therefore have to be regarded with caution.

The integral intensity of $J_{\text{ENA}} = 10^5\ \text{cm}^{-2}\ \text{sr}^{-1}\ \text{s}^{-1}$ in Figure 7 is what one expects for a stream of solar wind protons that are neutralized in the Mars hydrogen exosphere:

$$J_{\text{ENA}} = \sigma \int_{\text{LOS}} ds\ n_{\text{H}}(r) J_{\text{p}} = 2 \times 10^5 \text{cm}^{-2}\text{sr}^{-1}\text{s}^{-1}. \tag{7}$$

For Σ_{H} we have chosen $10^{16}\ \text{m}^2$ according to a UV limb emission measurement in April 2004 (Galli *et al.*, this issue); for the proton beam $J_{\text{p}} = 10^8\ \text{cm}^{-2}\ \text{sr}^{-1}\ \text{s}^{-1}$, assuming a solar wind speed of $420\ \text{km s}^{-1}$, and a proton density of $5\ \text{cm}^{-3}$. The thermal spread of the ENA spectrum shown in Figure 7 is somewhat too high for undisturbed solar wind protons and ENAs from undisturbed solar wind are unlikely to scatter in an angle of 60 around the Sun direction. This ENA signal probably is another example of the subsolar ENA jet of shocked solar wind protons, first described by Futaana *et al.* (2006b).

The typical spectrum of nightside ENAs (Equation (3)) is much broader than the spectrum of dayside solar wind ENAs in Figure 7, and it cannot be parameterized by the Maxwell-Boltzmann distribution of Equation (6). The width of the spectrum rather favors acceleration processes of planetary ions as possible explanation. There are, however, other examples of neutralized solar wind on the dayside whose spectra look similar to the one shown in Figure 3. The energy spectrum cannot be used as evidence against solar wind ENAs because the detected ENA spectrum is a convolution of charge-exchange processes along a LOS that intersects several regions of varying temperature, density and flow directions. Inside the IMB, there is no directed flow of solar wind protons anyway and the simple Equation (5) no longer applies. To our knowledge, there is only the model of Lichtenegger *et al.* (2002) that separately derives energy spectra of planetary and solar wind ENAs for a few special locations inside the magnetosheath, but these locations cannot be directly compared to our observations. Without further model work we can only assume that the high energy part of the spectrum above 1 keV is due to ENAs that originate from planetary pick-up protons in front of the bow shock but we generally cannot distinguish between the contributions of solar wind protons and of planetary protons to the typical ENA spectrum (Equation (3)). In the following Section 4.3 we will show the only example for which we can actually separate a low-energy and a high-energy ENA component, and in Section 4.4 we will see if we can at least compare the measured integral intensities to theoretical models.

4.3. Magnetosheath Measurements

On April 25, 2004, NPD was switched on while the spacecraft was still inside the magnetosheath, far in the downwind region (Figure 8). There, a distinct bimodal spectrum of ENAs abruptly appeared and vanished again half an hour later when the spacecraft was still outside the IMB. Afterwards the intensity decreased

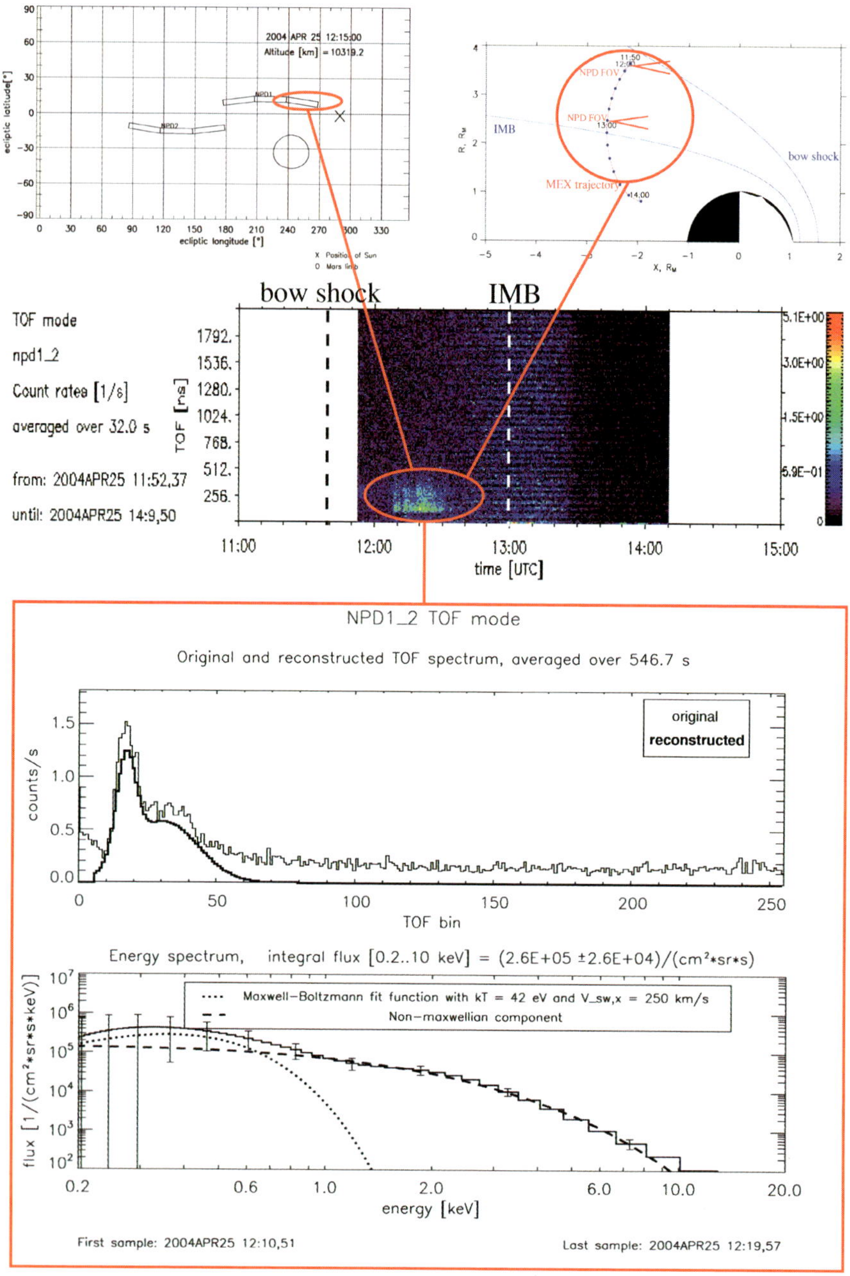

Figure 8. Hydrogen ENAs measured on April 25, 2004. The two uppermost panels show the observation configuration, in the middle follows the color-coded TOF measurement for the entire observation period of 2.25 h. The two lower panels show the TOF signal, averaged over 10 minutes from 12:10 to 12:20 UT, when the intensity reached its maximum. The dashed and the dotted line in the lowermost panel are fit curves for a maxwellian and a two-component power law distribution. The position of the IMB and of the bow shock refer to the plasma model of Kallio *et al.* (1997).

below 10^5 cm^{-2} sr^{-1} s^{-1}, and the usual ENA tail spectrum, as shown in Figures 3 and 6, was measured. Contrary to ion and electron data, the crossing of the IMB is not seen in NPD data, ENA imaging does not allow to map local plasma boundaries.

The ENA signal measured on April 25 is the only example of a bi-modal spectrum in the data base, and it is the only observation made on the nightside of Mars inside the magnetosheath, far away from the IMB. The spectrum shows a peak at 0.35 keV and a smeared out roll-over at 2.8 keV. Neither of the two components can be ENAs originated from undisturbed solar wind; the one with the roll-over at 2.8 keV is much too broad for typical solar wind temperatures, whereas energies of 0.35 keV are much too small for typical solar wind speeds between 400 and 700 km s^{-1}. The IMA data of H^+ and He^{2+} show that during the NPD measurement the local solar wind bulk velocity in the outer magnetosheath was still 0.9 keV (M. Fraenz, personal communication, 2006). The angle between the field-of-view to the Sun direction is much larger than on March 22 (see Figure 7), and yet the integral intensity again reaches several 10^5 cm^{-2} sr^{-1} s^{-1}. If we try to parameterize the bi-modal spectrum with the sum of two independent maxwellian distributions according to Equation (6), we find that the peak at 0.35 keV can be reproduced as slowed down, neutralized solar wind protons with a thermal spread of 20 to 200 eV and with a bulk velocity, $v_{sw,x}$, from 0 to 300 km s^{-1}. This is the dotted curve in the energy spectrum of Figure 8; the error bars at energies below 0.3 keV make a more accurate estimation impossible. The other component of the energy spectrum in Figure 8, represented by the dashed line, cannot be fitted by a maxwellian distribution with physically meaningful parameters without assuming an unreasonably high temperature. A two-component power law fit see Equation (2) for this component gives a roll-over of 2.8 ± 0.2 keV.

Ion measurements done with IMA (Dubinin *et al.*, 2006a,b,c) have proven the existence of planetary oxygen and hydrogen ions that are accelerated up to several keV within the magnetosheath. Oxygen ENAs can be excluded because O^+ ions would have to be accelerated to over 30 keV to reproduce the corresponding peak in the TOF spectrum. We conclude that the ENA component with the roll-over at 2.8 keV is due to planetary pick-up protons, whereas the low-energy component is due to decelerated solar wind protons. This conclusion is confirmed by the model of Lichtenegger *et al.* (2002) who predict that the planetary ENAs created upstream of the bow shock and in the outer magnetosheath have higher energies than the solar wind ENAs. The ratio of the energies of the two components found in Figure 8 (2.8 keV/0.35 keV) is higher than the factor of 4 we would expect if the stream of slow solar wind protons that produced the low energy peak had been the place where the high energetic pick-up ions were accelerated (Dubinin *et al.*, 2006c). The reason is that the NPD field-of-view did not cover the subsolar point, missing the streamlines of the fast solar wind. NPD detected only ENAs from the hot, decelerated solar wind protons and the planetary pick-up ENAs, which have a broader angular distribution than the solar wind itself. The integral intensity of the high-energy component, on the other hand, is astonishingly high: Lichtenegger *et al.* (2002) predict that the planetary H-ENA flux reaches at most 20% of the solar wind ENA flux, since the charge-exchange between solar wind protons and hydrogen neutrals is an important source for the planetary H^+ that subsequently create planetary H-ENAs. In contrast, we find for the spectrum in Figure 8 that the broad component with a roll-over at 2.8 keV accounts for as much as 60% of the entire ENA intensity J_{ENA} between 0.2 and 10 keV. To produce a beam of planetary H-ENA with an intensity of 2×10^5 cm^{-2} sr^{-1} s^{-1} it takes (according to Equation 7) a stream of planetary protons of 10^8 cm^{-2} sr^{-1} s^{-1} at some place in the magnetosheath. Up to now, no such intense pick-up proton streams with sufficient energies have been found in IMA data (Dubinin *et al.*, 2006c).

The energy spectrum and the integral intensity of the broad component are similar to the typical spectrum of tailward ENAs seen inside the IMB except for the place of the roll-over. For any other measurement in the data base the roll-over lies at energies well below 2 keV (Figure 5). We do not have other measurements from the magnetosheath on the nightside to assess if April 25, 2004, was a singular case or if ENAs measured in the outer magnetosheath always would show a bi-modal distribution.

4.4. Flux Statistics

The integrated ENA intensities can be organized into maps to be compared with simulation results. We combine all those measurements where the solar zenith angle and the distance to Mars are similar to one single plot. All measurements of tailward flowing ENAs were taken into account, including those weak ENA streams whose energy spectra are poorly constrained. During the NPD observations (see Figure 2) the solar zenith angle varied only between 135° and 160°. Because the orbit and the operational phase did not evolve much from April to May 2004 the entire data set (with the exception of the three observations in February) of 10 hours of NPD measurements may be presented by just four maps at different distances from the planet; they are shown in Figure 9. The top left panel (a) illustrates the measurement on April 25, 2004, in the magnetosheath (see Section 4.3), the other three panels (b, c, d) show the integral intensities measured inside the IMB (discussed in Section 4.1) at distances of 2.0, 1.0, and 0.6 R_{M}. We observe a homogenous picture of integral intensities inside the IMB. The signals from the Mars disk itself are weak, bordering to the detection limit of 10^4 cm^{-2} sr^{-1} s^{-1}, the more intense signals are seen around the limb close to the Sun direction. They reach values of a few 10^5 cm^{-2} sr^{-1} s^{-1} at most, corresponding to the yellow areas in Figure 9.

To map the global production rate of ENAs it would be desirable to combine the H-ENA signals that were observed close to the Sun direction in an additional image as one expects the maximum intensity of tailward ENAs to be parallel to the proton streamlines. The images in Figure 9 show all available data but they include

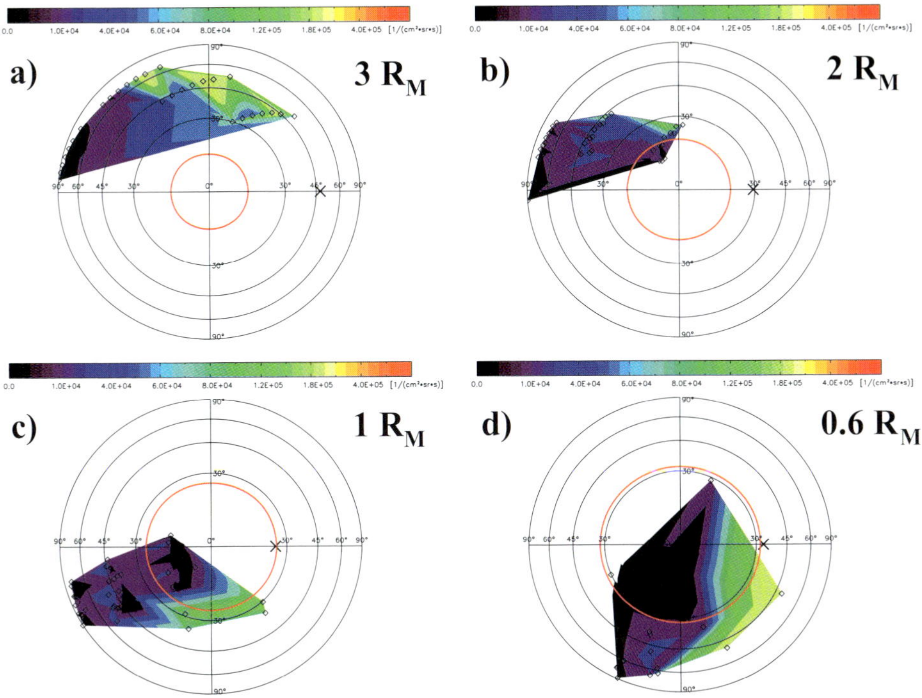

Figure 9. Images of integral H-ENA intensities in polar coordinates as seen from MEX. The spacecraft follows the trajectory shown in Figure 2, the distance to the Mars surface (red bold circle) decreases from 3 R_M (a) over 2 R_M (b) and 1 R_M (c) to about 0.6 R_M (d). The solar zenith angle is roughly 150° for all plots (X indicates the Sun position). The tiny diamonds denote the boresight directions of the NPD channels during the single observations. Image 9d can directly be compared to the model prediction shown in Figure 10.

directionality effects since the angle between the NPD field-of-view and the Sun varies from 30° to 120°. The ENA intensity drops quickly as the field-of-view is directed away from the Sun. Unfortunately, there are only two data points of 10 minutes each for which the angle between the NPD field-of-view and the Sun is smaller than 30° (the yellow area in Figure 9d). The reason for evading the Sun is the UV sensitivity of NPD. To estimate the upper limit of the global ENA production rate in Section 6 we will assume that the ENA intensities of 2×10^5 cm^{-2} sr^{-1} s^{-1} measured at these occasions apply to observations all around the Mars limb as long as the field-of-view is directed to the sub-solar point.

As comparison to the measurements shown in Figure 9d, Figure 10 shows the predictions for solar wind ENAs of a model from Gunell *et al.* (2006), and Figure 11 shows the predictions for planetary ENAs of the model from Lichtenegger *et al.* (2002), which has to be compared to Figure 9a. The ENA image shown in Figure 10 has been calculated for the observation configuration of the ENA measurements shown in Figure 9d. In the published version (Gunell *et al.*, 2006)

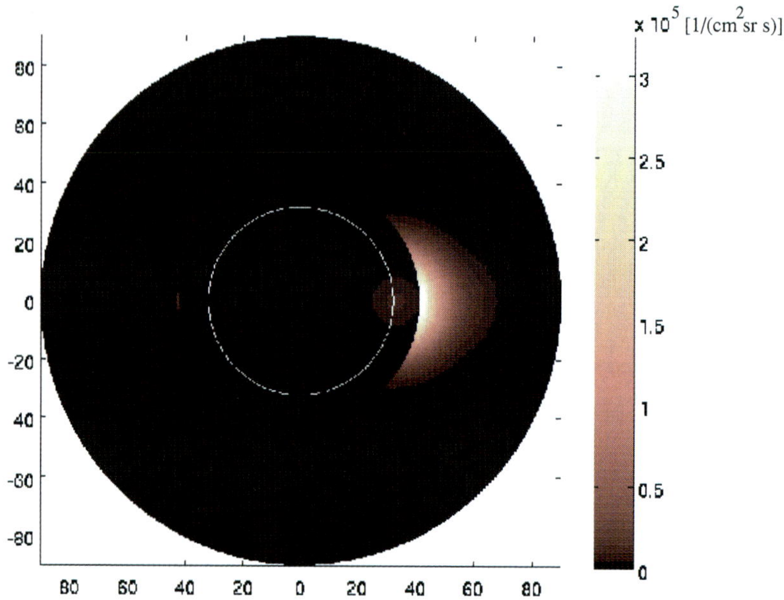

Figure 10. Model of solar wind ENA intensities (Gunell *et al.*, 2006) that has been adapted to the NPD measurement conditions in Figure 9d. The modeled solar wind ENA intensity increases to a few 10^5 cm^{-2} sr^{-1} s^{-1} close to the Sun direction. The axes are plotted in polar coordinates in units of degrees.

had taken the hydrogen exosphere parameters from the model of Krasnopolsky and Gladstone (1996). For the calculation presented here the values for the exobase density and temperature were replaced by recent measurements (Galli *et al.*, this issue). Krasnopolsky and Gladstone (1996) predicted for low solar activity $n_H = 10^{12}$ m^{-3} for the hydrogen exobase density and $T = 200$ K for the temperature. However, the UV Lyman-α limb emission (Galli *et al.*, this issue) measured during the orbit of April 25, 2004, indicates a much thinner hydrogen exosphere, $n_H = 10^{10}$ m^{-3}, with a very hot component above the exobase, $T \geq 600$ K. The hydrogen column densities are an order of magnitude lower than predicted by the model (Krasnopolsky and Gladstone, 1996) and the ENA intensities are bound to be lower than formerly calculated. The spatial distribution and the integral intensity of the solar wind ENA model (Figure 10) now match the observations (Figure 9d), whereas the cool and dense hydrogen exosphere model used in the published version (Gunell *et al.*, 2006) leads to H-ENA intensities that are an order of magnitude too high.

The planetary ENA fluxes shown in Figure 11, on the other hand, are taken directly from the publication of Lichtenegger *et al.* (2002) who also applied the model values published by Krasnopolsky and Gladstone (1996). This might explain why the measured H-ENA intensities are one or two orders of magnitude lower than Figure 11 suggests. If we assume that the high energetic H-ENA component seen

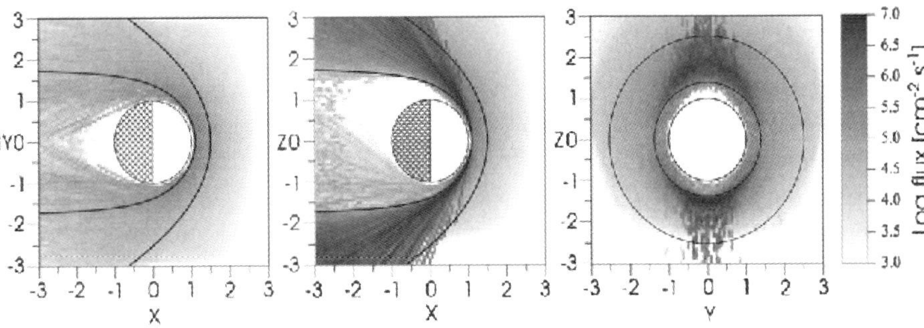

Figure 11. Modeled planetary ENA fluxes for solar minimum, taken from the work of Lichtenegger *et al.* (2002). The regions of intense production of planetary ENAs (black areas) are predicted to be aligned with the solar wind electric field (Z-axis in their reference frame). There, the modeled ENA flux exceeds 10^6 cm^{-2} s^{-1}, whereas NPD observed only 10^5 cm^{-2} s^{-1} (see Figure 9a and Equation (8)).

in the magnetosheath (Figure 9a) is due to planetary hydrogen atoms only we find for the measured ENA flux F_{pl} of planetary hydrogen atoms at most

$$F_{pl} \leq 2 \times 10^5\,\mathrm{cm}^{-2}\,\mathrm{s}^{-1}\,\mathrm{sr}^{-1} \times 0.5\,\mathrm{sr} = 10^5\,\mathrm{cm}^{-2}\,\mathrm{s}^{-1}, \tag{8}$$

where we have treated the ENA signal as a beam cone of roughly 0.5 sr aperture. This is about one order of magnitude below the predicted maximum of 10^6 cm^{-2} s^{-1} for the tailward H-ENA fluxes in Figure 11 calculated by Lichtenegger *et al.* (2002).

This comparison can only be qualitative because at every spacecraft position measurements of incoming ENAs from all directions (solid angle of 4π sr) would be required to transform the measured ENA intensities into a map of integral fluxes as presented in Figure 11. Equation (8) gives a valid estimate if the detected stream of ENAs is much more intense than ENA signals from any other direction at that place in the magnetosheath. According to the model of Lichtenegger *et al.* (2002) this assumption is justified: The dark region of maximum outflow in Figure 11 is aligned with the solar wind electric field. In this picture planetary ENAs originate from ionized hydrogen atoms that have been accelerated in the electromagnetic field of the solar wind outside the IMB before undergoing charge-exchange. MEX lacks a magnetometer but the proxies for the IMF direction derived from MGS data (2006) indicate that the magnetic field of the solar wind was directed dawnward on April 25, 2004. We can therefore assume that the measured maximum of ENA outflow seen above Mars (Figure 9a) really is coaligned with the direction of the electric field $\mathbf{E} = -\mathbf{v} \times \mathbf{B}$ at the given date and that this ENA hot spot corresponds to one of the two dark spots of planetary ENAs in Figure 11. We cannot exclude, however, that the IMF direction varied on a timescale smaller than the two hour interval provided by MGS magnetometer measurements (Brain, 2006). Although we are not certain about the IMF direction we establish a discrepancy between measured and predicted fluxes. The model of Lichtenegger *et al.* (2002) predicts solar wind ENA fluxes around the Mars limb that are even higher than the planetary ENAs by a factor of five.

Except for the one measurement in the magnetosheath (Section 4.3) the presented H-ENA signals can be understood as solar wind ENAs. The energy spectra do not allow us to deduce the source of the observed ENA signals measured inside the IMB. However, a solar wind generated ENA stream is expected to be strongly concentrated around the Mars limb in the Sun direction (see Figure 10), whereas a planetary ENA signal should be aligned with the electric field of the solar wind (see Figure 11). We do not know the exact direction of this field during the observations, but within the time span of roughly three weeks that covers the ENA images in Figure 9 we expect it to be varying randomly in the plain perpendicular to the Sun direction. It is therefore unlikely that regions with potentially larger ENA production always escaped our attention. The only region where we find high ENA intensities around the Mars limb is correlated with the direction to the Sun. This is the strongest argument that the majority of the H-ENAs seen inside the IMB are due to solar wind protons, and the intensity of a few 10^5 cm^{-2} sr^{-1} s^{-1} is consistent with solar wind ENAs, too. The measured ENA image fits to the predicted image of solar wind ENAs (compare Figures 9d and 10) except for the ENA streams from the planetary surface itself. These signals either are due to planetary ENAs, or the sharp obstacle boundary that is impenetrable to solar wind ENAs in the model is an oversimplification. According to calculations done by Kallio *et al.* (2006) solar wind ENAs are expected to spread into the eclipse because of the thermal spread of the parent solar wind protons and because the ENAs are scattered in the martian exosphere. Closer than 2 R_{M} above the planet in deep eclipse this signal of solar wind ENAs should be orders of magnitude lower than seen at Mars limb (Kallio *et al.*, 2006). The NPD measurements do not allow us to test this prediction as the spacecraft was never in deep eclipse.

Because the intensity as well as the spatial distribution of the ENA signals are consistent with model predictions for solar wind ENAs (Figure 10) we conclude that the majority of the ENAs measured inside the IMB are due to solar wind protons. Model calculations (Lichtenegger *et al.*, 2002) also indicate that the ratio of planetary to solar wind ENA fluxes should not exceed 20% at any time of the solar cycle. We cannot exclude that we have overlooked a hot spot of ENA production on the nightside of Mars (such as the dark region shown in the model plot in Figure 11) because our database is limited to a few months in 2004 and the pointing directions do not fully cover the magnetosheath or even the IMB. The narrow field-of-view covers only $6(30^\circ \times 4^\circ)$ of the entire sphere, which makes a comparison to modeled ENA fluxes difficult. Moreover, the electromagnetic configuration of the solar wind is not known either for most measurements. Nonetheless, the various observations on different days show a homogeneous picture of the tailward ENA flow. Inside the IMB the only hot spot of ENA production is related to the Sun direction; in the magnetosheath the burst of ENAs coincides with the predicted hot spot of planetary ENA production. For the following discussions we shall assume that the NPD ENA maps give us a representative image of the ENA flow at the martian nightside for solar minimum conditions and that no ENA hot spot has been overlooked.

5. Where are the Oxygen ENAs?

In analogy to hydrogen ENAs simulations predict significant oxygen ENAs in the tailward flow. Figure 13 shows the differential intensity of tailward O-ENAs according to the model of Barabash *et al.* (2002) for solar zenith angles 135° and 180°. The maximum energy to which an oxygen ion can be accelerated within the IMB region before charge-exchanging with neutral hydrogen or oxygen is predicted to be 1.7 keV. The bright spot of O-ENAs in the right panel, e.g., corresponds to an integral intensity of $J_{\mathrm{ENA}} \approx 10^6$ cm^{-2} sr^{-1} s^{-1} for the energy range between 0.1 and 1 keV. On the other hand, Gunell *et al.* (2006) compare the results of an empirical, a hybrid, and an MHD model; depending on the model they find maximum O-ENA intensities varying between 10^4 and 10^6 cm^{-2} sr^{-1} s^{-1} for the same observation conditions as plotted in Figure 13. These O-ENA simulations obviously are model-dependent. The exosphere parameters used in all models of Barabash *et al.* (2002) and Gunell *et al.* (2006) are taken from Krasnopolsky and Gladstone (1996).

O-ENA intensities as high as predicted by Barabash *et al.* (2002) should easily be detected by NPD, which has a detection limit of about 10^4 cm^{-2} sr^{-1} s^{-1} for neutral oxygen between 0.3 to 10 keV. Figure 12 shows the NPD instrument response to a monoenergetic oxygen beam of 0.7 keV from the calibration. The signature is flatter than for hydrogen ENAs but it is still well recognizable in the TOF spectrum. Because of the higher atomic mass, the peak in the TOF spectrum lies at much higher bins than for hydrogen ENAs of comparable energies (see for instance the TOF signal of neutralized solar wind protons in Figure 7).

In contrast to theoretical predictions there is not a single occasion in the entire data set of 10 hours where an oxygen ENA signal is unambiguously detected. In

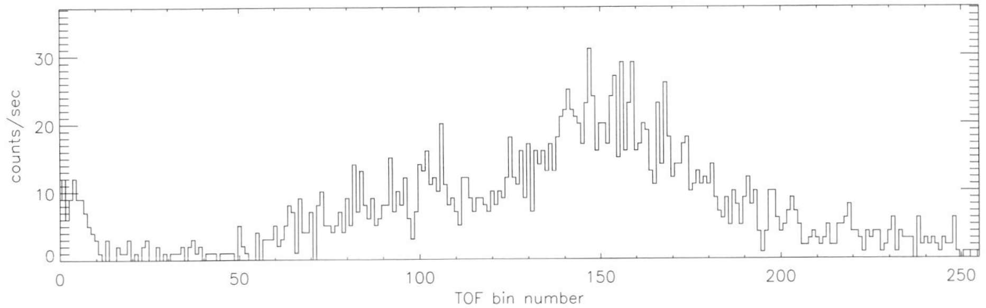

Figure 12. Response of the NPD2_1 channel to a beam of monoenergetic oxygen ENAs of 0.7 keV from the calibration. Incoming O-ENAs produce a peak in the TOF spectrum at higher TOF bins than H-ENAs and can thus be distinguished.

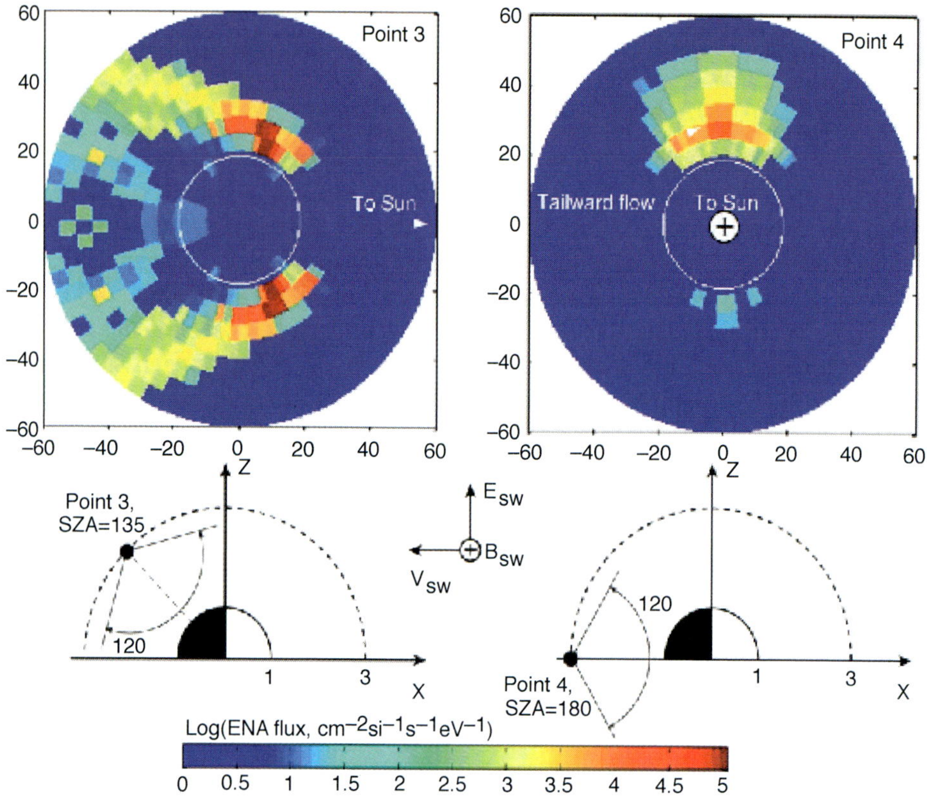

Figure 13. Model predictions for the oxygen ENA intensity on the nightside of Mars, taken from Barabash *et al.* (2002). Note that the intensity is given in eV^{-1} instead of keV^{-1}! In the right panel the ENA stream concentrates to the direction parallel to the electric field of the solar wind. Planetary oxygen ions are convected and accelerated to the same direction as planetary hydrogen atoms are (see right top panel of Figure 11).

most cases the TOF spectrum shows only UV photon noise in the TOF bins from 100 to 250 that would correspond to O-ENAs, and when there is a peak in the spectrum it is too flat to be significant. Even if we assume for all measured TOF spectra that all counts above the photon noise level in the bins from 100 to 250 were produced by O-ENAs the integral intensity never exceeds $J_{\text{ENA}} = 2 \times 10^4$ cm^{-2} sr^{-1} s^{-1}, integrated from 0.4 to 1 keV.

In summary, model calculations predict for comparable observation conditions oxygen ENA intensities that are one or two orders of magnitude higher than the upper limit of 10^4 cm^{-2} sr^{-1} s^{-1} of potential O-ENAs that we find in spring 2004. As we already stated in the discussion of the H-ENA integral fluxes (Section 4.4) it is improbable that an ENA hot spot as shown in Figure 13 was not discovered because the electric field of the solar wind was perpendicular to the NPD field-of-view for every observation. It is more plausible that O-ENA streams above the detection limit do not exist for solar minimum conditions. This is understandable because in 2004 the neutral hydrogen exosphere is much thinner than previously thought: Barabash *et al.* (2002) also used the neutral density profiles from Krasnopolsky and Gladstone (1996), which predict a hydrogen exobase density two decades higher than actual measurements for solar minimum (Galli *et al.*, this issue).

There is increasing evidence from IMA that also the Mars ionosphere was thinner during the MEX mission than predicted by models. Dubinin *et al.* (2006b) show that oxygen ions accelerated to several keV inside the IMB do exist, and a recent evaluation of IMA data (Barabash *et al.*, 2006) indicates that the integral flux of O^+ and O_2^+ is two orders of magnitude lower than the previous estimate from 1989 derived by Lundin *et al.* (1989) from Phobos 2 measurements. The O-ENA model of Barabash *et al.* (2002) was scaled to reproduce a global oxygen ion loss rate of $10^{25}\ \mathrm{s}^{-1}$, whereas the integration of O^+ fluxes, measured with IMA at solar minimum from 2004 to 2006, leads to a total ionospheric loss rate of only $10^{23}\ \mathrm{s}^{-1}$. The oxygen ion intensities inside the IMB are found to reach $J_\mathrm{p} = 10^6\ \mathrm{cm}^{-2}\ \mathrm{sr}^{-1}\ \mathrm{s}^{-1}$ at most (Barabash *et al.*, 2006). We lack recent measurements of the neutral oxygen exosphere to predict O-ENA intensities from these ion intensities. According to recent models (Krasnopolsky, 2002; Lichtenegger *et al.*, this issue) the radial column density of neutral oxygen, Σ_O, is less than $4 \times 10^{15}\ \mathrm{m}^{-2}$ for altitudes above 600 km. Since planetary oxygen ions typically reach energies ≥ 0.5 keV only above 1000 km (Dubinin *et al.*, 2006b), the production of O-ENAs is therefore dominated by the neutral hydrogen exosphere. We expect:

$$J_\mathrm{ENA} \approx \sigma \Sigma_H J_\mathrm{p}(s) = 10^3\ \mathrm{cm}^{-2}\ \mathrm{sr}^{-1}\ \mathrm{s}^{-1}, \qquad (9)$$

with the cross section $\sigma = 10^{-15}\ \mathrm{cm}^2$ for the charge-exchange between O^+ and H (Figure 1) and a hydrogen column density of $\Sigma_H = 10^{16}\ \mathrm{m}^{-2}$. Obviously, the nonexistence of oxygen ENA signals above the NPD detection threshold is consistent with the measured O^+ fluxes and the neutral hydrogen exosphere that are both one or two orders of magnitude thinner than assumed in the models of Barabash *et al.* (2002) and Gunell *et al.* (2006). The NPD observations are consistent with recent oxygen exospheric models (Krasnopolsky, 2002; Lichtenegger *et al.*, this issue), but do not rule out that the oxygen exosphere was also thinner in 2004 than theoretically predicted.

6. Global ENA Production Rates of Hydrogen and Oxygen

There are two problems if we try to estimate the global production rate of hydrogen and oxygen ENAs. First there are not enough measurements to cover the entire magnetosheath. Second, we have not clearly identified any O-ENAs at all, whereas for H-ENAs the ratio between planetary and solar wind ENAs remains unclear. At least we can give upper estimates of the global production rates of ENAs.

First we assume that the planetary ENA fluxes outside the magnetosheath are negligible. This assumption is justified because outside the magnetosheath only the subsolar ENA jet can contain notable fluxes of planetary ENAs. Futaana *et al.* (2006b) find that the integral intensity of the ENA jet amounts to only 5×10^5 cm^{-2} sr^{-1} s^{-1}, including solar wind and planetary ENAs. Because the source region of this jet is typically not more than a few 100 km in diameters the resulting loss rate amounts to 10^{20} s^{-1} at most. Second, we assume that no ENA hot spot has escaped our observations. This is plausible, but for future planetary ENA imaging missions it will be important to better map the magnetosheath as the ENA flows in the magnetosheath dominate the estimates of global production rates. Third, we integrate the H-ENA intensities shown in Figure 9 over aperture angle and cross section of the IMB and of the bow shock at the terminator. For this estimation we have approximated the IMB and the bow shock shape as axis-symmetric cones with a constant aperture angle of 26° for the IMB cone and 52° for the bow shock cone (Kallio *et al.*, 1997). If all H-ENA signals were entirely due to planetary hydrogen we then obtain 4×10^{23} s^{-1} as an upper limit of the global production rate of planetary ENAs. For this estimate we have assumed (see Section 4.4) that the tailward flow inside the IMB reaches 2×10^5 cm^{-2} sr^{-1} s^{-1} everywhere around the Mars limb for viewing directions close to the Sun and that the high ENA intensities of 3×10^5 cm^{-2} sr^{-1} s^{-1} in the magnetosheath (Figure 9a) are restricted to the two spots shown in Figure 11. If the higher ENA intensities around the Mars limb in Figure 9 are due to solar wind protons and only the weak signals from the Mars surface are due to planetary ions the global production rate reduces by an order of magnitude. In summary, we estimate the global production rate of planetary hydrogen ENAs to range between

$$Q = 10^{22} \ldots 4 \times 10^{23}\,\mathrm{s}^{-1}. \tag{10}$$

This is more than a factor of 10 below the estimate of Lichtenegger *et al.* (2002) of $Q = 10^{24} \ldots 10^{25}\,\mathrm{s}^{-1}$. It is also less than the production rate of solar wind ENAs on the dayside of Mars. For the ENA dayside albedo, reported by Futaana *et al.* (2006a), we find with $J_{\mathrm{ENA}} = 10^6$ cm^{-2} sr^{-1} s^{-1}

$$Q = J_{\mathrm{ENA}} \pi R_M^2 2\pi \approx 2 \times 10^{24}\,\mathrm{s}^{-1} \tag{11}$$

for the entire hemisphere. The tailward flowing ENAs, be they planetary or not, are a rather inconspicuous feature of the solar wind interaction with the martian atmosphere. The total H-ENA production rate of Mars, integrated over both hemispheres, including solar wind and planetary ENAs, thus amounts to

$$2 \ldots 3 \times 10^{24}\,\mathrm{s}^{-1}, \tag{12}$$

which again is a factor of 10 ... 30 times smaller than predicted by the three models published in Gunell *et al.* (2006) who apply the exospheric model of Krasnopolsky and Gladstone (1996).

Since the intensity of oxygen ENAs is at least an order of magnitude less than the intensity of H-ENAs we can set an upper limit of 10^{22} s^{-1} for the global loss rate of oxygen atoms from the atmosphere due to ENA production. Both for planetary hydrogen and oxygen ENAs the global production rates correspond to atmospheric loss rates of less than 1 g s^{-1}.

7. Conclusions

1. ENA measurements are a viable means to get a global image of the solar wind interaction with a planetary atmosphere, but to interpret the image theoretical models and knowledge of the exospheric density profile and of the solar wind parameters are mandatory. For future ENA imaging experiments we advise to make more observations in the magnetosheath. The ENA fluxes in the magnetosheath dominate the global ENA loss rate due to the large dimensions of the magnetosheath compared to the space inside the IMB or to the size of the planet. We have also learned that ENA images obtained from inside the IMB are difficult to interpret. The ENA data from the subsequent ASPERA-4/NPD experiment in Venus orbit are expected to be easier to interpret since the Venus Express spacecraft has a magnetometer.
2. On the nightside of Mars hydrogen ENAs up to a few 10^5 cm^{-2} sr^{-1} s^{-1} have been measured in the tailward flow, but no oxygen ENA signals have been detected.
   It remains unclear to which extent planetary hydrogen contributes to the measured H-ENA intensities. The presence of solar wind ENAs is clearly seen at the dayside (Futaana *et al.*, 2006a, b). On the nightside where the observation configuration should be more favorable to detect planetary ENAs (Lichtenegger *et al.*, 2002) the typical energy spectrum shows only a weak roll-over and does not allow the discrimination of two different populations. It is possible that the weak H-ENA signals from the Mars disk are due to planetary protons (Lichtenegger *et al.*, 2002), the more intense ENA streams of a few 10^5 cm^{-2} sr^{-1} s^{-1} close to the Sun direction probably are neutralized solar wind protons. The bimodal distribution seen on April 25 in the magnetosheath looks promising, the high energetic component probably consisting of planetary ENAs, but this is the only case where we can clearly identify two different components in the energy spectrum.
3. The spatial distribution of tailward flowing H-ENAs around the Mars limb is consistent with the solar wind ENA model of Gunell *et al.* (2006), when modified for a much thinner and hotter hydrogen exosphere. Likewise, the planetary H-ENA signal seen on April 25, 2004, matches the model of Lichtenegger *et al.* (2002), except that the measured intensities are an order of magnitude lower than predicted. The maximum intensities of hydrogen and oxygen ENAs have been found to be generally one or two decades lower than predicted by

all model calculations (Barabash *et al.*, 2002; Gunell *et al.*, 2006; Holmström *et al.*, 2002; Kallio *et al.*, 1997; Lichtenegger *et al.*, 2002) that use parameters for the martian hydrogen exosphere from the model of Krasnopolsky and Gladstone (1996). There, a very dense neutral hydrogen exosphere is predicted for solar minimum conditions whilst NPD observations of the Lyman-α airglow suggest a hydrogen surface density that is 20 times lower (Galli *et al.*, this issue). Since the predicted ENA intensity directly depends upon the density of neutral atoms along the LOS, the lower exospheric densities could explain the discrepancy both for hydrogen and oxygen ENAs. The exospheric densities of other species, such as O, influence the production of pick-up ions, but they are of minor importance (Kallio *et al.*, 1997) for the production of hydrogen and oxygen ENAs because of the low scale height of the oxygen corona. The neutral hydrogen density exceeds the neutral oxygen density for altitudes above 500 km (Lichtenegger *et al.*, this issue) where charge-exchange reactions of the previously accelerated oxygen ions need to take place to produce O-ENAs.

4. To whatever extent planetary atoms have contributed to the measured ENA signals, the derived global production rates correspond to atmospheric loss rates of less than 1 g s^{-1}, both for hydrogen and oxygen. This is much less than the total atmospheric loss rate of 1 kg/s for hydrogen and for oxygen according to Lammer *et al.* (2005).
5. Ionospheric escape and dissociative recombination induced escape were thought to be the dominating loss processes for atmospheric oxygen on Mars (for an overview see the work of Chassefière and Leblanc (2004)). An oxygen ion loss rate of 10^{25} s^{-1} has been estimated from measurements during the Phobos 2 mission in 1989 (Lundin *et al.*, 1989). Kim *et al.* (1998) and Lammer *et al.* (2005) estimate that the global oxygen loss rate due to dissociative recombination reaches 10^{25} s^{-1} as well. At least the ionospheric loss rate needs to be revised in the light of the recent NPD and IMA observations. The missing evidence of O-ENAs alone would not yet urge us to suggest a lower oxygen ion loss rate, but parallel to this work IMA data from 2004 to 2006 have been examined: Barabash *et al.* (2006) find that at solar minimum the O^+ escape rate is a factor of 100 below the value derived by Lundin *et al.* (1989). The charge-exchange reactions seem to be of minor importance for the oxygen loss of the martian atmosphere than previously thought. The NPD data, however, do not allow us to decide whether the oxygen exosphere itself is much thinner than modeled (Krasnopolsky and Gladstone, 1996) (in which case all oxygen loss rates would have to be downscaled) or if the extraction of planetary oxygen (Dubinin *et al.*, 2006a; Lundin *et al.*, 2004) is less efficient than assumed.
6. The actual loss rate of hydrogen of 10^{26} s^{-1} (Lammer *et al.*, 2005) must be re-examined as well because it is dominated by thermal escape, which has not been directly measured yet. An overestimation of the hydrogen exobase

density not only results in too high predicted ENA intensities, it also leads to an overestimation of the thermal escape flux F_{esc}:

$$F_{esc} = 0.5\frac{n_{\mathrm{H}}(r_{exo})U}{2\sqrt{\pi}}(\lambda + 1)\exp(-\lambda), \text{ where} \tag{13}$$

$$U = \sqrt{\frac{2kT}{m_{\mathrm{H}}}}, \text{ and } \lambda = \frac{GMm_{\mathrm{H}}}{kTr_{exo}}. \tag{14}$$

For the two widely used pairs of density and temperature of the exospheric hydrogen, namely $T = 350$ K, $n_H = 3 \times 10^{10}\,\mathrm{m}^{-3}$ according to the UV limb emission measurements done by Anderson and Hord (1971), and $T = 200$ K, $n_{\mathrm{H}} = 10^{12}\,\mathrm{m}^{-3}$ according to the model of Krasnopolsky and Gladstone (1996) for solar minimum conditions, Equation (13) yields in both cases a global loss rate of $10^{26}\,\mathrm{s}^{-1}$. The same thermal escape rate is calculated for the parameters $T = 1000$ K, $n_{\mathrm{H}} = 6 \times 10^{9}\,\mathrm{m}^{-3}$ derived by Galli *et al.* (this issue). The effects of the lower density and of the increased temperature compared to the model of Krasnopolsky and Gladstone (1996) cancel each other. The ionospheric escape rate of hydrogen was predicted to reach $10^{26}\,\mathrm{s}^{-1}$ as well (Lichtenegger and Dubinin, 1998) but probably this number has to be downscaled since in 2004 the upper limit of the total H-ENA production rate, as well as the surface density of the neutral hydrogen exosphere, are found to be one order of magnitude below the model estimate of Lichtenegger *et al.* (2002) or Gunell *et al.* (2006). Recent IMA measurements from 2005 of planetary pick-up protons (Dubinin *et al.*, 2006c) show H^+ fluxes between 10^5 and $10^6\,\mathrm{cm}^{-2}\,\mathrm{s}^{-1}$ in the outer magnetosheath, which is consistent with a neutral hydrogen exobase density $n_{\mathrm{H}} \leq 10^{11}\,\mathrm{m}^{-3}$, but an estimate of the global escape rate of H^+ from IMA data is not yet available. In any case the hydrogen loss rate due to the production of ENAs is orders of magnitude lower than the thermal escape rate.

7. It was never suggested that the production of ENAs might be the dominant escape process for oxygen or hydrogen at Mars, but given the ENA measurements done with NPD we can even conclude that it is completely negligible compared to other loss processes. Charge-exchange processes in general are of minor importance for the atmospheric loss of hydrogen. ENA measurements combined with UV limb emission measurements and independent ion measurements show a consistent picture of a thin martian exosphere at solar minimum.

Acknowledgments

The ASPERA-3 experiment on the European Space Agency (ESA) Mars Express mission is a joint effort between 15 laboratories in 10 countries, all sponsored by their national agencies. We thank all these agencies as well as the various departments/institutes hosting these efforts. This work is supported by the Swiss National Science Foundation.

References

Anderson, D. E., and Hord, C. W.: 1971, *JGR* **76**(28), 6666.

Barabash, S., Kallio, E., Lundin, R., and Koskinen, H.: 1995, *JGR* **100**(A11), 21307.

Barabash, S., Holmström, M., Lukyanov, A., and Kallio, E.: 2002, *JGR* **107**(A10), 1280.

Barabash, S., *et al.*: 2004, ASPERA-3: analyser of space plasmas and energetic ions for Mars Express, 121–139, in Mars Express: the scientific payload. ESA SP-1240, ed. by Andrew Wilson, Noordwijk, Netherlands.

Barabash, S., and Lundin, R.: 2006, *Icarus* **182**, 301.

Barabash, S., Fedorov, A., Lundin, R., and Sauvaud, J.-A.: 2006, *Science*, submitted.

Brain, D.: 2006, IMF Draping Direction at Mars, http://sprg.ssl.berkeley.edu/brain/rsrch/drapingdirxn-.html.

Carlsson, E., *et al.*: 2006, *Icarus* **182**, 320.

Chassefière, E., and Leblanc, F.: 2004,*P&SS* **52**, 1039.

Dubinin, E., *et al.*: 2006a, *Icarus* **182**, 337.

Dubinin, E., *et al.*: 2006b, *Icarus* **182**, 343.

Dubinin, E., Fränz, M., Woch, J., Barabash, S., Lundin, R., and Yamauchi, M.: 2006c, Hydrogen exosphere at Mars, pickup protons and their acceleration at the bow shock, GRL, in press.

Futaana, Y., *et al.*: 2006a, *Icarus* **182**, 424.

Futaana, Y., *et al.*: 2006b, *Icarus* **182**, 413.

Galli, A., *et al.*: 2006, *ApJ* **644**, 1317.

Galli, A., Wurz, P., Lammer, H., Lichtenegger, H. I. M., Lundin, R., Barabash, S., *et al.*: *Space Sci. Rev.*, this issue, doi: 10.1007/s11214-006-9089-7.

Gunell, H., Holmström, M., Barabash, S., Kallio, E., Janhunen, P., Nagy, A.F., *et al.*: 2006, *P&SS* **54**, 117.

Holmström, M., Barabash, S., and Kallio, E.: 2002, *JGR* **107**(A10), 1277.

Kallio, E., Luhmann, J. G., and Barabash, S.: 1997, *JGR* **102**(A10), 22183.

Kallio, E., *et al.*: 2006, *Icarus* **182**, 448.

Kim, J., Nagy, A. F., Fox, J. L., and Cravens, T. E.: 1998, *JGR* **103**(A12), 29339.

Krasnopolsky, V. A., and Gladstone, G. R.: 1996, *JGR* **101**(A7), 15765.

Krasnopolsky, V. A.: 2002, *JGR* **107**(E12), 5128.

Lammer, H., Selsis, F., Penz, T., Amerstorfer, U. V., Lichtenegger, H. I. M., Kolb, C., *et al.*: 2005, in Tokano, T. (ed.), *Advances in Astrobiology and Biogeophysics*, Springer-Verlag, Berlin, Germany, p. 25.

Lichtenegger, H. I. M., and Dubinin, E.: 1998, *EP&S* **50**, 445.

Lichtenegger, H. I. M., Lammer, H., and Stumptner, W.: 2002, *JGR* **107**(A10), 1279.

Lichtenegger, H. I. M., Lammer, H., Kulikov, Yu. N., Kazeminejad, S., Molina-Cuberos, G. H., Rodrigo, R., *et al.*: *Space Sci. Rev.*, this issue, doi: 10.1007/s11214-006-9082-1.

Lundin, R., Zakharov, A., Pellinen, R., Borg, H., Hultqvist, B., Pissarenko, N., *et al.*: 1989, *Nature* **341**, 609–612.

Lundin, R., *et al.*: 2004, *Science* **305**, 1933.

Macias, A., Riera, A., and Yañez, M.: 1983, *Phys. Rev. A* **27**(1), 213.

Nier, A. O., and McElroy, M. B.: 1977, *JGR* **82**(28), 4341.

Wurz, P.: 2000, in Scherer, K., *et al.* (ed.), *Detection of Energetic Neutral Particles, in The Outer Heliosphere: Beyond the Planets*, Katlenburg-Lindau: Copernicus-Gesellschaft e.V., 251.
Yamauchi, M., Futaana, Y., Fedorov, A., Dubinin, E., Lundin, R., Sauvaud, J.-A., *et al.*: *Space Sci. Rev.*, this issue, doi: 10.1007/s11214-006-9090-1.

OBSERVATIONS OF THE MARTIAN SUBSOLAR ENA JET OSCILLATIONS

A. GRIGORIEV[1,*], Y. FUTAANA[1], S. BARABASH[1] and A. FEDOROV[2]
[1]*Swedish Institute of Space Physics, Box 812, SE-981 28 Kiruna, Sweden;* [2]*Centre d'Etude Spatiale des Rayonnements, BP-4346, F-31028 Toulouse, France*
*(*Author for correspondence: E-mail: aug@irf.se)*

(Received 15 August 2006; Accepted in final form: 17 November 2006)

Abstract. The Neutral Particle Detector (NPD) of the ASPERA-3 experiment (Analyser of Space Plasmas and Energetic Atoms) on board the Mars Express (MEX) spacecraft observed an intense flux of H ENAs (energetic neutral atoms) with average energy of about 1.5 keV emitted anisotropically from the subsolar region of Mars. The NPD detected the ENA jet near the bow shock at radial distances of about 1 R_M from the Martian surface as the spacecraft moved outbound, while the NPD continuously pointed towards the subsolar region. The jet intensity shows oscillative behavior. These intensity variations occur on two clearly distinguishable time scales. The majority of the identified events have an average oscillation period of about 50 sec. The second group consists of events with long-scale variations with a time scale of approximately 300 sec. The *fast* oscillations of the first group exhibit a periodic structure and are detected in every orbit, while the *slow* variations of the second group are identified in ~40% of orbits. The intensity of the *fast* oscillations have a peak-to-valley ratio about 20 to 30% of the peak intensity. One of the possible mechanisms to explain *fast* oscillations is the formation of the low frequency ion waves at the subsolar region of Mars. *Slow* variations may be explained by either temporal variations in the ENA generation source or by a specific structure of the ENA generation source, in which hair-like ENA subjets can be present.

Keywords: Mars, solar wind, ENA

1. Introduction

The potential ability of ENA-imaging to diagnose plasma processes on the global scale has been recognized for a long time. A number of missions have carried out ENA measurements in order to investigate the dynamics of the terrestrial plasma environment under interaction with the solar wind (SW) (Barabash, 1995; Barabash *et al.*, 1998; C:son Brandt *et al.*, 2001; Mitchell *et al.*, 2000; Pollock *et al.*, 2000; Moore *et al.*, 2000). Modern planetary missions include ENA-detectors together with the plasma packages. The European Space Agency (ESA) missions towards Mars and Venus, namely Mars Express and Venus Express, carry the plasma and neutral particle packages ASPERA-3 and ASPERA-4 (Analyser of Space Plasma

DOI: 10.1007/s11214-006-9121-y

and Energetic Atoms) among the scientific payload. The packages are comprised of four instruments each: two ENA sensors, an electron spectrometer (ELS) and an ion mass analyser (IMA) (Barabash *et al.*, 2004, this issue). The two ENA sensors of ASPERA-3 and 4 are the Neutral Particle Imager (NPI) and the Neutral Particle Detector (NPD). The NPI is designed to measure ENAs with a relatively high angular resolution of $4.5^\circ \times 11.25^\circ$, but with no mass or energy analysis. The NPD has a lower angular resolution ($\sim 5^\circ \times 40^\circ$), but mass resolution is sufficient to separate H and O. A time-of-flight (TOF) section makes it possible to resolve particle velocities as well.

Mars does not possess a global intrinsic magnetic field, but has highly magnetized local regions (Acuña *et al.*, 1998). The lack of a magnetic field leads to the direct interaction of the SW with the upper part of the neutral atmosphere. The boundary upstream of the ionosphere is the magnetic pileup boundary (MPB) or the induced magnetosphere boundary (IMB) (Lundin *et al.*, 2004), identified and modeled by Vignes *et al.* (2000). Above the MPB, the solar wind protons are thought to be the dominant ion species, while below the MPB, planetary heavy ions prevail. A bow shock (BS) appears upstream the MPB, where the SW plasma flow is slowed from supersonic to subsonic and begins to divert around the obstacle. The region in-between the BS and MPB, containing both shocked SW ions and planetary heavy ions, is called the magnetosheath. Charge-exchange of the magnetosheath plasma with the exospheric gases, mainly H, of a density of 10^4–10^5 cm^{-3}, results in ENA generation.

The initial ASPERA-3/NPD observations revealed an ENA flux emitted anisotropically from the subsolar region of Mars (Futaana *et al.*, 2006). The ENA flux has been detected in the energy range 0.3–3 keV/amu when the FOVs of the NPD pointed toward the subsolar exosphere. Different mechanisms of ENA generation around Mars have been suggested, namely charge-exchange of the solar wind protons, hydrogen backscattered from the martian exosphere (Kallio and Barabash, 2001) and accelerated planetary ions, neutralized by charge-exchange in the exosphere (Lichtenegger *et al.*, 2002).

NPD observations revealed other important features of the ENA flux. First, it depends highly on the position and attitude of the satellite. So the ENA flux was being detected only under suitable SC location even though the FOV of the NPD was viewing toward the subsolar region. This was explained in terms of a highly directional ENA emission around the subsolar region: a subsolar ENA jet. Secondly, the ENA count-rate was found to decrease very rapidly (on the order of several tens of seconds) as long as the SC moved across the jet. This is mostly due to the spatially confined source of the ENA jet. Strictly speaking, one has to talk in terms of an ENA cone but statistical analysis of all observations will be conducted at a later time. The ENA jet flux intensity variations have been frequently observed.

The present study is focused on the observations and analysis of the ENA flux oscillations.

2. NPD Instrumentation and Data Set

A detailed description of the NPD instrument is available elsewhere (e.g. Futaana *et al.*, 2006; Barabash *et al.*, this issue). Here we present only information necessary in order to easier understand the data used as well as the observation geometry.

The NPD sensor consists of two detectors, NPD-1 and NPD-2, which are fully identical, differing only in their direction of the field of views (FOV). Each detector has a $9^\circ \times 90^\circ$ intrinsic FOV divided into three pixels (Dir-0, -1 and -2). The angular resolution per single pixel is $\sim 5^\circ \times 40^\circ$ of full width at half maximum (FWHM). Hence, a slight overlap of neighboring pixels FOV exists. The NPD detects ENA differential fluxes within the energy range 100 eV to 10 keV and is capable of resolving H and O by TOF measurement or pulse-height analysis.

As the NPD sensor possesses an open architecture, UV photons and ENA cannot be separated by the entrance collimator. UV photons entering the instrument interact with solid surfaces in a similar way as particles, causing the electron emission, and creating a lot of non-correlated events in the START and STOP detection systems. These non-correlated signals are random and occasionally they can correlate within the coincidence time gate, thus producing a false "ENA event". Since the TOF window is long enough, distribution of the false TOF signal (noise background) is basically constant over the entire TOF window. Hence the noise background level in the TOF spectrum can be estimated (see Appendix). Afterwards, during the data analysis, it was subtracted from the recorded signal.

In this study we used a data set obtained by the NPD-1, since NPD-2 did not view toward the subsolar region. The data set consists of 38 orbits recorded during the period May 24 – July 1, 2004 when the orbit and the attitude of the Mars Express spacecraft (MEX) were optimal to investigate ENA emissions from the subsolar region of Mars.

During this observation period, the NPD sensor was switched into so-called *BINning matrix* mode (shortly *BIN* mode). In this mode, the TOF of the correlated signals (valid START and valid STOP) is stored in logarithmically divided 16 TOF bins. Full spectrum accumulation time is 1 sec.

3. Data Analysis

The NPD-1 sensor registered anisotropic H ENA flux emitted from the martian subsolar region (Futaana *et al.*, 2006). The flux is an ENA jet/cone propagating outward from the subsolar region of Mars and spatially expanding. Unfortunately, the observation region, constrained within a limited range of angles, relative to the ecliptic plane, does not allow investigation of the spatial geometry of the detected ENA flux, whether it is just a jet or if it possesses a cone-like structure. The basic characteristics of the jet are as follows: averaged energy $\sim$1.5 keV/amu,

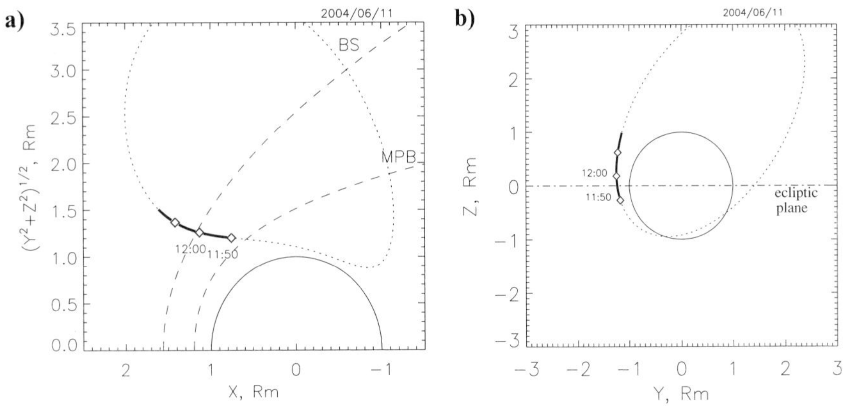

Figure 1. (a) MEX trajectory on June 11, 2004 (orbit 499). The cylindrical coordinate system based on the MSO reference frame is used. Thick line depicts the measurement interval. Dashed lines are the modeled bow shock (BS) and magnetic pileup boundary (MPB) (Vignes *et al.*, 2000). (b) MEX trajectory, projected in $YZ_{\rm MSO}$ plane (seeing from the Sun direction).

energy range is from 0.34 to 3.0 keV/amu. ENA differential flux is estimated to be $j \sim (4\text{–}7)\times 10^5 \mathrm{cm}^{-2}\mathrm{s}^{-1}\mathrm{sr}^{-1}$.

3.1. Observation Geometry

A typical MEX spacecraft orbital position (the orbit 499, on June 11, 2004) during the measurement cycle is shown in Figure 1. The orbital period during this time was about 6.5 hr. The NPD operations started ~20 minutes after the periapsis and last for ~30 minutes. The sensors were operational while the spacecraft was moving on the dayside outbound through the magnetosheath region through the bow shock and the region of undisturbed solar wind.

The plot on the left side (Figure 1(a)) is in the cylindrical coordinate system based on the Mars solar orbital (MSO) reference frame, scaled by martian radii ($R_{\rm M}$). In this frame, the Mars-Sun line is defined as the $+X$ direction, the $+Z$ axis is perpendicular to the orbital plane of Mars and the $+Y$ axis completes the right-hand orthogonal set and therefore $+Y$ is approximately opposed to the orbital motion of Mars. The vertical axis is a radial distance from the Mars-Sun line ($R = \sqrt{(Z^2+Y^2)}$). Dashed curved lines show the magnetic pileup boundary (MPB) and the bow shock (BS) locations modeled by Vignes *et al.* (2000). A dotted line shows the spacecraft (SC) trajectory during the observation period. The orbital part where the NPD sensor was switched on (11:50–12:17 UT) is depicted as a thick line, diamonds along this line represent time intervals of 10 min. Figure 1(b) shows the SC location projected on a $YZ_{\rm MSO}$ plane. MEX crosses the ecliptic plane, shown by a dot-dashed line, at the beginning of every measurement session.

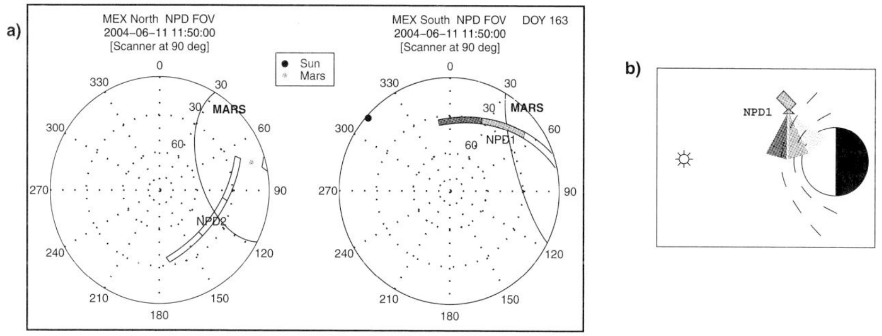

Figure 2. (a) NPD sensor observation geometry in the ecliptical coordinate system taken for the epoch 2000, with the SC in its center. Martian limb is shown by an ellipse. The elongated fans correspond to the both NPD FOV. A black and a grey polygon depict NPD-1 viewing directions Dir-1 and -2. The dark and light dots show the location of the centers of Sun and Mars correspondingly. (b) A sketch (not in scale) emphasizes the MEX location and the viewing direction of the NPD-1 during the measurements. The black and grey numbered sectors correspond to the NPD-1 channels Dir-1 and -2 FOV. A sun sign shows the Sun direction.

The MEX was nadir pointing during the observations near the perigee. The plane where ASPERA-3 was mounted was perpendicular to the nadir, facing the planet. In this configuration, the symmetry axis between the channels DIR-0 of NPD-1 and NPD-2 detectors was pointing towards the martian center (see Figure 1 in Futaana *et al.* (2006)). The NPD-1 FOV was covering the subsolar region of the Mars. Figure 2(a) shows the sensor observation geometry. In this plot, the SC-centered ecliptical coordinate system of the epoch 2000 is employed. The northern hemisphere is shown on the left side of the figure and the southern hemisphere is shown on the right side. The vantage point is the origin of the coordinate system. Martian limb is shown by an ellipse. It is partially visible on both hemispheres. The elongated fans show the NPD-1 and NPD-2 FOV. A black and a grey polygon depict NPD-1 viewing directions Dir-1 and -2. The dark and light dots show the location of the centers of Sun and Mars. Figure 2(b) represents a sketch that clarifies the NPD-1 observation geometry, emphasizing the sensor's location and attitude during the measurement. The black and grey sectors correspond to the detector channels Dir-1, -2 FOV. The channels DIR-1 and -2 of the NPD-1 sensor were observing the martian subsolar region and therefore suited for the subsolar ENA jet emission investigation.

3.2. ENA Jet Fluctuation Observation

During the observation period described in Section 2, the NPD sensor was operating in the BIN mode. The energy-time spectrograms obtained by the sensor channels DIR-1 and DIR-2 are shown on upper and middle panels of Figure 3. The data

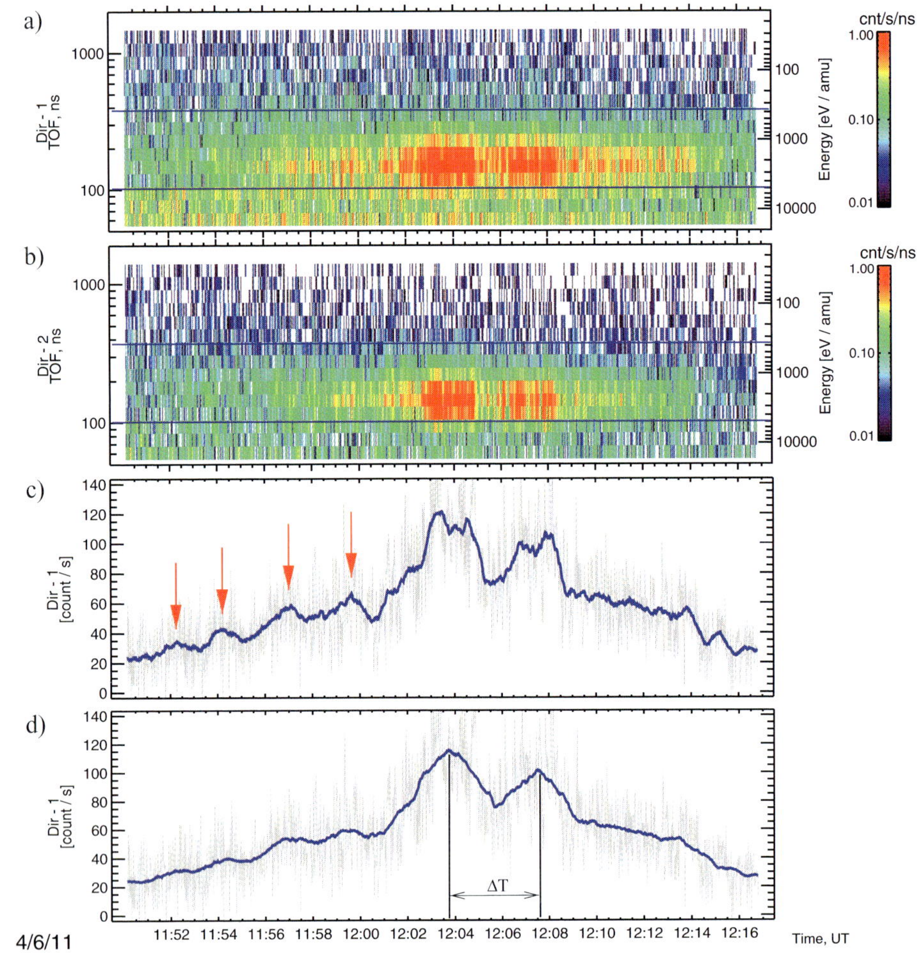

Figure 3. Panels a,b: TOF spectra of NPD-1 DIR-1 and -2 channels. The corresponding energy is shown in the right axis assuming the mass 1 amu. Horizontal blue lines specify the TOF window ξ = 99–374 ns, which covers the ENA jet energy range. Panels c,d: integrated over the TOF window ξ correlated count rate. *Fast* variations are identified by red arrows. *Slow* variations are depicted by thin vertical black lines. Different low pass filter window is used: 40 s – panel c, 90 s - panel d. The peak-to-peak period is marked as ΔT.

was obtained during 11:50–12:17 UT on June 11, 2004 (orbit 499). The intensity is color coded. The colorbar, on the right side, is scaled in counts/read-out/ns or counts/s/ns, assuming 1 ns TOF resolution of the sensor's electronics. The left vertical axis gives the particles TOF in ns. The right vertical axis is the energy scale, assuming H species.

UV induced background noise has been subtracted from the spectra (for details on the UV background noise estimation, see Appendix).

The jet peak energy is ~2.0 keV/amu in this case.

In order to investigate the signal variations, we defined a TOF window ξ as follows: the ξ consists of 6 TOF bins (99–374 ns), covering entirely the ENA jet energy range (0.34–5.1 keV/amu). The ξ window is marked by a pair of blue horizontal lines in each panel.

Two panels, Figure 3(c,d), show the correlated count of the channel DIR-1, integrated over the selected TOF window ξ, with different low-pass filter windows applied. The vertical axis is shown in the unit of [count/s]. Grey thin lines show the NPD-1 DIR-1 count integrated over the TOF window ξ. They reveal large fluctuation of the signal from the averaged values. As any ENA instrument, NPD possesses a residual sensitivity to UV flux. It causes a background in the correlated count rate, which does not depend on the time-of-flight. Therefore, to obtain the signal one subtracts the UV induced background which is assumed to be equal to the count rate at the last TOF bin, where no ENAs are expected due to low sensitivity of the sensor to the low-energy ENA (see Appendix). The signal-to-noise ratio is around 0.4–1.0. Therefore, statistical variations in the UV background on a level of 30% result in the statistical variations of the signal on a level of 60% or higher.

Thick blue lines show the smoothed data using the low-pass filter with 40 s (Figure 3(c)) and 90 s (Figure 3(d)) running averages.

As a general behaviour, the ENA jet intensity is rising while the SC is entering the ENA jet region and decaying as the SC is leaving the ENA jet region. At the same time, some periodic fluctuations of a signal appear during first 10–12 minutes after the measurement has started. The zoomed-in view of these signal variations is shown on Figure 4. Peaks of oscillations are depicted as arrows. The peak-to-peak period is 150–180 sec with the FWHM ~50 sec. The FWHM of a peak is calculated between the peak maximum and a reference level, which is estimated separately for each peak as a mean value of the local minima, neighboring the peak. The ENA jet signal's slow rising and falling is thus taken into account to minimize its influence on the FWHM of the peak estimation. The variations of the signal reach 30% on the maximum value. These are referred to as *fast* oscillations.

As the SC moves further, another type of signal variation occurs. Two high-intensity peaks appeared at approximately 12:04 and 12:08, as seen in Figure 3. These variations are different from the *fast* oscillations described earlier. The peak-to-peak period, ΔT, is ~270 sec, and the FWHM is about 110–140 sec. The variation depth is about 50%. This type of long-term signal variation is referred to as a *slow* variation.

The peak-to-peak times of the *slow* variation events do not depend on the step length of running average taken in a range of 20 to 120 sec. That is reasonable, since a shorter running window remains statistical fluctuations while a longer than 120 sec window smoothes them away.

In case of the *fast* variations, change in the step of the running average length within a range of 20 to 50 sec does not affect the periodicity of the variations.

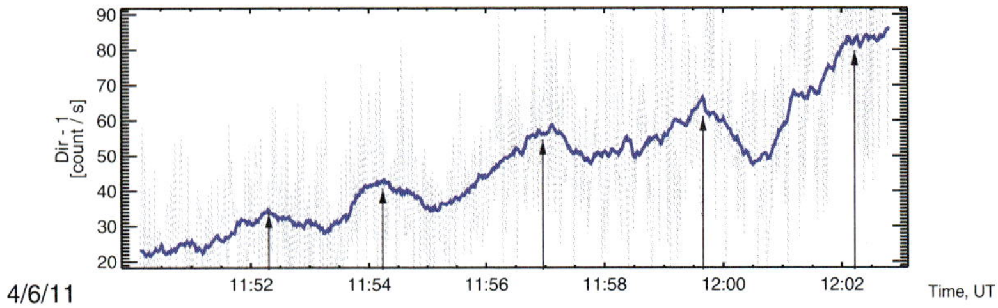

Figure 4. Typical flux variations due to *fast* oscillations. The picture is in the same format as Figure 3(c). The black arrows show the peaks of the *fast* oscillations.

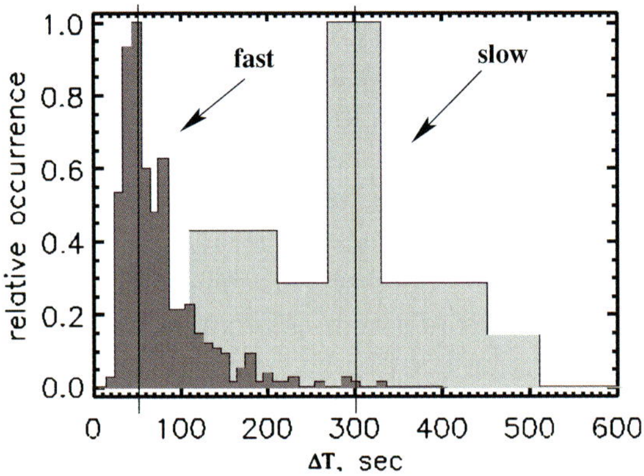

Figure 5. The peak-to-peak ΔT period distributions of both *fast* (dark-grey colored) and *slow* (light-grey colored) groups are shown.

3.3. Statistics on the Intensity Variations

The statistical distributions of the peak-to-peak times for two different groups of ENA jet flux intensity variations is shown in Figure 5. The histogram bin steps are equal to 20 sec and 60 sec for *fast* and *slow* oscillations respectively. Each distribution is normalized on a maximum of the corresponding distribution. The *fast* oscillations are described by the dark-grey distribution. The maximum of the distribution is at 50 sec and FWHM is ~30 sec. These variations are periodic and are detected in every orbit. The peak-to-valley ratio can reach 20–30 % of the max flux intensity magnitude value.

On the other hand, the *slow* variations are found in 15 of 38 orbits. The peak of the distribution is at ~300 sec and the FWHM is ~30–80 sec. The variations of the signal reach 60–80% on the maximum value.

The relative occurrence of FWHM of peaks in case of *slow* ENA jet variations is plotted in Figure 6. The histogram bin step equals to 40 sec. Evidently, the vast majority of the detected peaks has FWHM of ~160 sec. The peaks shape and location are highly variable from orbit to orbit. The number of peaks detected in

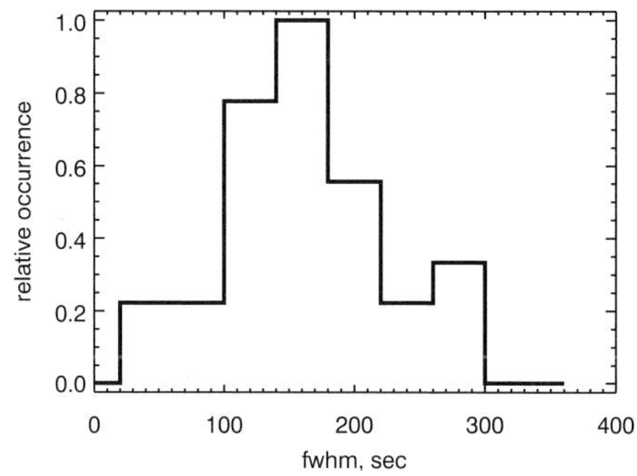

Figure 6. The distribution of FWHM of peaks in case of *slow* ENA jet variations. FWHM of a peak is expressed in seconds (X-axis). Y-axis marks the relative occurrence of cases. The histogram was calculated with the binning size 40 sec.

one orbit is in general two – they occur in 30% of all orbits. Multiple peaks (three and more) appear seldom, only in 10% of orbits.

Criteria for selecting *fast* and *slow* variation events are summarized as follows. In case of *fast* oscillations signal depth amplitude variation reach 20–30%. Number of peaks, observed during an observation period, is at least 3, with ΔT unchanged. Peak-to-peak period ΔT varies in a range 40–180 sec.

In case of *slow* variations signal depth amplitude change is ~50–80%. Number of peaks, detected during an observation period, is 2 or more, ΔT is ~300 sec ±15%.

Figure 7 shows the peak occurrence dependence on the angle between the $-Y$-axis and the MEX SC position, defined as φ (positive if counting clockwise). There is no preferable direction of the intensified ENA flux. The peaks are detected with equal probability from different vantage points within the NPD observation region.

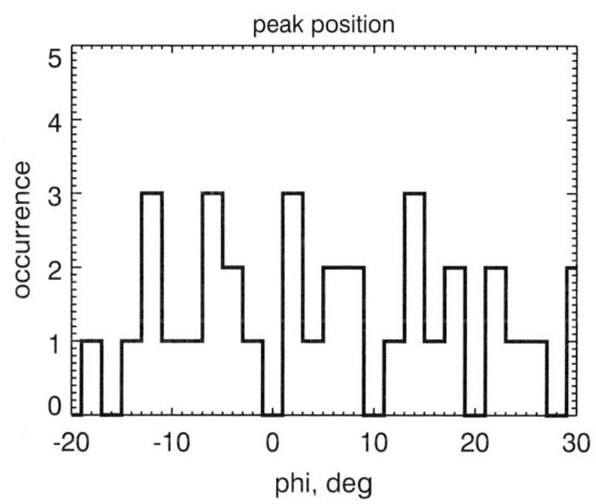

Figure 7. Location of the SC during ENA peak detection and its occurrence is shown. Angular distance between the $-Y$-axis and the MEX SC position (on Figure 1(b)) is shown by φ (see text for details). φ is in degrees (X-axis). Y-axis marks the peak occurrence.

4. Discussion

The NPD detected intense ENA fluxes with energies in the range 0.3–3.2 keV/amu when the FOVs of the NPD pointed toward the subsolar exosphere (Futaana *et al.*, 2006). In 100% of cases, the ENA flux intensity variations were detected. Two modes were identified, namely *fast* and *slow* fluctuations. As shown in observations, the *fast* variations occur with the typical frequency 0.005 to 0.035 Hz and are detected on each orbit. The *slow* fluctuations have the frequency 0.002 to 0.007 Hz and occur in 40% of the cases. Next, we discuss the possible generation mechanisms of the ENA jet intensity variations.

The MGS magnetometer team (Espley *et al.*, 2004) reported on the low-frequency (~0.04 Hz) magnetic field oscillations in the different plasma regions of Mars. One possible interpretation of the oscillations on the dayside of the magnetosheath is mirror mode instabilities. The important characteristic of the mirror modes is the anticorrelation between the magnetic field strength and the plasma density. Therefore, these kinds of magnetic field oscillations can represent one of the mechanisms of ENA jet *fast* oscillations generation. This is because the change of plasma density in the sheath is reflected by the change in ENA production-rate through the change in the occurrence of the charge-exchange process. However, the predominant period of the main fluctuation of magnetic field seems to be significantly shorter than the period of ENA *fast* oscillations, derived from the NPD observations. Therefore, the mirror-mode instability is a less plausible mechanism to create the *fast* oscillations.

Another possible mechanism is the global oscillation of the martian plasma obstacle. Futaana *et al.* (this issue) discussed an example of a global response of the martian plasma environment to the interplanetary shock. They used an ENA jet observation that shows an extremely intensive flux followed by a sudden decrease of the flux intensity. Using simultaneous plasma data, they concluded that the decrease is caused by the fast pile-up boundary displacement due to interplanetary shock. They also reported oscillations of the ENA jet flux (period ~1 min) just after the abrupt decrease, which are regarded in this study as the *fast* oscillations. However, since there have been no reports to show that the global oscillations of the martian plasma obstacle occur, this mechanism is also scarcely to create the ubiquitous *fast* oscillations.

Two frequency peaks in the ENA jet *fast* oscillations distribution are apparent in our data set: ~0.02 Hz and ~0.0125 Hz (Figure 5). Remarkably, the typical frequency of the electron flux intensity oscillations, observed by the ELS instrument of the ASPERA-3 package, is between 0.01 and 0.02 Hz (Winningham *et al.*, 2006). Such oscillations of electron fluxes have been observed in different regions of Mars, but the dayside magnetosheath is one of the active regions. The frequency range encompasses the typical O^+ ion gyrofrequency in the magnetosheath (Espley *et al.*, 2004). Cyclotron instability of the newly created planetary ion beams generates Alfvén waves with a frequency close to the local gyrofrequency (in the SC frame of reference). The associated plasma density variations result in ENA flux oscillation. While it seems that the observed *fast* ENA oscillations are connected to the detected electron flux oscillations (plasma is quasi-neutral), whether or not the wave energy is sufficient to cause 20–30% plasma density variations is unknown.

In contrast to the *fast* oscillations, the observed *slow* variations have the characteristic peak-to-peak time ~300 sec. And as shown in Section 3.3, the number of peaks in the case of *slow* ENA flux variations, detected in one orbit, is typically 2. Larger amount of peaks rarely occur. Such differences between the ENA jet flux fluctuation signatures implies that these two types of jet fluctuations have different production mechanisms.

The observed variations can reflect either temporal variations of the generation region or spatial structure of the jet itself. Because of the SC motion, we can not resolve this ambiguity. In the case of the temporal jet variations, such behavior can be explained by bursty intensification of the ENA source in consequence of the temporal changes in the upstream conditions, mostly the SW dynamic pressure. Holmström *et al.* (2002) investigated the deflected SW ENA flux dependence on the magnetic field magnitude and orientation, solar wind density and velocity. They concluded that the MPB position is the most important parameter to control the ENA flux. If the ENA generation region can move closer to the planet, then the SW ions can reach more dense parts of the martian exosphere, creating high ENA flux.

The waves on the magnetic pile-up boundary/induced magnetosphere boundary resulting from Kelvin-Helmholtz (KH) instability could, in principle, cause the observed oscillations. However, Penz *et al.* (2004) considered the development of the KH instability on the martian boundary in one-fluid approximation and concluded that the subsolar ionopause of Mars is stable with respect to the KH instability.

The observation points at the moments when the NPD-1 sensor detected peaks of *slow* ENA jet intensity variations are distributed uniformly over the whole observation region of the NPD, as shown on Figure 7. Since the observation period is long (~40 days), the distribution of IMF directions and magnitude, as well as SW parameters can be considered as random. This implies that the ENA generation region is very sensitive to the exterior and is voluble under unsteady solar wind dynamic pressure and IMF direction.

The signature supports temporal variations. However, there is a certain possibility that the *slow* variations of the ENA jet represents spatial structures of the jet itself. The majority of the detected peaks have the FWHM about 160 sec, corresponding to approximately 4–5° of an angular width of ENA stream, taking into account the SC velocity ~3 km/s and distance to the generation region $\sim 2R_{\mathrm{M}}$. If we assume the structure of "*subjets*" with an angular width of 4–5° (Figure 8), the observed *slow* variations can be explained. The ENA generation source under certain conditions can become patchy. What can cause such behaviour? Existence of small-scale magnetic field structures called magnetic flux ropes (e.g. Russell and Elphic, 1979; Elphic and Russell, 1983) can lead to such effects. The flux ropes existence at the martian ionopause has been reported by Vignes *et al.* (2004). The

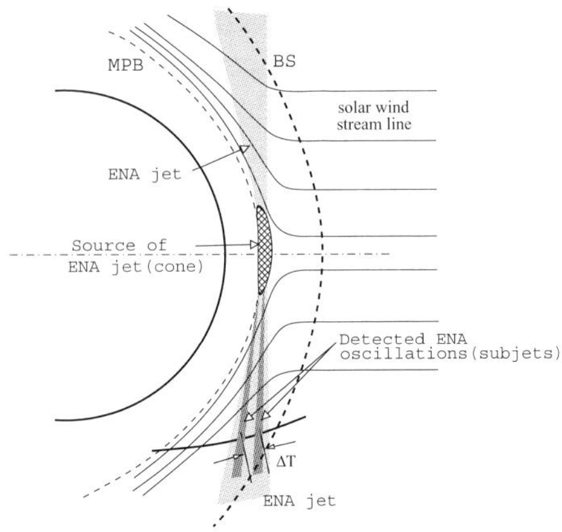

Figure 8. The schematic illustration of spatially resolved ENA subjets.

magnetic character of the flux ropes at Mars appears chaotic. At the same time, these magnetic structures are stationary in time relative to the speed of the SC. The width of the flux ropes is of the order of a few tens of kilometers. Most of the flux ropes have been detected at high solar zenith angles (SZA). The smaller number of flux ropes has been identified for SZA lower than 20° (Vignes *et al.*, 2004). Meanwhile, the size of the ENA generation region was reported to be within the SZA lower than 40° (Futaana *et al.*, 2006). Therefore, the existence of magnetic flux ropes can affect the transport of planetary protons to the magnetosheath and control the ENA production. The mechanism of ENA formation through charge-exchange of accelerated protons in the magnetosheath (Lichtenegger *et al.*, 2002) was reported by Futaana *et al.* (2006) to be one of the most probable ENA jet generation mechanisms.

More detailed study of the ENA jet behavior is a subject of future investigations. The next step would be to study the morphology of the ENA jet as a function of the upstream conditions such as SW dynamic pressure and IMF direction. But the key investigation to resolve the nature of the ENA jet oscillations and related dynamics of the Martian induced magnetosphere would be multipoint measurements at Mars. Those would allow to correctly correlate *in-situ* measurements of the local plasma parameters and remote observations of the ENA flux emerging from the interaction region.

5. Summary

We presented observations made by NPD of the martian subsolar ENA jet intensity oscillations. We categorized the flux intensity variations into two groups. First group is *fast* oscillations with the characteristic peak-to-peak time ~50 sec. These flux variations possess a periodic structure and are being observed in all orbits. The second group is *slow* variations, whose time scale is on the order of ~300 sec. These variations appear in 40% of the ENA jets.

The *fast* oscillations are possibly formed by the low frequency ion waves (0.01–0.02 Hz) at the subsolar region. The observed period correlates with electron oscillations detected by ELS which, in turn, is close to the oxygen gyroperiod. However, the energetics of the process is not clear.

The origin of the *slow* variations is a mystery. They might be related to temporal variations of the SW dynamic pressure. On the other hand, the *slow* oscillations can also be explained by a subjet structure as shown in Figure 8. Such structure could appear due to the existence of the magnetic flux ropes in the martian ionosphere.

Appendix: Background Count Rate Estimation

UV induced background noise is estimated in two different ways.

One way is called random coincidence method. This method is based on the probability distribution between two time series of random signal. The other way is using the instrumental response in the low energy range. No ENAs with energies <100 eV were detected, therefore the counts recorded in this energy range are expected to be the background. We discuss these two methods using three observation periods.

'Random Coincidence' method

The coincidence scheme (valid START and valid STOP) rejects most of the UV related signals. A valid ENA event requires signals from both START and STOP detectors within the TOF window (1900 ns). Background photons and stray particles would produce detector noise count rates proportional to the instrument geometrical factor and the square of the incoming ENA flux. The noise counts are random and the signals from START and STOP detectors can sometimes fit into defined TOF window and give a valid coincidence pulse. Nevertheless, as the UV flux causes random count detection on both Start and Stop units, the resulted TOF distribution can be approximated as a linear function over an entire TOF window. Strictly speaking, the coincident TOF distribution between two randomly distributed START and STOP signals obeys the exponential law, described as:

$$C_{\rm rc} = C_{\rm start} \times C_{\rm stop} \times \Delta T_{\rm tof} \times e^{-C_{\rm start} \times T_{\rm tof}}, \qquad (1)$$

where $C_{\rm rc}$ is the random coincidence expected rate, $C_{\rm start}$, $C_{\rm stop}$ are the START and STOP count rates, $\Delta T_{\rm tof}$ is the TOF window size (1900 ns in this case) and $T_{\rm tof}$ is a TOF value. Since the TOF window is short in comparison to the typical time between START events, the difference in the predicted noise count rate for $T_{\rm tof} = 0$ and $T_{\rm tof} = 1900$ ns is less than 10%. The random coincidence can be approximated by a linear function (constant) with an insignificant error in the estimations.

We examined three cases of the measurements in order to examine this method. The first two cases represent 'quiet' measurement cases, when the detected ENA signal was extremely low. The last case is the one examined in Section 3. One important fact is that in all cases, the NPD was observing the subsolar region of

Mars. Taking START and STOP not correlated count rates recorded onboard we can obtain the expected noise count rate C_{rc}, as shown in Table I:

TABLE I

The background count rates estimated by two different methods. Three cases are evaluated. C_{rc} and C_{tof} are the background count rates estimated by using random coincidence mode and TOF distribution respectively.

Case	UT	C_{rc} [c/s/ns]	C_{tof} [c/s/ns]
1	04/06/29 23:34–23:54	0.31	0.26
2	04/06/28 18:06–18:13	0.28	0.22
3	04/06/11 11:50–12:10	0.43	0.36

'TOF distribution' method

The TOF distribution, covering the TOF range 1514–1900 ns, can be considered as a background count rate (Futaana *et al.*, 2006), because it corresponds to ENAs with an energy of 14–22 eV assuming H, and 0.22–0.35 keV assuming O. The detection efficiency of the sensor for that low energy oxygen is less by a factor of 100 than that of H of energy 1.4 keV. There is no evidence of very intense fluxes of low energy oxygen in the observation region.

The background count rate C_{tof}, estimated by using the TOF distribution recoded onboard, is shown in Table I. Evaluated for the first two cases, noise is comparable, as the external conditions in those cases have been similar. The noise count rate, estimated for the case 3 is higher than those of other cases because of variations in UV background. UV background noise, estimated from detectors counts, is higher than that estimated from TOF distributions. This is due to the fact that values, derived from the random coincidence method, take into account both randomly correlated (false) and valid events. The TOF distribution method considers only false events and those valid events with energy in the range of 14–22 eV/amu, i.e. negligible part. Also, the theoretical estimation gives rather approximate noise evaluations while the TOF distribution method takes into account the sensor response. Hence, the TOF distribution method is thought to be more accurate. We applied the TOF distribution method in the data processing for this paper.

Acknowledgements

The ASPERA-3 experiment on the European Space Agency (ESA) Mars Express is a joint effort among 15 laboratories in 10 countries, all supported by their national agencies. We thank all these agencies, as well as the various departments and institutes hosting these efforts.

Dr. Y. Futaana has been supported by a JSPS Research Fellowships for Young Scientists.

References

Acuña, M. H., Connerney, J. E. P., Wasilewski, P., Lin, R. P., Anderson, K. A., Carlson, C. W., *et al.*: 1998, *Science* **279**, 1676.

Barabash, S.: 1995, *Satellite Observations of the Plasma-Neutral Coupling near Mars and the Earth.* Ph.D. thesis, Swedish Institute of Space Physics.

Barabash, S., Norberg, O., Lundin, R., Olsen, S., Lundin, K., Brandt, P. C., *et al.*: 1998, in R. F. Pfaff, J. E. Borovsky, and D. T. Young, (eds.), *Measurement Techniques in Space Plasmas, Field*, AGU Geophysical Monograph 103, pp. 257–262. American Geophysical Union, Washington, DC.

Barabash, S., Lundin, R., Andersson, H., Gimholt, J., Holmström, M., Norberg, O., *et al.*: 2004, in A. Wilson, (ed.), *Mars Express: The Scientific Payload*, volume SP-1240, pp. 121–139. ESA Special Publication.

Barabash, S., Lundin, R., Andersson, H., Brinkfeldt, K., Grigoriev, A., Gunell, H., *et al.*: *Space. Sci. Rev.*, this issue, doi: 10.1007/s11214-006-9124-8.

C:son Brandt, P., Barabash, S., Roelof, E. C., and Chase, C. J.: 2001, *J. Geophys. Res.* **106**(A11), 24663.

Elphic, R., and Russell, C.: 1983, *J. Geophys. Res.* **88**, 2993.

Espley, J. R., Cloutier, P. A., Brain, D. A., Crider, D. H., and Acuña, M. H.: 2004, *J. Geophys. Res.* **109**(A0).

Futaana, Y., Barabash, S., Grigoriev, A., Holmström, M., Kallio, E., C:son Brandt, P., *et al.*: 2006, *Icarus*.

Futaana, Y., Barabash, S., Grigoriev, A., Winningham, D., Frahm, R., and Lundin, R.: *Space. Sci. Rev.*, this issue, doi: 10.1007/s11214-006-9026-9.

Holmström, M., Barabash, S., and Kallio, E.: 2002, *J. Geophys. Res.* **107**(A10).

Kallio, E., and Barabash, S.: 2001, *J. Geophys. Res.* **106**(A1), 165.

Lichtenegger, H., Lammer, H., and Stumptner, W.: 2002, *J. Geophys. Res.* **107**(A10).

Lundin, R., Barabash, S., Andersson, H., Holmström, M., Grigoriev, A., Yamauchi, M., *et al.*: 2004, *Science* **305**, 1933.

Mitchell, D. G., Jaskulek, S. E., Schlemm, C. E., Keath, E. P., Thompson, R. E., Tossman, B. E., *et al.*: 2000, *Space Sci. Rev.* **91**(1–2), 67.

Moore, T. E., Chornay, D. J., Collier, M. R., Herrero, F. A., Johnson, J., Johnson, M. A., *et al.*: 2000, *Space Sci. Rev.* **91**(1–2), 155.

Penz, T., Erkaev, N., Biernat, H., Lammer, H., Amerstorfer, U., Gunell, H., *et al.*: 2004, *Planet. Space Sci.* **52**, 1157.

Pollock, C. J., Asamura, K., Baldonado, J., Balkey, M. M., Barker, P., Burch, J. L., *et al.* (2000). *Space Sci. Rev.* **91**(1–2), 113.

Russell, C., and Elphic, R.: 1979, *Nature* **279**, 618.

Vignes, D., Mazelle, C., Rème, H., Acuña, M. H., Connerney, J. E. P., Lin, R. P., *et al.*: 2000, *Geophys. Res. Lett.* **27**(1), 49.

Vignes, D., Acuña, M., Connerney, J., Crider, D., Rème, H., and Mazelle, C.: 2004, *Space Sci. Rev.* **111**, 223.

Winningham, J. D., Frahm, R. A., Sharber, J. R., Coates, A. J., Linder, D. R., Soobiah, Y., *et al.*: 2006, *Icarus*.

GLOBAL RESPONSE OF MARTIAN PLASMA ENVIRONMENT TO AN INTERPLANETARY STRUCTURE: FROM ENA AND PLASMA OBSERVATIONS AT MARS

Y. FUTAANA[1,2,*], S. BARABASH[2], A. GRIGORIEV[2], D. WINNINGHAM[3], R. FRAHM[3], M. YAMAUCHI[2] and R. LUNDIN[2]

[1]*Institute of Space and Astronautical Science, Japan Aerospace Exploration Agency, Yoshinodai 3-1-1, Sagamihara, 229-8510 Kanagawa, Japan*

[2]*Swedish Institute of Space Physics, Box 812, SE-981 28 Kiruna, Sweden*

[3]*Southwest Research Institute, San Antonio, TX 7228-0510, USA*

*(*Author for correspondence, E-mail: yoshifumi.futaana@irf.se)*

(Received 20 March 2006; Accepted in final form 5 July 2006)

Abstract. As a part of the global plasma environment study of Mars and its response to the solar wind, we have analyzed a peculiar case of the subsolar energetic neutral atom (ENA) jet observed on June 7, 2004 by the Neutral Particle Detector (NPD) on board the Mars Express satellite. The "subsolar ENA jet" is generated by the interaction between the solar wind and the Martian exosphere, and is one of the most intense sources of ENA flux observed in the vicinity of Mars. On June 7, 2004 (orbit 485 of Mars Express), the NPD observed a very intense subsolar ENA jet, which then abruptly decreased within ~10 sec followed by quasi-periodic (~1 min) flux variations. Simultaneously, the plasma sensors detected a solar wind structure, which was most likely an interplanetary shock surface. The abrupt decrease of the ENA flux and the quasi-periodic flux variations can be understood in the framework of the global response of the Martian plasma obstacle to the interplanetary shock. The generation region of the subsolar ENA jet was pushed towards the planet by the interplanetary shock; and therefore, Mars Express went out of the ENA jet region. Associated global vibrations of the Martian plasma obstacle may have been the cause of the quasi-periodic flux variations of the ENA flux at the spacecraft location.

Keywords: Mars, ASPERA-3, interplanetary shock, energetic neutral atoms, ENA jet, unmagnetized planet, interaction with solar wind

1. Introduction

The Martian plasma environment has been explored since the 1960s by the Mariner 4, and Mars 2, 3 and 5 spacecraft (e.g. see review by Vaisberg, 1992). Phobos 2 (1989) was the first mission to carry a complete set of modern plasma experiments (e.g. Sagdeev and Zakharov, 1989 and other articles in the Phobos-2 special issue [*Nature, 341*, pp. 581–618, 1989]). In 1998, the Mars Global Surveyor (MGS) spacecraft carried a magnetometer and an electron reflectometer (MAG/ER) to investigate the magnetic properties of Mars (e.g. see review by Nagy *et al.*, 2004). In 2003, the Mars Express spacecraft arrived at Mars. The Analyser of Space Plasma and Energetic Atoms (ASPERA-3) on board Mars Express is the first

Space Science Reviews (2006) 126: 315–332
DOI: 10.1007/s11214-006-9026-9

comprehensive plasma and neutral particle package capable of measuring ions, electrons and energetic neutral atoms (ENAs) to explore the vicinity of Mars (Barabash *et al.*, 2004).

Mars has no global intrinsic magnetic field, but locally magnetized areas distributed globally. The MGS MAG/ER found strong magnetic fields of crustal origin, especially in the southern hemisphere (Acuña *et al.*, 1998, 1999). Owing to the lack of the global magnetic field and the existence of the strong and localized magnetic field, the interaction of Mars with the solar wind is much different than that of the Earth.

Our knowledge of the global Martian plasma environment is an average view. Temporal changes of the global Martian plasma environment are not well understood, particularly over short time scales. It is generally difficult to investigate temporal variations of global structure from *in situ* observations by a single spacecraft. One way to investigate the global Martian plasma environment from a single spacecraft is to analyze accumulated data statistically. For example, Vignes *et al.* (2002) used 553 bow shock crossings to investigate boundary locations from MGS MAG/ER data relative to solar wind conditions.

Energetic Neutral Atom (ENA) imaging techniques have developed rapidly during the last decade. ENA imaging has become a powerful means to remotely investigate the plasma environment and the neutral exosphere of planets. Several Earth-orbiting spacecraft have carried ENA imagers to investigate the dynamics of the terrestrial plasma environment, such as in the auroral ionosphere, the cusp, the radiation belt, and the plasmasheet (e.g. Roelof *et al.*, 1985; Barabash *et al.*, 1998; Mitchell *et al.*, 2000; Pollock *et al.*, 2000; Moore *et al.*, 2000).

Although ENA generation in the vicinity of Mars is quite different from that of the Earth (Barabash *et al.*, 2004 and references therein), ENA imaging is still a powerful tool in order to investigate the spatial structures and temporal variations of the global plasma environment of Mars. Understanding of the Martian ENA environment is one of the main scientific goals for the ASPERA-3 experiment. The ASPERA-3 comprises four instruments: two ENA sensors, an electron spectrometer and an ion mass analyzer (Barabash *et al.*, 2004). The two ENA sensors are called the Neutral Particle Imager (NPI) and the Neutral Particle Detector (NPD). Note that Mars Express does not carry a magnetometer. The initial results of ENA imaging have, in general, confirmed theoretical predictions as summarized below.

Kallio and Barabash (2001) has calculated the flux of ENAs emitted from the dayside Martian exobase. In their calculations, the solar wind ENAs (charge-exchanged solar wind protons in the upstream region of the bow shock) can reach very low altitudes within the Martian atmosphere, where elastic and inelastic collisions become dominant. As a result, some of the solar wind ENAs are expected to be scattered back into the space. Futaana *et al.* (2006a) confirmed the existence of such backscattered ENAs from the Martian upper atmosphere by analyzing the NPD data. The backscattered ENAs are emitted globally with the flux of $\sim 10^7 \mathrm{cm}^{-2}\,\mathrm{s}^{-1}$. The observed signatures are consistent with calculations, while the only difference was that the flux of the backscattered ENAs was higher than in theoretical calculations. They concluded that the flux of the backscattered ENAs which originated from the direct input of the solar wind protons is significant at Mars.

Kallio *et al.* (2006) suggested an "ENA occultation" technique: part of the so-called solar wind ENAs, which are produced by charge exchange between solar wind protons and extended exosphere, are expected to be scattered in the near-Mars exosphere around the planetary limb, penetrating into the Martian tail. The ENA occultation is similar to stellar, solar wind and radio occultation measurements. In this ENA occultation technique, the ENAs penetrating into the tail region are used to infer information about the Martian exosphere. Kallio *et al.* (2006) also simulated the solar wind ENAs penetrating into the Martian tail as a separate ENA population. They used a Monte-Carlo model to simulate the interaction of these ENAs with the Martian atmosphere. Although their simulation predicted that the ENA flux was too low to be observed by existing instruments, it is worthwhile examining whether these ENAs are detectable by ASPERA-3. Brinkfeldt *et al.* (2006) reported on NPI analysis of signals from the direction around the limb during transversals of the Martian optical umbra. By comparing with simulations, they concluded that the NPI signals can also be explained by a ~20 eV perpendicular temperature of the solar wind protons. However at present, there are no reference observations of the solar wind temperature, and the above discussion still remains an open question.

Mars Express also found a substantial flux of ENAs in the direction tangential to the solar wind flow direction. Gunell *et al.* (2006) reported ENA signals coming from the dayside magnetosheath observed by NPI. By comparing NPI data with an ENA generation model in the shocked solar wind (Kallio *et al.*, 1997), they concluded that these observations are ENAs of shocked solar wind origin. On the other hand, Futaana *et al.* (2006b) showed ENA emission from the subsolar region detected by the NPD instrument. The flux is denoted as a subsolar ENA jet (or cone) because the emitted flux is highly directional from the subsolar region. The question whether these ENA observations by different instruments are of the same origin is still under investigation.

As discussed in Futaana *et al.* (2006b), we can consider two possible source for the subsolar ENA jet: shocked solar wind protons and protons of planetary origin. The first source was discussed by Holmström *et al.* (2002), and is responsible for the ENA flux detected by NPI (Gunell *et al.*, 2006). When the shocked solar wind protons charge exchange with exospheric particles, they are converted to ENAs and form the subsolar ENA jet. The second source is described in Lichtenegger *et al.* (2002). After photoionization of the exospheric cold neutral atoms, the resulting ions are accelerated via energy or momentum exchange (Peréz-de-Tejada, 1987) with the shocked solar wind, eventually reaching the same energy or momentum as the shocked solar wind. When such accelerated ions experience the charge exchange reaction, they are observed as a subsolar ENA jet.

No matter what the subsolar ENA jet generation mechanism is, we can conclude that the jet is generated in the vicinity of the subsolar region of the Martian upper atmosphere as a result of the interaction between the solar wind and the Martian upper atmosphere. This means that detailed investigations of subsolar ENA jets would advance the understanding of the dynamics involved in the interaction between the planet and the solar wind.

In this paper, we analyze a peculiar subsolar ENA jet event which was recorded by the NPD on June 7, 2004 (orbit 485) and discuss the global Martian plasma environment. For this event, the NPD measured an extremely high flux of ENAs from a subsolar ENA jet. This strong ENA flux decreased abruptly over a time scale of ~10 sec. Following the decrease, periodic enhancements of ENA fluxes were observed. We show the NPD data together with *in situ* plasma data to investigate global signatures of the interaction between the Martian upper atmosphere and the solar wind.

2. Instrumentation and Data

2.1. Neutral Particle Detector

The Mars Express spacecraft carried the first ENA instrument to Mars as a part of the Analyser of Space Plasma and Energetic Atoms (ASPERA-3) experiment (Barabash *et al.*, 2004). The ENA instrument is composed of two sensors: the Neutral Particle Imager (NPI) and the Neutral Particle Detector (NPD). The NPI is designed to measure ENAs with high angular resolution ($\sim 4.5^\circ \times 11.25^\circ$) and with a total field of view of 360° (32 directions), but without mass and energy resolution. The NPD can resolve particle velocities and masses, but possesses a lower angular resolution ($\sim 5^\circ \times 40^\circ$) and only has total field view of ~180° (6 directions). For the analysis in this paper, we used only NPD data as a measure of the ENA flux.

The NPD sensor consists of two identical detectors, NPD-1 and NPD-2, each containing three directions (Dirs-0, -1 and -2) that form approximately a 90° fan. We used only two directions of NPD-1 during this analysis. The NPD measures ENA differential flux over the energy range 100 eV–10 keV, resolving H and O. The velocity is obtained by measuring the time-of-flight (TOF) of each ENA (Barabash *et al.*, 2004). The nominal operation mode of the NPD is the 'bin-matrix' mode. In this mode, each TOF signal is accumulated into 16 logarithmically-divided TOF bins (50–1900 nsec). In order to analyze TOF spectra of ENAs, the recorded count rate in each TOF bin [counts/sec] is normalized by dividing by the TOF window width [nsec]. The time resolution for ENA detection is 1 sec (Futaana *et al.*, 2006b).

Since the NPD is an open system, there exist background counts due to ultraviolet (UV) photons which enter into the detector. UV photons can stochastically generate correlated signals, which are not ENA-originated signals. This background level is not constant, but has temporal changes due to the change in the local UV flux, i.e. from the spacecraft location and the direction of instrument aperture. We have regarded the correlated counts recorded in the TOF channel of 1514–1900 ns (corresponding to 14–22 eV/amu) as the background level. Since the MCP response to ENAs is extremely low in this energy range (∼20 eV), all of this channel's counts are considered as UV-originated counts. Theoretically, the background level has a TOF dependence, but the difference is not large (less than 10%) in the TOF range of the NPD. Therefore, we have assumed a constant background level over the TOF range. The background level has been subtracted from the TOF data before analyses.

2.2. Plasma Spectrometers

The ASPERA-3 observes *in situ* plasma velocity distribution functions by using two plasma sensors: the Electron Spectrometer (ELS) and the Ion Mass Analyzer (IMA) (Barabash *et al.*, 2004). The ELS is a spherical top-hat analyser with the aperture of $4^\circ \times 360^\circ$. There are 16 anodes corresponding to the directions of the incident electrons, and each has an angular resolution of $\sim 4^\circ \times 22.5^\circ$. The ELS provides two-dimensional electron velocity distribution functions over the energy range of ∼0.4 eV–20 keV with 128 energy steps. The time resolution (i.e. the time for one energy sweep of the top-hat analyzer) is 4 sec.

The IMA can perform quasi-three-dimensional ion observations within the energy range of ∼10 eV–30 keV in 96 energy steps. The total field of view is $90^\circ \times 360^\circ$ with an angular resolution of $\sim 5.6^\circ \times 22.5^\circ$. The IMA possesses an electrostatic deflection system (elevation analyzer) in front of a top-hat analyzer to obtain 3-D ion velocity distribution functions. By sweeping the voltage of the deflector, the view angle can be changed by $\pm 45^\circ$ with respect to the aperture plane of the top-hat analyzer. Each energy sweep takes 12 sec, and a 3-D distribution function is obtained by completing an angular sweep with the time resolution of 192 sec.

3. Observations

Figure 1 shows the Mars Express orbit in the cylindrical Mars-Sun orbit (MSO) coordinate system from 11:00 to 14:30 UT on June 7, 2004 (orbit 485). The pericenter was at 13:22:26 UT with an altitude of ∼260 km from the Martian surface. In this figure, the dotted lines show the average locations of the modeled bow shock (BS) and the magnetic pileup-boundary (MPB) determined by Vignes *et al.* (2000). The bow shock is the region created in front of the Martian plasma obstacle where the solar wind is decelerated from supersonic to subsonic speed. The MPB is the boundary where the magnetic field becomes strong and the electron flux is decreased (e.g. Nagy *et al.*, 2004). These models were derived from statistical studies from Mars Global Surveyor (MGS) observations, but it is known that these boundaries are highly variable depending on upstream conditions. Therefore, we need to identify these boundaries using the *in situ* plasma data for analysis of this single event.

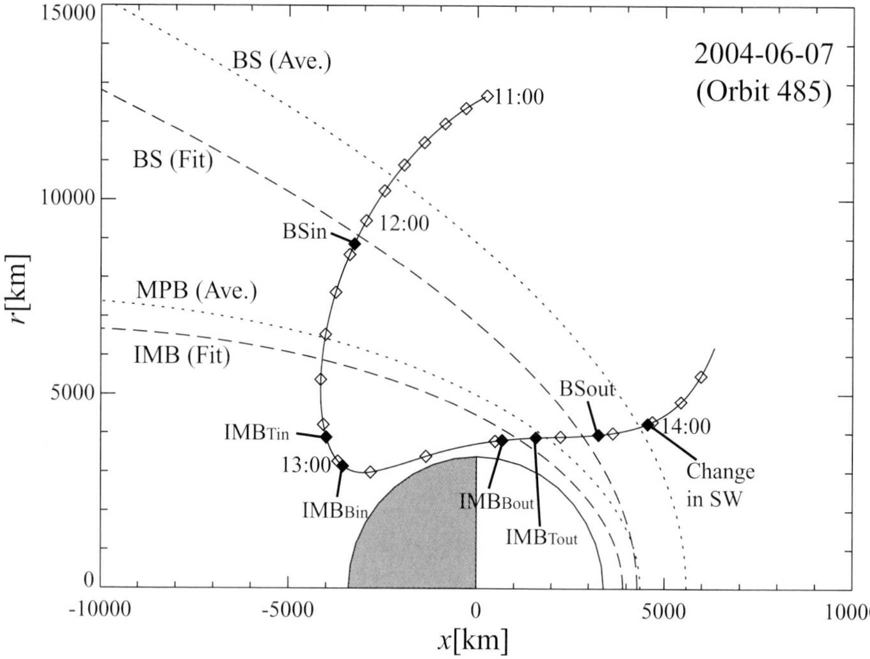

Figure 1. The orbit of Mars Express on June 7, 2004 (orbit 485) in the cylindrical coordinate system. The periapsis occurred at 13:22:26 UT. The dotted lines are the averaged locations of the bow shock (BS) and the magnetic pileup boundary (MPB) described in Vignes *et al.* (2000). The bow shock and the induced magnetosphere boundary (IMB) crossings are shown by filled symbols. The satellite position when the change in solar wind conditions was observed (13:58:30 UT) is also indicated. The dashed lines are the calculated positions of bow shock and induced magnetosphere boundary assuming the scaling law.

Figure 2 shows the plasma observations between 11:00 and 14:15 UT. From top to bottom, the energy-time spectra for ELS (CH-3 and -7) and IMA are displayed. The periodic change ($\sim$3 min) within the IMA data is due to the angular sweep of the electrostatic deflection system (see Section 2).

Mars Express was in the upstream solar wind region when the plasma observations started ($\sim$11:15 UT), and approached Mars from its nightside. At $\sim$12:07 UT, Mars Express crossed the bow shock as indicated by sharp increase of the electron and the ion temperatures. The crossing position is drawn by the filled symbol (BS_{in}) in Figure 1, which shows that the bow shock was located much closer to Mars than predicted by the model. The ion count gradually decreased between 12:53–13:02 UT. These changes in solar wind ions were due to the crossing of the stopping boundary of the solar wind. Lundin *et al.* (2004) named this solar wind stopping boundary the induced-magnetosphere boundary (IMB). Because the IMB has a finite thickness in general, we used two definitions of the IMB: the topside IMB (IMB_T) and the bottomside IMB (IMB_B). The IMB_T is the boundary where the

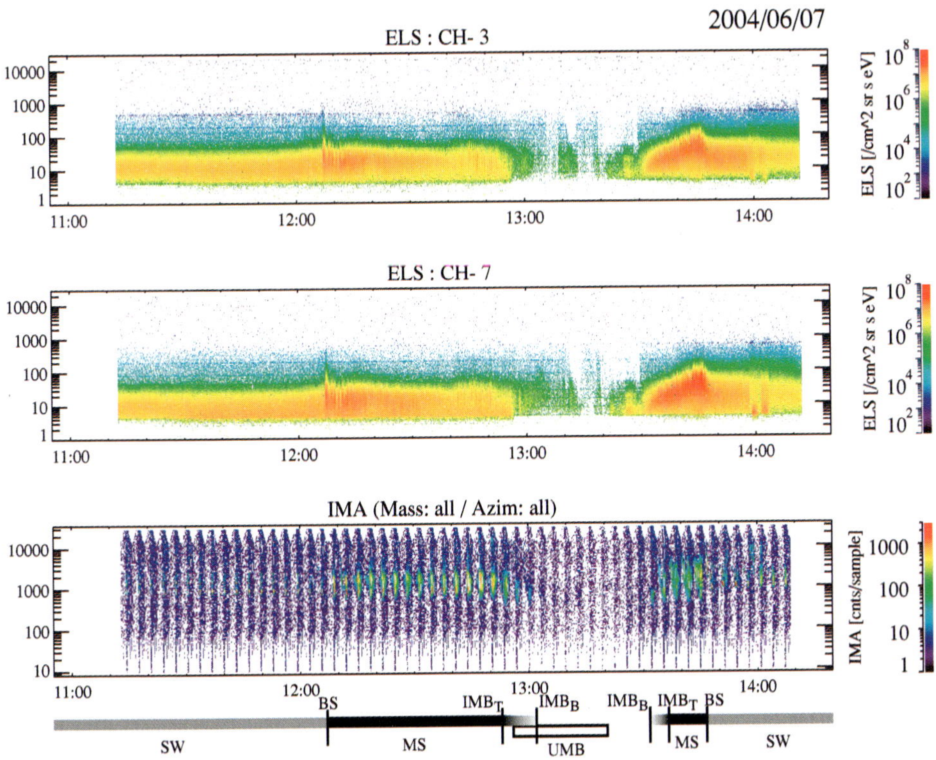

Figure 2. The electron and ion observations throughout the observation for orbit 485. The first and second panel shows the ELS energy-time spectrogram (Ch-3 and -7), and the third panel shows the IMA energy-time spectrogram summed over all view directions and mass channels. At the bottom, plasma boundaries and domains are denoted, where BS is the bow shock, IMB is the induced magnetosphere boundary, SW is the solar wind, MS is the magnetosheath, and UMB is the umbra.

shocked solar wind flux and velocity start to decrease (12:53 UT, $\mathrm{IMB}_{T\mathrm{in}}$), and the IMB_B is the boundary where shocked solar wind disappears (13:02 UT, $\mathrm{IMB}_{B\mathrm{in}}$). Note that Mars Express was in the umbra region between 12:58–13:20 UT. The outbound IMB_B and IMB_T crossings are identified as 13:31 ($\mathrm{IMB}_{B\mathrm{out}}$) and 13:36 UT ($\mathrm{IMB}_{T\mathrm{out}}$), respectively. At around 13:47 UT, Mars Express crossed the bow shock ($\mathrm{BS}_{\mathrm{out}}$) again and exited to the solar wind region. One can also see a change of the solar wind conditions at around 13:58:30 UT.

The bow shock and the IMB shape and location during the observation can be fitted by the boundary crossing positions identified above. Here we applied the scaling law to the boundary location models by Vignes *et al.* (2000), which are the average locations of the bow shock and the MPB calculated by 290 orbits of Mars Global Surveyor. For fitting the bow shock shape, we used the positions of inbound ($\mathrm{BS}_{\mathrm{in}}$) and outbound ($\mathrm{BS}_{\mathrm{out}}$) bow shock crossing. For fitting the IMB shape, we used the outbound IMB crossing positions ($\mathrm{IMB}_{B\mathrm{out}}$ and $\mathrm{IMB}_{T\mathrm{out}}$) under the assumption

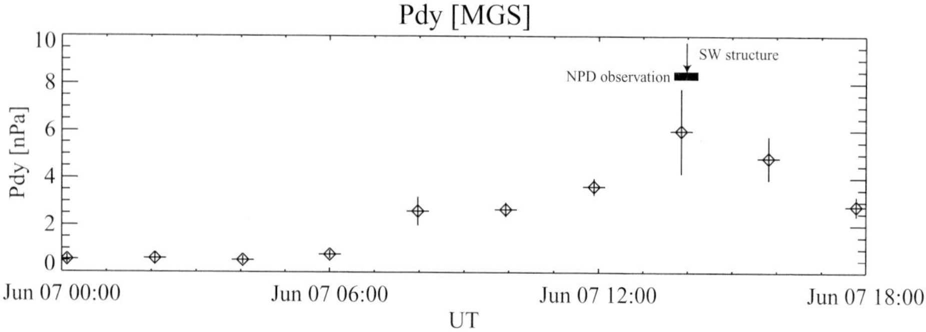

Figure 3. Solar wind dynamic pressure estimated by MGS observations (Crider *et al.*, 2003). The estimates were performed by fitting the magnetic field data obtained in the dayside northern hemisphere (~30 min). The NPD observation time (13:42–14:13UT) is shown by a filled rectangle, and the arrival time of the solar wind structure detected by Mars Express (~13:58:30UT) is shown by the arrow.

that the IMB and the MPB are identical. The fitted boundaries are shown by dashed lines in Figure 1. The scaling factors are ~0.77 and ~0.90 for the bow shock and the IMB, respectively. We can conclude that during orbit 485, the dayside Martian plasma environment was substantially compressed compared to its average size.

Figure 3 shows MGS dynamic pressure proxy data between 0:00–18:00 UT on June 7 at Mars. The magnetic field data observed by MGS were converted to the solar wind dynamic pressure by assuming pressure balance. The time resolution of the data is 2 hours (corresponding to the MGS orbital period), and each estimate has been generated by fitting the magnetic field data from the dayside northern hemisphere (~30 min) (Crider *et al.*, 2003). The dynamic pressure (Pdy) was a typical value of ~0.5 nPa before 6:00 UT. Figure3 also shows that there are two increases of Pdy: at 8:00 UT to ~2.5 nPa and at 14:00 UT to ~6 nPa. This solar wind pressure is extremely large in the vicinity of Mars.

The NPD-1 observation during this orbit is shown in Figure 4a and b. The instrument was on between 13:42 and 14:13 UT on June 7, 2004. The vertical axis is the TOF of the ENAs, which is converted to the hydrogen ENA energy as shown on the right axis. The background levels have been subtracted from the observed TOF spectra as described in Section 2. Only directions 1 and 2 (Dirs-1 and -2) observe the ENA jet flow. Figure 4c shows the instrument count rate integrated over the range 50–1900 ns (corresponding to 14 eV–20 keV). The maximum count rate was ~1000 counts/s, which corresponds to a flux of $J = (0.6\text{–}1) \times 10^7\ \text{cm}^{-2}\ \text{sr}^{-1}\ \text{s}^{-1}$, where we employ the geometric factor (G_0) and efficiency(ϵ) of $\epsilon \cdot G_0 = (9.78\text{–}17.1) \times 10^{-5} \text{cm}^2\text{sr}$ for 1 keV hydrogen atoms. This flux is approximately 5 times higher than the typical flux of $(1\text{–}2) \times 10^6\ \text{cm}^{-2}\ \text{sr}^{-1}\ \text{s}^{-1}$ (Futaana *et al.*, 2006b). The observed energy of this ENA jet (0.8-3.0 keV corresponding to ~390–760 km/s for hydrogen ENAs) is consistent with the typical subsolar ENA jet.

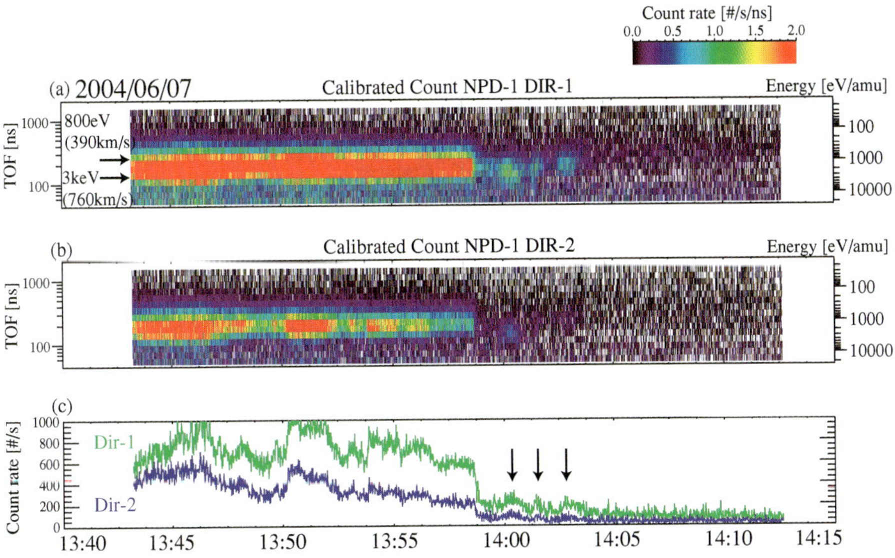

Figure 4. Time of flight spectrum of NPD-1 for (a) DIR-1 and (b) DIR-2. The energy corresponding to the time of flight (assuming mass =1 amu) is shown on the right axis. The intense signal with energies 0.8–3.0 keV is the subsolar-ENA flux. (c) The count rate of the subsolar ENA flux integrated over the time of flight. Abrupt depletion of ENA fluxes occurred at ~13:58:40 UT. Following variations in flux are depicted by arrows.

Figure 5 shows a fish-eye projection of geometry of Mars with respect to the field of view (FOV) of the NPD. The dark and light gray rectangles correspond to the Dirs-1 and -2 FOVs, respectively. Mars and the fitted IMB are shown by the black and the gray curves. The Mars-sun line is also indicated by a black hair line with filled symbols at 0, 500, 1000 and 3397 km above the subsolar point. The observation geometry, the intensity, and the energy of the ENAs suggest that the intense ENA signal before 13:58:40 UT comes from the subsolar region, which is the subsolar ENA jet reported by Futaana *et al.* (2006b).

A notable signature in this observation is the abrupt decrease of the subsolar-ENA flux at 13:58:40 UT within ~10 sec, i.e. a distance of 30 km. Moreover, after the ENA flux has decreased, the NPD observed quasi-periodic enhancements with three peaks in the ENA flux (indicated by three arrows in Figure 4c) up to 200 counts/sec (which is the typical count rate of subsolar ENA jets) . The time interval between each peak is ~1 min.

4. Discussion

As described in the introduction, the subsolar ENA jet is a result of the interaction between the solar wind and the Martian upper atmosphere. Stationary characteristics

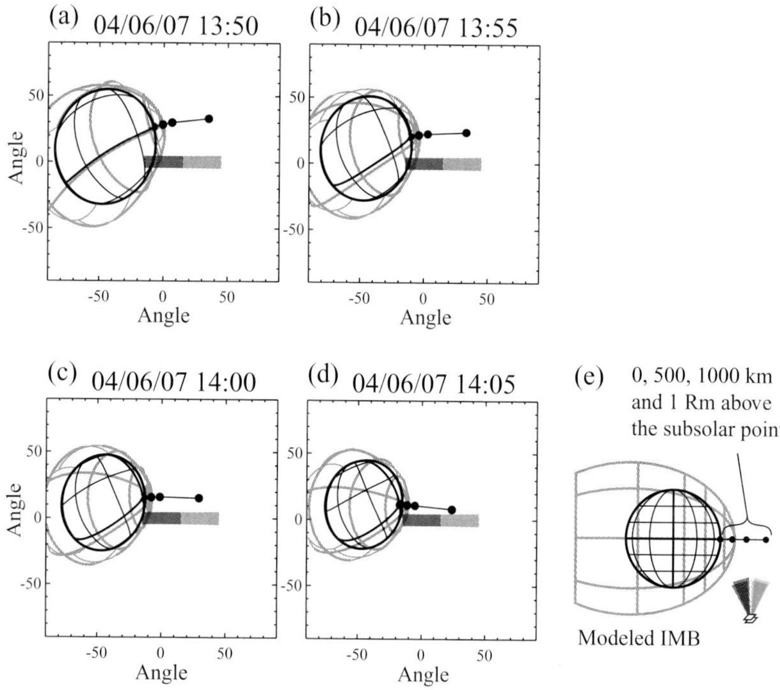

Figure 5. (a)–(d) Time series of the geometry of Mars with respect to the fields of view of the Dir-1 (dark gray) and Dir-2 (light gray). The cartoon (e) is the legend for the projection of the panels (a)–(d). The projection is determined with respect to the center of Dir-1. The FOVs of the NPD are fixed in this projection. Mars and the fitted IMB (see Figure 1) are shown by the black and the light gray curves. The Mars-sun line is indicated by a black hair line and the point at 0, 500, 1000, and 3397 km above the subsolar point are identified by black symbols.

of the observed subsolar ENA jet were investigated by Futaana *et al.* (2006b). In 38 orbits with favorable FOV configurations of the NPD, the subsolar ENA jet can be detected in 36 orbits. Moreover, one third of the outer boundaries of the subsolar ENA jets exhibit clear edges. The typical time scale for the crossing this boundary is about 1 min, which corresponds to a thickness of 200 km for the outer boundary.

However, the ENA jet observed during the orbit 485 shows three peculiarities: the flux is ~5 times higher than the nominal jet flux, the outer boundary is extremely clear (corresponding to the abrupt decrease within ~10 sec at ~13:58:40 UT), and there are quasi-periodic enhancements (~1 min) just after the boundary crossing. Since these characteristics are quite unique, it is difficult to interpret them by using nominal solar wind modulations.

One possible interpretation is that the change is caused by crustal magnetized regions of Mars (Acuña *et al.*, 1998, 1999). A magnetic anomaly can reconfigure the MPB shape and the location of the subsolar region (Crider *et al.*, 2002), and the change in the MPB may explain the peculiarities. However, this idea is not feasible because the subsolar point at the time of the observations was ~(45°E, 17°N), where crustal magnetization is very weak. It is also difficult to reconfigure the MPB within the time scale of 10 sec (corresponding to the 0.04° rotation of Mars).

Another idea is that these signatures are caused by temporal variations in the Martian plasma environment triggered by an change in the solar wind dynamic pressure. This idea well explains these signatures, and is consistent with simultaneous plasma observations.

The first peculiarity of the extremely high ENA flux observed before the abrupt decrease can be interpreted as the result of a compressed Martian plasma obstacle. The bow shock and the IMB were closer to Mars than the average locations of these boundaries, as described in the previous section. This means that the Martian plasma obstacle was more compressed and smaller than the average (Figure 1). This compression might be a result of the increase in solar wind dynamic pressure at 8:00 UT (Figure 3). Under a compressed configuration, the magnetosheath solar wind ions can reach lower altitudes, where the neutral particle density is higher. The scale height at the height of the exobase is about 30 km (Fox and Dalgarno, 1979). So therefore, the intensity of the ENA jet is expected to be sensitive to the compression of the IMB. Assuming exponentially decreasing exosphere with height, 5 times higher density corresponds to a decrease of 48 km in altitude, which is not an unreasonable value.

The second peculiarity, the abrupt decrease in ENA flux at ~13:58:40 UT, is interpreted as a result of a clear solar wind structure crossing as indicated by the ion and the electron observations (13:58:30 UT). This structure may be correlated with the second increase in the dynamic pressure which occurred sometime between ~13:30 and 14:00 (Figure 3).

Figure 6 shows detailed plots of the simultaneous plasma observations obtained by the ELS and the IMA after the bow shock crossing (13:47 UT). From top to bottom, (a) flux of the NPD, (b) the energy-time (E-t) spectrogram for the ELS, (c) the time series of the electron count rate for 5 energy ranges, (d) the E-t spectrogram for the IMA, (e) the energy mass spectrograms for the IMA are displayed. Figure 6(e-1) and (e-2) display the energy-mass spectra integrated over all the viewing direction of IMA before and after the solar wind conditions changed at ~13:58:30 UT. The thick lines show the mass per charge (M/q) profiles for selected mass species.

The bow shock crossing was at ~13:47 UT, and afterward Mars Express was in the solar wind. By using the electron data, it can easily be identified that the solar wind conditions changed at ~13:58:30 UT (indicated by arrows). Features are also observed in the electron flux enhancements (<20 eV) between 13:59–14:04 UT; however these flux enhancements are artifact signatures, which may be due to photoelectrons of satellite or instrument origin. These fluxes are always observed when special viewing configurations occur. The change in the solar wind at ~13:58:30 UT is not the artifact; the undisturbed electron fluxes at 13:50-13:55

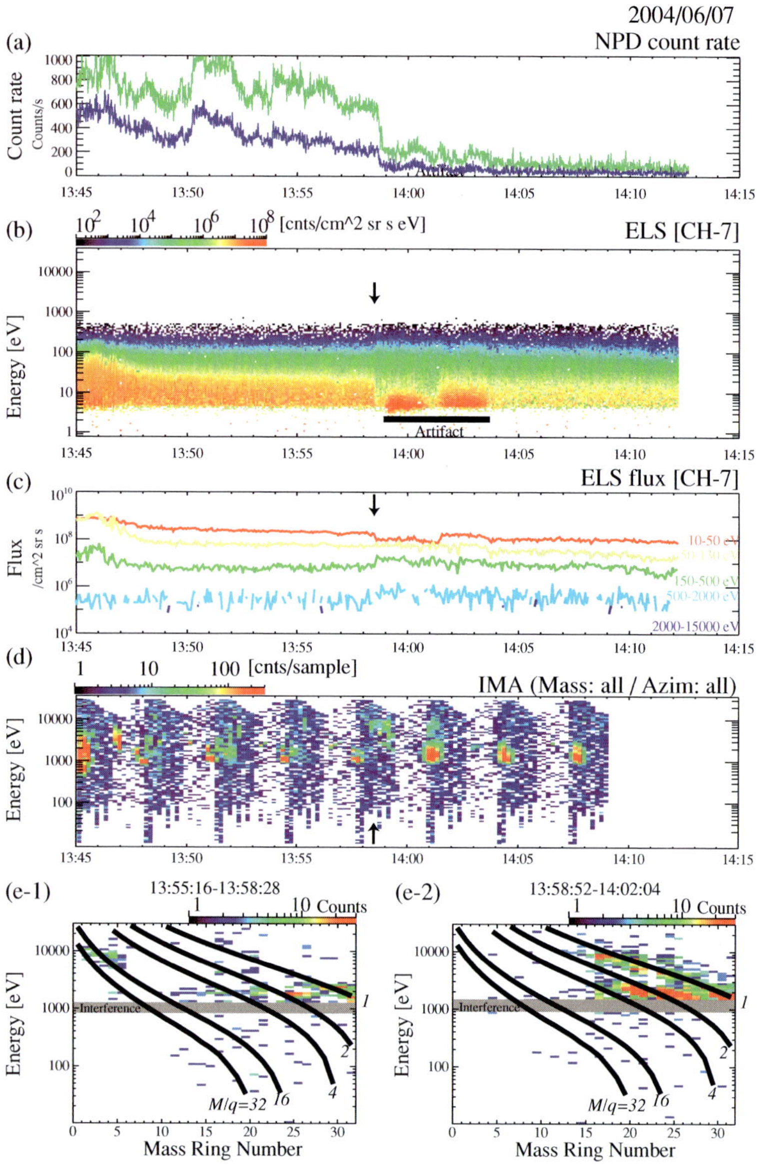

Figure 6. The simultaneous observations from IMA and ELS. The top panel (a) is the integrated ENA flux (identical to Figure 4(c)). The panel (b) shows the energy-time spectrogram observed by ELS. The panel (c) shows the time series of counts with the energy range 10–50, 50–130, 150.0–500, 500–2000 and 2000–15000 eV from ELS. The panel (d) shows the IMA energy-time spectrogram between 13:45 and 14:15 (Operation ended at 14:09 UT). The bottom panels (e-1) and (e-2) display the energy-mass spectra as measured by IMA. The thick lines correspond to the mass per charge profiles of the observed ions for $M/q = 1$, 2, 4, 16 and 32.

and 14:05–14:10 are clearly changed. Such a change in spectra is not observed in data from other orbits.

Lacking magnetic field data, it is difficult to tell whether this abrupt flux change is caused by a shock or a discontinuity. However, there are several characteristics of the plasma observations which strongly suggesting that this solar wind structure was indeed an interplanetary shock. First, the electron flux with energies larger than 150 eV was increased while the flux with energies less than 50 eV was decreased (Figure 6b and c). This signature can be interpreted as the superthermal electron heating by an interplanetary shock (e.g. Feldman *et al.*, 1983; Treumann and Terasawa, 2001). Simultaneously, IMA data in Figure 6(d) indicate that the solar wind ions were heated and the velocity distribution function was broadened. We also see nonthermal ions with an energy range of 2–10 keV around the time of the solar wind structure crossing ($\sim$13:59 UT). The energy-mass spectrum (Figure 6(e-2)) indicates that the ions are protons ($M/q = 1$). Since we cannot see such nonthermal protons before the solar wind structure crossing (Figure 6(e-1)), these nonthermal protons are generated by the solar wind structure. We can conclude that the nonthermal ions are solar wind protons reflected at the supercritical shock surface (e.g. Thomsen, 1985) since such nonthermal ions are observed as well in the vicinity of shock surfaces. Moreover, such reflected ions do not theoretically exist in the vicinity of discontinuities. From the above investigations, the solar wind structure is most likely an interplanetary shock.

The explanation of how the interplanetary shock surface (even if the interface is not the shock, but just the increase of the dynamic pressure) results in the abrupt decrease of the ENA jet flux is illustrated in Figure 7. As shown in Figure 7a, Mars Express was in the ENA jet. After the interplanetary shock surface hit the Martian plasma obstacle, the obstacle was compressed due to the higher dynamic pressure of the downstream medium of the interplanetary shock (Figure 7b). This is analogous to the sudden commencement at the Earth's magnetosphere (e.g. Araki, 1994). Under the compressed situation, the obstacle moved closer to Mars and the solar wind streamlines changed. The reconfiguration of the obstacle shape and location caused the ENA jet generation region to move closer to the planet. The reconfiguration also change the shape of the generation region. As a result, the satellite exited from the jet region. Even though the FOV looked toward the generation region, the plasma streamlines at the subsolar region did not point toward the satellite. As a result, no flux could not be detected.

The third peculiarity of this observation is the quasi-periodic enhancements of ENA flux occurring just after the abrupt decrease. The period is $\sim$1 min and we observed at least three peaks in the data (13:59–14:05 UT, arrows in Figure 4c). The generation mechanism of these enhancement is still an open question, but one possible candidate is that there were global vibrations of the Martian plasma obstacle triggered by the interplanetary shock. Such global vibrations cause the jet generation region to move back and forth resulting in variations of the ENA flux at the satellite location. From *in situ* observations conducted by the Phobos 2

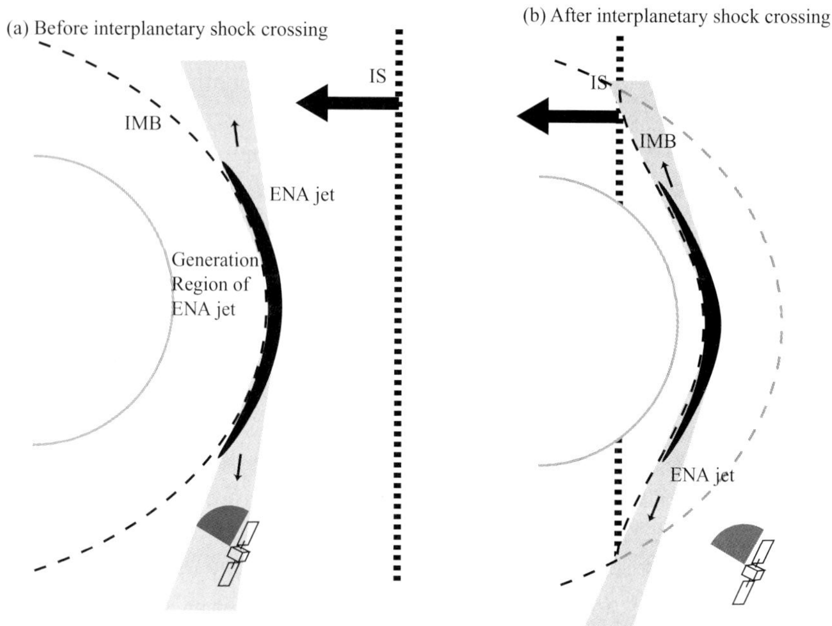

Figure 7. Illustration of the interpretation (see text for details): (a) before and (b) after the shock crossing of the planetary ENA jet source. The source region of the subsolar ENA jet, which is very close to the IMB, was pushed toward the planet due to the high dynamic pressure of the interplanetary shock. The satellite went out of the subsolar ENA jet very quickly due to this reconfiguration of the Martian plasma obstacle.

and Mars Global Surveyor spacecraft, ultra low-frequency magnetic oscillations in the Martian magnetosheath have been observed while the mechanism that causes such the oscillations is not yet known (Espley *et al.*, 2004). This kind of global vibration of the plasma obstacle can provide one possible explanation for the quasi-periodic enhancement in jet flux just after its abrupt decrease, but careful investigations by global hybrid simulations are necessary and are yet to be performed.

We should note that it is possible for the global structure of the ENA jet to be influenced by the interaction with an inclined interplanetary shock. If the interplanetary shock had an inclination (defined by the angle between the Mars-Sun line and the normal direction of the interplanetary shock), the interplanetary shock would hit the flank-side of the Martian plasma obstacle. This may introduce an observable reconfiguration of the ENA jet as well, which in turn could result in the decrease of the ENA jet flux at the satellite location.

However, the simple calculation below shows that the interplanetary shock was nearly perpendicular to the Mars-Sun line during this event. Figure 8 shows the 7-min observation of the ELS flux and the NPD count rate around the interplanetary shock crossing. The time of the shock arrival of the satellite position is ~13:58:30

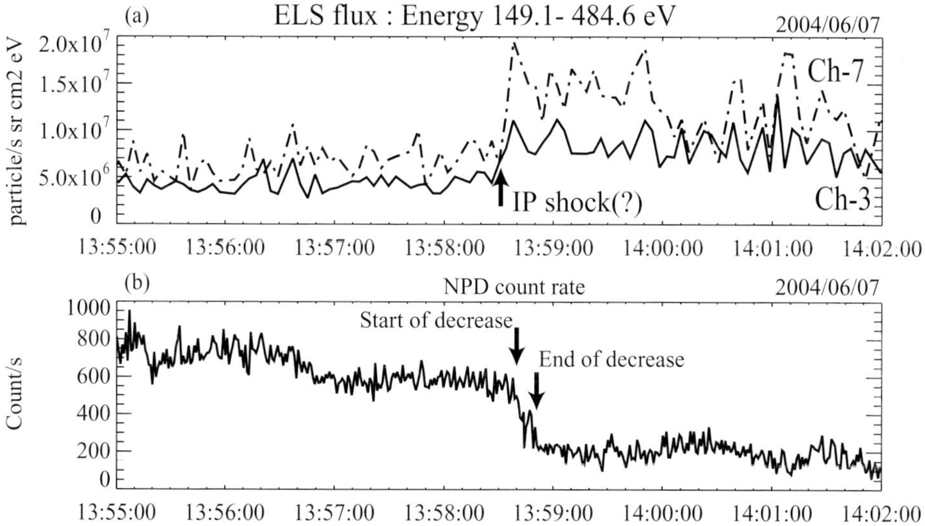

Figure 8. Electron and ENA flux around the shock crossing (13:55–14:02 UT): (a) The electron flux with energies between 149.1–484.6 eV (identical to the green line in Figure 6(c)). (b) The ENA count rate (identical to the green line in Figure 4(c)).

UT as determined from the ELS observation (Figure 8(a)), while the ENA decrease starts at ~13:58:40 UT (Figure 8(b)). The distance between the Martian subsolar region and Mars Express was $L \sim 4400$ km (see Figure 1). The ENAs (average velocity was $v \sim 575$ km/s) took $T \sim 7.7$ sec to arrive at the spacecraft location after their generation in the subsolar region. This means that the interplanetary shock hit the Martian plasma obstacle at around 13:58:32 UT, which was approximately the same time as when the shock was detected by Mars Express. Since the x-coordinate of Mars Express at the interplanetary shock crossing was almost the same as that of the Martian magnetic obstacle as seen in Figure 1, we conclude that the interplanetary shock could have a small inclination.

From this peculiar event, we are also able to discuss the extent of the subsolar jet generation region by the investigation of the travel time for the ENAs which forms the jet. The extent of the generation region along the line of sight, ΔL, can be estimated by the decrease time of the ENA jet flux at the satellite position, ΔT, and the dispersion of the ENA velocity. The relationship is described as

$$\left(\frac{\Delta L}{L}\right)^2 \sim \left(\frac{\Delta T}{T}\right)^2 + \left(\frac{\Delta v}{v}\right)^2,$$

where T and L is the average travel time and the distance between the generation region and the satellite, v is the velocity of the ENA, and Δv is the deviation of the ENA velocity. Figure 8b shows that it took $\Delta T \sim 10$ sec for the ENA flux decrease (13:58:40–13:58:50 UT). T and L are about 7.7 sec and

4400 km from the above discussion. The observed ENA energies are 800 eV-3 keV (Figure 4), which corresponds to the hydrogen ENA velocities of 390–760 km/s, i.e., $v \sim 575$ km/s and $\Delta v \sim 370$ km/s. Using these values, the extent ΔL is calculated as 6400 km. This extent is approximately the same order as the Martian diameter.

5. Summary

The NPD data obtained on June 7, 2004 (orbit 485) exhibited an extremely high flux of a subsolar ENA jet. The observed flux was approximately 5 times higher than the typical ENA jet flux, and abruptly decreased over a very short time (less than 10 sec), when the spacecraft crossed the outer boundary of the ENA jet. We interpret this peculiar event as a result of a quick reconfiguration of the Martian plasma environment due to the arrival of a solar wind structure. A simultaneous structure showing a pressure increase was also observed in the MGS dynamic pressure data. The simultaneous *in situ* ion and electron observations showed a clear boundary in the solar wind at the time of the decrease of the ENA jet. The solar wind ions and electrons were heated after the solar wind boundary crossing and generation of the reflected ions was observed. Because these signatures are commonly observed in the vicinity of shock surfaces, the change in the interplanetary medium was most likely due to an interplanetary shock.

When the interplanetary structure hit the Martian plasma obstacle, the IMB was pushed toward the planet and the generation region of the subsolar ENA jet moved toward Mars (Figure 7). This global reconfiguration could cause the sudden decrease of the subsolar ENA jet observed at the Mars Express location. After the sudden decrease, the ENA flux exhibited quasi-periodic enhancements with a period of ~1 min. This periodic flux may be interpreted as a result of the global vibration of the Martian plasma obstacle with a characteristic frequency of ~1 min.

This NPD observation provided information about the response in the Martian plasma environment to a change in solar wind conditions. The Martian plasma obstacle responds to the solar wind change on a very short time scale (~10 sec). Moreover, the response appears to be elastic, and a solar wind dynamic pressure pulse can induce a global vibration mode of the Martian plasma environment with ~1 min period.

The event reported in this paper is unique. When the interplanetary shock reached the Martian plasma obstacle, the spacecraft was located in the ENA jet, and we have plasma and neutral particle data during this event. Unfortunately, no statistical analysis is possible at present.

We would like to emphasize that the observations of subsolar ENA jets can be used as a global monitoring tool to investigate the dynamics of the Martian plasma obstacle. Such a remote observation of the plasma environment is one advantage of ENA imaging. By comparing *in situ* plasma observations and remote ENA observations, one can investigate global signatures of the interaction between the Martian upper atmosphere and the solar wind.

Acknowledgments

Y. Futaana has been supported by a Postdoctoral Fellowship for Research Abroad program from the Japan Society for the Promotion of Science and by a JSPS Research Fellowships for Young Scientists. We would like to acknowledge Dr. Mario Acuña at NASA/GSFC and the MGS MAG/ER team for providing the dynamic pressure data at Mars and thank Dr. D. Crider at the Catholic University of America for fruitful discussion concerning the MGS data. The work at SwRI was supported by NASA contract NASW0003. The ASPERA-3 experiment on the European Space Agency (ESA) Mars Express is a joint effort among 15 laboratories in 10 countries, all supported by their national agencies. We thank all these agencies, as well as the various departments and institutes hosting these efforts.

References

Acuña, M. H., Connerney, J. E. P., Wasilewski, P., Lin, R. P., Anderson, K. A., Carlson, C. W. *et al.*: 1998, *Science* **279**, 1676–1680.

Acuña, M. H., Connerney, J. E. P., Ness, N. F., Lin, R. P., Mitchell, D., Carlson, C. W. *et al.*: 1999, *Science* **284**, 790–793.

Araki, T.: 1994, in: M. J. Engebretson, K. Takahashi, and M. Scholer (eds.), *Solar Wind Sources of Magnetospheric Ultra-Low-Frequency Waves, Geophysical Monograph*, vol. 81, pp. 183–200.

Barabash, S., Norberg, O., Lundin, R., Olsen, S., Lundin, K., Brandt, P. C. *et al.*: 1998, in: R. F. Pfaff, J. E. Borovsky, and D. T. Young (eds.), *Measurement Techniques in Space Plasmas, Field*, AGU Geophysical Monograph 103, American Geophysical Union, Washington, DC, pp. 257–262.

Barabash, S., Lundin, R., Andersson, H., Gimholt, J., Holmström, M., Norberg, O. *et al.*: 2004, in: A. Wilson, (ed.), *Mars Express: The Scientific Payload*, vol. SP-1240, ESA Special Publication, pp. 121–139.

Brinkfeldt, K., Gunell, H., Brandt, P., Barabash, S., Frahm, R. A., Winningham, J. D. *et al.*: 2006, *Icarus* **182**(2), 439–447.

Crider, D. H., Acuña, M. H., Connerney, J. E. P., Vignes, D., Ness, N. F., Krymskii, A. M. *et al.*: 2002, *Geophys. Res. Lett.* **29**(8), 1170, doi: 10.1029/2001GL013860.

Crider, D. H., Vignes, D., Krymskii, A. M., Breus, T. K., Ness, N. F., Mitchell, D. L. *et al.*: 2003, *J. Geophys. Res.* **108**(A12), 1461, doi: 10.1029/2003JA009875.

Espley, J. R., Cloutier, P. A., Brain, D. A., Crider, D. H., and Acuña, M. H.: 2004, *J. Geophys. Res.* **109**, A07213, doi: 10.1029/2003JA010193.

Feldman, W. C., Anderson, R. C., Bame, S. J., Gosling, J. T., Zwickl, R. D., and Smith, E. J.: 1983, *J. Geophys. Res.* **88**, 9949–9958.

Fox, J. L., and Dalgarno, A.: 1979, *J. Geophys. Res.* **84**(A12), 7315–7333.

Futaana, Y., Barabash, S., Grigoriev, A., Holmström, M., Kallio, E., C:son Brandt, P., *et al.*: 2006a, *Icarus* **182**(2), 424–430.

Futaana, Y., Barabash, S., Grigoriev, A., Holmström, M., Kallio, E., C:son Brandt, P. *et al.*: 2006b, *Icarus* **182**(2), 413–423.
Gunell, H., Brinkfeldt, K., Holmström, M., Brandt, P., Barabash, S., Kallio, E. *et al.*: 2006, *Icarus* **182**(2), 439–447.
Holmström, M., Barabash, S., and Kallio, E., 2002, *J. Geophys. Res.* **107**(A10), 1277, doi: 10.1029/2001JA000325.
Kallio, E., and Barabash, S.: 2001, *J. Geophys. Res.* **106**(A1), 165–177.
Kallio, E., Luhmann, J. G., and Barabash, S.: 1997, *J. Geophys. Res.* **102**(A10), 22,183–22,197.
Kallio, E., Barabash, S., Brinkfeldt, K., Gunell, H., Holmström, M., Futaana, Y. *et al.*: 2006, *Icarus* **182(2)**, 448–463.
Lichtenegger, H., Lammer, H., and Stumptner, W.: 2002, **107**(A10), 1279, doi: 10.1029/2001JA000322.
Lundin, R., Barabash, S., Andersson, H., Holmström, M., Grigoriev, A., Yamauchi, M. *et al.*: 2004, *Science* **305**, 1933–1936.
Mitchell, D. G., Jaskulek, S. E., Schlemm, C. E., Keath, E. P., Thompson, R. E., Tossman, B. E. *et al.*: 2000, *Space Sci. Rev.* **91**(1–2), 67–112.
Moore, T. E., Chornay, D. J., Collier, M. R., Herrero, F. A., Johnson, J., Johnson, M. A. *et al.*: 2000, *Space Sci. Rev.* **91**(1–2), 155–195.
Nagy, A. F., Winterhalter, D., Sauer, K., Cravens, T. E., Brecht, S., Mazelle, C. *et al.*: 2004, *Space Sci. Rev.* **111**, 33–114.
Peréz-de-Tejada, H.: 1987, *J. Geophys. Res.* **92**, 4713–4718.
Pollock, C. J., Asamura, K., Baldonado, J., Balkey, M. M., Barker, P., Burch, J. L. *et al.*: 2000, *Space Sci. Rev.* **91**(1–2), 113–154.
Roelof, E. C., Mitchell, D. G., and Williams, D. J.: 1985, *J. Geophys. Res.* **90**, 10,991–11,008.
Sagdeev, R. Z., and Zakharov, A. V.: 1989, **341**(6243), 581–585.
Thomsen, M. F.: 1985, in: B. T. Tsurutani, and R. G. Stone (eds.), *Collisionless Shocks in the Heliosphere: Reviews of Current Research*, American Geophysical Union, pp. 253–270.
Treumann, R. A., and Terasawa, T.: 2001, *Space Sci. Rev.* **99**, 135–150.
Vaisberg, O. L.: 1992, in: J. G. Luhmann, M. Tatrallyay, and R. O. Pepin (eds.), *Venus and Mars Atmospheres, Ionospheres, and Solar Wind Interactions*, *AGU Geophysical Monograph*, vol. 66, pp. 311–326.
Vignes, D., Mazelle, C., Rème, H., Acuña, M. H., Connerney, J. E. P., Lin, R. P. *et al.*: 2000, *Geophys. Res. Lett.* **27**(1), 49–52.
Vignes, D., Acuña, M. H., Connerney, J. E. P., Crider, D. H., Rème, H., and Mazelle, C.: 2002, *Geophys. Res. Lett.* **9**, 1328, doi: 10.029/2001GL014513.

AURORAL PLASMA ACCELERATION ABOVE MARTIAN MAGNETIC ANOMALIES

R. LUNDIN[1,*], D. WINNINGHAM[2], S. BARABASH[1], R. FRAHM[2], D. BRAIN[11], H. NILSSON[1], M. HOLMSTRÖM[1], M. YAMAUCHI[1], J. R. SHARBER[2], J.-A. SAUVAUD[3], A. FEDOROV[3], K. ASAMURA[4], H. HAYAKAWA[4], A. J. COATES[5], Y. SOOBIAH[5], C. CURTIS[6], K. C. HSIEH[6], M. GRANDE[7], H. KOSKINEN[8], E. KALLIO[8], J. KOZYRA[9], J. WOCH[10], M. FRAENZ[10], J. LUHMANN[11], S. MCKENNA-LAWLER[12], S. ORSINI[13], P. BRANDT[14] and P. WURZ[15]

[1]*Swedish Institute of Space Physics, Box 812, S-98 128, Kiruna, Sweden*
[2]*Southwest Research Institute, San Antonio, TX 7228-0510, USA*
[3]*Centre d'Etude Spatiale des Rayonnements, BP-4346, F-31028 Toulouse, France*
[4]*Institute of Space and Astronautical Science, 3-1-1 Yoshinodai, Sagamichara, Japan*
[5]*Mullard Space Science Laboratory, University College London, Surrey RH5 6NT, UK*
[6]*University of Arizona, Tucson, AZ 85721, USA*
[7]*Rutherford Appleton Laboratory, Chilton, Didcot, Oxfordshire OX11 0QX, UK*
[8]*Finnish Meteorological Institute, Box 503 FIN-00101 Helsinki, Finland*
[9]*Space Physics Research Lab., University of Michigan, Ann Arbor, MI 48109-2143, USA*
[10]*Max-Planck-Institut für Sonnensystemforschung, D-37191 Katlenburg-Lindau, Germany*
[11]*Space Science Lab., University of California in Berkeley, Berkeley, CA 94720-7450, USA*
[12]*Space Technology Ltd., National University of Ireland, Maynooth, Co. Kildare, Ireland*
[13]*Instituto di Fisica dello Spazio Interplanetari, I-00133 Rome, Italy*
[14]*Applied Physics Laboratory, Johns Hopkins University, Laurel, MD 20723-6099, USA*
[15]*University of Bern, Physikalisches Institut, CH-3012 Bern, Switzerland*
(**Author for correspondence: E-mail: rickard.lundin@irf.se*)

(Received 4 April 2006; Accepted in final form 25 October 2006)

Abstract. Aurora is caused by the precipitation of energetic particles into a planetary atmosphere, the light intensity being roughly proportional to the precipitating particle energy flux. From auroral research in the terrestrial magnetosphere it is known that bright auroral displays, discrete aurora, result from an enhanced energy deposition caused by downward accelerated electrons. The process is commonly referred to as the auroral acceleration process. Discrete aurora is the visual manifestation of the structuring inherent in a highly magnetized plasma. A strong magnetic field limits the transverse (to the magnetic field) mobility of charged particles, effectively guiding the particle energy flux along magnetic field lines.

The typical, slanted arc structure of the Earth's discrete aurora not only visualizes the inclination of the Earth's magnetic field, but also illustrates the confinement of the auroral acceleration process. The terrestrial magnetic field guides and confines the acceleration processes such that the preferred acceleration of particles is frequently along the magnetic field lines. Field-aligned plasma acceleration is therefore also the signature of strongly magnetized plasma.

This paper discusses plasma acceleration characteristics in the night-side cavity of Mars. The acceleration is typical for strongly magnetized plasmas – field-aligned acceleration of ions and electrons. The observations map to regions at Mars of what appears to be sufficient magnetization to support magnetic field-aligned plasma acceleration – the localized crustal magnetizations at Mars (Acuña *et al.*, 1999). Our findings are based on data from the ASPERA-3 experiment on ESA's Mars Express, covering 57 orbits traversing the night-side/eclipse of Mars. There are indeed strong similarities

Space Science Reviews (2006) 126: 333–354
DOI: 10.1007/s11214-006-9086-x

between Mars and the Earth regarding the accelerated electron and ion distributions. Specifically acceleration above Mars near local midnight and acceleration above discrete aurora at the Earth – characterized by nearly monoenergetic downgoing electrons in conjunction with nearly monoenergetic upgoing ions. We describe a number of characteristic features in the accelerated plasma: The "inverted V" energy-time distribution, beam vs temperature distribution, altitude distribution, local time distribution and connection with magnetic anomalies. We also compute the electron energy flux and find that the energy flux is sufficient to cause weak to medium strong (up to several tens of kR 557.7 nm emissions) aurora at Mars.

Monoenergetic counterstreaming accelerated ions and electrons is the signature of field-aligned electric currents and electric field acceleration. The topic is reasonably well understood in terrestrial magnetospheric physics, although some controversy still remains on details and the cause-effect relationships. We present a potential cause-effect relationship leading to auroral plasma acceleration in the nightside cavity of Mars – the downward acceleration of electrons supposedly manifesting itself as discrete aurora above Mars.

Keywords: aurora, plasma acceleration, Mars magnetic anomalies

1. Introduction

The magnetized planets in our solar system, such as Earth, Jupiter and Saturn, are all known to accommodate aurora in their magnetic circumpolar regions. The existence of a strong planetary magnetic dipole appears to be the main condition for the occurrence of aurora, whether caused by an external/solar (e.g. the Earth) or internal (e.g. Jupiter) energy source. The distinguishing difference between "aurora" and airglow lies not in the physical atomic/molecular exitation processes, but rather in their macroscopic/structural appearance. In essence aurora may be defined as light emissions occurring within strongly magnetized plasmas. Aurora is highly structured light emissions, while airglow represents diffuse emissions.

The Earth's polar aurora and its related ionospheric and magnetospheric phenomena, is well explored and reasonably well understood. The first measurements establishing its relation with charged particle precipitation into the Earth's atmosphere was made in the early 1960s (MacIlwain, 1960). Intense aurora in the form of extended bright arcs in the sky, sometimes referred to as discrete aurora, results from downward electron acceleration at higher altitudes above the Earth. Hannes Alfvén proposed already in 1958, that magnetic field-aligned quasi-static electric fields should be an important acceleration process related with bright aurora. Field-aligned electric fields, accelerating electrons downward, will both increase the electron energy and enhance the downward electron fluxes leading to brightened auroral displays in the topside atmosphere. The first evidence for electrostatic acceleration came from sounding rockets measurements (Albert, 1967; Evans, 1968). They showed that intense fluxes of nearly monoenergetic precipitating electrons were present above auroral arcs. Subsequent low-altitude satellite measurements (Frank and Ackerson, 1971) provided a global view of the electron precipitation, demonstrating that nearly monoenergetic electron structures denoted "inverted V's" were a common feature above the auroral oval. The notion "inverted V" referred to the spectral shape in an energy-time spectrogram. A connection between "inverted V's" and electric fields (Gurnett and Frank, 1973) supported the hypothesis of electron acceleration in an electrostatic potential structure. Evans (1974) analyzed the distribution of accelerated electrons below a field-aligned electrostatic potential, considering in detail the interaction between accelerated primary electrons and the atmosphere. The interaction leads to electron energy spectra above the aurora that contain primary/accelerated electrons as well as backscattered electrons (degraded primary- and secondary electrons).

A consequence of electrostatic acceleration is the separation of charges. If electrons are accelerated downward one expects positive ions to be accelerated upward. Indeed, narrowly collimated upward beams of accelerated ionospheric ions were eventually observed (Shelley *et al.*, 1976) at high-altitudes (4000–8000 km) above the aurora, results subsequently reconfirmed by several auroral satellites (see Moore *et al.*, 1999 for a review). Simultaneous upward accelerated ion beams and downward accelerated electron beams are therefore evidence for quasi-electrostatic acceleration of plasma along magnetic field lines.

Field-aligned electrostatic acceleration is generally discussed in connection with dipole magnetic field geometries, such as that of the Earth. Electrostatic acceleration is not expected to occur in weakly magnetized plasma environments such as those near comets, Venus and Mars. The Mars Global Surveyor (MGS) findings of strong crustal anomalies at Mars (Acuña *et al.*, 1999) have considerably changed the picture. We now expect to find diverging magnetic field "cusps" above Mars (Krymskii *et al.*, 2002) and "minimagnetospheres" (Mitchell *et al.*, 2001) with local magnetic conditions similar to those found above the Earth's polar region. A distinctive difference, though, is that the crustal magnetization is very weak compared to the Earth's dipole field. It was therefore not obvious that the plasma would be sufficiently magnetized to accommodate processes known to us from the Earth as being associated with aurora. Furthermore, the crustal field is characterized by a set of multipoles centered at specific longitudes and latitudes near the equator of Mars. The aurora, if existent, was therefore expected to be highly dependent on local time, visible only near the equatorial nightside at specific longitudes and latitudes.

The first evidence for structured emissions that could be aurora on Mars was indeed related with a crustal magnetization at Mars (Bertaux *et al.*, 2005). They also inferred the emissions to be related with particle precipitation. Publications of ongoing work followed and we now have ample evidence that intense electron precipitation occurs over magnetized regions at Mars (Lundin *et al.*, 2006a; Brain *et al.*, 2006). Evidence from ASPERA-3 data (Lundin *et al.*, 2006a) showed the signature of field-aligned plasma acceleration that suggests the existence of parallel electric field acceleration – accelerating electrons downward and ions upward. The acceleration regions, "inverted V's", are connected to semi-open flux tubes (cusps/clefts) associated with crustal magnetization regions. The latter might have been proof for a strongly magnetized plasma, if it wasn't for the lack of magnetic field measurements on Mars Express. However, Brain *et al.* (2006a) found field-aligned currents, up to 1 microamp per square meter, associated with accelerated electrons, another signature associated with magnetic field-aligned plasma acceleration in a "strongly" magnetized plasma. Therefore, we have ample evidence for two important aspects:

1. Aurora occurs on the nightside of Mars – connected with crustal magnetization regions.
2. Auroral plasma acceleration, usually reserved for strongly magnetized planets with dipole fields, may also occur in localized magnetic cusps/clefts extending into space from weakly magnetized planets such as Mars. A solar wind / boundary layer driven generator/dynamo feeds energy to the plasma accelerator and the planetary ionosphere via field-aligned electric currents.

In this study we expand on the results of Lundin *et al.* (2006a), going into more detail on the acceleration process, its latitude and local time dependence, its altitude dependence and the energy flux expected in the nightside of Mars leading to what one may term the *aurora equatorialis*. The reason for its location so close to local midnight will be discussed. Finally, we suggest a simple concept of a current circuit that couples to the energy source – the dynamo propelled by solar wind forcing. We also discuss where and how the indirect forcing is coupled to the accelerator/load, in and above the nightside ionosphere of Mars.

2. ASPERA-3 Results

The ASPERA-3 experiment (Barabash *et al.*, 2004) has two plasma instruments, an ELectron Spectrometer (ELS) and an Ion Mass Analyzer (IMA). ELS provides electron measurements in the energy range 0.001–20 keV with 8% energy resolution. The intrinsic field of view is $4 \times 360°$. The 360° aperture is divided into 16 sectors. The sensor consists of a top hat electrostatic analyzer in a very compact design.

IMA is an improved version of the ion mass spectrographs TICS/Freja, IMIS/Mars-96, IMI/Planet-B, and almost identical to the Ion Composition Analyzer, ICA, (Nilsson *et al.*, 2006) on ESAs Rosetta mission. IMA provides ion measurements in the energy range 0.01–30 keV/e for the main ion components H^+, He^{++}, He^+, O^+, and the group of molecular ions ($20 < M/q <\sim 80$). IMA has a $4.6 \times 360°$ field of view. Electrostatic sweeping provides elevation ($\pm 45°$) coverage for a total field of view of about 2π. The elevation sweep shows up as periodic "pulses" in the ion energy-time spectrogram (FigureS 2 and 9).

The total material forming the basis of this study comprises over 130 orbits of ASPERA-3 data during 2004 and 2005, where we have analyzed in detail data from 57 traversals of the Martian mid-tail eclipse. During this period the spacecraft traversed the midnight sector close to the central tail. An important reason for

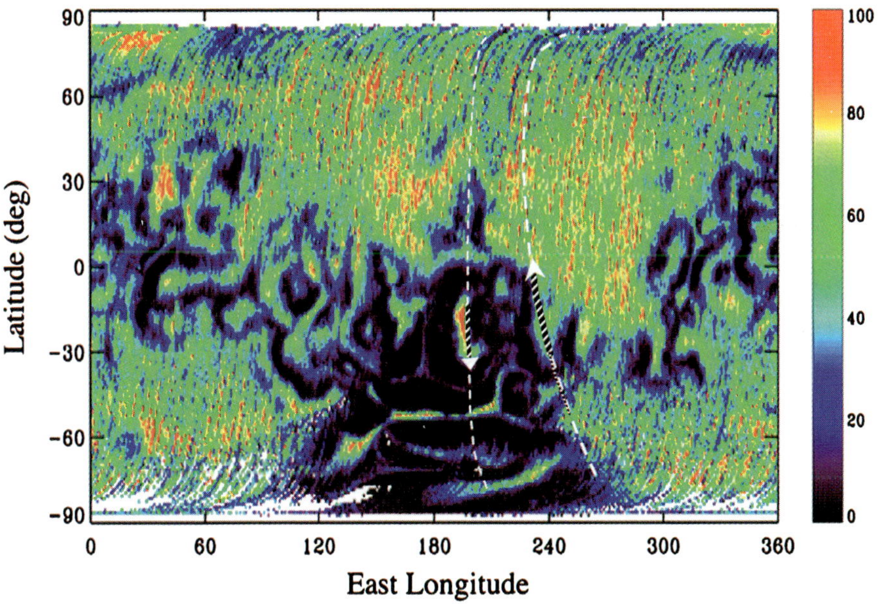

Figure 1. Nightside map of magnetic anomalies. The scale indicates the percentage of open magnetic field lines at MGS altitudes (≈400 km), red indicating 100% open and black fully closed magnetic flux tubes (after Brain *et al.*, 2005). Two satellite ground tracks are marked by white dashed lines, black-and white part corresponding to the "inverted V" cases of Figure 2 (right) and 3 (left) respectively. Arrows marks the spacecraft track of the inverted V traversal.

selecting these periods was that they were characterized by coincident energization of ions and electrons in eclipse. In a number of traversals we find that the tail cavity represents a void of particles. These are all cases when the eclipse contained no passes over magnetic anomalies. In the statistics of acceleration structures presented we note that eclipses frequently contain even longer periods without-, than with plasma acceleration. There is a general lack of plasma acceleration when the spacecraft traverse regions of fully "closed" or fully open magnetic field lines (Lundin *et al.*, 2006a; Brain *et al.*, 2006).

Eclipse cavity passes facilitates a direct geographic mapping along magnetic field lines connecting to the magnetic cusps/clefts at Mars. This geometry provides more radial extension of the field lines into the near Mars tail. A geographic latitude and longitude projection to the 400 km MGS open/closed magnetic field map determined by Brain *et al.* (2005) is used. Figure 1 shows such a map, with the open (red) and closed (black) magnetic field lines at 400 km color-coded. Closed field lines (black) implies shielding against penetration of low energy particles, while open field lines (red) enables direct particle access to the atmosphere of Mars. On top of this map, trajectories of the Mars Express spacecraft showing orbits from two "inverted V" cases (Figures 2 and 3) are indicated, the "inverted V" marked out by hatched lines. The right dashed line corresponds to the orbit in Figure 2 and

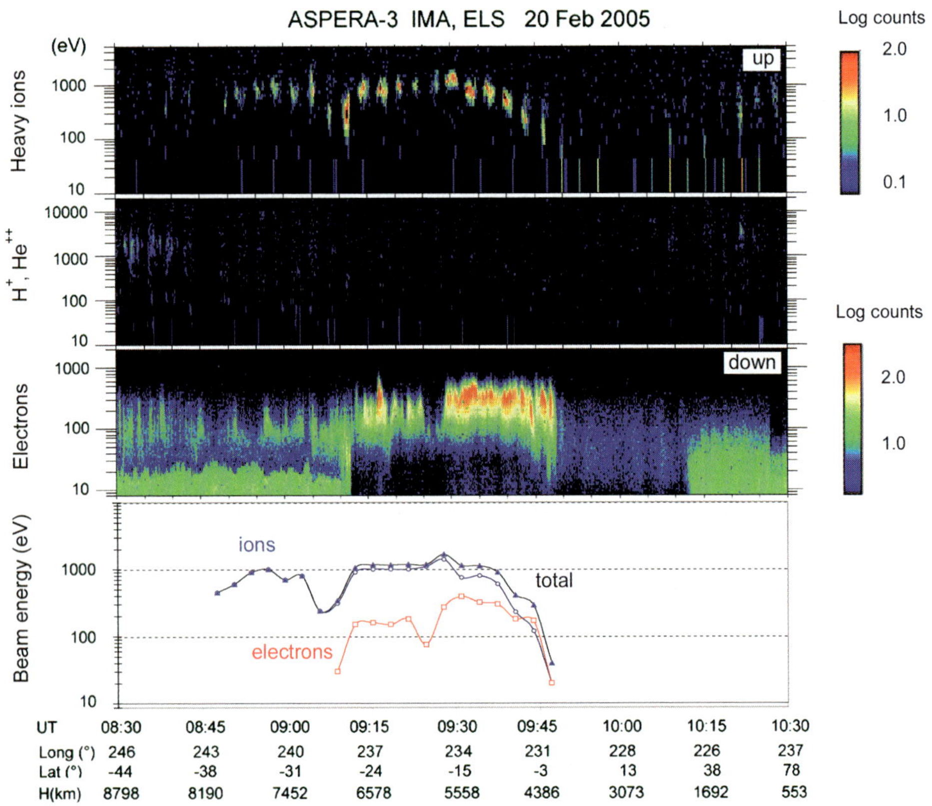

Figure 2. ASPERA-3 energy-time spectrogram and flux peak data for ions and electrons, the spacecraft traversing a high-altitude "Inverted V" event in the tail-eclipse (09:13–10:12 UT). Top panel shows spectra for upgoing heavy ions (O^+, O_2^+, CO_2^+), second panel spectra for antisunward solar wind ions, and third panel spectra for downgoing electrons. The fourth panel depicts acceleration energies (peak energy) for ions (blue), electrons (red), and total acceleration (electrons + ions). The coordinates are in Mars East longitude and latitude (Figure 1).

the left hatched line passing over the "island" of open flux tubes corresponds to the orbit in Figure 3.

Figure 2 shows ion and electron energy-time spectrograms for the "inverted V" event on 20 Feb., 2005 (adapted from Lundin *et al.*, 2006a). The upper panel shows energy spectra for upgoing heavy ions using counts per readout (0.125 s). Narrow beams of heavy ionospheric ions (O^+, O_2^+, and CO_2^+) are observed flowing upward/tailward while substantial fluxes of electrons move in the opposite direction (downward/sunward). Second panel show energy spectra of solar wind ions (H^+ + He^{++}). The low/vanishing fluxes of solar wind ions demonstrate that the spacecraft was inside the induced Mars magnetosphere during this time period. The inbound and outbound crossing of IMB as determined by strong decrease/increase of sheath fluxes occurred at 08:30 UT and 10:42 UT respectively. The third panel shows energy spectra for downgoing electrons (using counts/second), the data accumulated

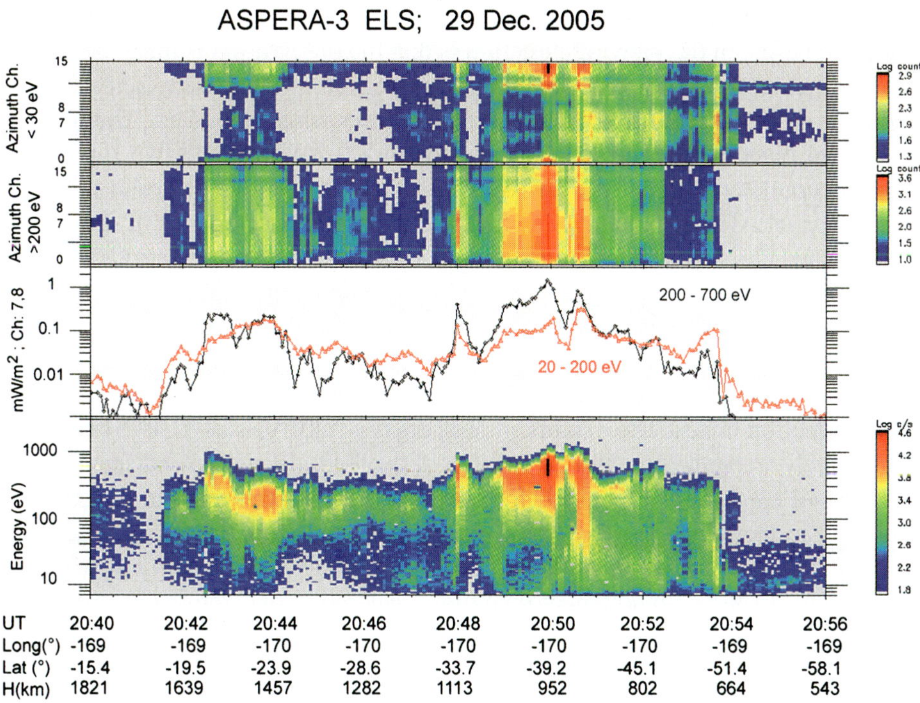

Figure 3. An example of a low-altitude "inverted V" event. The ASPERA-3 electron data is taken during a tail eclipse period (20:32–20:52 UT). The first two panels show integral counts from 16 azimuth sectors. The upper panel for <30 eV electrons, the second panels for >200 eV electrons. The third panel shows the downward energy flux (mW/m^2) in two energy intervals measured by sectors 7 and 8. Bottom panels shows energy-time spectra averaged over sectors 6–11 (downward directed). Coordinates are in Mars East longitude and latitude.

from sectors 7–10. The fourth panel shows electron and ion peak energies determined from the energy-time spectra, and the combined (total) acceleration. Notice the general characteristics of these "inverted V" events: the coincident existence of narrow upgoing ionospheric ion beams and energized downgoing electrons; the coincident acceleration of ions and electrons on what appears to be a common magnetic flux tube. This corroborates the analogy with auroral acceleration near the Earth. The MEX geographic mapping of the "inverted V" (09:10–09:50 UT) is illustrated in Figure 1. Notice that the trajectory maps to semi-open magnetic field lines – in the boundary between a magnetic anomaly and open field lines. It does not map directly to the closed flux tubes from the magnetic anomaly (dark region). This is another analogy with the Earth's auroral zone, the "inverted V's" are usually found in the boundary region between open and closed magnetic field lines.

Figure 3 shows a low-altitude "inverted V" case, this time with no data indicating ion acceleration. The energy-time spectrogram (bottom panel) is for approximately downward directed electrons (Sectors 7 and 8). The lack of observable ion fluxes may be due to spacecraft shadow or incomplete coverage, but a more likely explanation based on the viewing direction is that the observation is made below the acceleration region. The electron energy flux peaked in this case exactly at local midnight, at an altitude of 900–1000 km above the surface of Mars. The integral downward electron energy flux corresponds to 7.0 mW/m^2. The two top panels show accumulated counts from all 16 ELS sectors for <30 eV (first) and >200 eV (second panel). The two panels illustrate the general characteristics of downward field-aligned electron acceleration, i.e. higher fluxes (counts) of electrons with energies near or above the electron peak energy (bottom panel) and enhanced fluxes of degraded primaries and backscattered electrons well below the electron peak energy.

Figure 4 shows a series of 6 ELS electron spectra taken around the region of peaked electron fluxes. The maximum peak energy, 520 eV, is at 20:49:51 UT. One immediately recognizes the typical features of the field-aligned auroral acceleration process (see e.g. Moore *et al.*, 1999 for a review). The spectral shape is evidence for "auroral" acceleration in a quasi-static electric potential drop. Based on the acceleration model by Evans, 1974 we identify three categories in the energy distribution of electrons related with downward/parallel acceleration of electrons in a potential drop V_0: the peak energy of the accelerated primaries ($E_P = eV_0$);

(1) accelerated primary electrons ($E \geq E_P$);
(2) degraded primaries and backscattered electrons ($E < E_P$);
(3) secondary electrons ($E \ll E_P$).

Degraded primaries, backscattered and secondary electrons originate from a combination of wave-particle interaction, electron back-scattering, and secondary low energy electrons emerging as a result of impacting primary electrons. The angular distribution in the downward to perpendicular direction is rather isotropic, in the sense that a downward flow is less obvious. However, this is consistent with field-aligned acceleration in a diverging magnetic field, some of the electrons magnetically mirroring before precipitating into the atmosphere (e.g. Chiu and Schulz, 1978). The precipitating electron flux, and the associated field-aligned electric current, is determined from the loss-cone. Lacking magnetic field measurements we are unable to determine the loss cone, but we may on basis of the low-energy fluxes infer that, at least part of the time, it was covered by sector 13. The reason for this is the enhanced fluxes of <100 eV electrons that stands out (20:49–20:50 UT) compared to the fluxes in other sectors. Our interpretation is that these are backscattered secondary electrons produced by energetic electron precipitation into the atmosphere of Mars (category 3)).

It is evident from Figure 4 that the peak electron energy is higher than the thermal energy of the accelerated electrons. Lundin *et al.* (2006a,b) discussed this observational fact for both ions and electrons in the eclipse cavity of Mars. Figure 5 shows an updated version of the observations by Lundin *et al.* (2006b), i.e. the relation between ion and electron beam energy versus beam temperature, respectively.

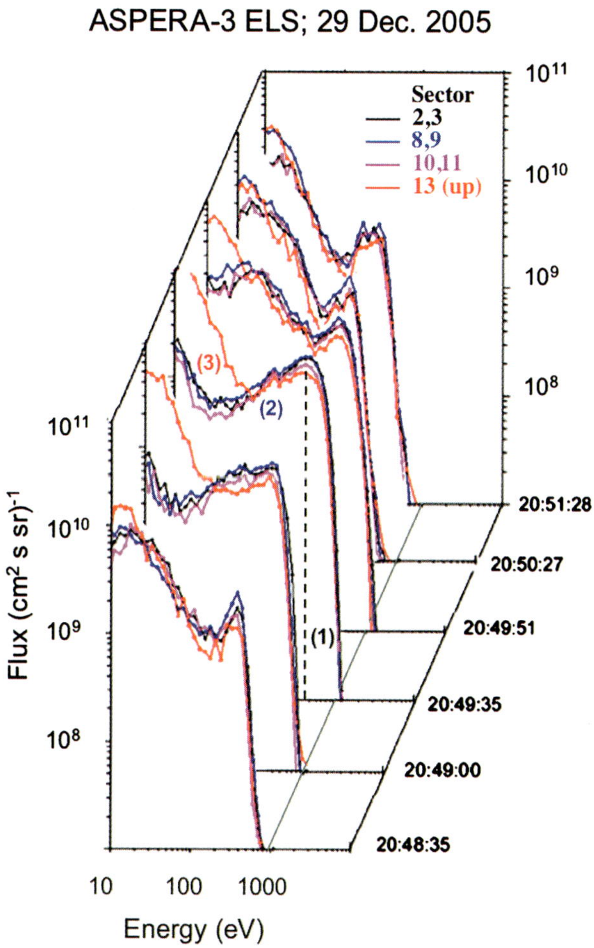

Figure 4. ASPERA-3; ELS electron energy spectra taken from the low-altitude "inverted V" of Figure 3. The spectra shows the typical characteristics of field-aligned acceleration of a primary distribution (1) and the degraded primary (2) + secondary (3) electron distribution below the peak energy. The spectra are for look directions 3+4 (black), 8+9 (blue), 10+11 (magenta), and 13 (red), the latter corresponding to upgoing electrons.

Figure 5 demonstrates a good correlation between ion beam energy and ion beam temperature. The linear relation $y = 8.5x + 623$ (eV) has a correlation coefficient of 0.76. Studies of upward flowing H^+ ions, near the Earth, by the Viking satellite (Moore *et al.*, 1999; Figure 2.28) gives a linear relation $y = 4.9x + 210$ (eV) with a correlation coefficient 0.74. The relations illustrate the coupling between parallel acceleration and ion heating, with low heating rate for low parallel ion acceleration. The ion beams are cool below $\approx$1 keV, the temperature increasing proportional to the parallel energy above $\approx$1 keV, but the beam energy remain at least a factor of 10 higher than the beam temperature. In analogy with the Earth, the increased ion

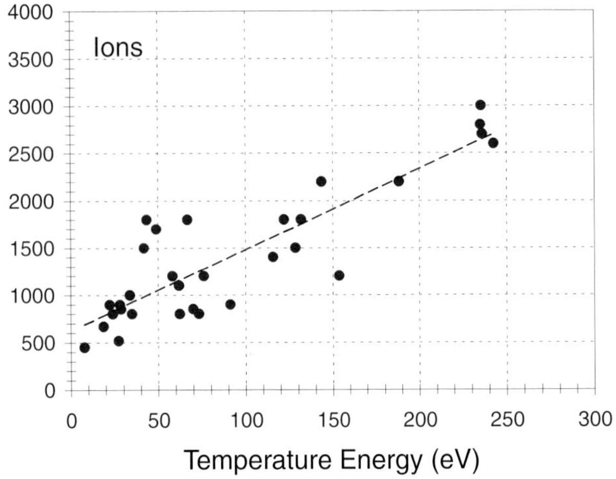

Figure 5. Ion "inverted V" peak energy versus temperature in the nightside of Mars. Oblique line marks a linear fit with function $y = 8.5x + 623$ (eV). Correlation coefficient = 0.76.

beam temperature may be due to a transverse (to the magnetic field) acceleration process (Sharp *et al.*, 1977). Transverse acceleration/heating are governed by a multitude of wave-particle energization processes above the Earth's polar region (Moore *et al.*, 1999). The ion acceleration as implied from Figure 5 may be due to a combined/bimodal (transverse + parallel) acceleration of ions parallel and perpendicular to the local magnetic field as suggested by Klumpar *et al.* (1984). The result of a bimodal acceleration is that the ion beam energy depends on ion mass. This fact was discovered from mid-altitude orbiting auroral satellites already in the early 1980s (Collins *et al.*, 1981). Lundin and Hultqvist (1989) presented a concept for a combined field-aligned electrostatic acceleration and low frequency wave acceleration that could explain the mass-dependent, bimodal, acceleration. They noted that low-frequency waves interacting with plasma in a diverging magnetic field leads to a velocity dependent forcing denoted "magnetic moment pumping", MMP. Guglielmi and Lundin (1999) provided a theoretical background for ponderomotive acceleration, including MMP, induced by Alfvén waves and/or ion cyclotron waves. A combination of ponderomotive forcing (MMP) and electrostatic acceleration will lead to an essentially electrostatic acceleration for electrons and a combined electrostatic and velocity-dependent (wave-induced) parallel acceleration of ions.

The ion mass-energy distribution shown in Figure 6 illustrates the mass-dependent energization of upgoing ions from Feb 20, 2005 (Figure 1). The ion mix in the beam was 47±10% O_2^+, 37±8% O^+, and 16±3% CO_2^+. The average peak energy during the time interval is: O_2^+ ≈0.93 keV, O^+ ≈0.62 keV and CO_2^+ ≈1.16 keV, suggesting a mass dependent energization process. From $v = \sqrt{2E/Nm_p}$, where N represents the number of nucleons, we find $v(O^+)$ ≈85 km/s, $v(O_2^+) \approx$ 71 km/s and

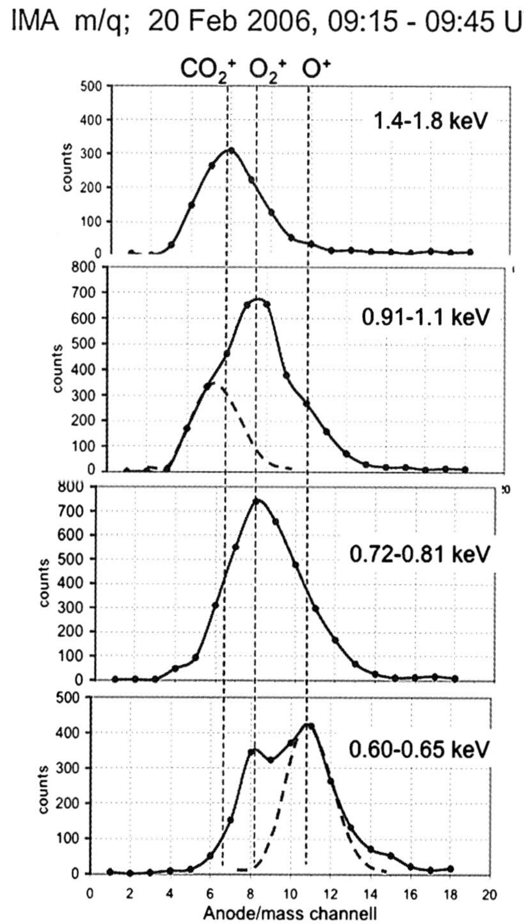

Figure 6. Ion mass spectra for 4 energy intervals for the ion beams in Figure 1, integrated over the time 09:15–09:45. Dotted lines mark the three major heavy ion species. Dashed lines shows the response for a single ion species. The average composition for the time interval is O_2^+ −49%, O^+ – 30%, and CO_2^+ −21%. Average peak energy: O_2^+ ≈0.93 keV, O^+ – ≈0.62 keV and CO_2^+ – 1.16 keV.

$v(CO_2^+) \approx 67$ km/s. These values are quite close, indicating a velocity dependent component of the field-aligned acceleration process.

Assuming now a bimodal acceleration mechanism providing an energy dependent (electrostatic) as well as velocity dependent energization we may use:

$$E = E_i + eV_0 + \frac{Nm_p v_x^2}{2} \tag{1}$$

where E_i is the particle initial energy, V_0 the (electrostatic) acceleration voltage, N is the number of nucleons, m_p is the proton mass, and v_x is the velocity increase from a velocity-dependent acceleration.

Assuming that the initial particle energy is zero, the equation can be solved by making at least two simultaneous measurements for two different ions species, assuming that both species are affected by the same acceleration voltage V_0. From the energy E_1 and E_2 for the two ion species, having the number of nucleons N_1 and N_2, we get:

$$eV_0 = \frac{N_1 E_2 - N_2 E_1}{N_1 - N_2} \tag{2}$$

Taking the peak energies for O_2^+, O^+, and CO_2^+ and introducing their corresponding N (32, 16, 44) we obtain three solutions for eV_0. They all fall close to $eV_0 \approx 310$ eV. This implies that about half of the O^+ energization is due to a velocity increase; the remaining half is due to electrostatic acceleration (eV_0). For higher masses the ratio between velocity increase and electrostatic acceleration is even higher ($O_2^+ \Rightarrow$ about 2/3). By the same token we expect that electrons, having much lower mass than the ions, will be accelerated by primarily electrostatic acceleration.

The above relations represent useful tool to distinguish, by means of ion composition measurements, energy dependent acceleration from velocity dependent acceleration. The fact that a velocity-dependent acceleration appears may be responsible for half of the field-aligned acceleration is an interesting aspect that requires further considerations. It is important to note that the velocity dependent acceleration is directed, maintaining an order of magnitude lower temperature than the beam energy. It is a directed acceleration with minute ion heating, in agreement with MMP ponderomotive forcing by Alfvén waves (Lundin and Hultqvist, 1989; Guglielmi and Lundin, 1999). Intense low-frequency waves measured by MGS (Espley *et al.*, 2004) and MEX (Winningham *et al.*, 2006) may be the wave energy source for the observed velocity dependent acceleration. The analogy to the inflow of waves (e.g. Chaston *et al.*, 2005) and the corresponding wave energization observed above the Earth (e.g. André *et al.*, 1998) is striking.

Regarding the electron acceleration we find from Figure 7 that the electron beam energy and electron thermal energy display some (weak) correlation (correlation coefficient 0.58). In this case we have to infer an exponential fit for maximum correlation coefficient. The lower electron beam energy versus beam temperature ratio (2–5) compared to the ions is probably related with the high altitude origin of the electrons, most likely from the sheath and tail boundary layer. For low peak energies the electron temperatures are well in the range of typical magnetosheath electron temperatures (20–50 eV). The trend of increasing electron temperature with increasing beam energy implies a combined heating and field-aligned acceleration process. Electron heating is in general connected with wave activity along Terrestrial auroral field lines. Wave activity inferred from high-time resolution ELS data in connection with ion and electron acceleration (Winningham *et al.*, 2006; Lundin *et al.*, 2006b) implies similar heating processes near Mars. An ongoing study using orbit conjunctions between MGS and MEX, combining magnetic field, electron and ion data will elaborate further on this topic.

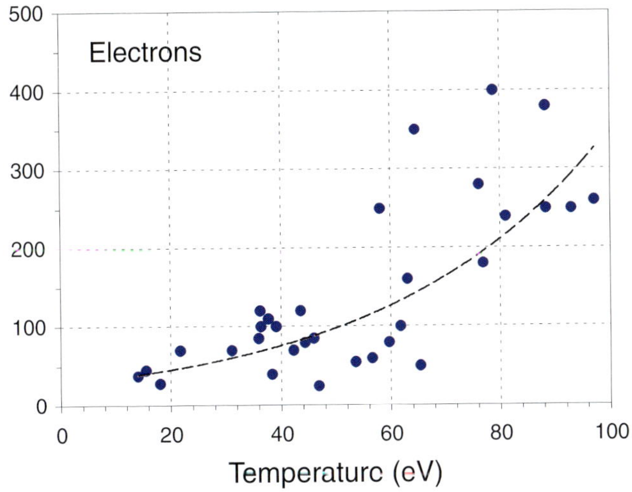

Figure 7. Electron "inverted V" peak energy versus temperature in the night-side of Mars. Dashed line marks an exponential fit with function $y = 27.2\exp(0.026x)$ (eV). Correlation coefficient = 0.58.

Having concluded that acceleration process affect ions and electrons slightly differently, an imminent question is where? We noted already the close connection to the crustal magnetic field in the midnight sector of Mars, so relevant issues are therefore the altitude, local time and geographic distribution.

Figure 8 shows the relative contribution of ions to the total (electron + ion) acceleration. The basis is a downward acceleration of electrons and upward acceleration of ions, the total acceleration in a flux tube given by the sum of the ion and electron acceleration. The total/maximum acceleration achievable corresponds to the ion beam energy in the tail, above the acceleration region, and the electron beam energy below the acceleration region in the upper ionosphere of Mars. This is corroborated in Figure 8. The altitude dependence suggests a gradual change/turnover of the acceleration, the acceleration taking place between the ionosphere and some 10 000 km above Mars. The figure also suggests that a large fraction of the acceleration takes place at altitudes below 2000 km.

The relative contribution of ion acceleration to the total acceleration (ion acceleration/total acceleration), support the hypothesis that the acceleration is at least in part due to altitude dependent electric field acceleration. Dashed line gives a logarithmic fit to the data points, the correlation coefficient being $R^2 = 0.78$. The logarithmic fit gave in fact the best correlation of all fits. The lack of accelerated electrons observed above $\approx$8000 km implies a height limited process. This is expected if the acceleration process is coupled to the crustal magnetic field at Mars. The confined regions of crustal magnetization imply a rapid decrease of the magnetic field intensity with height, but the field lines may yet extend far into the tail of the Martian umbra.

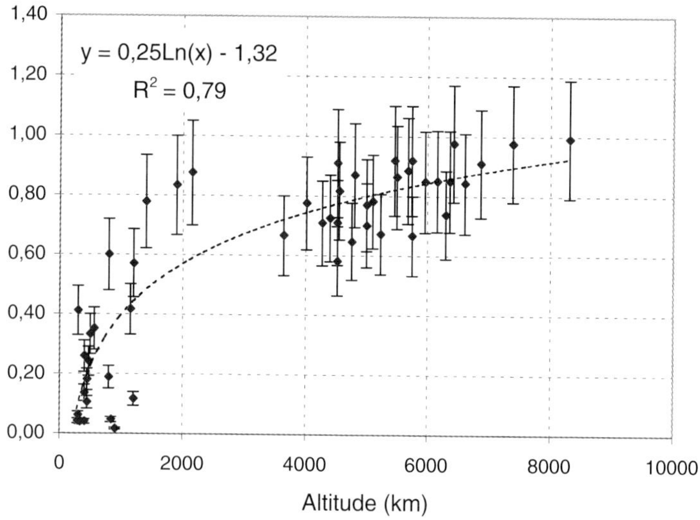

Figure 8. Relative ion contribution to the total acceleration versus altitude. Dashed curve shows a logarithmic fit to the data with correlation coefficient 0.79. Error bars marks estimated errors in determining the relative ion contribution of the total ion+electron acceleration.

While Figure 8 illustrates a statistical/average situation, Figure 9 shows a case of the instantaneous coupling between ion and electron acceleration, i.e. the ion acceleration decreases simultaneously with an increase of the electron acceleration. The total/average acceleration stays rather constant, indicating that the spacecraft encounters a gradient of the acceleration region, where the downward acceleration of electrons briefly extends to higher altitudes. In analogy with field-aligned electrostatic acceleration over the Earth's auroral oval, one may envisage this case as an apparent dip towards the center of the field-aligned electric field region. Notice that the electrons display a similar variability versus energy as that observed in Figures 2 and 3, again suggesting a simultaneously operating wave induced velocity dependent acceleration.

The final issue concerns the mapping of nightside/eclipse plasma acceleration to Martian magnetic anomalies. Figure 10 shows a projection of the 57 "inverted V" cases in solar ecliptic latitude and longitude. There is a clear tendency for "inverted V's" to occur near local midnight, the observations clustering around an average local time, latitude entry (LT = 23.1, lat = −5.3°) and exit (LT = 23.4, lat = −11.3°) of the "inverted V's". The clustering of observations close to local midnight is in part due to the selection criteria (eclipses), but it can neither explain the clear shift of the encounters towards the evening sector, nor can it explain the strong clustering within three hours of local midnight.

In Figure 11 we have instead plotted the "inverted V" footprints versus geographic latitude and longitude. This provides a more dispersed picture compared to Figure 10. Notice that we have transformed the background map (Figure 1) such

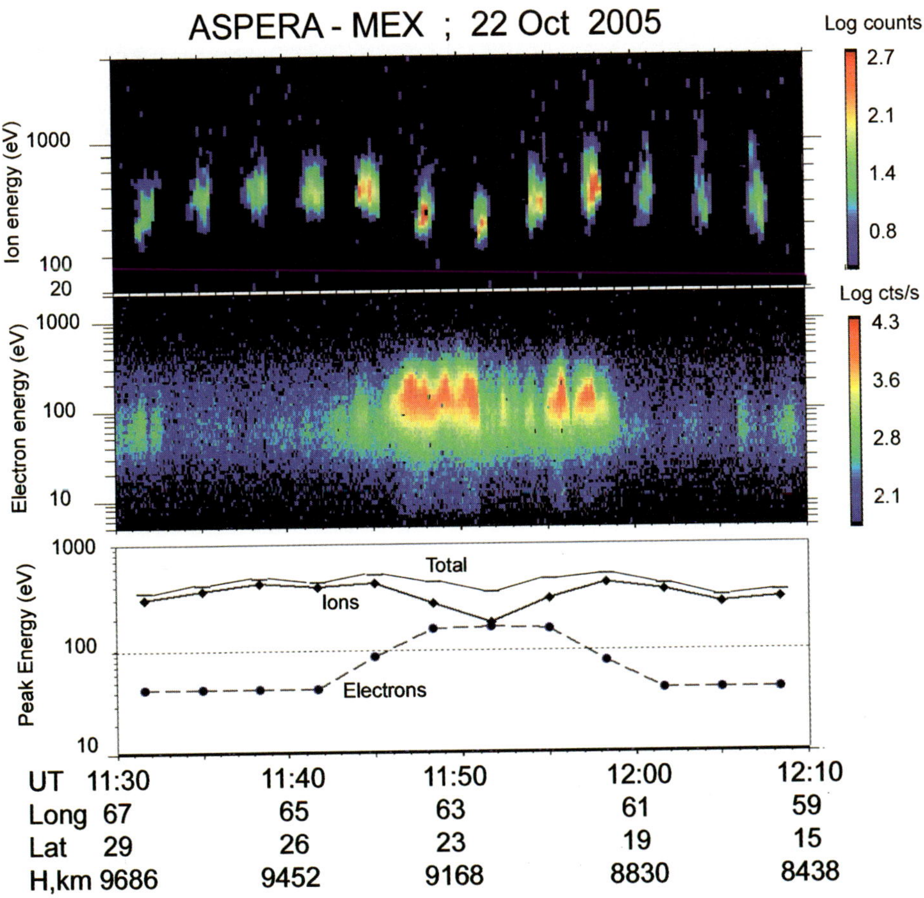

Figure 9. Energy time spectra for upgoing ions (upper panel) and downgoing electrons (middle panel) in the acceleration region. The bottom panel shows peak energies for ions, electrons, and sum peak energy (ions-electron). The event illustrates the strong coupling between electron and ion acceleration.

that the extremes – open field lines and closed field lines, are colored black. White lines mark the latitude and longitude mapping of the "inverted V's" above Mars. Clearly there is a clustering of data points in transition regions between open and closed field lines. Few data points fall within larger areas of open (black) or closed (black) magnetic field lines. We therefore conclude that the "inverted V's" are associated with boundary regions between open and closed field lines. The precipitation of electrons and the corresponding acceleration and escape of ionospheric ions appear to take place in either cusps interfacing magnetic anomalies or near the outer boundary interfacing the large-scale magnetic anomaly region with the non-magnetized region of Mars.

To complete the analogy between Terrestrial discrete aurora and Martian aurora the local downward energy flux of electrons is determined as well as the

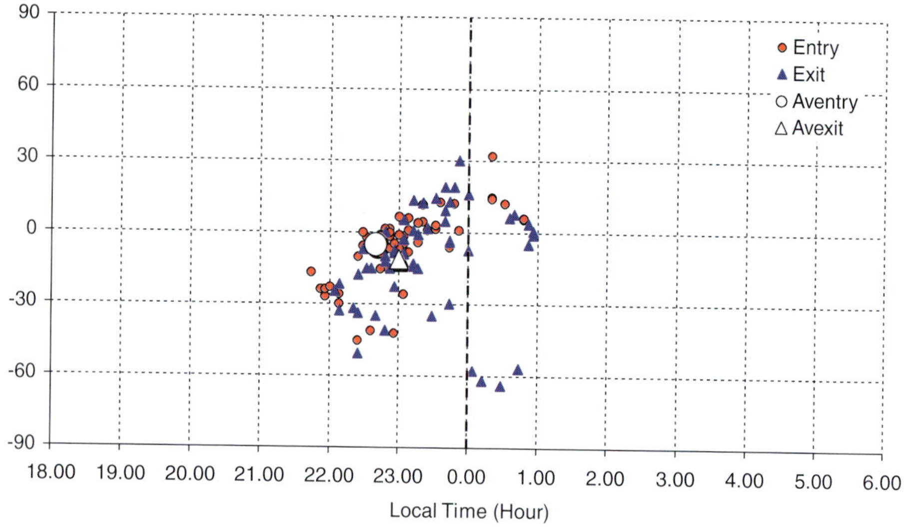

Figure 10. The distribution of nightside ion "inverted V's" plotted versus solar ecliptic (SE) coordinates. Notice the agglomeration of data points near local midnight. Entry and exit of the "inverted V's" are marked by different symbols. Average entry (LT = 23.1, lat = −5.3°) and exit (LT = 23.4, lat = −11.3°) are also marked.

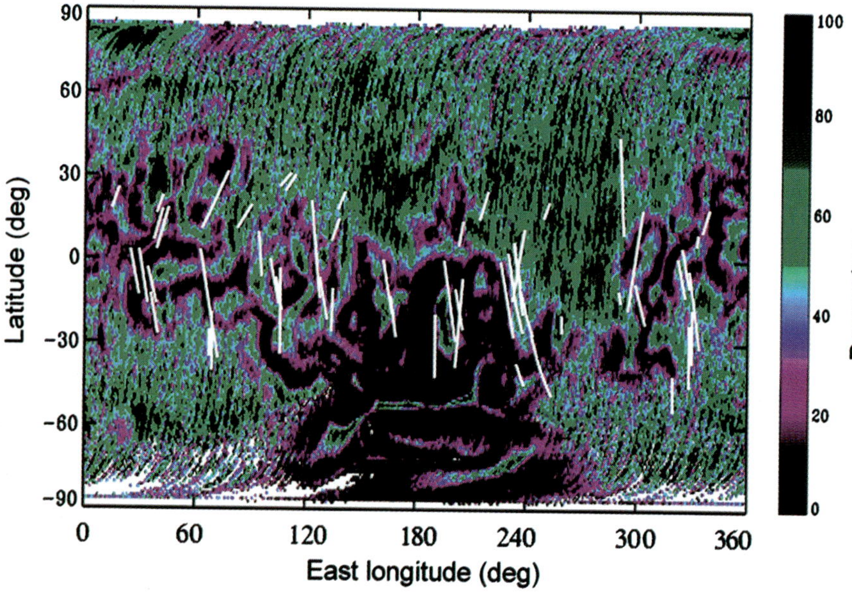

Figure 11. Ground track projections of nightside ion "inverted V's" and their relation to the percentage of open field lines over crustal magnetic regions at Mars.

electron energy flux expected on top of the atmosphere from the acceleration of electrons below the spacecraft. The latter was computed using the acceleration voltage (eV_0) inferred from the almost monoenergetic outflowing ionospheric ion beams (Figures 1 and 5). As noted from Equations (1, 2) and the related text, ponderomotive acceleration by waves may correspond to about half of the ion acceleration. Therefore, a much more detailed analysis is required to infer correctly the electrostatic part of the field-aligned acceleration. Nevertheless, we will here for the sake of simplicity assume that the ion and electron acceleration is not mass (velocity) dependent, enabling an "electrostatic scaling" as described below. The assumption is quite reasonable considering the finding in Figure 9, i.e. the electron and ion peak energy scales almost equally. For electrostatic acceleration along a unit magnetic flux tube the total energy flux gain is given by:

$$P_a = \left(\frac{B_a}{B_i}\right) eV_o\Phi_i \tag{3}$$

where Φ_i represents the local downward electron flux, V_0 the acceleration voltage, and B_a/B_i is the ratio of the magnetic field along the accelerating flux tube, the subscript a marking top of atmosphere (also bottom of acceleration structure) and subscript i marking the locus of observation above eV_0. Assuming for the sake of simplicity that the acceleration takes place in a narrow altitude region ($B_a/B_i \approx 1$) we obtain: $P_a \approx eV_o\Phi_i$, i.e. the downward electron energy flux above the atmosphere is given by:

$$\Theta_a = \Theta_i + eV_0\Phi_i \tag{4}$$

where Θ_i is the locally measured downward electron energy flux. In Figure 12 we have plotted Θ_a versus Θ_i for 28 different cases. We note here that the locally measured energy flux is in the range 0.01–10 mW/m^2 while the accelerated precipitating energy flux into the upper nightside atmosphere of Mars is in the range 1–25 mW/m^2. The latter is indeed a substantial electron precipitation, the highest value corresponding to rather bright aurora at the Earth ($\approx$40 kR of 557.7 nm). The naked eye catches an aurora in the range a few kR in full darkness, and the aurora stands out clearly above 10 kR. The auroral light is produced from excitation of atoms and molecules in the upper atmosphere, the lines reflecting the composition of the gas. The auroral light intensity is related to how emission lines are stimulated by particle precipitation, the emissions discovered in 150–300 nm band (CO and O) by Bertaux *et al.* (2005) being less referred to in connection with terrestrial aurora. More data about the upper atmosphere composition is therefore needed to make a more detailed comparison between Martian and Terrestrial aurora.

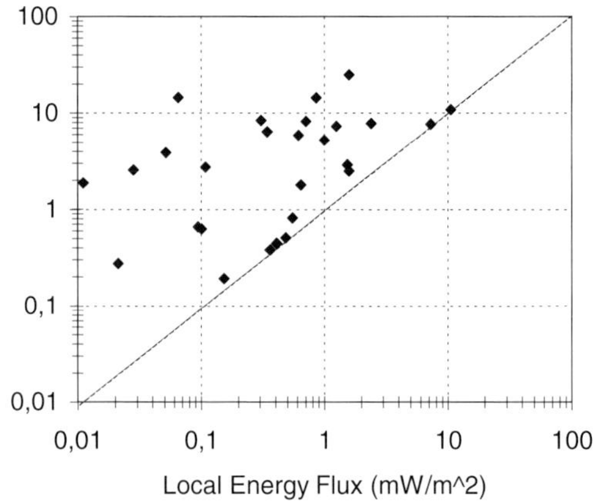

Figure 12. In-situ measured electron energy flux plotted versus model energy flux determined from an electrostatic parallel acceleration below the s/c. Low-altitude acceleration (eV_0) is determined by the peak energy of upgoing ion beams. The diagonal line, indicating no acceleration below the s/c, shows that parallel acceleration amplifies the downward energy flux by of the order a factor of ten. Measurements points falling along the diagonal line were all taken at altitudes below 900 km.

3. Discussions and Conclusions

Plasma acceleration with "inverted V" like ion features have been reported from the Phobos-2 spacecraft orbiting Mars (Lundin *et al.*, 1989) and more recently by Mars Express (Lundin *et al.*, 2006a,b). A connection with aurora, and the corresponding auroral plasma acceleration process, could be made following the discovery of localized strong magnetization regions in the Martian crust (Acuna *et al.*, 1999). This lead to speculations about magnetic topologies and ionosphere processes above Mars similar to those above the Earth (e.g. Mitchell *et al.*, 2001; Krymskii *et al.*, 2002). It also lead to experiments pinpointing towards the interesting magnetic field regions and to the identification of auroral-like emissions (Bertaux *et al.*, 2005) and the occurrence of auroral acceleration process in regions connected to the crustal magnetic field at Mars (Lundin *et al.*, 2006a).

The plasma acceleration leads to an inflow of external (solar wind) plasma as well as an outflow of ionospheric plasma at Mars. Both inflow and outflow lead to losses, the inflow (of energetic electrons) causing ionization in the nightside upper atmosphere, thereby promoting plasma acceleration and escape. While ionospheric plasma losses are effective on the solar EUV illuminated dayside exposed to a fierce bombardment of solar wind plasma (e.g. Perez-de Tejada, 1987; Lundin *et al.*, 1989; Luhmann and Bauer, 1992), it is supposedly less effective in the deep nightside of Mars. Direct nightside losses will require processes in the tail cavity that can channel energy down to the (tenuous) nightside ionosphere.

Results from ASPERA-3 on MEX (Lundin *et al.*, 2004) suggest that energization and outflow of plasma may initiate at fairly low altitudes, on the dayside potentially below the MEX pericenter ($\approx$270 km). Similarly, the nightside energization and outflow may also reach down to low altitudes, perhaps even lower in view of the low nightside ionization rate.

The auroral acceleration process is usually connected with strongly magnetized plasmas interacting with a planetary ionosphere, such as the hot plasma contained in the magnetospheres of Earth, Jupiter and Saturn. It is therefore not an obvious case for Mars. The solar wind interaction with Mars is essentially Venus like or comet like, the crustal magnetic field regions mainly perturbing the solar wind interaction with Mars. However, the process evidently operates in the tail shadow zone of Mars where the field lines from the crustal magnetizations may extend out and contain sufficiently tenuous plasma. The latter enables stronger magnetic control, leading to morphologies and an ionosphere-magnetosphere interaction that resembles that taking place in the Earth's distant magnetotail. Besides commonalities presented here between the Earth's auroral plasma acceleration and the accelerated plasma in the nightside cavity of Mars, the findings by Brain *et al.* (2006a) adds another aspect into this – the field-aligned electric currents (FACs, Iijima and Potemra, 1976). It is well known from the terrestrial magnetospheric physics that FACs play an important role in the magnetosphere-ionosphere interaction (see e.g. Moore *et al.*, 1999). The FACs are connected to the ionosphere (load) in one end and to a dynamo/generator in the other. In this way the interaction may be described by a simple electric current circuit analogy. The interaction is in reality much more complex, involving waves, current limitations and instabilities etc. However, the electric current circuit analogy is yet an important aid into understanding the cause-effect relationship.

Figure 13 is a diagrammatic representation of a dynamo driven current circuit connecting to the nightside ionosphere at Mars. The upper panel is a view perpendicular to the ecliptic plane showing how solar wind plasma forcing sets up a dynamo, producing transverse/dynamo currents ($\underline{j} \times \underline{B} = \nabla \underline{P} > 0$) closing in the ionosphere (load) via FACs. The field-aligned potential drop on the negatively polarized side of the dynamo (lower panel) acts as an accelerator for particles, upward for ions and downward for electrons. Downward accelerated electrons represent the energy source for the aurora. The dynamo itself is governed by the external solar wind forcing, i.e. as long as a plasma dynamic pressure gradient ($\nabla \underline{P}$) can be maintained perpendicular to B. That situation is quite unstable, though, in view of the geometrical constraints of a magnetic flux tube attached to the surface at Mars in a relatively confined region. A "flapping" oscillatory motion may be induced by the external flow, the dynamic pressure gradient varying with the eigenfrequency of the flux tube. The corresponding acceleration of electrons may vary with the frequency of a standing Alfvén wave. Such variable signatures are in fact observed in the accelerated electron distribution. Notice for instance in Figures 2 and 9 that the electron energy peaks are modulated with a periodicity ranging between 1–2 min.

In this report we have analyzed in more detail the auroral acceleration process in the nightside/eclipse of Mars above magnetic anomalies (Lundin *et al.*, 2006a). The observational data is extended to lower altitudes and we have elaborated on the

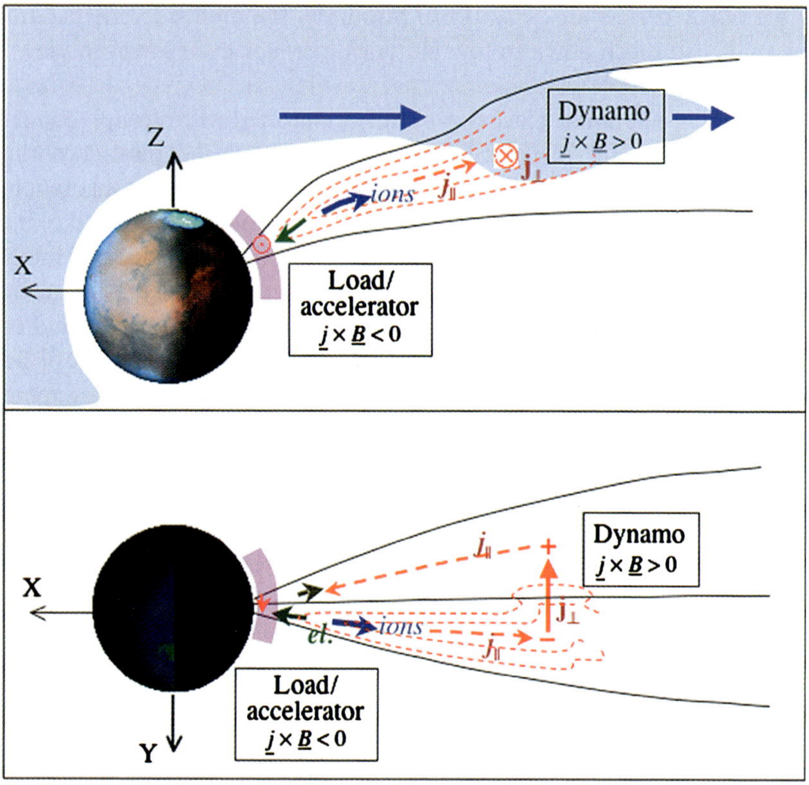

Figure 13. Diagram of the field-aligned auroral plasma acceleration region above a magnetic cusp/cleft at Mars. The dynamo is governed by the cross-field flow of solar wind plasma. The current circuit, driven by the dynamo, consists of a pair of field-aligned currents and an ionospheric closure current.

potential energy source, the dynamo driving the auroral acceleration process above Mars. The energy and mass sources are aspects of the dynamo-accelerator that requires further studies. The outflowing ions clearly originate from the ionosphere. On the other hand the high altitude source, the solar wind plasma access of magnetic flux tubes connected to Mars, remains an issue for future studies. The intense fluxes of upgoing ionospheric ions from a tenuous nightside ionosphere suggest the formation of auroral plasma density cavities, like in the Earth's nightside ionosphere (Calvert, 1981). A combination of parallel electric fields and waves deepens the cavity and promotes a bimodal acceleration process (e.g. Klumpar *et al.*, 1984). A combined velocity dependent and energy dependent field aligned acceleration will manifests itself as a mass-dependent acceleration process (Lundin and Hultqvist, 1989) leading to different peak energies for different masses as first reported on by Collin *et al.* (1984). Velocity dependent acceleration in a non-magnetized planetary environment is also the characteristics of a pick-up process as described Luhman and Schwingenshuh (1990). However, we report here on nearly antisunward ion outflow extending from magnetic anomalies projected to the planetary ionosphere – at low altitudes near local midnight. Considering the draped tail magnetic field of Mars we may therefore exclude contributions from a transverse (to B) pickup process in the solar wind electric- and magnetic field. One may of course interpret the velocity-dependent part of the acceleration as an "ion pickup" process, but then in a strong diverging crustal magnetic field, similar to that in the Earth's dipole field above the auroral oval.

The overall situation in the tail umbra of Mars and the Earth therefore show a number of similarities, in particular with respect to the acceleration of plasma, eventually leading to aurora in their respective topside atmosphere. The "inverted V" electron and ion energy-time characteristics, the mass dependence of the ion acceleration, the electron energy distribution, the altitude distribution, and the close connection to the planetary magnetic field are all in support for such a conclusion. A more enigmatic feature is the concentration of events to midnight, with a significant shift towards premidnight at Mars. An external solar wind electric field forcing may in principle cause such an organization, but only if it is unidirectional and likewise for the Martian magnetic field. Anything else would lead to a spread around midnight. Another possible explanation is that the Mars rotational speed can cause a "Parker spiral" effect of the escaping ionospheric plasma. However, estimates lead to a time shift of no more than 10 minutes, while the average shift is about 1 hour. Moreover, the mapping always falls close to the "expected" crustal magnetic field regions. Thus, there is no apparent shift of the magnetic field lines towards dawn or dusk. This implies that it is the source region, the dynamo that is being shifted towards dusk. For what reason, that is the question.

References

Acuña, M. J., Connerey, J., Ness, N., Lin, R., Mitchell, D., Cralsson, C., *et al.*: 1999, *Science* **284**, 790.

Albert, R. D.: 1967, *Phys. Rev. Lett.* **18**, 368.

André, M., Norqvist, P., Andersson, L., Eliasson, L., Eriksson, A. I., Blomberg, L., *et al.*: 1998, *J. Geophys. Res.* **103**, 4199.

Barabash, S., Lundin, R., Andersson, H., *et al.*: 2004, The Analyzer of Space Plasmas and Energetic Atoms (ASPERA-3) for the Mars Express Mission, In Mars-Express – The Scientific Payload, ESA-SP-1240.

Bertaux, J.-L., Leblanc, F., Witasse, O., Quemerais, E., Lilensten, J., Stern, S.A., *et al.*: 2005, *Nature* **435**, 9.

Brain D., Luhmann, J., Mitchell, D., and Lin, R.: 2005, Expected influence of crustal magnetic fields on the ASPERA3 ELS observations: Lessons learnt from MGS. Paper presented at 1st mars Express Science conference, 21–25 Feb, 2005.

Brain, D. A., Halekas, J. S., Peticolas, L. M., Lin, R. P., Luhmann, J. G., Mitchell, D. L., *et al.*: 2006, *Geophys. Res. Lett.*, 10.1029/2005GL024782.

Calvert, W.: 1981, *Geophys. Res. Lett.* **8**, 919.

Chaston, C. C., Peticolas, L. M., Carlson, C. W., McFadden, J. P., Mozer, F., Wilber, M., *et al.*: 2005, *J. Geophys. Res.* **110**, A02211, doi:10.1029/2004JA010483.
Chiu, Y. T., and Schulz, M.: 1978, *J. Geophys. Res.* **83**, 629.
Collin, H. L., Sharp, R. D., and Shelley, E. G.: *J. Geophys. Res.* **89**, 2185.
Dubinin, E., Lundin , R., Koskinen, H., and Pissarenko, N.: 1993, *J. Geophys. Res.* **98**, 3991.
Evans, D. S.: 1968, *J. Geophys. Res.* **73**, 2315.
Evans, D. S.: 1974, *J. Geophys. Res.* **79**, 2853.
Espley, J. R., Cloutier, P. A., Crider, D. H., Brain, D. A., and Acuña, M. H.: 2004, *J. Geophys. Res.*, 2004AGUFMSA13A1120E.
Frank, L. A. and Ackerson, K. L.: 1971, *J. Geophys. Res.* **76**, 3612.
Guglielmi, A. and Lundin, R.: 2001, *J. Geophys. Res.* **106**, 13219.
Gurnett, D. A. and Frank, L. A.: *J. Geophys. Res.* **78**, 145.
Iijima, T. and Potemra, T. A.: 1976, *J. Geophys. Res.* **81**, 2165.
Kallio, E., Barabash, S., Luhmann, J. G., Koskinen, H., Lundin, R., and Norberg, O.: 1994, *Geophys. Res. Lett.* **99**, 23547.
Klumpar, D. M., Peterson, W. K., and Shelley, E. G.: 1984, *J. Geophys. Res.* **89**, 10779.
Krymskii, A. M., Breus, T. K., Ness, N. F., Acuña, M. H., Connerney, J. E. P., Crider, D. H., *et al.*: 2002, *J. Geophys. Res.* **107**(A9), 1245, doi:10.1029/2001JA000239.
Luhman, J. G. and Schwingenshuh, K.: 1990, *J. Geophys. Res.* **95**, 939.
Luhmann, J. G. and Bauer, S. J.: 1992, AGU monograph **66**, 417.
Lundin, R., Zakharov, A., Pellinen, R., Hultqvist, B., Borg, H., Dubinin, E. M., *et al.*: 1989, *Nature* **341**, 609.
Lundin, R. and Hultqvist, B.: 1989, *J. Geophys. Res.* **94**, 6665.
Lundin, R., Barabash, S., Andersson, H., Holmström, M., *et al.*: 2004, *Science* **305**, 1933.
Lundin, R., Winningham, D., Barabash, S., *et al.*: 2006a, *Science* **311**, 980.
Lundin, R., Winningham, D., Barabash, S., Frahm, R., and the ASPERA-3 team: 2006b, *ICARUS*, April 2006.
Lyons, L. R., Koskinen, H. E. J., Blake, J. B., Egeland, A., Hirahara, M., Øieroset, M., *et al.*: 1999, *Space Sci. Rev.* **88**, 85.
MacIlwain, C. E.: 1960, *J. Geophys. Res.* **65**, 2727.
Mitchell, D. L., Lin, R. P., Mazelle, C., *et al.*: 2001, *J. Geophys. Res.* **106**, 23419.
Moore, T. E., Lundin, R., Alcayde, D., Andre, M., Ganguli, S. B., Temerin, M., *et al.*: 1999, *Space Sci. Rev.* **88**.
Pérez-de Tejada, H.: 1987, *J. Geophys. Res.* **92**, 4713.
Russell, C. T., Luhmann, J. G., Schwingenshuh, K., Riedler, W., and Yeroshenko, Ye: 1990, *Geophys. Res. Lett.* **17**, 897.
Sharp, R. D., Johnson, R. G., and Shelley, E. G.: 1977, *J. Geophys. Res.* **82**, 3324.
Shelley, E. G., Johnson, R. G., and Sharp, R. D.: *Geophys. Res. Lett.* **3**, 654.
Winningham, J. D., Frahm, R. A., Sharber, J. R., Coates, A. J., Linder, D. R., Soobiah, Y., *et al.*, and the Aspera-3 Team: 2006, *ICARUS*, April issue.

INVESTIGATION OF THE INFLUENCE OF MAGNETIC ANOMALIES ON ION DISTRIBUTIONS AT MARS

H. NILSSON[1,*], E. CARLSSON[1,2], H. GUNELL[1], Y. FUTAANA[1], S. BARABASH[1], R. LUNDIN[1], A. FEDOROV[3], Y. SOOBIAH[4], A. COATES[4], M. FRÄNZ[5] and E. ROUSSOS[5]

[1]*Swedish Institute of Space Physics, P.O. Box 812, SE-981 28 Kiruna, Sweden*
[2]*Luleå University of Technology, Luleå, Sweden*
[3]*Centre d'Etude Spatiale des Rayonnements, Toulouse, France*
[4]*Mullard Space Science Lab, Imperial College, London, UK*
[5]*MPI für Sonnensystemforschung, Katlenberg-Lindau, Germany*
*(*Author for correspondence, E-mail: hans.nilsson@irf.se)*

(Received 7 May 2006; Accepted in final form 18 August 2006)

Abstract. Using data from the Mars Express Ion Mass Analyzer (IMA) we investigate the distribution of ion beams of planetary origin and search for an influence from Mars crustal magnetic anomalies. We have concentrated on ion beams observed inside the induced magnetosphere boundary (magnetic pile-up boundary). Some north-south asymmetry is seen in the data, but no longitudinal structure resembling that of the crustal anomalies. Comparing the occurrence rate of ion beams with magnetic field strength at 400 km altitude below the spacecraft (using statistical Mars Global Surveyor results) shows a decrease of the occurrence rate for modest (<40 nT) magnetic fields. Higher magnetic field regions (above 40 nT at 400 km) are sampled so seldom that the statistics are poor but the data is consistent with some ion outflow events being closely associated with the stronger anomalies. This ion flow does not significantly affect the overall distribution of ion beams around Mars.

Keywords: plasma, Mars, ions

1. Introduction

The solar wind interaction with the near-Mars space environment has been studied mainly by the Phobos-2 spacecraft [e.g. (Lundin *et al.*, 1989, 1991; Breus *et al.*, 1991; Barabash *et al.*, 1991; Trotignon *et al.*, 1996)], the Mars Global Surveyor (MGS) [e.g. (Mitchell *et al.*, 2000, 2001; Vignes *et al.*, 2000; Crider *et al.*, 2002; Krymskii *et al.*, 2003; Bertucci *et al.*, 2005; Brain *et al.*, 2005)], combinations of these two data sets (Trotignon *et al.*, 2006) and the, at the time of writing, most recently arrived spacecraft Mars Express [e.g. (Lundin *et al.*, 2004; Fränz *et al.*, 2005; Soobiah *et al.*, 2005)]. Much of the picture emerging from the first two spacecraft has been summarized in Nagy *et al.* (2004). The solar wind interaction with the near-Mars space results in several distinctive regions, mainly the bow-shock, the magnetosheath and the magnetic pile-up region. These regions are dominated by the solar wind magnetic field which is draped around the obstacle. However

Space Science Reviews (2006) 126: 355–372
DOI: 10.1007/s11214-006-9030-0

MGS data shows clearly that the crustal magnetic fields [e.g. (Acuña *et al.*, 1998; Connerney *et al.*, 1999)] of Mars significantly affect the distribution of electrons in near-Mars space, in particular at the magnetic pile-up boundary (Vignes *et al.*, 2000; Crider *et al.*, 2002; Brain *et al.*, 2005) and the ionopause [e.g. (Mitchell *et al.*, 2001; Fränz *et al.*, 2005)].

The magnetic field of the magnetic pile-up region (MPR) is the interplanetary magnetic field draped around the planetary obstacle. The outer boundary towards the magnetosheath is termed the magnetic pile-up boundary (MPB) and is characterized from MGS measurements by an increase in magnetic field strength (Crider *et al.*, 2002) and a decrease in suprathermal electron fluxes and a decrease in magnetic field variability and wave activity (Brain *et al.*, 2005). The decrease in suprathermal electrons is consistent with energy loss of the magnetosheath electrons due to impact ionization of exospheric neutrals (Crider *et al.*, 2000). The MPB is thus not a pressure balance boundary, nor an impenetrable obstacle, at least not for magnetosheath electrons and magnetic fields. The ions of the magnetic pile-up region are expected to be mainly of planetary origin but the more extensive MGS data set lacks ion data.

The lower boundary of the magnetic pile-up region is characterized by a further reduction of the electron fluxes of magnetosheath origin, and below the MPR planetary origin photo-electron fluxes dominate. Mitchell *et al.* (2000, 2001) identify this as the Martian ionopause. The many strong crustal magnetic anomalies in the southern hemisphere stands off solar wind electrons up to higher altitudes in both the boundary regions.

The crustal magnetic fields also affect the ionosphere at altitudes well below the ionopause and even the neutral atmosphere. Krymskii *et al.* (2003) reported increased electron temperatures inside the "mini-magnetospheres" created by strong crustal magnetic fields, through confinement of photo-electrons, as well as a cooler neutral atmosphere which is shielded from additional heating by the solar wind interaction. Ness *et al.* (2000) reported an influence of magnetic fields on the ionospheric scale height, where horizontal fields inhibit vertical diffusion as compared to vertical or magnetic field-free regions. Mitchell *et al.* (2001) showed similar results at higher altitudes where strong crustal fields allowed the ionosphere to extend to higher altitudes, resulting in regions with enhanced photo-electron fluxes at an altitude of 400 km in the dayside. On the other hand photo-electron drift from day- to nightside and magnetosheath origin electron access were inhibited in the closed crustal fields on the nightside resulting in "void" regions with very low electron fluxes. Series of plasma void regions were often separated by electron flux-spikes. This tends to occur where the radial magnetic field is near a local maximum. The presence of magnetosheath-like electrons on such field-lines suggests that they are or were once connected to the magnetosheath, and the situation is thus similar to the cusps in the Earth's magnetosphere but on a much smaller scale. Brain *et al.* (2006) took the similarities with the Earth further, showing that peaked electron spectra, resembling the accelerated electron spectra associated with aurora on Earth, were frequently observed near strong radial crustal fields in the Martian nightside.

Thus the MGS results have firmly established the importance of the crustal fields for a number of electron plasma processes and structures at Mars. What about the influence of crustal fields on the ions? MGS lacks an ion spectrometer and we must now turn to Mars Express measurements.

We first tie the measurements from the two spacecraft together by looking at the reported electron observations from Mars Express. The work of Fränz *et al.* (2005) confirmed the crustal field influence on the statistical distribution of magnetosheath electron stand-off distance and the work of Soobiah *et al.* (2005) compared Mars Express electron spectrometer results with those obtained from MGS by Mitchell *et al.* (2001) and the Mars crustal magnetic field model of Cain *et al.* (2003) to investigate the influence of magnetic anomalies on the electron fluxes. They found that the presence of plasma voids in the nightside and flux enhancements in the dayside were well ordered by the Cain magnetic field model.

As Mars Express does not carry a magnetometer it is customary to call the planetary boundary towards the magnetosheath the Induced Magnetosphere Boundary (IMB) rather than the MPB, but it has been shown that on a large scale these are the same (Lundin *et al.*, 2004; Vignes *et al.*, 2000). The only works which so far have discussed ion observations in relation to magnetic anomalies are those by Lundin *et al.* (2004, 2005, 2006). These works report ion outflow as observed by the ASPERA-3 Ion Mass Analyzer (IMA). It is suggested that ion energization frequently involves acceleration by field-aligned electric fields and low frequency waves (as determined from electron flux variations, Winningham *et al.* (2005)). These can involve induced or draped magnetic fields just as well as crustal fields, but in Lundin *et al.* (2006) only deep nightside tail events were studied in an attempt to avoid the influence of non-crustal fields. Evidence of large scale field-aligned electric fields was found in the form of accelerated beam-like outflowing ionospheric ions observed simultaneously with precipitating electrons with peaked energy spectra, similar to what is observed in the auroral region on Earth. Mapping these events to crustal sources indicated that they were associated with magnetic cusps. The altitude of the observations was fairly high (several thousand km) and the mapping thus somewhat uncertain, but the association with magnetic anomalies is strengthened by the fact that the observations reported by Brain *et al.* (2006) clearly show that the peaked electron spectra observed by MGS at 400 km altitude are associated with strong radial crustal magnetic fields.

There thus seem to be cases when the magnetic anomalies may also be of importance for the ions. For the large scale distribution of ions this should mainly be for low energy ions or for field-aligned acceleration events because the crustal fields are relatively weak at altitudes where more energetic ions can be expected. The gyro-radii of ions quickly become large compared to the scale size of the anomalies when they are energized to energies observable by IMA (lower limit between 10 and 100 eV, see discussion in Section 2). However just as at Earth the outflow of planetary ions is essentially a two-step process where the flow may either be regulated at the source (ionosphere, availability of ions) or by the energization process which typically occurs at higher altitudes. The purpose of this paper is to examine the potential role of crustal magnetic fields on the distribution of ions in near-Mars space. This has been done in 4 steps:

(1) We have examined the clearest of the electron events reported by Soobiah *et al.* (2005) which were associated with magnetic anomalies and examined the corresponding ion data.
(2) We have also examined the orbits containing the ion events used by Lundin *et al.* (2005) and compared the data with the Cain magnetic field model (Cain *et al.*, 2003) on a case basis.
(3) We have studied the distribution of all energetic planetary ion beam events reported by Carlsson *et al.* (2006), including an extended study of similar events also for the year 2005.
(4) We have gone through all the data when IMA was used in a non-entrance deflection scanning mode to improve time resolution and the ability to observe low energy ions (see Section 2).

2. Instrument Description

The Ion Mass Analyzer (IMA) is a mass resolving ion spectrometer, part of the ASPERA-3 instrument onboard Mars Express (Barabash and The ASPERA-4 Team, 2006). IMA:s twin ICA on the Rosetta spacecraft is described in detail in Nilsson *et al.* (2006). IMA consists of an electrostatic acceptance angle filter, an electrostatic energy filter, and a magnetic velocity analyzer. Particles are detected using large diameter (100 mm) microchannel plates and a two-dimensional anode system. The energy range of the instrument is nominally from 10 eV to 36 keV and an angular field-of-view of $360^\circ \times 90^\circ$ is achieved through electrostatic deflection of incoming particles. This field of view is partially obstructed by the spacecraft body and the solar panels. IMA is mounted on the spacecraft $-Z$ side, facing towards spacecraft $-Y$ (i.e. the instrument symmetry axis is along spacecraft Y), see Figure 1. The basic field-of-view of the instrument is the spacecraft X–Z plane, particles are brought in from $\pm 45^\circ$ out of this plane through the electrostatic deflection system. The deflection system does not have high enough voltage to reach all angles for the highest energies and not enough voltage resolution to reach all deflection angles for low energies. Above 15 keV the field-of-view is restricted towards the central viewing plane. For energies below 100 eV the angular resolution is degraded. Turning off the entrance deflection scan and using the instrument in a 2D mode removes the resolution problems at low energies and improves the instrument's ability to measure low energy ions as well as the time resolution. The time for one full energy scan is 12 s. and for one full measurement of 16 different

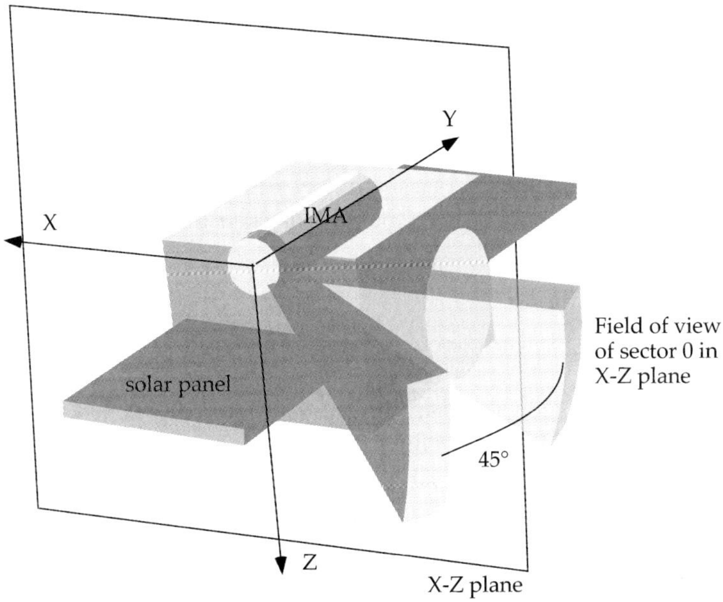

Figure 1. Schematic figure of the IMA Ion Mass Analyzer on the Mars Express spacecraft. Indicated are the spacecraft coordinate system and the field-of-view of one sector of the instrument at no deflection and at 45° away from the spacecraft.

deflection angles the time resolution becomes 192 s. The no entrance deflection mode may therefore be necessary to catch any finer structure of the ion distribution, in particular at low altitude where high time resolution is more important. IMA may run in different spatial and mass resolution modes to save telemetry. In practice almost all data is in the full resolution mode; no binning of data from different acceptance angles or binning of mass anodes is made (instrument mode 24).

Mass resolution is obtained through the magnetic velocity analyzer, where particles with the same energy but different mass will hit the micro-channel plate in different locations due to the analyzer magnetic field. The range of masses observable and the mass resolution can be influenced by adding energy to the incoming particles through a post-acceleration voltage. This voltage is applied between the electrostatic energy filter and the magnetic velocity analyzer and is controlled by a 3-bit reference value (0–7), corresponding to post-acceleration voltages between 0 and 4.3 kV.

3. Observations

3.1. Electron Events Associated with Magnetic Anomalies

The clearest and most pronounced electron signatures associated with magnetic anomalies reported by Soobiah *et al.* (2005) were investigated to see if any ion signatures were found. This corresponded to 20 events selected from a total of 57 events identified in data from 144 orbits. The result was negative. Usually no ions at all were detected and when ions were detected they were not exactly coincident with either electron signatures or magnetic anomalies as determined from the Cain model (Cain *et al.*, 2003). Care was taken to determine that the IMA instrument was looking downward during at least some of the events. However the poor angular coverage at low energies means that there may still be low energy ions associated with the magnetic anomalies (there must be at least thermal ions due to charge neutrality).

3.2. Magnetic Fields Around Clear Ion Observation Events

Having failed to find good ion data in step 1 described above we proceeded to check the magnetic field as determined from the Cain model around some clear ion signatures, those reported by Lundin *et al.* (2005). A total of 30 events were plotted and investigated in detail. Typically ion beams were observed at the lowest altitude and some cases occurred at magnetic anomalies. However ion beams clearly existed even when no magnetic anomaly was nearby or the extrapolated Cain model field was very weak at the altitude of observation. No general similarity in the fine structure of ions and the magnetic field model was found though the temporal resolution may have been too poor to allow such a comparison. We report this part of the study for completeness, but will show data only from the cases when the IMA instrument was run in the "no entrance deflection" mode in Section 3.4). Then we also make a comparison with the magnetic field at a fixed altitude to avoid the risks inherent in extrapolating the Cain model to higher altitudes than the data from which the model was obtained.

3.3. The Distribution of Planetary Origin Ion Beam Events

Here we used the data base of the ion observations used by Carlsson *et al.* (2006). It consists of all heavy ion beams (O^+, CO_2^+, CO^+/O_2^+) as identified from manual inspection of data from inside the nominal Induced Magnetosphere Boundary (IMB). A sample ion beam (in high time resolution "no entrance deflection mode") is shown in Figure 2. The same event is marked with number 1 in Figure 7. The observation altitude was in the range 2000–3000 km, and the solar zenith angle was 136°–140°.

This database has been updated with all ion beam events observed up to 22 October 2005, likewise determined from visual inspection of all IMA data obtained inside the nominal IMB. In Carlsson *et al.* (2006) only post-acceleration level 1 (out of three, 0 (none), 1 (reference value 1–4) and 2 (reference value 5–7)) was used, but for the subsequent data all identified events regardless of post-acceleration setting have been used (a total of 818 events). Before proceeding to investigate a possible influence on the distribution of planetary origin ion beams from crustal

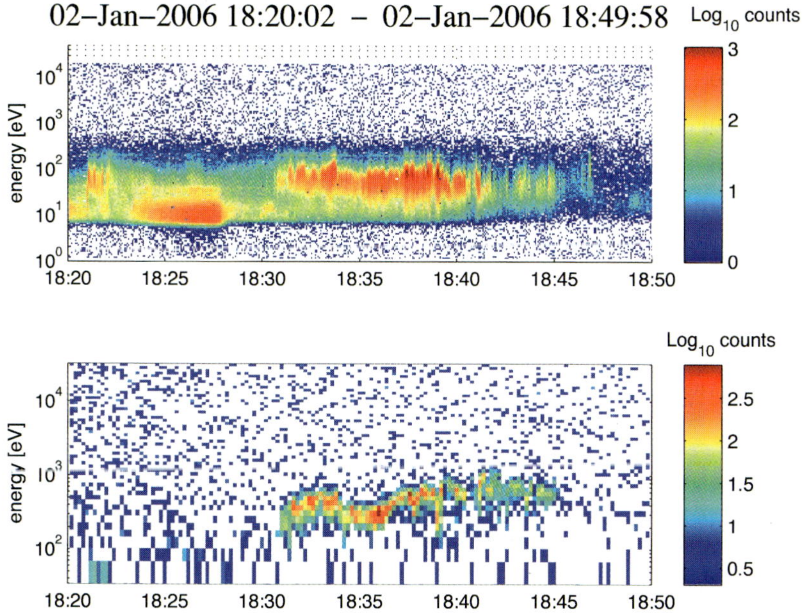

Figure 2. Sample energy spectra of a heavy ion beam from 20060102 when no entrance deflection scanning was used. The upper panel shows the corresponding electron spectra summed over all sectors (the electrons are typically rather isotropic). The lower panel shows the ion counts, summed over all sectors. One sector dominates and only a few sectors (of $22.5^\circ \times 5^\circ$) detect any ions at all. The Y-axis shows particle energy in eV for both panels, and the x-axis time (UT).

magnetic fields, we show in Figure 3 the ion beam occurrence rate (panel a) and the spacecraft coverage (number of passes in bin, panel b) as a function of solar zenith angle (x-axis) and altitude (y-axis). As can be seen, there is a clear dependence in the sense that dayside beams are observed at low altitude and nightside beams at high altitude. A lack of coverage at the lowest nightside altitudes is also evident, caused by restricted operation in spacecraft eclipse. Essentially the distribution follows what we expect from the induced magnetosphere boundaries and we can say that we do not have a strong dependence on solar zenith angle.

In order to search for an influence on the distribution of these ion beam events from magnetic anomalies, we have plotted their occurrence rate as a function of latitude and longitude, using 20×20 bins, i.e. a resolution of $18^\circ \times 9^\circ$ in longitude-latitude space. The data was also binned in altitude, and the normalized result for four different altitude bins (up to 1000 km, 1000–2000, 2000–3000 and 4000–10000 km) is shown in Figure 4. The distribution was calculated such that each "event" (continuous presence of an ion beam) was counted only once inside each latitude, longitude and altitude bin. The same type of distribution was then obtained for all cases when IMA was on in full resolution mode (mode number 24), post-acceleration setting was 1 for the 2004 data (all according to the housekeeping data) and Mars Express was inside the nominal IMB. This result was used to normalize

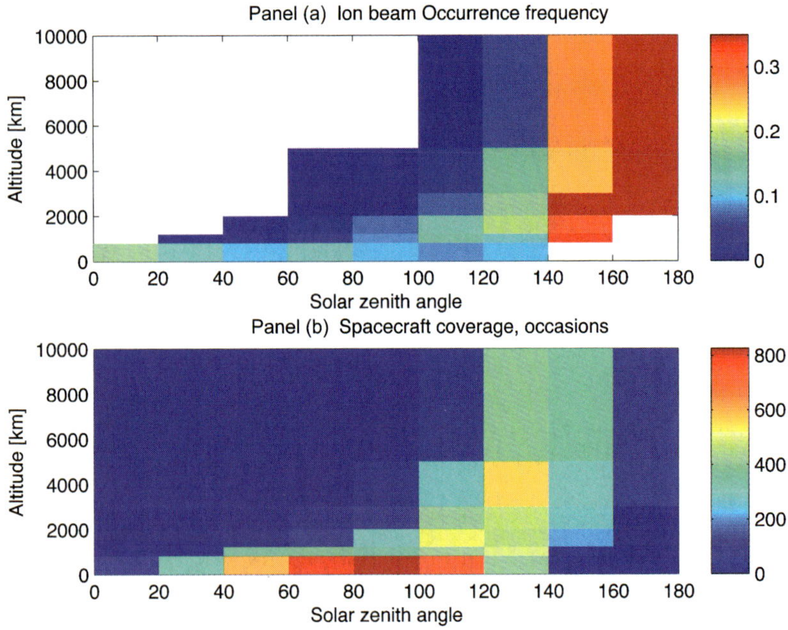

Figure 3. Panel (a) Distribution of occurrence rate of ions beams as a function of solar zenith angle (x-axis, degrees) and altitude (y-axis, [km]). Panel (b) Number of satellite passes through each statistical bin when IMA was operating in an appropriate mode.

the beam occurrence. What is shown in Figure 4 is the normalized occurrence frequency. The number of events is rather small and the plot in Figure 4 therefore rather noisy. It can nevertheless clearly be noted that the events occur over all locations on Mars. There seems to be some preference for northern latitudes for low altitudes (below 1000 km, panel a) and some preference for southern latitudes just below the equator at high altitude (above 3000 km, panel d). There is also a relatively low occurrence frequency for the southernmost latitude bins. Possibly this could indicate a large-scale influence from magnetic anomalies as these are stronger in the southern hemisphere. However latitude distributions are very sensitive to the orbit characteristics which causes an ambiguity between altitude and latitude dependence. The perigee is drifting so in due time all latitudes will be sampled at different altitudes but this is not true for a data set from a limited time interval such as the three-month data with entrance deflection turned off discussed in Section 3.4.

There is no longitudinal distribution (which is much less sensitive to orbit characteristics) resembling that of magnetic anomalies which are shown in Figure 7. To further investigate the significance of the observed north-south asymmetry we plot in Figure 5 the number of orbits when IMA was on in the right mode when Mars Express passed above the indicated latitude and longitude bin. Clearly there are a significant amount of samples in the south for low altitudes as well as for the

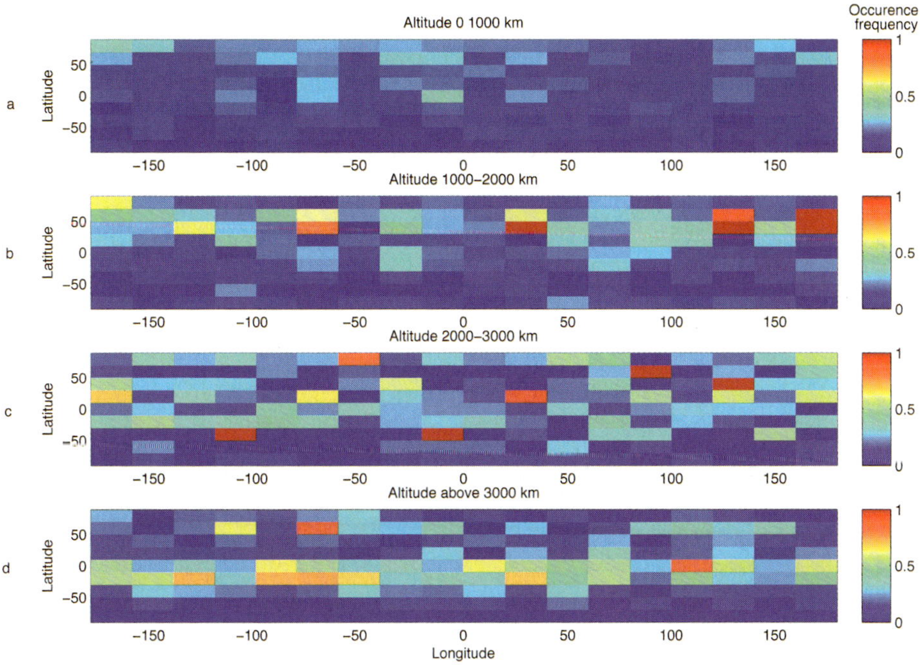

Figure 4. Distribution of occurrence rate of ions beams. Panel (a) shows the altitude interval up to 1000 km, (b) shows the interval 1000 to 2000 km, (c) 2000 to 3000 km and (d) 3000 up to 10000 km altitude.

southernmost latitudes at high altitude. Finally one may note that there is an increasing occurrence rate at higher altitude. As IMA cannot detect thermal ions this is consistent with an extended altitude range where ion energization is significant. This can be both thermal ionospheric origin ions and newly created pick-up ions.

One may also study the presence of ion beam events as a function of the magnetic field as determined from MGS statistics. We have used the data for 400 km altitude as provided by Connerney *et al.* (2001). Interpolating the MGS data (we used linear interpolation) at each measurement point yields rather few points above significant anomalies. We show in Figure 6 the occurrence frequency of ion beams for some altitude intervals as a function of radial magnetic field at 400 km altitude. Radial magnetic field was used because ion outflow can be expected along strong radial fields. Strong transverse fields can be expected to inhibit vertical plasma transport so we have made the same study with the transverse and total magnetic field as well, but with no significant differences in the result. Almost all data points are located above magnetic field values below 40 nT, as shown by the grey bars in the plot (number of data points on right y-axis). We can now discern an influence from magnetic anomalies on ion beams at the lowest altitudes. Up to 2000 km the likelihood of observing an ion beam decreases in the presence of even rather low

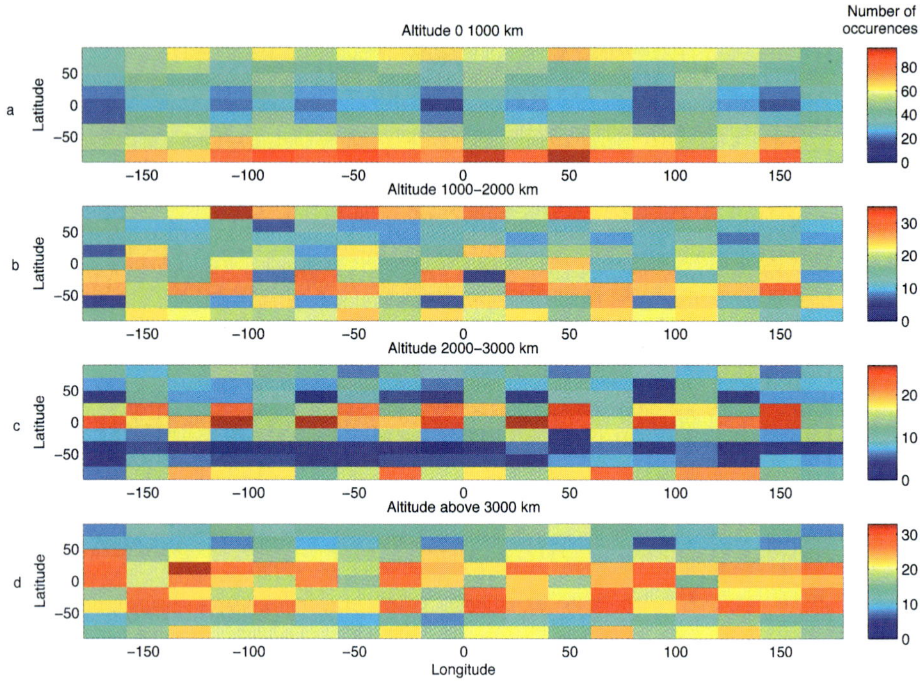

Figure 5. Distribution of number of different orbits of Mars Express passing through different latitude, longitude and altitude bins, when IMA was on in the right mode. Panel (a) shows the altitude interval up to 1000 km, (b) shows the interval 1000 to 2000 km, (c) 2000 to 3000 km and (d) 3000–10000 km altitude.

magnetic field strengths. The statistics are very poor for the data at higher magnetic field values, but there are a number of events there also for the lowest altitudes below 2000 km. In the altitude interval above 2000 km there is no influence on the occurrence of ion beams from the magnetic field for the values below 40 nT where statistics are relatively good. The north-south asymmetry seen in the latitude distribution could in principle cause a spurious dependence on the magnetic field as the magnetic fields are in general stronger in the southern hemisphere. We have therefore performed the same calculations for the southern hemisphere only but with no significant change in the result. The large scatter with some very high occurrence frequencies for higher magnetic field values is consistent with significant magnetic anomalies playing a particular role in some outflow events (like the field-aligned acceleration events reported by Lundin *et al.* (2006)) but the net contribution is small. We have made the same plot also for a magnetic map where we used the maximum magnetic field value within up to $\pm 5^\circ$ of each grid of the magnetic field map (which has a resolution of $1^\circ \times 1^\circ$) to allow for a less precise mapping to the closest strong anomaly. This caused a more even spread of the data and confirmed the lack of influence of magnetic fields on the large scale distribution for altitudes

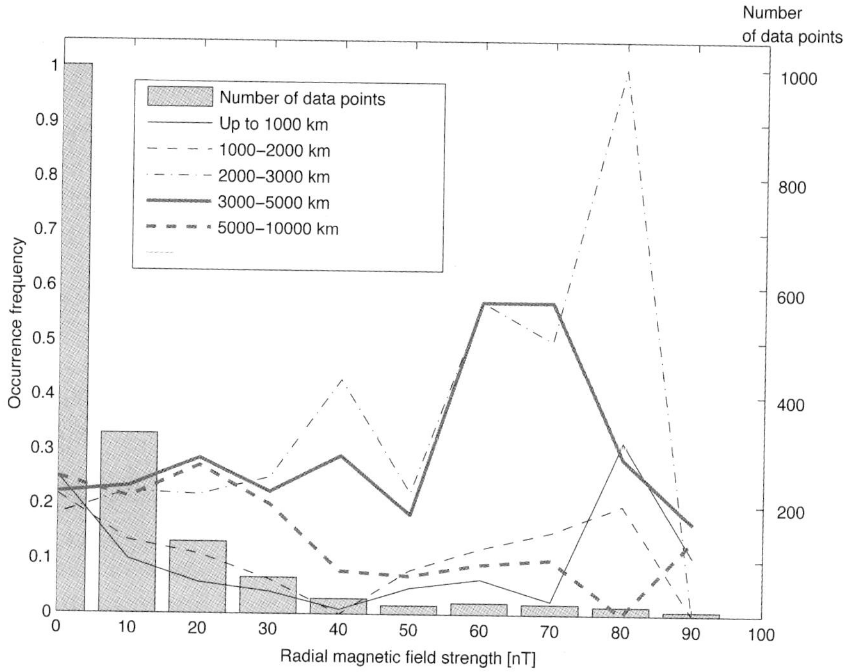

Figure 6. Occurrence rate of ions beams vs. radial crustal magnetic field at 400 km altitude below the spacecraft for 5 different altitude intervals as indicated in the figure. Grey bars in the background indicate the number of ion beam data points (summed over all altitude intervals) in each statistical bin.

above 2000 km. We also tried the transverse and total magnetic fields which resulted in somewhat larger scatter but otherwise little difference in the result. The "noisy" peaks for higher magnetic field values were most pronounced for the original radial magnetic field data used in Figure 6, indicating that these particular events indeed map rather precisely to the strong radial field regions.

We have in the discussion above used a radial mapping of the location of the spacecraft down to our reference magnetic field model at 400 km. The obvious alternatives are to compare to the magnetic field values of an extrapolated mathematical model (i.e. the Cain model) at the actual altitude of observation or do field-tracing along such a model, which must be coupled to an IMF/draped field-line model to justify the effort of such a precise mapping. The latter is desirable but outside the scope of our current work. We have tried the former but it suffers from the fact that all extrapolated model fields are weak at high altitudes whereas the radial field can be stretched out and therefore stronger during certain events. Relatively strong crustal fields at high altitude always correspond to strong crustal magnetic fields at low altitude as well.

3.4. Distribution of Events with no Entrance Deflection

The small data set from December 2005 to March 2006 (when IMA was run with entrance deflection off) is particularly important both because of its higher time resolution and its better ability to measure low energy ions. We have visually inspected all such orbits and picked out ion beam events in the lower altitude part. The heavy ion counts summed over all energies are plotted along the satellite track as shown in Figure 7 (background count levels subtracted). Also shown in Figure 7 is a grayscale map of the radial crustal magnetic field at 400 km altitude (Connerney *et al.*, 2001) so that the fine structure of the ion counts can be compared with the fine structure of the radial field. About half of the data points were obtained below 1500 km altitude, and altitudes up to 4000 km have been used. Clearly many of the events occur where there are no strong magnetic anomalies straight below the spacecraft. Fine structures also occur when no magnetic anomaly is nearby. A number of events with significant structure do occur close to magnetic anomalies, and the structure could possibly arise either from an interaction between pick-up

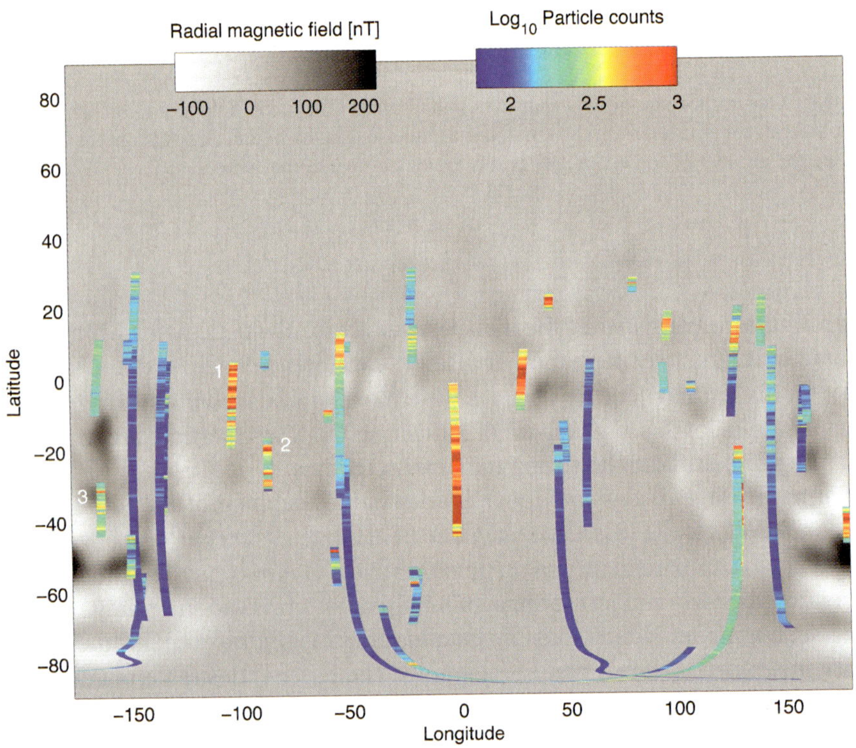

Figure 7. Counts in the heavy ion mass channels mapped radially to the planet, superposed on a map of the radial crustal field at 400 km as obtained from the MGS spacecraft. Three cases are marked with a number, these are discussed in more detail in the text.

ions and the anomalies or because the ions emanate from the anomalies as in the cases reported by Lundin *et al.* (2006). We leave a closer investigation of this to future case studies but show two sample cases here. The first, from 2006-02-16, numbered 2 in Figure 7, is a sample of a beam that appears very structured and occurs away from any anomaly, as shown in Figure 8. The other sample is shown in Figure 9 and was taken in the close proximity of a significant crustal magnetic anomaly. Both figures have four panels, where the first shows electron counts from the ELS electron spectrometer, the second shows proton counts, the third shows oxygen ion counts and the fourth shows altitude (black line, left y-axis) and radial magnetic field at 400 km altitude, interpolated from the statistical MGS results (red line, right y-axis). The oxygen ion counts dominated but may contain contributions from heavier ions (see Carlsson *et al.*, 2006, for details). Just as the impression one gets from the overview shown in Figure 7, there is indeed significant structure in the heavy ion counts in both cases. These are often correlated with variability

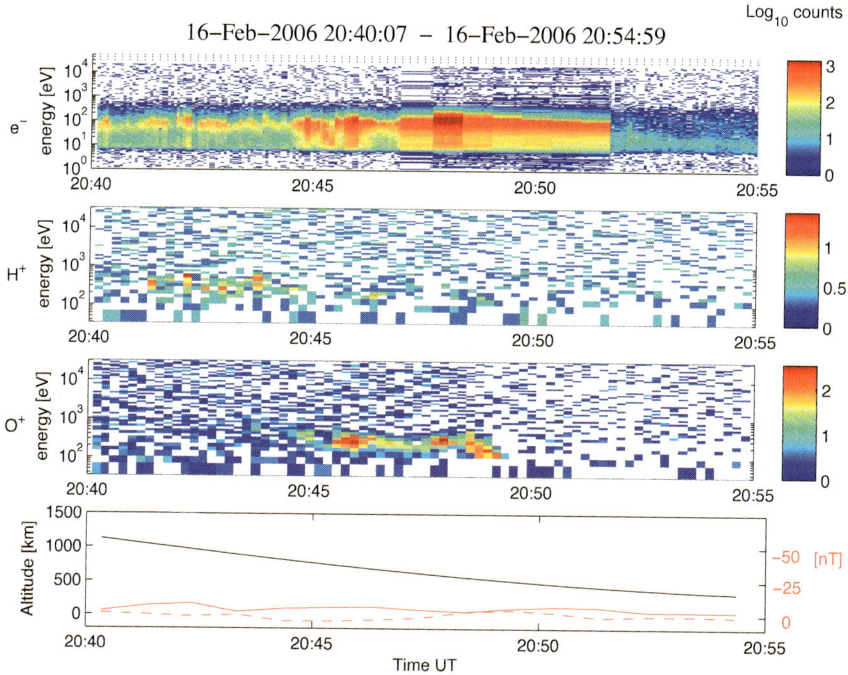

Figure 8. Sample energy spectra of a heavy ion beam from 2006-02-16 when no entrance deflection scanning was used. The first panel show the corresponding electron spectra summed over all sectors (the electrons are typically rather isotropic). The two consecutive panels show the H^+ and O^+ ion counts, summed over all sectors. Only a few sectors (of $22.5^\circ \times 5^\circ$) detect any ions at all. The Y-axis shows particle energy in eV for all of the first three panels, and the x-axis time (UT). The bottom panel shows the altitude of the spacecraft (black line, left y-axis [km]) together with the radial (red dashed line) and transverse (red solid line) magnetic field at 400 km altitude, radially below the spacecraft (right y-axis, [nT]).

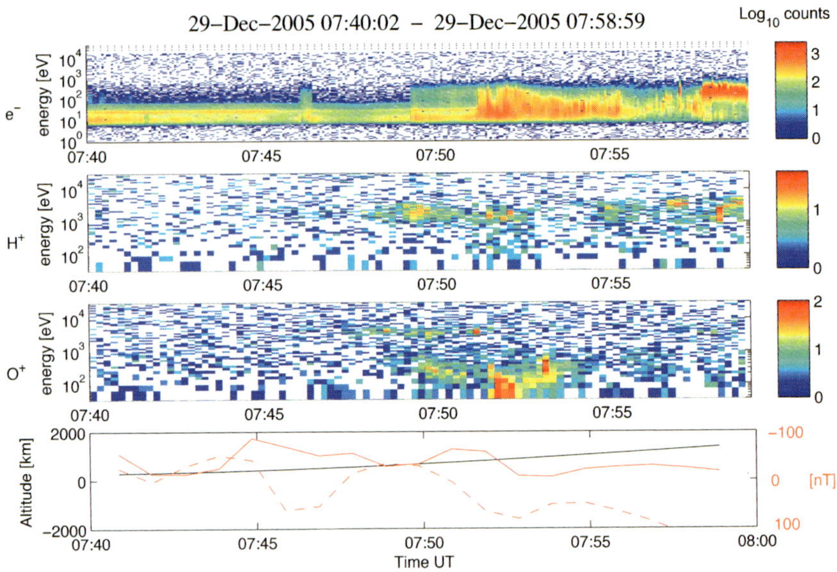

Figure 9. Sample energy spectra of a heavy ion beam from 2006-12-29 when no entrance deflection scanning was used. The first panel show the corresponding electron spectra summed over all sectors (the electrons are typically rather isotropic). The two consecutive panels show the H^+ and O^+ ion counts, summed over all sectors. Only a few sectors (of $22.5^\circ \times 5^\circ$) detect any ions at all. The Y-axis shows particle energy in eV for all of the first three panels, and the x-axis time (UT). The bottom panel shows the altitude of the spacecraft (black line, left y-axis [km]) together with the radial (red dashed line) and transverse (red solid line) magnetic field at 400 km altitude, radially below the spacecraft (right y-axis, [nT]).

in the electron counts but we leave the detailed comparison for future studies. In the case observed close to an anomaly it turns out that low energy ions are observed at a peak in the radial magnetic field (dashed red line). On both sides of the low energy ions we observe beams with a narrow energy distribution. Low energy ions are those most likely to be affected by the magnetic fields, and such low energy ionospheric ions are usually not observed in the ion beam events, (e.g. the samples shown in Figures 2 and 8). The low energy ions were not observed at the lowest point in the orbit, nor were low energy ions observed at the peak in the radial field which the spacecraft passed at a lower altitude at about 7:46 UT. The observed structure is therefore consistent with an influence due to the anomaly, but this is not a clear proof that this is really the case and anomalies are clearly not associated with such enhancements all the time. Rather the opposite is true according to our statistical results discussed in Section 3.3, where ion beams were less common above moderately strong (10–40 nT) crustal fields than over regions with the lowest crustal fields.

The most important observation is that the events occur everywhere regardless of the presence of anomalies and there is a considerable amount of fine structure everywhere, not only close to anomalies. One may note that fine structure in the "no entrance deflection" cases can be due to flow direction changes, not necessarily a particle flux modulation, but it still represents a small scale structure in the plasma characteristics.

4. Discussion and Conclusions

The data shown in this paper clearly indicate that heavy ion beams (and thus of planetary origin) occur over most locations above Mars. This is a very expected result as planetary ions are expected to dominate in the magnetic pile-up region which is draped around the planet. There are, however, three interesting findings in our data set.

(1) There is some north-south asymmetry in the data. Ion beams are somewhat less common at low altitudes (up to 1000) over the southern hemisphere and more common just below the equator in the southern hemisphere for high altitude (above 3000 km). North-south asymmetries are sensitive to orbit characteristics but the data we show have been normalized to take into account the number of observations in the different latitude-longitude-altitude intervals used in our study. There is no longitudinal variation resembling those of the magnetic anomalies. Our conclusion is that the latitudinal asymmetry is not directly caused by magnetic anomalies. One may have to compare with the average solar wind electric field direction and possibly planet rotation axis tilt to explain the difference. As was shown by Dubinin *et al.* (2005), Fedorov *et al.* (2005) the heavy ion flux distribution is well organized by the solar wind electric field. Furthermore the ion beam occurrence rate increases somewhat with altitude. This is consistent with an extended altitude region where ion energization up to the beam energies of 100 eV and above occurs.

(2) A study of the dependence of ion beam occurrence rate vs. radial magnetic field at 400 km altitude revealed no dependence on ion beam occurrence at altitudes above 2000 km. For altitudes below 2000 km a dependence on ion beam occurrence could be seen. The occurrence frequency was highest for the lowest magnetic field region (0–10 nT) and decreased for the moderately strong crustal fields (about 40 nT). For higher magnetic field values the variability was large, mainly because of poor statistics but the decrease seen from about 0 to 40 nT magnetic field strength can clearly not be extrapolated to higher magnetic field cases. There was no signature of magnetic anomalies discernible in the longitude distribution of the ion beams, and the latitude asymmetry discussed above seems not to be directly related to magnetic anomalies. This is different from what is the case for electrons where the ionopause and magnetic-pileup boundary as detected from electron data are clearly modulated by the presence of magnetic anomalies (see references in introduction), and the strongest

fields have a clear and pronounced effect. Typical electron signatures associated with magnetic anomalies have been demonstrated (Mitchell *et al.*, 2001; Soobiah *et al.*, 2005). The reason for the discrepancy between ion and electron observations is most likely finite gyro radii effects of the ions for the energies observable by the IMA instrument. The beams observed by IMA typically have energies of several 100 eV. In a magnetic field of 50 nT a 300 eV O^+ ion has a gyro radius of 200 km. When the field is down to 10 nT the gyro radius is 1000 km. Therefore ions may be tied to the magnetic fields at low altitudes where the magnetic field is strong and the ion energy typically low. As the ions gain energy they will still be affected by magnetic anomalies, but in a dynamical way, they will not stay in place the way the electrons do. The ions may be accelerated along the field-line, in which case the mirror force will keep the outflowing ions beam-like and tied to the original field-line (Lundin *et al.*, 2006) which is indeed the only clear ion-magnetic anomaly association reported). The influence may also be through small scale ripples in the draped field-lines, causing non-adiabatic drift and acceleration through the centrifugal force mechanism (Cladis, 1986; Cladis *et al.*, 2000).

It could, despite gyro-radius considerations, be possible that the actual number flux and ion composition would be influenced by magnetic anomalies. This would then mainly concern ionospheric upflow and escape, not pick-up ions. At Earth the ionospheric escape is a two-stage process consisting of initial upflow in the ionosphere observable for example by incoherent scatter radar [e.g. (Nilsson *et al.*, 1996; Ogawa *et al.*, 2003)] and subsequent energization to escape velocity at higher altitudes. The initial upflowing ions are typically gravitionally bound and flow down again unless further heating processes take place at higher altitudes (as is the case at Earth, from the topside ionosphere and throughout the magnetosphere, e.g see discussion and references in Nilsson *et al.* (2004)). The lower altitude processes regulate the number flux of planetary origin ions and, if something similar occurs at Mars, would most likely be strongly affected by the magnetic anomalies as these regulate ionospheric scale height and heating rates [e.g. (Krymskii *et al.*, 2003)]. A possible explanation for our results of decreasing ion beam occurrence for intermediate strength crustal fields is that most field-lines close below the spacecraft and reduce the vertical transport of ionospheric ions. It would therefore be worthwhile to study the number flux and detailed composition of the planetary origin fluxes as a function of geographic location above Mars. This is, however, a rather demanding task as the detailed ion composition requires a manual inspection of every mass spectrogram [e.g. (Carlsson *et al.*, 2006)] and is thus beyond the scope of this report.

(3) There is considerable small scale structure in many of the ion beams observed. Some of these structures may indeed be caused by magnetic anomalies which show variations on the proper spatial scale. It would be of interest from a fundamental plasma physics point-of-view to identify some such cases and study

them in detail, but if magnetic anomalies disperse or further energize already picked-up ions passing through them this should not be of major importance for the total outflow. Outflow caused more directly by processes associated with the crustal fields could have some influence on the total outflow. If that is the cause of small scale structures then due to the small size it cannot have a large overall impact on the ion escape from Mars. Inhibition of vertical transport and therefore lower escape is rather more in line with the results obtained in this study (see point 2 above). The data we presented in Figure 7 contained cases with considerable fine structure also when no magnetic anomalies were nearby so clearly there are other plasma structuring processes at work as well.

Acknowledgments

This project was supported by the Swedish National Space Board and the funding agencies of our co-investigators. The MGS magnetic field data was downloaded from the Goddard Space Flight Center web pages, and we gratefully acknowledge the work of GSFC, JPL and all other institutes and individuals involved in the Mars Global Surveyor MAG-ER experiment.

References

Acuña, M. H., Connerney, J. E. P., Wasilewski, P., Lin, R. P., Anderson, K. A., Carlson, C. W., *et al.*: 1998, *Science* **279**(5357), 1676, doi: 10.1126/science.279.5357.1676.

Barabash, S.: 2006, *Plan Space Sci.*, submitted.

Barabash, S., Dubinin, E., Pisarenko, N., Lundin, R., and Russell, C.: 1991, *Geophys. Res. Lett.* **18**, 1805.

Bertucci, C., Mazelle, C., Acuña, M., Russel, C., and Slavin, J.: 2005, *J. Geophys. Res.* **110**(A01209), doi: 10.1029/2004JA010,592.

Brain, D., Halekas, J. S., Lillis, R., Mitchell, D., Lin, R., and Crider, D.: 2005, *Geophys. Res. Lett.* **32**, doi: 10.1029/2005GL023,126.

Brain, D., Halekas, J. S., Peticolas, L., Lin, R., Luhmann, J., Mitchell, D., *et al.*: 2006, *Geophys. Res. Lett.* **33**(L01201), doi: 10.1029/2005GL024,782.

Breus, T., Krymskii, A., Lundin, R., Dubinin, E., Luhmann, J., Yeroshenko, Y., *et al.*: 1991, *J. Geophys. Res.* **96**(A7), 11165–11174.

Cain, J., Ferguson, B. B., and Mozzoni, D.: 2003, J. geophys. Res. **108**(E2), doi 10.1029/2000JE001, 487.

Carlsson, E., Fedorov, A., Barabash, S., Budnik, E., Grigoriev, A., Gunell, H., *et al.*: 2006, *Icarus*, **182**(2), 320.

Cladis, J. B.: 1986, *J. Geophys. Res.* **13**, 893.

Cladis, J. B., Collin, H. L., Lennartsson, O. W., Moore, T. E., Peterson, W. K., and Russell, C. T.: 2000, *Geophys. Res. Lett.* **27**, 915, doi: 10.1029/1999GL10737.

Connerney, J. E. P., Acuna, M. H., Wasilewski, P. J., Ness, N. F., Réme, H., Mazelle, C., *et al.*: 1999, *Science* **284**(5415), 794.

Connerney, J. E. P., Acuna, M. H., Wasilewski, P. J., Kletetschka, G., Ness, N. F., Réme, H., *et al.*: 2001, *Geophys. Res. Lett.* **28**(21), 4015.

Crider, D., Cloutier, P., Law, C., Walker, P., Chen, Y., Acuña, M., *et al.*: 2000, *Geophys. Res. Lett.* **27**(1), 45.

Crider, D., Acuña, M., Connerney, J., Vignes, D., Ness, N., Krymskii, A., *et al.*: 2002, *Geophys. Res. Lett.* **29**(8), doi: 10.1029/2001GL013,860.

Dubinin, E., Lundin, R., Fränz, M., Woch, J., Barabash, S., Fedorov, A., *et al.*: 2006, *Icarus*, **182**(2), 337.

Fedorov, A., Budnik, E., Sauvaud, J. A., Mazelle, C., Barabash, S., Lundin, R., *et al.*: 2006, *Icarus*, **182**(2), 329.

Fränz, M., Winningham, J., Dubinin, E., Roussos, E., Woch, J., Barabash, S., *et al.*: 2006, *Icarus*, **182**(2), 406, doi: 10.1016/j.icarus.2005.11.016.

Krymskii, A., Breus, T., Ness, N., Hinson, D., and Bojkov, D.: 2003, *J. Geophys. Res.* **108**(A12), doi: 10.1029/2002JA009,662.

Lundin, R., Zakharov, A., Pellinen, R., Hultqvist, B., Borg, H., Dubinin, E., *et al.*: 1989, *Nature* **341**, 609.

Lundin, R., Dubinin, E. M., Koskinen, H., Norberg, O., Pissarenko, and N., Barabash, S. W.: 1991, *Geophys. Res. Lett.* **18**(6), 1059.

Lundin, R., Barabash, S., Andersson, H., Holmström, H., Grigoriev, A., Yamauchi, M., *et al.*: 2004, *Science* **305**, 1933.

Lundin, R., Winningham, D., Barabash, S., Frahm, R., Andersson, H., Holmström, M., *et al.*: 2006, *Icarus*, **182**(2), 308, doi: 10.1016/j.icarus.2005.10.035.

Lundin, R., Winningham, D., Barabash, S., Frahm, R., Holmström, M., Sauvaud, J. A., *et al.*: 2006, *Science* **311**(5763), 980.

Mitchell, D., Lin, R., Rème, H., Crider, D., Cloutier, P., Connerney, J., *et al.*: 2000, *Geophys. Res. Lett.* **27**(13), 1871.

Mitchell, D., Lin, R., Mazelle, C., Rème, H., Cloutier, P., Connerney, J., *et al.*: 2001, *J. Geophys. Res.* **106**(E10), 23419.

Nagy, A., Winterhalter, D., Sauer, K., Cravens, T., Brecht, S., Mazelle, C., *et al.*: 2004, *Space Sci. Rev.* **111**(1–2), doi: 10.1023/B:SPAC.0000032,718.47,512.92.

Ness, N., Acuña, M., Connerney, J., Kliore, A., Breus, T., Krymskii, A., *et al.*: 2000, *J. Geophys. Res.* **105**(A7), 15991.

Nilsson, H., Yamauchi, M., Eliasson, L., Norberg, O., and Clemmons, J.: 1996, *J. Geophys. Res.* **101**, 10947.

Nilsson, H., Joko, S., Lundin, R., Réme, H., Sauvaud, J. A., Dandouras, I., *et al.*: 2004, *Ann. Geophys.* **22**, 2497.

Nilsson, H., Lundin, R., Lundin, K., Barabash, S., Borg, H., Norberg, O., *et al.*: 2006, *Space Sci. Rev.*, doi: 10.1007/s11214-006-9031-z.

Ogawa, Y., Fujii, R., Buchert, S. C., Nozawa, S., and Ohtani, S.: 2003, *J. Geophys. Res.* **108**, doi: 10.1029/2002JA009,590.

Soobiah, Y., Coates, A., Linder, D., Kataria, D., Winningham, J., Frahm, R., *et al.*: 2006, *Icarus*, **182**(2), 396, doi: 10.1016/j.icarus.2005.10.034.

Trotignon, J., Dubinin, E., Réjan, G., Barabash, S., and Lundin, R.: 1996, *J. Geophys. Res.* **101**(A11), 24965.

Trotignon, J. G., Mazelle, C., Bertucci, C., and Acuna, M. H.: 2006, *Plan Space Sci.* **54**, 357.

Vignes, D., Mazelle, C., Réme, H., Acuña, M., Connerney, J., Lin, R., *et al.*: 2000, *Geophys. Res. Lett.* **27**(1), 49.

Winningham, J., Frahm, R., Sharber, J., Coates, A., Linder, D., Soobiah, Y., *et al.*: 2006, *Icarus*, **182**(2), 360, doi: 10.1016/j.icarus.2005.10.033.

OBSERVATIONS OF VERTICAL REFLECTIONS FROM THE TOPSIDE MARTIAN IONOSPHERE

E. NIELSEN[1,*], H. ZOU[1,6], D. A. GURNETT[2], D. L. KIRCHNER[2], D. D. MORGAN[2], R. HUFF[2], R. OROSEI[3], A. SAFAEINILI[4], J. J. PLAUT[4] and G. PICARDI[2]
[1]*Max Planck Institute for Solar System Research, 37191 Katlenburg-Lindau, Germany*
[2]*Department of Physics and Astronomy, University of Iowa, Iowa City, IA 52242, USA*
[3]*Istituto di Astrofisica Spaziale e Fisika Cosmica (NAF), 00133 Rome, Italy*
[4]*Jet Propulsion Laboratory, Pasadena, CA 91109, USA*
[5]*Infocom Department, "La Sapienza" University of Rome, 00184 Rome, Italy*
[6]*School of Earth and Space Sciences, Peking University, Beijing 100871, P. R. China*
(**Author for correspondence, E-mail: nielsen@mps.mpg.de*)

(Received 30 March 2006; Accepted in final form: 2 November 2006)

Abstract. The Martian ionosphere has for the first time been probed by a low frequency topside radio wave sounder experiment (MARSIS) (Gurnett *et al.*, 2005). The density profiles in the Martian ionosphere have for the first time been observed for solar zenith angles less than 48 degrees. The sounder spectrograms typically have a single trace of echoes, which are controlled by reflections from the ionosphere in the direction of nadir. With the local density at the spacecraft derived from the sounder measurements and using the lamination technique the spectrograms are inverted to electron density profiles. The measurements yield electron density profiles from the sub-solar region to past the terminator. The maximum density varies in time with the solar rotation period, indicating control of the densities by solar ionizing radiation. Electron density increases associated with solar flares were observed. The maximum electron density varies with solar zenith angle as predicted by theory. The altitude profile of electron densities between the maximum density and about 170 km altitude is well approximated by a classic Chapman layer. The neutral scale height is close to 10 to 13 km. At altitudes above 180 km the densities deviate from and are larger than inferred by the Chapman layer. At altitudes above the exobase the density decrease was approximated by an exponential function with scale heights between 24 and 65 km. The densities in the top side ionosphere above the exobase tends to be larger than the densities extrapolated from the Chapman layer fitted to the measurements at lower altitudes, implying more efficient upward diffusion above the collision dominated photo equilibrium region.

Keywords: mars, ionosphere, electron densities, top side sounder

1. Introduction

The Earth's upper ionosphere is protected by strong geomagnetic fields from direct interaction with the solar wind. Only at very high latitudes is the ionosphere connected by field aligned currents to interplanetary space. Contrary to this Mars has a rather weak crustal magnetic field over parts of the surface, and the rest of the planet is essentially non-magnetic. The absence of a strong global magnetic field allows the solar wind to directly interact with Mars's upper atmosphere and

Space Science Reviews (2006) 126: 373–388
DOI: 10.1007/s11214-006-9113-y

ionosphere. For strong solar wind pressure a magnetic field may be induced in the ionosphere in order to hold off and deflect the solar wind. The different magnetic characteristics of Earth and Mars lead to significant differences between the top side ionosphere on the planets. The lower ionosphere plasma is expected to be dominated by photo-chemical equilibrium between solar electron/ion production and local losses (Chapman, 1931; Budden, 1966). Apart from solar zenith angle effects, there may be further spatial variations owing to crustal magnetic fields (Nagy *et al.*, 2004). However, at higher altitudes, above the photo chemical equilibrium region, vertical diffusion and horizontal plasma transport processes as well as plasma wave activity induced owing to planet rotation, magnetic fields, and solar wind interactions lead to modifications of the plasma at Mars (Luhman *et al.*, 1992; Shinagawa, 2000; Acuna *et al.*, 1998, 1999; Ness *et al.*, 2000; Wang and Nielsen, 2004).

The Martian ionosphere electron densities have been observed before using the radio occultation technique by (Mars 2, 3, 4 and 6, Mariner 4, 6, 7, and) Mariner-9 (Kliore, 1992), by Viking orbiter and Viking landers (Hanson *et al.*, 1977), by Mars Global Surveyor (Hinson *et al.*, 1999; Bougher *et al.*, 2001), and by Mars Express (Paetzold *et al.*, 2005). Because Mars is an exterior planet the radio occultation technique allows observations only in a zenith angle range of roughly ± 45 degrees around the terminator. Some results presented in this work are new in the sense that such observations have not been presented before: observations of the Martian ionosphere at small solar zenith angles, direct control of the electron density peak on solar rotation (radiation), and presentation and discussion of virtual and real range obtained by a topside sounder operating at Mars. Other results are new in the sense that we have used a new kind of experiment on Mars and used the measurements to confirm (or dispute) earlier results. Even if we only confirm old results it is worth while to show that the same physics results from different experimental techniques.

MARSIS (= Mars Advanced Radar for Subsurface and Ionosphere Sounding) is a top side sounder, which has been operated on board the ESA spacecraft Mars Express (Nielsen, 2004) since July 2005. The observations are aimed to study the Martian ionosphere and its interaction with the sun (Gurnett *et al.*, 2005). The MARSIS experiment is described by Picardi *et al.* (1999).

In this work we demonstrate, that the sounder yields convincing evidence that at the lower altitudes the altitude profile of the electron densities follow the predictions for a Chapman layer, both with respect to zenith angle variations as to control by solar radiation. At higher altitudes the densities are variable and are typically larger than the densities extrapolated from the Chapman layer dominating the densities closer to the density maximum.

2. The Sounder

The radio wave sounder technique has been used extensively to explore the Earth's ionosphere (Budden, 1966; Kenneth, 1965). Sounders have been deployed both on the ground (Rishbeth and Garriott, 1969) and on orbiting satellites (Nelms *et al.*, 1966). Sounder measurements provide detailed information on variations of the ionospheric electron density with distance from the sounder (Warren, 1963; Hagg, 1967).

Ionosphere soundings rely on the fact that electromagnetic waves can not propagate in a plasma when the wave frequency, f, is below the electron plasma frequency, f_p, and that the wave is reflected where $f = f_p$. The plasma frequency is controlled by the electron density, N_e,

$$f_p^2 = \frac{e^2 N_2}{4\pi^2 m \varepsilon_0} \tag{1}$$

$$f_p = 8.97\sqrt{N_e[\mathrm{el/cm^3}]}\ [\mathrm{kHz}] \tag{2}$$

The basic sounding measurement consists of transmitting a radio wave pulse at a discrete frequency. If the radio frequency is larger than the local plasma frequency the radio wave propagates away from the radar. Specular reflection of the radio signal occurs from regions where the ionosphere is stratified in directions perpendicular to the radar wave vector, and where the plasma frequency (Equation (2)) is equal to the radar wave frequency. The sounder detects the pulse return as a function of time following transmission, the delay time.

The antenna used for transmitting and receiving the radio waves is a dipole, which has a significant gain in all directions (except in directions parallel to the dipole elements). Thus, reflected signals can be received from all directions.

The MARSIS sounder operates with radio waves in the frequency range from 0.1 to 5.4 MHz. This means that electron densities between 125 and 3.8×105 el/cm^3 in the top side ionosphere are probed by the radar. Varying the sounder frequency in steps, a spectrogram of the received signals is build up in a coordinate system of frequency versus delay time. A spectrogram for normal operations displays the intensity of reflected echoes as a function of time for each of 160 discrete frequencies. The time resolution of the measurements is the same as the transmitter pulse width and is 91.4 micros, corresponding to a spatial resolution of $\sim$13.7 km. After transmission of the pulse and a further 162.5 micros dead time, the echo intensity is measured 80 times, covering delay times up to 8 ms or a range of $\sim$1200 km.

3. Single Trace Spectrogram – The Vertical Echo

An example of a spectrogram with a single trace obtained with the top side sounder MARSIS is shown in Figure 1.

This is a typical Martian spectrogram of reflections from the ionosphere in nadir. For this event reflections were observed at frequencies between 1.0 and 3.4 MHz. At low frequencies the echo merges out of background determined by the transmitter power, antenna gain in direction of nadir, and absorption in the ionosphere. The

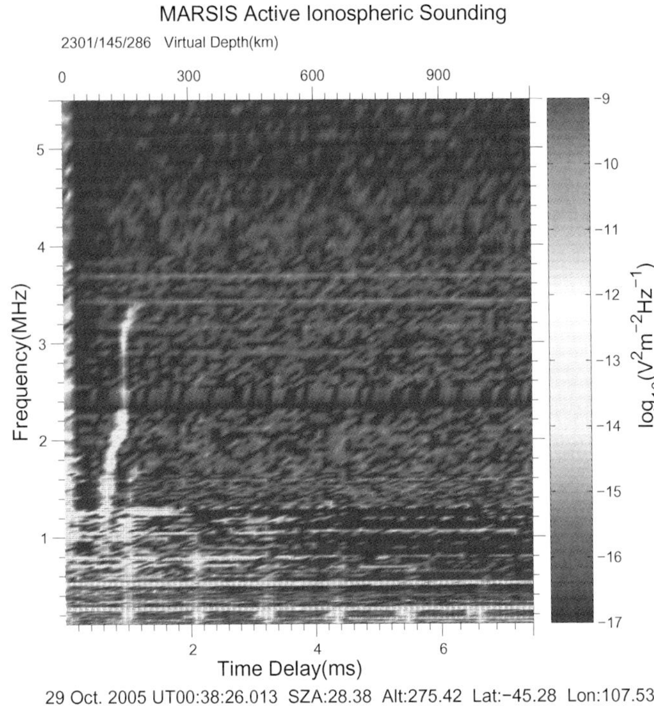

Figure 1. MARSIS spectrogram with frequency versus delay time covering 7.54 ms. The horizontal line near 2.4 MHz is instrumental in origin. A total frequency sweep lasts 1.23 s, and is repeated every 7.5 s.

echoes are reflected at electron densities between 1.24×10^4 and 1.44×10^5 el/cm^3. Delay times are from 0.8 to 1.44 ms. The delay times result from the combined effects of the electron density altitude profile and the altitude of the sounder.

The maximum frequency reflected from the ionosphere is a measure of the maximum plasma frequency (maximum electron density). The maximum plasma frequency may be up to ~1 MHz on the night side and ~5 MHz on the dayside.

Since the electron density in the Martian ionosphere is generally decreasing with altitude, low radar frequencies are reflected from the upper parts of the ionosphere, and therefore associated with relatively short delay times. With increasing frequency the reflections occur at larger distances and longer delay times. Radar signals with a frequency exceeding the maximum plasma frequency of the target will penetrate the ionosphere and reach the surface. There it will be reflected/scattered back towards the radar, and if the ionosphere absorption is low enough the ground wave may be detected at the radar (Nielsen *et al.*, 2006). In this particular case no ground wave is observed.

In Figure 1 notice several strong (resonance) horizontal echoes with a separation between nearest neighbors of 265 kHz. These echoes are higher harmonics of the local plasma frequency at the spacecraft excited by the radar transmissions. The harmonics are caused by nonlinear distortion in the receiver (Gurnett *et al.*, 2006). Using Equation (2) the observed frequency separation translates into a local density of 860 electrons/cm^3. This value of the local density is used later in the inversion process of the spectrograms which yields the vertical electron density profile.

In the following are discussed the maximum electron densities, solar control of the densities, and the altitude profile of the electron densities.

4. The Electron Density Maximum

In connection with ephemeris data the spectrograms yield the density maximum and associated solar zenith angle as they vary during an orbit. We select every 5 minutes the maximum density and zenith angle from all orbits between July and October, 2005. This ensures good coverage in zenith angle and yield a large amount of observations. The Chapman theory predicts that the density maximum (N_m) depends on the sub-solar density (N_0) and zenith angle (θ) as given by Budden (1966)

$$N_m = N_0 \cos^n(\theta) \tag{3}$$

where the simple theoretical value of the exponent is $n = 0.5$. Equation (3) is expected to be valid for zenith angles less than ~85 degrees. Taking the logarithm on both sides a linear least square fit yields $n = 0.48$ and $N_0 = 1.79 \times 105[\text{el/cm}^3]$,

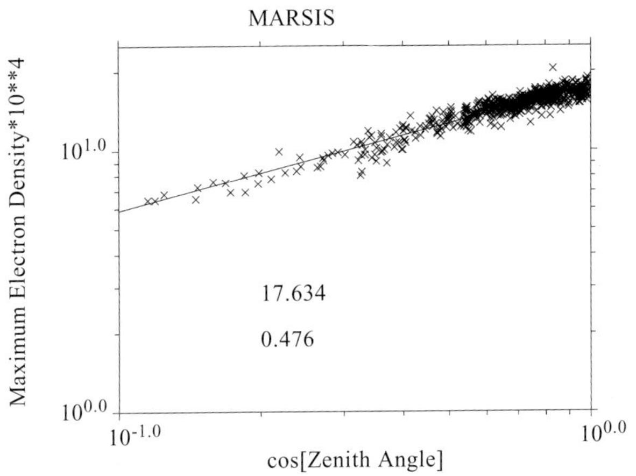

Figure 2. Least squares fit of Equation (3) to observations of maximum density versus zenith angle. For the best fit the exponent is 0.48 consistent with the theoretical prediction, and the maximum sub-solar density is 1.8×10^5 [el/cm^3]. The data are from the latitude interval +75 (north) to −85 degrees.

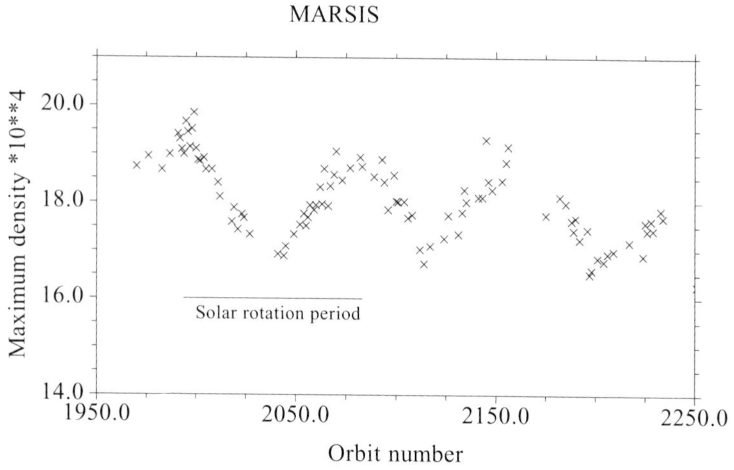

Figure 3. Time variation of the maximum sub-solar electron density (from July 22 to October 14, 2005).

see Figure 2. (This relation is similarly shown to be valid in Figure 3 of Gurnett *et al.* (2005)).

The solid line is the best fit. For zenith angles <85 degree the function sqrt(cos(θ)) is a realistic approximation. For $\theta > 85$ degree ($\cos\theta > 0.1$) the predicted zenith angle dependence require the cos-function in Equation (3) be replaced by the function $1/Ch(x, \theta)$, which takes into account the complications of radiative energy transfer to the atmosphere near the terminator (Chapman, 1931; Rishbeth and Garriott, 1961). The parameter $x = (R_M + h)/H$ is about 350 for Mars ($R_M = 3398$ km, h(height of density maximum) ~130 km, and H(the neutral scale height) ~10 km). Use of the Ch-function improves the fit for large zenith angles. But the function still tends to be a lower limit for densities for zenith angles >90 degrees (Gurnett *et al.*, 2005). Just behind the terminator the electron densities are larger than predicted by photochemical equilibrium. This is likely an effect of transport by winds of dayside plasma across the terminator into the nightside. These results are on line with Zhang *et al.* (1990a,b), Kliore (1992), and Gurnett *et al.* (2005).

It was tested if there were time variations of the electron density taking place during the time interval covered by the observations. For each orbit and for observations with zenith angle $\theta < 85$ degrees the sub-solar density was estimated using Equation (3). Thus, for each orbit the subsolar density was derived for each zenith angle and the average subsolar density calculated. The average of these densities is displayed as a function of orbit number (~time) in Figure 3.

The quite smooth sinusoidal variation of the density with time implies a real time variation of the sub solar densities. That the density variations have a period, which is comparable to the solar rotation period of 26 (Earth-) days points to solar

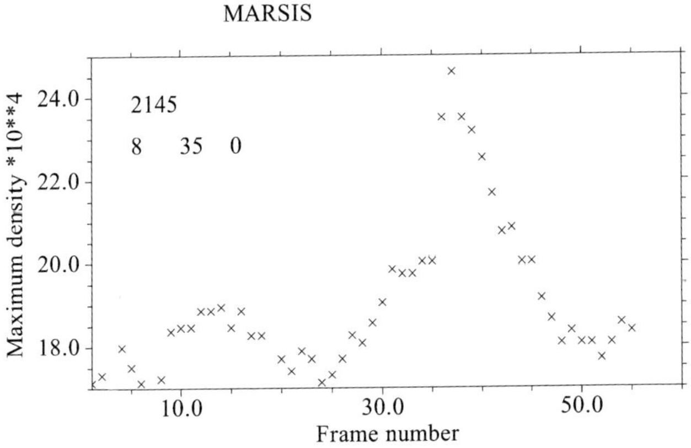

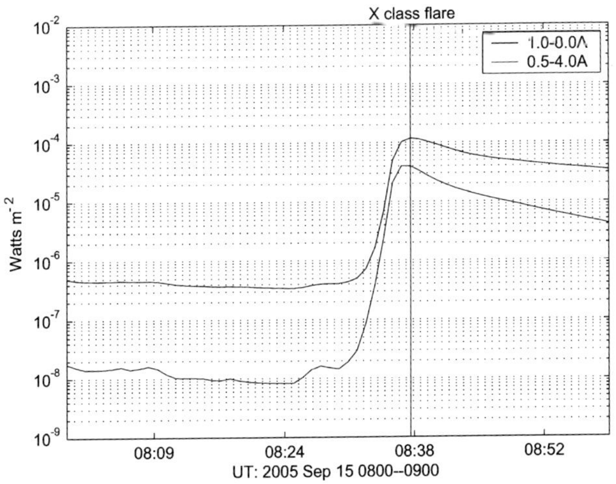

Figure 4. Sudden increase of the electron density maximum (top panel) nearly simultaneous to an increase in solar X-ray flux of wavelength between 0.5 and 3 Angstrom (red curve) and between 1 and 8 Angstrom (blue curve) (bottom panel). The X-ray data are from the GOES spacecraft at the Earth (Bornmann *et al.*, 1996). The top panel covers 7 minutes from 0835 to 0842 UT. The bottom panel covers from 0800 to 0900 UT. (All times are on September 15, 2005 (see also Gurnett *et al.* (2005), Figure 3).

radiation as the factor controlling the Chapman layer electron density. This result is consistent with Mendillo *et al.* (2003) and Withers and Mendillo (2005).

During orbit 2145 (September 15, 2005) at ~0839UT (frame 36 in Figure 4) the electron density maximum in the Martian ionosphere suddenly increased from 1.8×105 to 2.4×105 (Gurnett *et al.*, 2005). This coincides closely in time with an increase in solar X-ray fluxes (Bornmann *et al.*, 1996) measured onboard GOES spacecraft at the Earth (Figure 4). The X-rays precipitating into the atmosphere are an additional agent of ionization. These X-rays have an ionizing effect at the altitude of the density maximum sufficient to increase the equilibrium maximum electron density. Similar results on the effects of solar flares on the electron densities were reported recently by Mendillo *et al.* (2006).

5. The Electron Density Altitude Profile

The (typical) echo trace in Figure 1 results from reflections of the sounder waves from horizontally stratified layers of constant electron density in the ionosphere. The trace displays the delay time as a function of sounder frequency.

If the radio waves travelled in vacuum, then the delay time multiplied by the speed of light would equal two times the distance between spacecraft and reflection region. Distances calculated this way are referred to as virtual (or apparent) ranges, r', and have the form,

$$r' = \int_0^{t_d} c dt \tag{4}$$

where c is the speed of light, t_d is delay time, and t is time. However, the radio waves actually propagate through plasma with a group velocity equal to the product of the speed of light and the refractive index of the plasma, n, given by

$$n = \sqrt{1 - \left(\frac{f_p}{f}\right)^2} \tag{5}$$

where n is real for $f > f_p$ (indicating propagation). The group velocity of the radio wave is,

$$v_g = c \cdot n = c \cdot \sqrt{1 - \left(\frac{f_p}{f}\right)^2} < c \tag{6}$$

Since f_p^2 is proportional to the electron density the speed with which the radio wave propagates is a function of the local electron density. The real distance travelled by the radio wave in time t_d is

$$r = \int_0^{t_d} cndt \tag{7}$$

Rewriting this equation we have an expression for the delay time,

$$t_d = \int dt = \int \frac{dr}{c \cdot n} \tag{8}$$

the integral is along the round trip, radar-reflection-radar.

For each signal frequency the observed delay time multiplied by the speed of light and divided by two is the virtual range to the reflection region where the plasma frequency equals the signal frequency. The virtual ranges depend on the spatial variations of the electron density between the spacecraft and the reflection

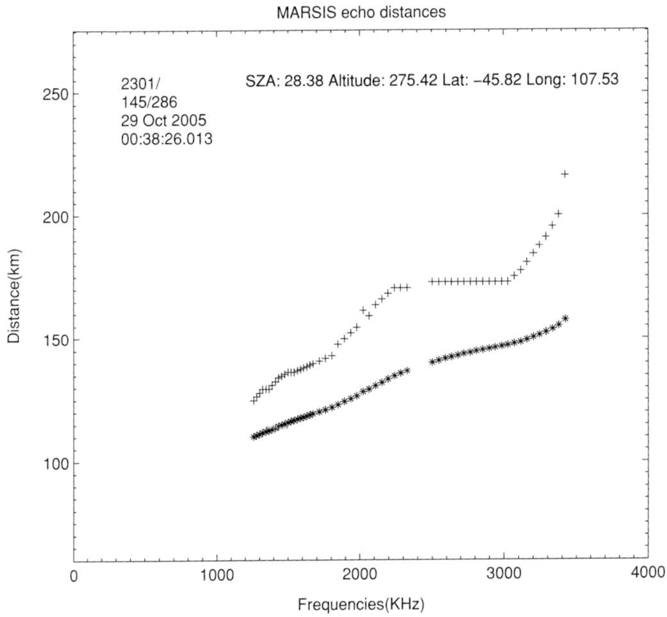

Figure 5. The upper curve is derived from the data in Figure 1 by multiplying the observed delay times with the speed of light (virtual range). This curve is known as an ionogram. The lower curve is the real range variations derived using the lamination technique (see text).

region. Since the propagation speed is reduced in the plasma the virtual range is an upper limit on the real range.

The echo trace in Figure 1 was digitized and the virtual ranges (or distances) versus the sounder frequencies are shown by the upper curve in Figure 5. The top-side sounder measurements cover that part of the ionosphere located below the spacecraft and higher than the altitude of the maximum plasma frequency. For this case the virtual distances vary from ~110 to 160 km. In this case the virtual distance increases with frequency in the whole interval. However, note that for some events the slope of the curve for the lower frequencies is negative, i.e. the virtual distance decreases with increasing frequency. If the distances were real this would not be physically possible; because it implies that lower electron densities are observed at larger distance than higher density layers. However the distances shown are virtual distances. In order to reduce the virtual distances to realistic real distances one must take into account the densities between the spacecraft and the region of reflection of the lower frequencies.

The next step is to invert the observed virtual ranges into a self consistent real electron density profile, with real ranges, such that Equation (8) is satisfied.

The ionogram can be inverted to a real height profile of electron densities using the lamination technique (see Zou and Nielsen, 2004). In this method the ionosphere between the spacecraft down to the maximum electron density is divided into a series of horizontal slabs (N), and it is assumed that the real range varies linearly with plasma frequency inside a slab such that dr/df = constant. For distances closer to the spacecraft than the distance to the lowest frequency reflection point the electron density is assumed to decrease exponentially with altitude with a scale height H_t.

The variable 'time' (in Equation (4)) is replaced by the variable 'real range', or distance. Using that $dt = dr/cn$ and that dr/df is constant within each slab, Equation (4) is modified to

$$r' = \int_0^{t_d} \frac{1}{n} \int_{f_0}^{f_{\max}} \frac{1}{n} \frac{dr}{df} \int_{f_0}^{f_{\max}} \frac{1}{n} df \tag{9}$$

where f_0 is the plasma frequency at the spacecraft and $f_{\max}$ is the plasma frequency at the maximum electron density. This form of the virtual height equation reveals that the virtual range rapidly increases where the curve r' versus f' is steep (dr/df large). This is the case near the peak of a density layer, as for example near the maximum plasma frequency. As noted above this effect can be seen in Figure 1.

Writing this equation for each of the slabs starting from the top ($i = 0$) to slab $i = j$ and summing over range, we have,

$$r' = \sum_i^j dr_i \left(\frac{1}{df_{i,i-1}} \int_{f_{i-1}}^{f_i} \frac{1}{n} df \right) = \sum_i^j dr_i M(j, i) \tag{10}$$

Here are N equations with N unknown real range intervals, dr_j, one for each slab. Recognizing that M is a matrix the set of equations can be inverted and solved for dr,

$$dr = M^{-1} r' \tag{11}$$

and finally the real range to slab j is,

$$r_j = \sum_{i=1}^{j} dr_i \tag{12}$$

First some general comments to the inversion process. If echoes were observed all the way from the spacecraft to the density maximum the inversion could be carried out self consistently. In this case the first slab is at the spacecraft and the N'th slab at the distance to reflection of the highest observed frequency. However, typically the range to the first reflection point is not zero but a considerable distance from the spacecraft. The densities within this distance are not known. In order to proceed with the inversion one must make an assumption about these densities.

We have carried out the inversion using the local density at the spacecraft derived from the plasma frequency oscillations as an 'anchor point'. We use the earlier determined density at the spacecraft and assume the density variation between the spacecraft and the first reflection point is exponential with a scale height H_t (which

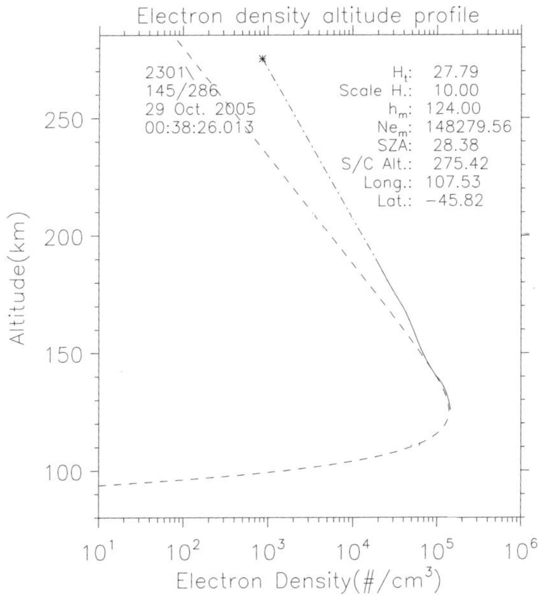

Figure 6. The solid and dash-dot curve is the estimated profile of the top side electron densities derived using the lamination technique. The solid curve covers the densities for which echoes were observed, and the dash-dot curve the assumed exponential decrease. Note, the dash-dot curve starts at the derived electron density (star) at the height of the spacecraft. The dashed curve is a Chapman layer fitted to the observations. In the plot is noted the associated maximum electron density (148279 el/cm^3) and altitude of the maximum (∼124 km) in the sub solar region, together with the neutral scale height (10.0 km), a top-side scale height (∼28 km), and solar zenith angle (∼28 degrees).

is determined by the analysis). This approach is only possible when the plasma frequency can be determined from the observations. That is not always possible.

This approach is only valid within the assumption of exponential decay. The actual profile may be non-exponential. If the profile is non-exponential that will influence the derived distance and densities between spacecraft and the first point of reflection. There seems to be a trend in the data that an exponential decay is a good first approximation but not a complete description. If the assumed exponential decrease is overestimating/under estimating the densities then the derived real distance from spacecraft to first reflection point will be an underestimate/overestimate, and the altitude of the density maximum will be overestimated/underestimated.

The inversion technique has been applied to the data in Figure 1 and Figure 5. Figure 5 also displays the derived real range curve (the lower curve). With the spacecraft height this real range curve has been transformed to the height versus electron density curve, solid in Figure 6. The exponential decrease (dash-dot curve) of electron densities at high altitudes (with $H_t \sim 28$ km) is also included in the figure; the curve starts at the altitude of the spacecraft at the observed local electron density (marked by a star).

Next we test how the real density profile fits to the predictions of the Chapman theory. The electron density profile in a Chapman layer is given by,

$$N_e(h) = N_o \exp\frac{1}{2}\left[1 - \frac{h - h_o}{H_1} - \sec(\theta)\exp\left(-\frac{h - h_0}{H_1}\right)\right] \quad (13)$$

where h_o is the sub solar maximum electron density and altitude of the density maximum. θ is the solar zenith angle. H_1 is the neutral atmosphere scale height. For a scale height of 10 km the layer is a good fit to the observed density profile. The predicted profile for a zenith angle of 28 degrees is displayed in Figure 6 (dashed). Clearly the observations and theory matches very well for altitudes around the electron density maximum up to an altitude of 160 km. The scale height is in the range determined in earlier Martian electron density measurements. It is concluded that up to an altitude of ~160 km, the observations are well represented by a Chapman layer.

To check the derivations we have used the derived real profile curve for all altitudes above the density peak to calculate the delay time versus frequency. We

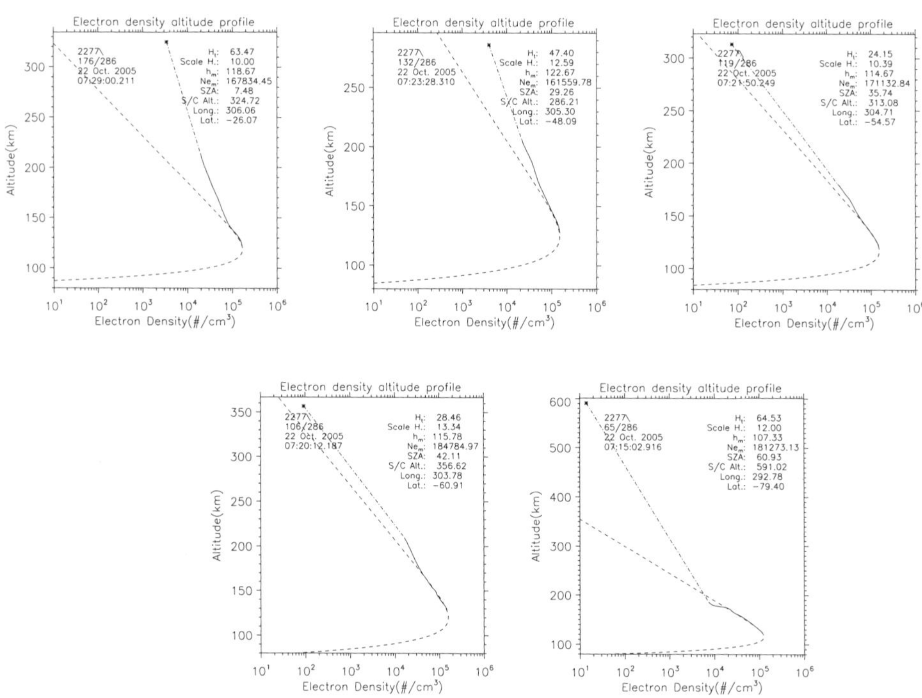

Figure 7. Five electron density profiles from the same orbit (2277) for solar zenith angles between ~7 and 60 degree, with fitted Chapman functions. Increased diffusion above ~180 km leads to densities which are relatively high when compared to the Chapman layer fitted to the observations at lower heights.

find the predicted time delays are in good agreement with the observed ones. This implies self consistency of the inversion process.

In Figure 7 are shown 5 panels with electron density profiles derived as described above in the solar zenith angle interval from 7 to 60 degrees. In all cases there is a good fit of the data to a Chapman layer from the density maximum up to about 160 km. The neutral scale height is from 10 to 13 km. At higher altitudes the scale height increases to values between 24 and 65 km. Thus the density in the top side ionosphere exceeds the densities in the Chapman layer extrapolated above 160 to 180 km. With decreasing collisions the plasma is apparently able to more effectively diffuse upward above 180 km.

It is concluded that the ionosphere at the density peak and in the altitudes above up to ~180 km is well described as a Chapman layer with a neutral scale height from 10 to 13 km. This means that the photochemical equilibrium region extends up to between 160 and 180 km in good agreement with earlier results. Above the equilibrium region the scale height is between 24 and 65 km also in the same framework as earlier results.

The measurements were obtained with a top side sounder radar, a new instrument of a kind not flown before at Mars. The measurements extend for the first time to zenith angles less than 48 degrees. Examples of MARSIS spectrograms have illustrated several points about the day side Martian ionosphere:

1. the zenith angle dependence of the electron density maximum is as predicted by the Chapman theory. This was shown in a statistical sense, in Figure 2;
2. the density maximum is controlled by solar radiation and varying with solar rotation;
3. solar flare X-ray burst causes increase of the electron density maximum;
4. the electron density altitude profile just above the density peak is well described by a Chapman function with a neutral atmosphere scale heights between 10 and 13 km,
5. the ionosphere above the photoequilibrium region, ending at between 160 and 180 km, is variable and was approximated by an exponential decrease corresponding to minimum scale heights between 24 and 65 km.

These results were partly reached by analyzing the echoes reflected from the ionosphere in the nadir using the lamination method to invert the observed ionograms.

6. Discussion

A top side ionospheric sounder measures the delay times for radio waves propagating from the sounder to a reflection region and back, as a function of frequency. MARSIS measures electron densities in the range from 125 to 3.8 $\times 10^5$ el/cm^3. The sounder transmit in a broad angular interval (essentially 4π) centered on the nadir. Specular reflections occur from the Fresnel zone which has a dimension of 23 km at a frequency of 2 MHz and a spacecraft altitude of 900 km. Altitude variations in the reflections over the Fresnel zone would tend to widen the spectrogram trace. Occasionally the width of the echo trace corresponds to a range interval of ~30 km indicating simultaneous reflections can occur in neighboring range bins.

Provided observations are available all the way from the spacecraft to the reflection point, Equation (10) can be inverted to yield the electron density as a function of radar range (or of height above the planet surface). However, normally the closest echo originates some distance from the sounder. Since the radio waves are also influenced by the electron densities in the 'blind region' between the spacecraft and the first echo, one must in order to proceed assume how the density varies in that region. The local density at the spacecraft can often be derived from resonance lines in the spectrogram. They occur with a separation between neighboring lines at the plasma frequency of the local plasma. The density profile in the blind region is assumed to start with the local density at the spacecraft altitude and to be exponentially decreasing with altitude below the spacecraft. This assumption allows the data to be inverted to yield the real electron density profile using the lamination technique.

In this report MARSIS observations between 7 and 60 degree solar zenith angle are presented.

The ionosphere is typically characterized by a spectrogram with a single trace of reflections. For each frequency there is only a reflection for one delay time. This implies a horizontally stratified ionosphere with an electron density gradient pointing towards the sounder.

The maximum plasma frequency corresponds typically to a maximum density in the range 10^4 to 10^5 el/cm^3, values consistent with earlier results. The solar zenith angle dependence of the density maximum is as predicted by theory: the density decreases as the square root of the cosine to the zenith angle (Equation (3)) for angles <85 degree and as the square root of the inverse Chapman function for angles >85 degree.

The density maximum is varying periodically in time with the solar rotation period implying that solar radiation controls the maximum. X-rays are known to cause excess ionization in the Earth's ionosphere, giving rise to Sudden Cosmic Noise Absorption. X-ray bursts were found to coincide with a density increase also in the Martian ionosphere. Such an effect of solar X-rays on Mars has been suggested in the past (Nielsen, 1998).

The observations further confirm that the density profile from the maximum and up to an altitude of 160 to 180 km is well described as a Chapman layer. The equivalent subsolar altitudes of the electron density maximum is from 107 to 123 km. The neutral air scale height in the Chapman layer varied between 10 and 13 km. At higher altitudes the density decreased exponentially with a scale height in the range 24 to 65 km. These parameter values are in general agreement with earlier results. Note that the altitude of the density maximum is dependent on the assumed exponential decrease of the densities in the 'blind region'. Deviations from the assumption could raise or lower the altitude of the density maximum. The assumption has however little influence on the values of the other parameters characterizing the Chapman layer.

Note, that these results are also valid for zenith angles <48 degree. This is outside the zenith angle range accessible to previous measurements.

Acknowledgements

MARSIS was built and is jointly managed by the Italian Space Agency and NASA. Mars Express was built and is operated by the European Space Agency. The research at the University of Iowa was supported by NASA through contract 1224107 with the Jet Propulsion Laboratory.

References

Acuna, M. H., Connerney, J. E. P., Ness, N. F., Lin, R. P., Mitchell, D., *et al.*: 1999, *Science* **284**, 790.

Acuna, M. H., Connerney, J. E. P., Wasilevsky, P., Lin, R. P., Anderson, K. A., Carlson, C. W., *et al.*: 1998, *Science* **279**, 1676.

Bornmann, P. L., Speich, D., Hirman, J., Mathison, L., Grubb, R., Garcia, H., *et al.*: 1996, *Proc. SPIE* **2812**, 291.

Bougher, S. W., Engel, S., Hinson, D. P., and Forbes, J. M.: 2001, *Geophys. Res. Lett.* **28**, 3091.

Budden, K. G.: 1966, *Radio Waves in the Ionosphere*, The University Press, Cambridge.

Chapman, S.: 1931, *Proc. Phys. Soc.* **43**, 483.

Gurnett, D. A., Huff, R. L., Morgan, D. D., Persoon, A. M., Averkamp, T. F., Kirchner, D. L., *et al.*: 2006, COSPAR.

Gurnett, D. A., Kirchner, D. L., Huff, R. L., Morgan, D. D., Persoon, A. M., Averkamp, T. F., *et al.*: 2005, *Science* **310**, 1929, 10.1126/science.1121868.

Hagg, E. L.: 1967, *Can. J. Phys.* **45**, 27.

Hanson, W. B., Sanatani, S., and Zuccaro, D. R.: 1977, *J. Geophys. Res.* **82**, 4351.

Hinson, D. P., Simpson, R. A., Twicke, J. D., Tyler, G. L., and Flasar, F. M.: 1999, *J. Geophys. Res.* **104**, 26997.

Kenneth, D.: 1965, *National Bureau of Standards Monograph* **80**.

Kliore, A. J.: 1992, In *Venus and Mars: Atmospheres, Ionospheres, and Solar Wind Interactions*, Geophysical Monograph 66, American Geophysical Union.

Luhmann, J. G., Tatrallyay, M., and Pepin, R. O.: 1992, in Luhmann, J. G., Tatrallyay, M., and Pepin, R. O. (eds.), *Geophysical Monograph* 66, American Geophysical Union.

Mendillo, M., Smith, S., Wroten, J., Rishbeth, H., and Hinson, D.: 2003, *J. Geophys. Res.* **108**(A12), 1432, doi:10.1029/2003JA009961.

Mendillo, M., Withers, P., Hinson, D., Rishbeth, H., and Reinisch, B.: 2006, *Science* **311**, 1135.

Nagy, A. F., Winterhalter, D., Sauer, K., Cravens, T. E., Brecht, S., Mazelle, C., *et al.*: 2004, *Space Sci. Rev.* **111**, 33.

Nelms, G. L., Barrington, R. E., Belrose, J. S., Hartz, T. R., McDiarmid, I. B., and Brace, L. H.: 1966, *Can. J. Phys.* **44**, 1419.

Ness, N. F., Acuna, M. H., Connerney, J. E. OP., Kliore, A. J., Breus, T. K., Krymskii, A. M., *et al.*: 2000, *J. Geophys. Res.* **105**, 15991.

Nielsen, E., Morgan, D. D., Kirchner, D. L., Plaut, J., and Picardi, G.: 2006, *Plan. Space Sci.*, in press.

Nielsen, E.: 2004, *Space Sci. Rev.* **111**, 245.

Nielsen, E.: 1998, in Attema, E., Schwehm, G., and Wilson, A. (eds.), *Proc. 32nd ESLAB Symp. 'Remote Sensing Methodology for Earth Observation and Planetary Exploration' ESTEC,* Noordwijk, *The* Netherlands, ESA SP-423, ESA Publ. Div., Noordwijk, pp. 215–223.

Pätzold, M., Tellmann, S., Häusler, B., Hinson, D., Schaa, R., and Tyler, G. L.: 2005, *Science* **310**, 837.

Picardi, G., Sorge, S., Seu, R., Fedele, G., Federico, C., and Orosei, R.: 1999, Mars advanced radar for subsurface and ionosphere sounding (MARSIS). Info-Com. Dept., Technical report N. MRS-001/005/99, version 2.0.

Rishbeth, H., and Garriott, O. K.: 1969, *Introduction to Ionospheric Physics*. International Physics Series, Volume 14, Academic Press.

Shinagawa, H.: 2000, *Adv. Space Res.* **26**(10), 1599.

Shinagawa, H., and Cravens, T. E.: 1989, *J. Geophys. Res.* **94**(A6), 6506.

Wang, J.-S., and Nielsen, E.: 2004, *Ann. Geophysicae* **22**(1–5), SRef-ID: 1432-0576/ag/2004-22-1.

Warren, E. S.: 1963, *Nature* **197**, 636.

Withers, P., and Mendillo, M.: 2005, *Planet. Space Sci.* **53**, 1401.

Zhang, M. H. G., Luhmann, J. G., Kliore, A. J., and Kim, J.: 1990a, *J. Geophys. Res.* **95**(B9), 14829.

Zhang, M. H. G., Luhmann, J. G., and Kliore, A. J.: 1990b, *J. Geophys. Res.* **95**(A10), 17095.

Zou, H., and Nielsen, E.: 2004, Methods for obtaining electron density profiles from MARSIS ionograms and derivation of parameters characterizing the profiles. MPAe-W-485-04-01.

LOCATIONS OF ATMOSPHERIC PHOTOELECTRON ENERGY PEAKS WITHIN THE MARS ENVIRONMENT

R. A. FRAHM[1,*], J. R. SHARBER[1], J. D. WINNINGHAM[1], P. WURZ[2], M. W. LIEMOHN[3], E. KALLIO[4], M. YAMAUCHI[5], R. LUNDIN[5], S. BARABASH[5], A. J. COATES[6], D. R. LINDER[6], J. U. KOZYRA[3], M. HOLMSTRÖM[5], S. J. JEFFERS[1], H. ANDERSSON[5] and S. MCKENNA-LAWLER[7]

[1]*Southwest Research Institute, 6220 Culebra Road, San Antonio, TX 78228, USA*

[2]*University of Bern, Physikalisches Institut, Sidlerstrasse 5, CH-3012 Bern, Switzerland*

[3]*Space Physics Research Laboratory, University of Michigan, 2455 Hayward Street, Ann Arbor, MI 48105, USA*

[4]*Finnish Meteorological Institute, Box 503, FIN-00101 Helsinki, Finland*

[5]*Swedish Institute of Space Physics, Box 812, S-98 128, Kiruna, Sweden*

[6]*Mullard Space Science Laboratory, University College London, London RH5 6NT, UK*

[7]*Space Technology Ireland, National University of Ireland, Maynooth, Co. Kildare, Ireland*

*(*Author for correspondence: E-mail: frahm@swri.edu)*

(Received 2 February 2006; Accepted in final form 14 November 2006)

Abstract. By identifying peaks in the photoelectron spectrum produced by photoionization of CO_2 in the Martian atmosphere, we have conducted a pilot study to determine the locations of these photoelectrons in the space around Mars. The significant result of this study is that these photoelectrons populate a region around Mars bounded externally by the magnetic pileup boundary, and internally by the lowest altitude of our measurements (~250 km) on the dayside and by a cylinder of approximately the planetary radius on the nightside. It is particularly noteworthy that the photoelectrons on the nightside are observed from the terminator plane tailward to a distance of ~3 R_M, the Mars Express apoapsis. The presence of the atmospherically generated photoelectrons on the nightside of Mars may be explained by direct magnetic field line connection between the nightside observation locations and the Martian dayside ionosphere. Thus the characteristic photoelectron peaks may be used as tracers of magnetic field lines for the study of the magnetic field configuration and particle transport in the Martian environment.

Keywords: Mars, photoelectrons

Introduction

On June 3, 2003, the European Space Agency (ESA) launched the Mars Express (MEX) spacecraft. The spacecraft reached Mars and was injected into orbit on December 25, 2003. One experiment on the MEX spacecraft is the Analyzer of Space Plasmas and Energetic Atoms-3 (ASPERA-3) (Barabash *et al.*, 2004), which measures *in situ* ions, electrons, and energetic neutral atoms at Mars. Ambient electrons are measured *in situ* by the Electron Spectrometer (ELS) of ASPERA-3.

Space Science Reviews (2006) 126: 389–402
DOI: 10.1007/s11214-006-9119-5

Prior to the launch of Mars Express, it was known that the Mars ionospheric electron spectrum contained photoelectron peaks resulting from the photoionization of atmospheric gases. Based on neutral mass spectrometer data obtained from the Mars Viking lander and a solar spectrum (Hinteregger, 1976), Mantas and Hanson (1979) calculated the electron spectrum below 100 eV in the Mars atmosphere, with both horizontal and vertical magnetic fields, demonstrating the presence of spectral peaks in the range of 21–24 eV and at 27 eV. These results were corroborated later that same year by a model of Fox and Dalgarno (1979) without a magnetic field, and later updated by Fox (2004). These electron peaks result primarily from the ionization of CO_2 by solar He 30.4 nm photons produced for the most part in the photochemical equilibrium region and transported to higher altitudes. Thus the photoelectron peaks detected at a remote location may be considered as tracers of the magnetic field line to the production region (Frahm *et al.*, 2006). Evidence of these peaks was observed in the spectral measurements of the Electron Reflectometer (ER) on Mars Global Surveyor (MGS) (Mitchell *et al.*, 2001), but they were not resolved by that instrument. The peaks were first resolved by spectral measurement with the electron spectrometer on Mars Express (Lundin *et al.*, 2004; Frahm *et al.*, 2006), which began routine operations in its orbit around Mars in early 2004.

The paper reporting the photoelectron peaks in the Martian ionosphere (Frahm *et al.*, 2006) described the early results from ELS, demonstrating the presence of the photoelectron peaks and their relationship to the sheath of shocked solar wind plasma. In that paper it was pointed out that the CO_2 photoelectron peaks are routinely observed on the dayside of the planet at and above the MEX periapsis altitude of ~250 km. The peaks were also reported on the night side of the terminator plane, sometimes very distant from Mars, even near apoapsis, at more than 10,000 km altitude. (During 2004 the MEX apoapsis changed from an altitude of ~11,600 km to ~10,100 km and periapsis changed from an altitude of ~270 km to ~250 km in mid year to ~300 km at the end of the year.) In an effort to understand how these photoelectrons reach the large distances, Liemohn *et al.* (2006) used ELS photoelectron observations and simulations with a global MHD code (Ma *et al.*, 2002, 2004) to study the likely paths of the electrons detected by the spacecraft. Two of the three cases studied indicated that the photoelectrons were produced on open field lines connecting to the planet in the afternoon/evening sector, while the third suggested that the photoelectrons were observed on interplanetary field lines draping the planet.

Based on the notion that photoelectrons may be used as tracers of the magnetic field configuration near Mars, we expand on our previous work by making use of the large body of observations obtained from the MEX electron spectrometer to determine the location of the CO_2 photoelectrons in the space around the planet. We report here the results of this pilot study, which used electron data for most of the year 2004. In some ways the results are quite surprising, in that a large number of photoelectrons were found far down the Martian tail and generally outside the umbral region. These findings provide clues about the magnetic field configuration and the transport of photoelectrons in the Martian environment. They may also be compared with simulations using global scale MHD modeling (see Liemohn *et al.*, this issue).

Instrument

The ELS is a spherical top-hat analyzer with a field of view of $360° \times 4°$. The 360° measurement plane is divided into 16 sectors, each sector is 22.5° wide. The ELS analyzer constant (the average value is 7.23 ± 0.05 eV/volt) and energy resolution (average value of $\Delta E/E = 0.083 \pm 0.003$) are slightly sector dependent and were determined by laboratory measurements at 10 keV. Measurements of variations in the analyzer constant and energy resolution as functions of energy indicated an energy-dependent, relative efficiency factor for each sector. Sector comparisons using a stable nickel-63 nuclear source determined these energy-dependent variations. The energy independent physical geometric factor was determined to be 5.88×10^{-4} cm^2 sr from simulation.

The ELS covers the energy range from 0.4 eV to 20 keV with a dual range deflection power supply. The deflection voltage ranges from 0 to 20.99 V for the low range and 0 to 2800.0 V for the high range. The energies selected are sector dependent, but have maximum values of approximately 150 eV and 20 keV. Each supply has a control resolution of 4096 linear voltage values within its full range. Of the 8192 possible deflection voltage values, 128 are selected to comprise the ELS energy sweep. The values are sequenced from highest to lowest voltage in a time of 4 s. The last sweep step is a fly-back step and is ignored in the data analysis.

Most of the ELS data examined in this paper were acquired under normal energy resolution and high time resolution; however, some of the data were acquired in high energy resolution and high time resolution modes. In a few cases, data were used when ELS was in its low energy resolution and low time resolution modes. *High energy resolution* means that data are acquired with energy steps less than the energy bandwidth (ΔE) of the analyzer, and the spectral values of each sweep are telemetered; at *normal energy resolution*, data are acquired with energy steps equal to the energy bandwidth (ΔE), and spectral values of each sweep are telemetered. In *low energy resolution* mode, data are acquired at energy steps equal to the energy bandwidth (ΔE), but the spectral measurement is degraded by addition of spectral values over 2, 4, or 8 energy steps before being telemetered. In *high time resolution*, a complete ELS energy sweep occurs within 4 s and is fully telemetered. *Lower time resolution* modes are degraded by addition of full energy sweeps (2, 4, 8, or 16 sweeps) before being telemetered.

The ELS instrument is mounted on the ASPERA-3 scan platform, which makes possible full 3D electron distributions. However, this platform was not activated until January of 2006 and its use has been limited since. Only ELS data from 2004 have been used in the pilot study.

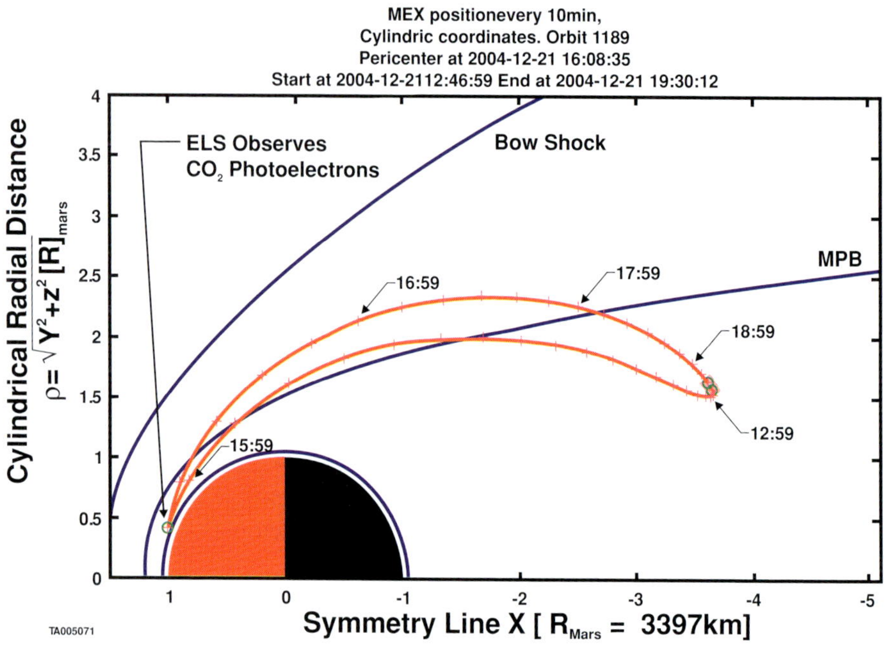

Figure 1. Mars Express orbit 1189 on December 21, 2004. As Mars Express traveled through the dayside ionosphere, photoelectron peaks were observed. Mars cylindrical coordinates with the Sun at the left, are used to express the spacecraft orbit as distance from the Mars-Sun line. The blue curves mark the average positions of the bow shock and MPB (Magnetic Pileup Boundary) as derived by Vignes *et al*. (2000).

Observations

A measurement of the photoelectrons in the Martian ionosphere on December 21, 2004 is illustrated in Figures 1, 2, and 3. Figure 1 describes the location of the MEX orbit in cylindrical coordinates, ρ and X of the Mars-centered Solar Orbital (MSO) system (unabberated). In the MSO system, X points toward the Sun, Z is perpendicular to the planet's velocity and is directed toward the northern ecliptic hemisphere, and Y completes a right handed, orthogonal system. The coordinate $\rho = \sqrt{Y^2 + Z^2}$. In Figure 1 the Sun is to the left, and the outer blue curves mark the average positions of the bow shock and magnetic pileup boundary (MPB) as determined by Vignes (2000) based on 290 orbits of MGS observations. The innermost blue circle marks 150 km altitude.

Measurements of ELS electrons in the minutes before and after periapsis on Day 356 of 2004 are shown in Figure 2. The plots are energy-time spectrograms of 17 minutes of data beginning at 1600:00 UT of Day 356 with periapsis at 1608:24 UT. Figure 2 shows measurements from four directions looking from the spacecraft: toward Mars, away from Mars, toward the East, and toward the West. Two enhancements of electron energy flux occurring at energies near 20 eV are observed

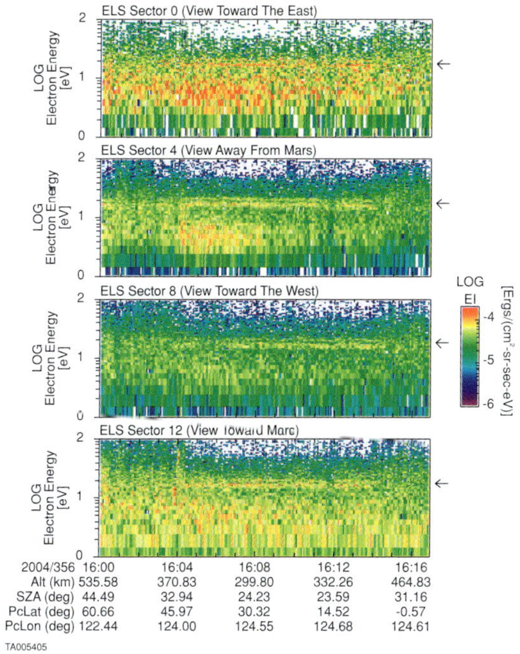

Figure 2. Energy-time spectrogram in the Martian ionosphere. Values of differential energy intensity are color coded. Spectrograms are marked in terms of view from the spacecraft and show photoelectron peaks measured in selected directions. Spacecraft location is given in terms of planetodetic altitude (Alt), solar zenith angle at the spacecraft (SZA), planetocentric latitude (PcLat), and planetocentric longitude (PCLon), which is measured toward the east. An arrow at the right of each spectrogram marks the approximate energy of the photoelectron peaks.

in each spectrogram between about 1604 and 1614 UT (their approximate energy is marked at the right of each panel with an arrow). They are not quite so easy to see in the first plot (toward East) because of higher fluxes below 20 eV. These enhancements are the major peaks in the photoelectron spectrum resulting from ionization of CO_2 in the Mars atmosphere. They are a unique signature of photoionization. On this pass they are measured from an altitude of 370 km on the inbound part of the orbit, through periapsis (300 km altitude), and out to 412 km on the outbound portion.

Further examination of the photoelectron spectra near periapsis is presented in Figure 3, which shows photoelectrons directed (a) away from Mars and (b) toward Mars. The spectra are plotted as energy intensity in units of ergs/(cm^2 s sr eV) and were obtained by averaging over a two-minute interval centered near periapsis. Also shown are the measurement uncertainties at the $+$ and -1σ level, which include Poisson counting statistics, errors resulting from the telemetry compression scheme, and instrumental uncertainties. The spectra show the signature of the photoelectron peaks at about 21 eV and about 17 eV. The theoretical locations of these peaks are 27 eV and 21–24 eV, which indicates that during this time, the spacecraft was charged to about −6 volts. It is important to note that these photoelectron features in the electron energy spectrum, although shifted in energy, are easily recognizable, because the electrons have been produced uniquely and in sufficient numbers for easy identification. Our ability to distinguish these atmospherically produced photoelectrons from those produced by the spacecraft is based on this identification of the peaks characteristic of the Mars atmospheric environment, as discussed earlier. These usually stand out easily even when spacecraft-produced photoelectrons and secondary electrons are present. Further information may be found in Frahm *et al.* (2006).

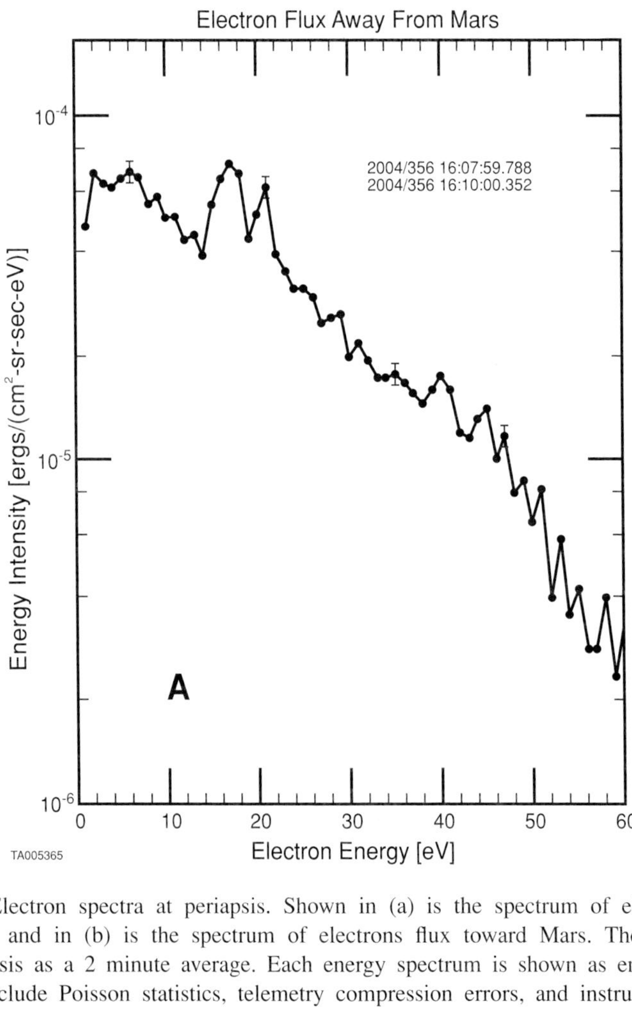

Figure 3. Electron spectra at periapsis. Shown in (a) is the spectrum of electrons flux away from Mars and in (b) is the spectrum of electrons flux toward Mars. These data are taken near periapsis as a 2 minute average. Each energy spectrum is shown as energy intensity. Error bars include Poisson statistics, telemetry compression errors, and instrument uncertainties. (*Continued on next page*)

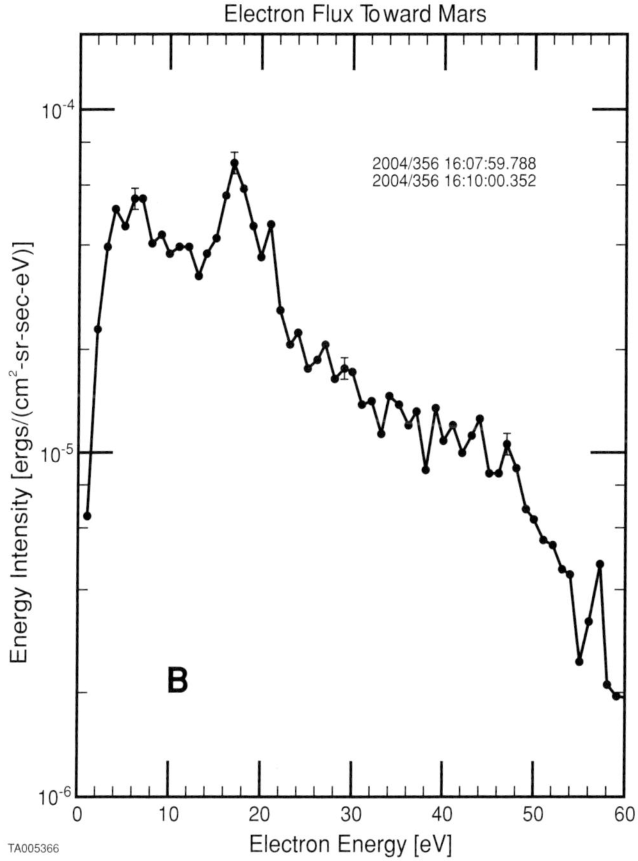

Figure 3. (*Continued*)

In the ELS data we see photoelectron peaks at essentially all altitudes sampled by the Mars Express orbit. Figure 4 shows eleven spectra measured at various altitudes over this range. The energy range of the photoelectron peaks in each spectrum is highlighted. There is some variation in the magnitudes of the signature peaks from one spectrum to another, but all are comparable. This has bearing on the source region and will be discussed later. The spectrum measured at the lowest altitude was obtained in the ionosphere near periapsis (260 km) in the sunlit atmosphere. The spectrum measured at the highest altitude was obtained near apoapsis where the spacecraft altitude was 10,100 km. In Table I we provide the date and time of each spectral measurement shown in Figure 4, along with altitude, planetodetic latitude and longitude, solar zenith angle, and local solar time of each observation. The ranges of each entry correspond to the beginning and end time of each spectral measurement. The table entries are arranged in order of increasing altitude.

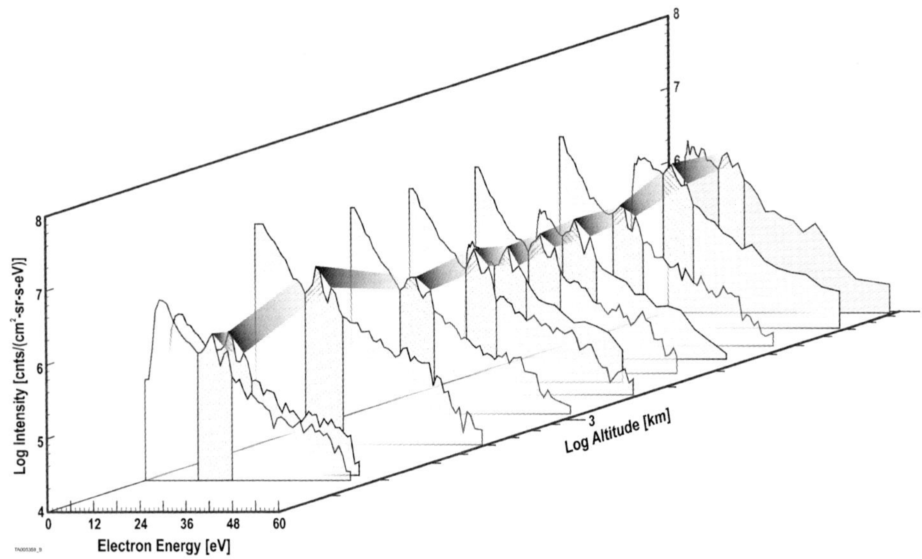

Figure 4. Electron spectra obtained at periapsis and apoapsis and selected altitudes in between, illustrating that the atmospheric photoelectron peaks can be found over the entire altitude range of the spacecraft. Energy bands containing the atmospheric photoelectron peaks are highlighted.

TABLE I

Orbit-altitude data for photoelectron spectra of Figure 4.

Year	Day	Universal Time (hh:mm:ss)	Altitude (km)	PD Latitude (deg)	PD Longitude (deg)	Solar Zenith Angle (deg)	Solar Time (hr)
2004	176	22:07:29–22:09:44	267.6–295.7	−86.56–−81.10	296.45–10.46	112.56–107.26	2.15–3.76
2004	156	04:46:27–04:47:35	299.1–327.9	−67.57–−63.03	98.23–99.51	92.21–88.68	5.75–6.09
2004	107	10:45:56–10:50:34	628.9–385.5	−85.69–−75.49	168.07–281.66	102.1–86.02	4.93–7.20
2004	187	19:56:38–19:57:46	1206.9–1109.2	−14.81–−17.86	345.57–345.49	123.37–124.18	20.33–20.42
2004	187	19:51:00–19:52:09	1696.0–1604.4	−02.05–−04.18	346.13–346.02	118.48–119.55	19.88–19.94
2004	178	00:30:01–00:33:24	2222.2–1908.7	+2.48–−3.58	187.54–187.09	121.42–124.12	20.11–20.28
2004	178	07:06:33–07:07:38	2863.7–2769.9	+13.15–+11.75	90.26–90.10	115.8–116.58	19.74–19.79
2004	172	03:04:59–03:07:14	3698.8–3488.6	+20.35–+17.77	94.49–94.10	114.71–116.24	19.18–19.79
2004	166	19:08:57–19:10:06	4832.6–4738.0	+29.26–+28.32	164.69–164.48	111.49–112.19	19.56–19.62
2004	115	07:00:01–07:03:04	7107.4–6900.8	+13.20–+11.68	242.65–242.02	147.11–148.08	21.91–21.95
2005	009	13:43:26–13:44:32	10113.3–10113.4	−40.34–−39.99	150.51–150.26	153.05–153.33	23.98–23.99

To determine the occurrence and locations of the CO_2 photoelectrons in an aggregate sense in the space around Mars, we conducted a pilot statistical study using ELS data taken between January 5 and November 13 of 2004. For simplicity only one ELS sector (sector 3) was used, and consequently, for this study, the statistics do not register the flow direction at each location. Statistics were built up by accumulating the occurrence and location in MSO coordinates of every spectrum containing the CO_2 photoelectron peaks. For two-dimensional presentations, the data are then binned into 100×100 km bins in that system.

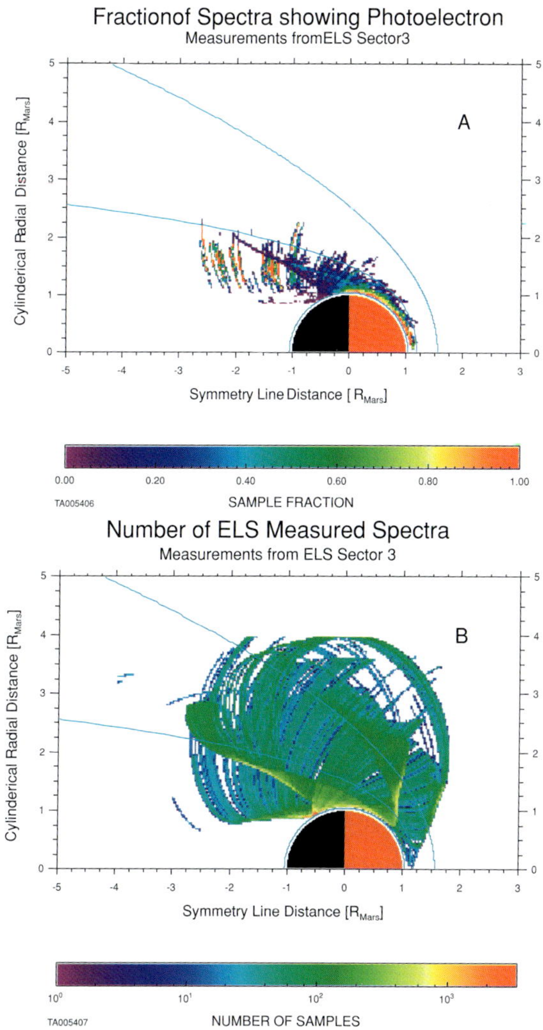

Figure 5. Fractional occurrence of atmospheric photoelectron peaks. The region near Mars is examined to show (a) the fraction of occurrence of atmospheric photoelectrons peaks and (b) number of samples made by the ELS. In both cases, the sun is at the right, and the average positions of the Martian bow shock and MPB, as determined by Vignes *et al.* (2000), are shown. Note that a zero-sample fraction is shown as the background color (see text).

The results of the study are presented in Figure 5. In Figure 5(a) each pixel of data represents a fractional occurrence of the CO_2 photoelectron peaks, color-coded by the scale below the plot. Figure 5(b) shows the regions sampled by Mars Express at times when ELS was on. Also shown in Figure 5(a) are the average positions of the bow shock and MPB as determined by Vignes *et al.* (2000) using MGS magnetometer and electron observations. It is clear from Figure 5a that most of the photoelectron signatures are found within the average MPB. In other words, the atmospherically produced photoelectrons, whether they were detected near or far from the planet, were inside the magnetic pileup region. We have checked the points outside the MPB curve of Figure 5a (an average); and, using ELS data, all of these lie below the altitude range over which the shocked solar wind electrons are absorbed. This region defines the MPB (Crider *et al.*, 2000; Vignes *et al.*, 2000); thus every point in the figure definitely lies within (below) the instantaneous magnetic pileup boundary. We note here that in the presentation of Figure 5(a), there is no distinction made between a "zero" that results because the region was not sampled and a "zero" that results when the region is sampled and no photoelectron peaks are observed. These may be distinguished with the help of Figure 5(b), which shows the regions sampled.

In Figure 5 one also observes that the frequency of occurrence of photoelectron peaks is largest on the dayside of the planet at low altitudes in the ionosphere. This is consistent with their copious production in the photochemical equilibrium layer below. We also note that the atmospheric photoelectron peaks were not seen in the solar wind; i.e., they are not seen anywhere near or outside of the statistical bow shock in spite of good sampling of the interplanetary medium.

Figures 6 and 7 show the same data (shown in Figure 5) in different perspectives. Figure 6 shows the data from the dayside of Mars ($X > 0$) projected onto the YZ-plane, while Figure 7 shows data from the night side ($X < 0$) projected onto the same plane. These figures show where the ELS measurements were obtained and also where the sampling was incomplete (shown in the (b) part of each figure). The dayside data (Figure 6) indicates good coverage at locations away from the planet,

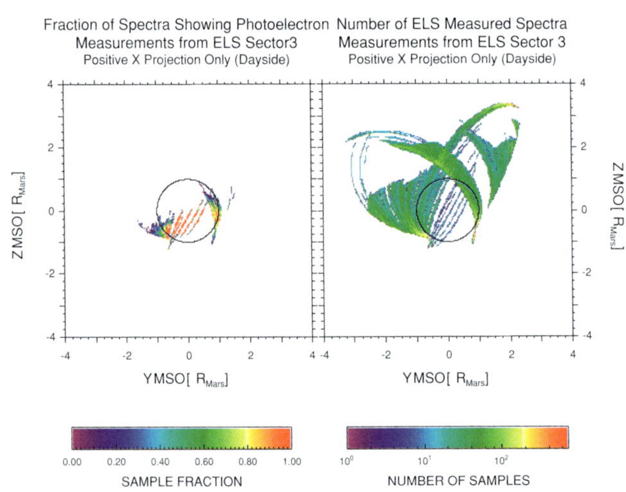

Figure 6. Dayside ELS measurements. The data of Figure 5 for $X > 0$ are projected on the YZ-plane. The fraction of times photoelectrons measured are shown in (a), while the total number of measurement points is shown in (b). Note that a zero-sample fraction is shown as the background color (see text).

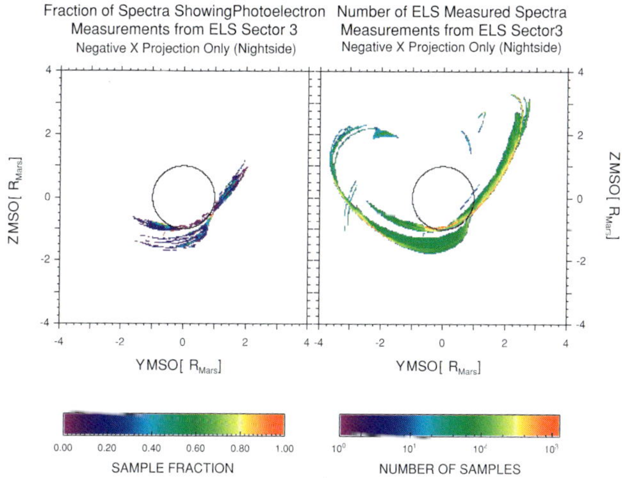

Figure 7. Nightside ELS measurements. The data of Figure 5 for $X < 0$ are projected on the YZ-plane. The fraction of times photoelectrons are measured are shown in (a) while the total number of measurement points is shown in (b). Note that a zero-sample fraction is shown as the background color (see text).

but the small number of measurements occurs at altitudes too high for detection of the photoelectron peaks. In Figure 7, the gap just below the outline of the planet in Figure 7(a) results from the similar sampling gap in Figure 7(b) indicating that there is no coverage at that location.

Usually when MEX is far from Mars, the measurement plane of ELS is parallel to the Mars orbital plane. As the spacecraft nears the planet, MEX is generally commanded into one of two observational modes. The first keeps the ELS measurement plane parallel to the Mars orbital plane (one sensor will look along the Mars-Sun line), and the second tilts the spacecraft such that the planet radial vector is parallel to the ELS measurement plane. Since this pilot study uses only one ELS sector (sector 3), it is possible that the difference in aspect resulting from different spacecraft observational modes might affect our statistics. It has been demonstrated that the measurement of flows can be very aspect sensitive at large distances from the planet (Frahm *et al.*, 2006; Liemohn *et al.*, 2006). The slightly reduced occurrence percentages in the region near $-0.75\ R_M < X < -0.25\ R_M$ of Figure 5 could be an indication of this, as this is the region where MEX often maneuvers into the ELS radial measurement configuration. At this point this effect is a possibility and will be investigated further in future work.

Discussion

The significant result of this study is that atmospherically produced photoelectrons populate a region around Mars bounded externally by the magnetic pileup boundary, and internally by the lowest altitude of our measurements (~250 km) on the dayside and by the a cylinder defined approximately by the edge of the planet's shadow on the nightside. On the nightside, the photoelectrons were observed as far from the terminator plane as ~3 R_M. The dayside results may be understood in terms of ionization of atmospheric carbon dioxide (with some ionization of atomic oxygen) by solar extreme ultraviolet radiation (Frahm *et al.*, 2006, and references therein). However, it is the observation of the CO_2 photoelectrons at large distances down the flanks of the magnetosheath that is the unexpected result of this study. How then might atmospherically generated photoelectrons be found at such distances?

The fact that the CO_2 peaks measured in the distant locations have magnitudes comparable to those measured on the dayside argues strongly for a direct connection, most likely along magnetic field lines, to the production region – in this situation the dayside ionosphere. Both observations and simulation studies suggest such a connection.

The solar wind interaction with Mars results in a "draping" magnetic field morphology (Nagy *et al.*, 2004). Measurements using the Magnetometer/Electron Reflectometer on Mars Global Surveyor (Bertucci *et al.*, 2005) have shown the draping effect to be easily measured within the magnetic pileup *region* (MPR), the region immediately interior to the MPB. Typically, as MGS moves from the shocked solar wind (magnetosheath), through the MPB and into the MPR, the magnetic field continues to pile up (strengthens), became more regular (contains progressively fewer fluctuations), and exhibits the draping quality. Draping is not detected within the shocked solar wind. The lower boundary of the MPR is within the ionosphere and may be as low as the exobase (Mitchell *et al.*, 2001; Nagy *et al.*, 2004). Based on our own observations (e.g., Figure 2), this is the region we know to be a source region for the CO_2 photoelectrons on the dayside of Mars. Thus the MGS observations place the dayside photoelectron source on draped field lines, some of which have a high likelihood of connecting with down-tail regions shown by this study to be populated by the photoelectrons.

Simulations have provided another means to study the particle and field environment. For example, Ma *et al.* (2004), using an MHD model run for maximum solar wind conditions and the presence of crustal fields, show a magnetic pileup region, draping of the magnetic field, and minimagnetospheres resulting from the presence of relatively strong crustal magnetic field sources (Harnett and Winglee, 2003). In the Ma *et al.* (2004) simulations, some field lines emanating from crustal source regions connect the dayside ionosphere to nightside locations.

In a study reported in this issue, Liemohn *et al.* (this issue) have used the Ma *et al.* (2004) code to look specifically at the question of magnetic field line connectivity between the dayside ionosphere and regions antisunward of the terminator plane. In the simulation, magnetic field lines were extracted from an array of starting points in the terminator plane connecting the dayside and nightside regions. As a part of the study, statistics were recorded on the fraction of field lines showing magnetic connectivity to the dayside ionosphere. This enabled a very direct comparison with the results of our study. The Liemohn *et al.* (this issue) results are strikingly similar to the ELS statistics shown in our Figure 5, making clear the connection between the dayside ionosphere and locations antisunward of the terminator plane, including the photoelectrons observed several R_M tailward of that plane. Liemohn *et al.* conclude that "the high-altitude photoelectrons are the result of direct magnetic connectivity to the dayside at the moment of the measurement, and no extra trapping or bouncing mechanisms are needed to explain the data." We completely agree with that conclusion.

Since atmospheric photoelectrons, originating from the dayside, are observed in the tail of Mars, by charge conservation there must also be low energy planetary ions at those locations. A study of the relation between low-energy planetary ions and photoelectrons will be a natural follow-on to this study. Low-energy planetary ions are confirmed to be observed with the photoelectrons at more than 10,000 km (the pass of the last entry in Table I).

Conclusions

By identifying peaks in the photoelectron spectrum produced by photoionization of CO_2, we have conducted a pilot study to determine the locations of these photoelectrons in the space around Mars.

The significant result of this study is that these photoelectrons populate a region around Mars bounded externally by the magnetic pileup boundary, and internally by the lowest altitude of our measurements ($\sim$250 km) on the dayside and by a cylinder of approximately the planetary radius on the nightside. It is particularly noteworthy that the photoelectrons on the nightside are observed from the terminator plane tailward out to a distance of $\sim$3 R_M, the MEX apoapsis.

The presence of the atmospherically generated photoelectrons on the nightside of Mars may be explained by direct magnetic field line connection between the nightside observation locations and the Martian dayside ionosphere. Thus the characteristic photoelectron peaks may be used as tracers of magnetic field lines for the study of the magnetic field configuration and particle transport in the Martian environment.

Acknowledgements

The ASPERA-3 experiment on the European Space Agency (ESA) Mars Express mission is a joint effort between 15 laboratories in 10 countries, all sponsored by their national agencies. We thank all these agencies as well as the various departments/institutes hosting these efforts. We wish to acknowledge support through the National Aeronautics and Space Administration (NASA) contract NASW-00003 in the United States, Particle Physics and Astronomy Research Council (PPARC) in the United Kingdom, and wish to thank those NASA officials who had the foresight to allow augmentation of the original ASPERA-3 proposal for ELS so that it would provide the additional capabilities which allowed the science described in this paper to be conducted. We also wish to acknowledge the Swedish National Space Board for their support of the main PI-institute and we are indebted to ESA for their courage in embarking on the Mars Express program, the first ESA mission to the red planet.

References

Barabash, S., *et al.*: 2004, in Wilson, A., and Chicarro, A. (eds.), *Mars Express: The Scientific Payload*, European Space Agency Special Report SP-1240, European Space Agency Research and Scientific Support, European Space Research and Technology Centre, Noordwijk, The Netherlands, p. 121.

Bertucci, C., Mazelle, C., and Acuña, M. H.: 2005, *J. Atmos. Sol. Terr. Phys.* **67**, 1797.

Crider, D., Cloutier, P., Law, C., Walker, P., Chen, Y., Acuña, M., *et al.*: 2000, *Geophys. Res. Lett.* **27**(1), 45.

Fox, J. L.: 2004, *J. Geophys. Res.* **109**, A11310, doi:10.1029/2004JA010380.

Fox, J. L., and Dalgarno, A.: 1979, *J. Geophys. Res.* **84**, 7315.

Frahm R., *et al.*: 2006, *Icarus* **182**, 371.

Harnett, E. M., and Winglee, R. M.: 2003, *Geophys. Res. Lett.* **30**(20), 2074, doi:10.1029/2003 GL017852.

Hinterreger, H. E.: 1976, *J. Atmos. Terr. Phys.* **38**, 791.

Liemohn, M. W., *et al.*: 2006, *Icarus* **182**, 383.

Liemohn, M. W., Ma, Y., Frahm, R. A., Fang, X., Kozyra, J. U., Nagy, A. F., *et al.*: *Space Sci. Rev.*, this issue, doi: 10.1007/s11214-006-9116-8.

Lundin, R., *et al.*: 2004, *Science* **305**, 1933.

Ma, Y., Nagy, A. F., Hansen, K. C., DeZeeuw, D. L., Gombosi, T. I., and Powell, K. G.: 2002, *J. Geophys. Res.* **107**(A10), 1282, doi:10.1029/2002JA009293.

Ma, Y., Nagy, A. F., Sokolov, I. V., and Hansen, K. C.: 2004, *J. Geophys. Res.* **109**, A07211, doi:10.1029/2003JA010367.

Mantas, G. P., and Hanson, W. B.: 1979, *J. Geophys. Res.* **84**, 369.

Mitchell, D. L., Lin, R. P., Mazelle, C., Réme, H., Cloutier, P. A., Connerney, J. E. P., *et al.*: 2001, *J. Geophys. Res.* **106**, 23419.

Nagy, A. F., *et al.*: 2004, *Space Sci. Rev.* **111**, 33.

Vignes, D., Mazelle, C., Réme, H., Acuña, M. H., Connerney, J. E. P., Lin, R. P., *et al.*: 2000, *Geophys. Res. Lett.* **27**, 49.

X-RAYS FROM MARS

KONRAD DENNERL
Max-Planck-Institut für extraterrestrische Physik, Giessenbachstraße, 85748 Garching, Germany
(E-mail: kod@mpe.mpg.de)

(Received 23 February 2006; Accepted in final form 16 August 2006)

Abstract. X-rays from Mars were first detected in July 2001 with the satellite Chandra. The main source of this radiation was fluorescent scattering of solar X-rays in its upper atmosphere. In addition, the presence of an extended X-ray halo was indicated, probably resulting from charge exchange interactions between highly charged heavy ions in the solar wind and neutrals in the Martian exosphere. The statistical significance of the X-ray halo, however, was very low. In November 2003, Mars was observed again in X-rays, this time with the satellite XMM-Newton. This observation, characterized by a considerably higher sensitivity, confirmed the presence of the X-ray halo and proved that charge exchange is indeed the origin of the emission. This was the first definite detection of charge exchange induced X-ray emission from the exosphere of another planet. Previously, this kind of emission had been detected from comets (which are largely exospheres) and from the terrestrial exosphere. Because charge exchange interactions between atmospheric constituents and solar wind ions are considered as an important nonthermal escape mechanism, probably responsible for a significant loss of the Martian atmosphere, X-ray observations may lead to a better understanding of the present state of the Martian atmosphere and its evolution. X-ray images of the Martian exosphere in specific emission lines exhibited a highly anisotropic morphology, varying with individual ions and ionization states. With its capability to trace the X-ray emission out to at least 8 Mars radii, XMM-Newton can explore exospheric regions far beyond those that have been observationally explored to date. Thus, X-ray observations provide a novel method for studying processes in the Martian exosphere on a global scale.

Keywords: Mars, X-rays, solar wind, charge exchange, X-ray scattering

1. Introduction

Since July 2001 we have known that Mars is an X-ray source. Mars is the fourth planet found to emit X-rays, after the Earth (e.g. Winckler *et al.*, 1958; Grader *et al.*, 1968), Jupiter (Metzger *et al.*, 1983) and Venus (Dennerl *et al.*, 2002). The first X-ray observations of Mars were made with the ROSAT satellite (Trümper, 1983) on three occasions during April 10–13, 1993, yielding a total exposure of 75 minutes. However, during two observations, Mars was unfavourably placed in the field of view of the ROSAT Position Sensitive Proportional Counter (Pfeffermann *et al.*, 1986), so that only 35 minutes of exposure were left to search for any X-ray emission, and no X-ray signal was detected (Dennerl, 2002).

The motivation for the ROSAT observation was to reveal the relative significance of Thomson scattering, fluorescent scattering and airglow in Mars'

Space Science Reviews (2006) 126: 403–433
DOI: 10.1007/s11214-006-9028-7

predominantly CO_2 atmosphere. In 1996, the discovery of comets as a new, unexpected class of bright X-ray sources (Lisse *et al.*, 1996; Dennerl *et al.*, 1997; Mumma *et al.*, 1997) led to an increased interest in X-ray studies of solar system objects. It revealed the importance of another process for the generation of soft X-rays which was overlooked for a long time: charge exchange between highly charged heavy ions in the solar wind and neutral gas in the solar system (Cravens, 1997).

Mars has a tenuous atmosphere, a weak magnetic field, and orbits at a heliocentric distance where comets exhibit significant activity. Mars can thus be considered as a planetary analog to a comet, and it is reasonable to expect that the same solar wind charge exchange processes which cause comets to emit X-rays also operate in the Martian exosphere. This possibility was investigated by Cravens (2000) and Krasnopolsky (2000). Holmström *et al.* (2001) made predictions about the X-ray morphology based on detailed computer simulations. The other likely source of X-ray radiation from Mars was scattering of solar X-rays, which was studied by Cravens and Maurellis (2001). All these model investigations were prior to the detection of X-rays from Mars.

2. First Detection of X-rays from Mars

On 4 July 2001, X-rays from Mars were detected for the first time (Dennerl, 2002). The observation was performed with the ACIS-I detector on the Chandra X-ray satellite. With its unprecedentedly high spatial resolution, Chandra was excellently suited for an observation of Mars. Furthermore, in contrast to other X-ray satellites which can observe only at directions which are approximately perpendicular to the direction of the Sun, Chandra was the first imaging X-ray satellite which was able to observe Mars around opposition, when it is closest to Earth.

Although the closest approach of Mars to Earth, with a minimum distance of 0.45 AU, had occurred already on 22 June 2001, the Chandra observation was postponed by two weeks. This decision was motivated by the fact that Mars was expected to be an X-ray source mainly due to fluorescent scattering of solar X-rays, and computer simulations of this process, already successfully tested on Venus (Dennerl *et al.*, 2002), had indicated that observing at a non-zero phase angle would result in a diagnostically more valuable image than observing at opposition: while a practically uniform X-ray brightness across the whole planet was expected for a phase angle close to zero, a phase angle of $\sim 15^\circ$ should already result in a characteristic X-ray brightening on the sunward limb. The decision to postpone the Chandra observation was also supported by the favorable fact that Mars was still approaching the perihelion of its orbit, so that its distance from the Earth would increase only slightly. Furthermore, the small loss of X-ray photons due to the reduced solid angle would be almost compensated by the fact that Mars would then be closer to the Sun and would intercept more solar radiation.

At the time of the observation, Mars was at a heliocentric distance of 1.45 AU and at a geocentric distance of 0.46 AU. Its diameter was 20.3″, and it was observed at a phase angle of 18.2° and at a solar elongation of 153.7°. The Chandra observation lasted for 9 hours and 13 minutes. During this time, the satellite did not track Mars, but kept its orientation fixed with respect to the sky. As seen from the orbit of Chandra, Mars was moving along a curved path across the sky, and this path was directly visible in the X-ray image (Figure 1). The photons were detected with X-ray CCDs which were read out every 3.2 s, so that a post-facto transformation of the individual photons into the rest frame of Mars was possible, resulting in the first X-ray image of this planet (Figure 2).

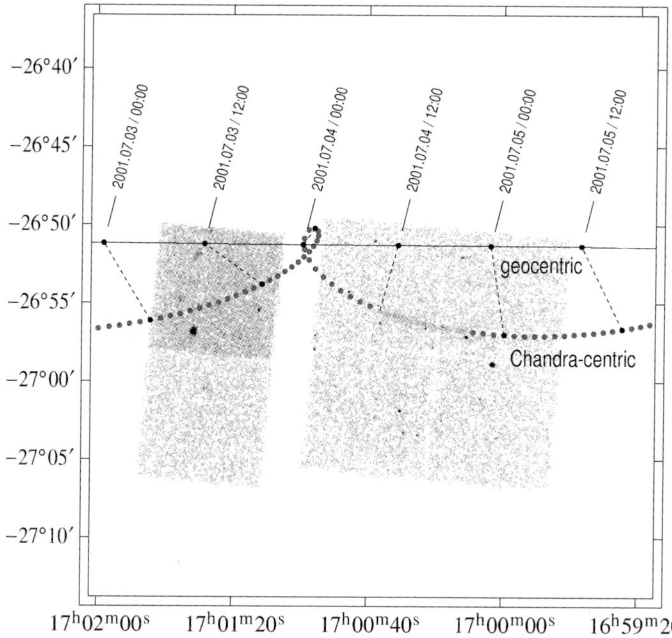

Figure 1. Apparent motion of Mars in celestial coordinates (J 2000), as seen from the Earth and Chandra. The solid line shows the geocentric path of Mars, with its position indicated every 12 hours. Dashed lines illustrate the parallactic shift at these times due to the Chandra orbit. Chandra-centric positions are shown every hour by dots, which match the apparent size of Mars. The loop at 4 July, 0 UT reflects the perigee passage of Chandra. Superimposed on this diagram is the Chandra image in celestial coordinates, obtained at $E < 1.0$ keV. The sequence of dots in the Chandra-centric curve is interrupted to show the curved soft X-ray trail of Mars on the X-ray CCDs. In addition to Mars, several point sources show up in the Chandra image. The photons from these sources were excluded in the subsequent analysis (from Dennerl, 2002).

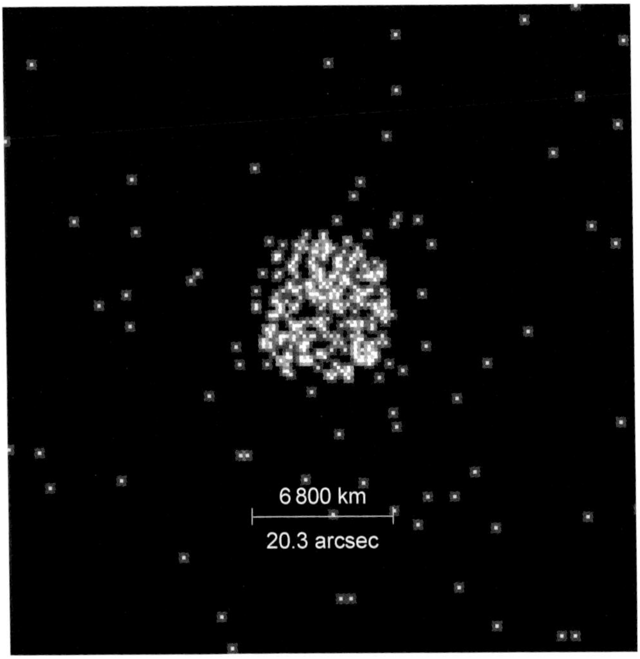

Figure 2. First X-ray image of Mars, obtained with Chandra ACIS-I on 4 July 2001. Only photons in the instrumental energy range $E = 0.40–0.73$ keV were selected and transformed into the rest frame of Mars. Trails of point sources were removed (from Dennerl, 2002).

3. Scattering of Solar X-rays

Mars appeared as an almost fully illuminated disk, with an indication of the phase effect predicted by the earlier computer simulations of the scattering of solar X-rays in its atmosphere. These simulations concentrated on fluorescent scattering, because this process was expected to be the dominant one. Cravens and Maurellis (2001) found that the X-ray intensity due to fluorescence of O and N alone exceeds that of elastic scattering in the broad spectral band 2–120 Å (0.1–6.2 keV) by a factor of 2.4.

In order to make a direct comparison of the observed X-ray image with that expected for fluorescent scattering of solar X-rays, these simulations were specifically tuned to the conditions of the Chandra observation (Dennerl, 2002). The ingredients to the simulation were the composition and density structure of the Martian atmosphere, the photoabsorption cross sections and fluorescence efficiencies of the major atmospheric constituents, and the incident solar spectrum.

Our knowledge about the Martian atmosphere is mainly based on spacecraft observations, in particular the Viking 1 and 2 missions (e.g. Nier and McElroy, 1977). More recently, the Mars Global Surveyor and Mars Odyssey accelerometer

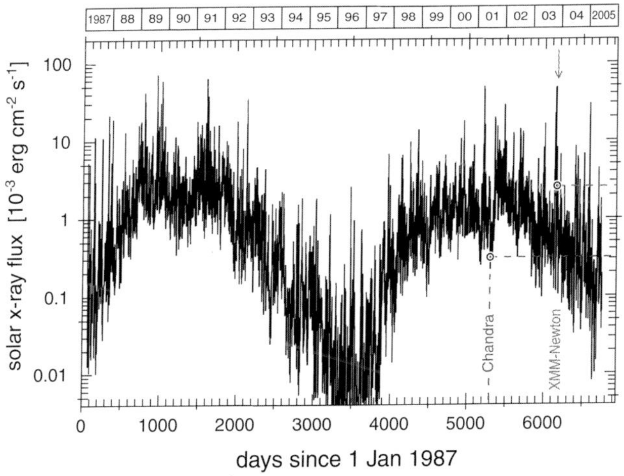

Figure 3. 1–8 Å (1.55–12.4 keV) solar X-ray flux at 1.0 AU, measured with the GOES satellites. The times of the Chandra and XMM-Newton observations are marked, together with the corresponding mean solar X-ray flux, which was highly variable during the XMM-Newton observation (from Dennerl *et al.*, 2006a).

measurements, obtained during aerobraking, revealed the large-scale and small-scale structure of the thermosphere in unprecedented detail (Withers, 2006). With this information, sophisticated models are being developed, like the Mars Global Reference Atmospheric Model (Mars-GRAM), an engineering-level Mars atmosphere model which is widely used for many Mars mission applications (e.g. Justus *et al.*, 2005).

For the purpose of modeling the scattering of solar X-rays in the Martian atmosphere, a simplified model was adopted, which describes the total density in the form of analytical expressions for different phases of the solar cycle (Sehnal, 1990a,b). This model is based on the COSPAR Reference Atmosphere of Mars together with Viking 1 and 2 measurements and theoretical considerations. Motivated by the general behaviour of the soft solar X-ray flux (Figure 3), solar maximum conditions were selected (Figure 4[1]). For simplicity it was assumed that the Martian atmosphere is composed of C, N, and O only, neglecting the ~1.6% contribution of other elements, mainly Ar, and the following composition was used: 64.9% oxygen, 32.4% carbon, and 2.7% nitrogen. As the main constituents, C and O, are contained in CO_2, this composition was assumed to be homogeneous throughout the atmosphere. Viking 1 and 2 measurements showed that the Martian atmosphere is mixed to heights in excess of 120 km (Nier and McElroy, 1977). Above the homopause,

[1]In Dennerl (2002), a definition was adopted where the exosphere starts at 100 km, while 180–250 km is a more realistic value; for practically all planetary exospheres, the exobase occurs near a density of $\sim 10^8$ cm^{-3}. The 100–180 km region is generally defined as the thermosphere. The labels in Figures 4, 6, and 7 have been changed accordingly.

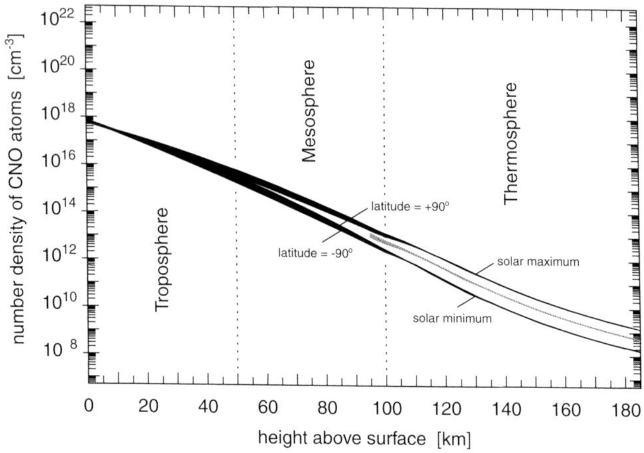

Figure 4. Number density $n_{CNO} = n_C + n_N + n_O$ of the sum of C, N, and O atoms in the Mars model atmosphere (Sehnal, 1990a,b) as a function of the height above the surface, for solar minimum, solar maximum, and the intermediate state. Below 100 km, the density depends also on the latitude (adapted from Dennerl, 2002).

at ~110 km, different species start separating out according to their mass: first O, then H_2 and H take over as the most abundant species.

From the photoabsorption cross sections (e.g. Reilman and Manson, 1979) and the C, N, and O contributions listed above, the effective cross section of the Martian atmosphere was computed (Figure 5a). This, together with the atmospheric density structure, yielded its optical depth, as seen from outside (Figure 6). It turned out that, at solar maximum, the Martian atmosphere becomes optically thick to photoabsorption of incident solar X-rays with $E = 0.1$–1.0 keV between 113 km and 100 km. The solar spectrum for the time of the Chandra observation was derived from SOLAR 2000 (Tobiska *et al.*, 2000). To improve the coverage towards energies above 100 eV, synthetic spectra were computed with the model of Mewe *et al.* (1985) and aligned with the SOLAR 2000 spectrum by adjusting the temperature and intensity. The resulting spectrum, scaled to the heliocentric distance of Mars, is shown in Figure 5b (upper curve).

For the simulation, the irradiated part of the Martian atmosphere was sampled with a grid of cubic volume elements with a side length of 1 km. Following the direction of the incoming solar X-rays, the absorbed radiation was then computed for each volume element. Figure 5b (lower curve) illustrates how the incident solar spectrum is modified by atmospheric photoabsorption. Only a small part of the absorbed energy, however, is emitted in the form of fluorescent photons, because of the small fluorescent yields for C, N, O (0.25%, 0.55%, and 0.85%, respectively; Krause, 1979). Figure 7 shows the resulting volume emissivities of fluorescence photons for the subsolar atmospheric column (zenith angle 0°) and for a column at the terminator (zenith angle 90°). The height of maximum emissivity rises with

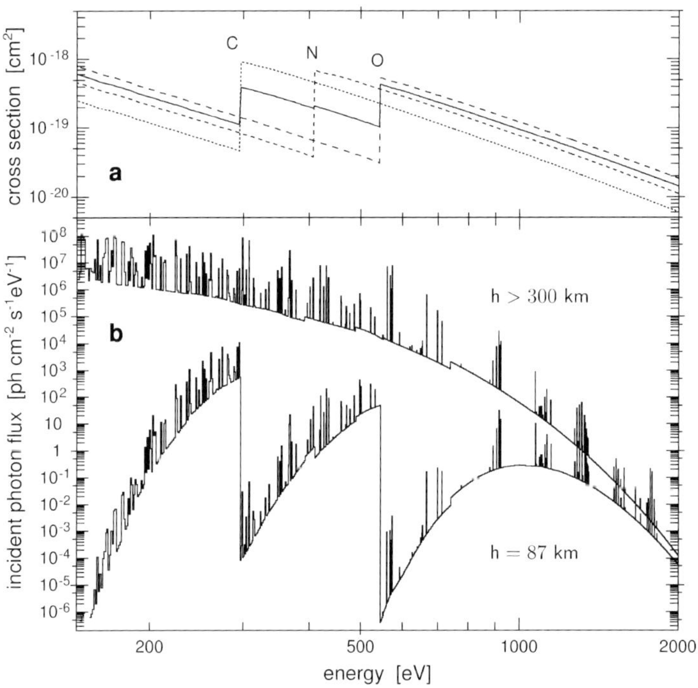

Figure 5. (a) Photoabsorption cross sections σ_C, σ_N, σ_O for C, N, and O (dashed lines), and σ_{CNO} for the chemical composition of the Mars atmosphere (solid line). (b) Incident solar X-ray photon flux on top of the Mars atmosphere ($h > 300$ km) and at 87 km height. The spectrum is plotted in 1 eV bins. At 87 km, it is considerably attenuated just above the K_α absorption edges, recovering towards higher energies (from Dennerl, 2002).

increasing solar zenith angle because of the increased path length and absorption along oblique solar incidence angles. By sampling the radiation in the volume elements along the line of sight, starting from the element which is farthest away from the observer, synthetic X-ray images of Mars were then accumulated in the fluorescence energies of C, N, O for the phase angle of the Chandra observation.

These images (Figure 8a–c) exhibited a fairly uniform glow of the disk accomanied by a pronounced limb brightening on the sunward side. This is due to the fact that the scattering of solar X-rays is most efficient in the upper atmospheric regions, at heights above ~100 km, and extends into the tenuous, optically thin parts of the thermosphere (Figure 7). There, the volume emissivities add up along the line of sight without considerable absorption, so that the observed brightness is mainly determined by the extent of the atmospheric column along the line of sight. As a result, the sun-lit hemisphere of Mars appears surrounded by an almost transparent luminous shell in X-rays, and Mars looks brightest at the sunward limb since more luminous material is there.

Detailed comparison of the simulated images (Figure 8a–c) shows that the amount of limb brightening is somewhat different for the three energies. According

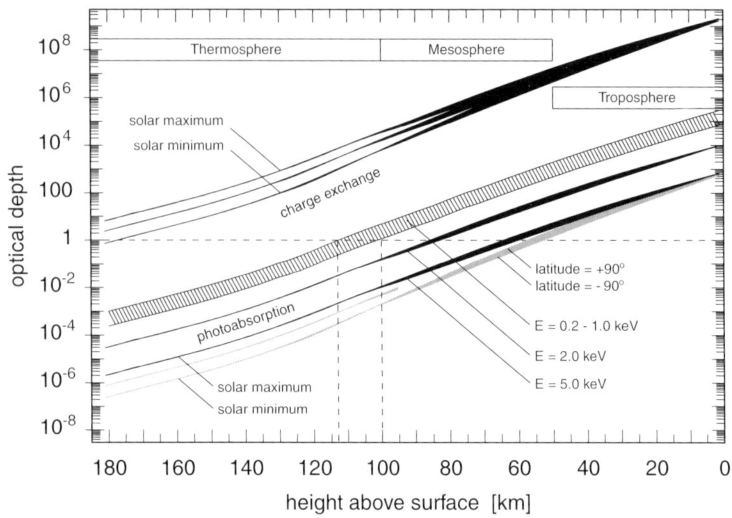

Figure 6. Optical depth $\tau = \tau_C + \tau_N + \tau_O$ of the Martian model atmosphere (Sehnal, 1990a,b) with respect to charge exchange (above) and photoabsorption (below), as seen from outside. The upper/lower boundaries of the hatched area refer to energies just above/below the C and O edges (cf. Figure 5a). For better clarity the dependence of the photoabsorption on the solar cycle is only shown for $E = 5.0$ keV; the curves for the other energies refer to the solar maximum. The dashed horizontal line, at $\tau = 1$, marks the transition between the transparent ($\tau < 1$) and opaque ($\tau > 1$) region. For charge exchange interactions, a constant cross section of $3 \cdot 10^{-15}$ cm^2 was assumed. Due to this high cross section, $\tau = 1$ is reached already at heights of 180 km and above; even for exospheric hydrogen, the opacity to charge exchange is of the order of 0.1 and not negligible. For photoabsorption at $E = 0.2$–1.0 keV, the atmosphere becomes opaque between 113 km and 100 km for solar maximum conditions. During solar minimum, this transition occurs ~10 km deeper in the atmosphere (adapted from Dennerl, 2002).

to the computer simulations, this brightening depends sensitively on the density and chemical composition of the Martian atmosphere. Thus, precise measurements of this brightening could provide a novel method of obtaining remotely direct information about the atmospheric structure in the mesosphere and thermosphere, for different phases of the solar cycle. A direct comparison of the simulated images with the Chandra image (Figure 8d), however, suffers from low photon statistics, as only ~300 photons were detected from Mars during the whole observation. Nevertheless, the predicted limb brightening can be seen in the surface profiles (Figure 11b), which show also indications for a fading on the opposite side, in agreement with the simulation.

Mars is an extremely faint X-ray source: its observed X-ray flux was only 5×10^{-10} of the optical flux. Taking into account that the energy of an X-ray photon exceeds that of an optical photon by two orders of magnitude, this means that there was on average only one X-ray photon among 2×10^{11} photons from Mars. This extremely low number of X-ray photons in the Mars spectrum is due

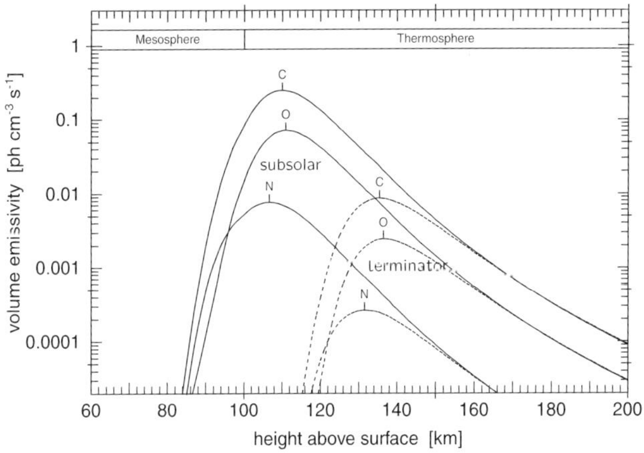

Figure 7. Volume emissivities of C, N, and O K_α fluorescent photons at zenith angles of 0° (subsolar, solid lines) and 90° (terminator, dashed lines) for the incident solar spectrum of Figure 5b. In all cases the maximum emissivity occurs in the thermosphere, where the optical depth depends also on the solar cycle (Figure 6; adapted from Dennerl, 2002).

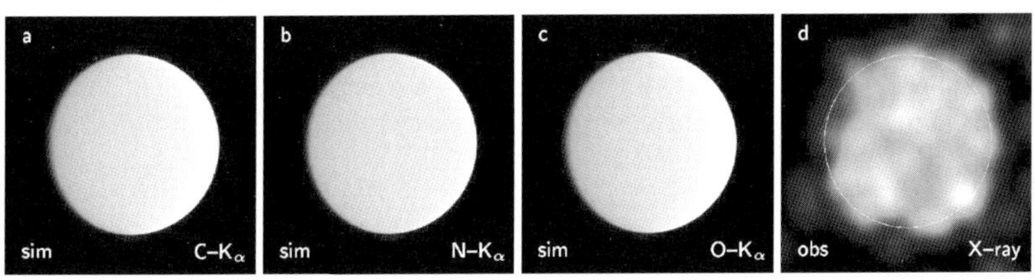

Figure 8. (a–c) Simulated X-ray images of Mars at C-K_α, N-K_α, and O-K_α, for the phase angle of the Chandra observation (18.2°). All images show some limb brightening, especially at C-K_α and O-K_α. (d) Observed X-ray image, accumulated in the energy range 0.4–0.7 keV and smoothed with a Gaussian function with $\sigma = 1.2''$. The circle indicates the geometric size of Mars. This image is dominated by O-K_α fluorescence photons. Although the brightness fluctuations are mainly caused by photon statistics and are not significant, there is evidence for limb brightening on the right-hand (sunward) side (cf. Figure 11b; from Dennerl, 2002).

to the low X-ray flux of the Sun and the low X-ray albedo of the Martian CO_2 atmosphere, which is in turn caused by the very low X-ray fluorescence yields of light elements.

In order to calculate the X-ray luminosity, it is necessary to know the angular distribution of the scattered solar X-rays, as these photons are not emitted isotropically. For this purpose, X-ray intensities were determined from synthetic images for different phase angles (Figure 9). By spherically integrating these intensities for the three fluorescence energies over phase angle, the following luminosities were obtained from the simulation: 2.9 MW for C, 0.1 MW for N, and 1.7 MW for O. These values, directly derived from the solar spectrum, the Mars model atmosphere and quantities from atomic physics, thus predicted a total X-ray luminosity of 4.7 MW.

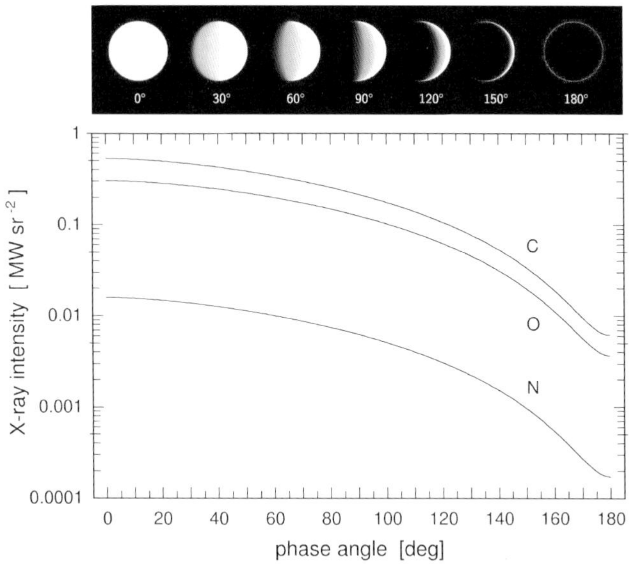

Figure 9. X-ray intensity of Mars as a function of phase angle, in the fluorescence lines of C, N, and O, for the conditions of 4 July 2001. The images at top, all displayed in the same intensity coding, illustrate the appearence of Mars at O-K_α for selected phase angles. Depending on the incident solar spectrum, there may be episodes when the O-K_α flux exceeds that of C-K_α (from Dennerl, 2002).

The corresponding value, obtained from the observed flux, was $\sim 3.5 \pm 0.6$ MW. Both values agreed well with the prediction of Cravens and Maurellis (2001), who had estimated a luminosity of 2.5 MW due to X-ray fluorescence, with an uncertainty factor of about two.

The first X-ray spectrum of Mars (Figure 10), obtained with the ACIS-I detector onboard Chandra, was dominated by a single narrow emission line. Although this line appeared at 0.65 keV, it was interpreted as the O-K_α fluorescence line at 0.53 keV. This conclusion was motivated by the fact that in the case of Venus a similar line, observed at 0.6 keV with the same detector, could be uniquely identified to be at 0.53 keV by an additional observation with an X-ray grating (LETG; Dennerl *et al.*, 2002). The apparent energy shift was most likely caused by optical loading, a superposition of the charges released by 0.53 keV photons and optical photons, during the 3.2 s exposure of each CCD frame. The other expected emission lines, from C and N fluorescence, were too close or even outside the sensitivity range of the ACIS-I detector. Thus, the observed X-ray morphology, the luminosity, and the X-ray spectrum were all in full agreement with fluorescent scattering of solar X-rays.

No evidence for temporal variability was found. This was also in agreement with fluorescent scattering of solar X-rays, because the solar X-ray flux was quite steady at that time. Dennerl (2002) noticed that the conditions encountered during

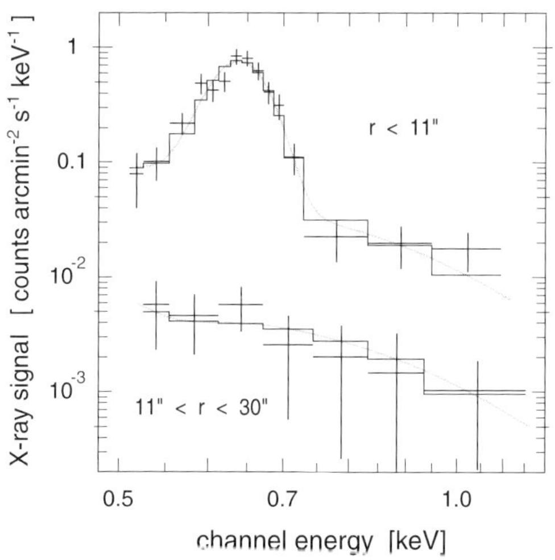

Figure 10. X-ray spectra (background subtracted) of the Martian disk (top) and halo (bottom), obtained with Chandra ACIS-I. The disk spectrum is dominated by a single narrow emission line, caused by O-K_α fluorescence and shifted to higher energies due to optical loading. At higher energies the presence of an additional component is indicated. The spectral shape of this component agrees with that of the halo (from Dennerl, 2002).

this observation were favorable for testing the hypothesis of dust-related X-ray emission: scattering of solar X-rays on very small dust particles was one of the early suggestions for explaining the X-ray emission from comets. Wickramasinghe and Hoyle (1996) had noted that X-rays could be efficiently scattered by dust particles, if their size is comparable to the X-ray wavelength. Such attogram dust particles ($\sim 10^{18}$ g) would be difficult to detect by other means. Their existence was postulated by Owens *et al.* (1998) in order to explain the X-ray properties of comet C/1995 O1 (Hale-Bopp). There was the possibility that such particles might be present in the upper Mars atmosphere, in particular during episodes of global dust storms.

Incidentally, on June 24 a local dust storm on Mars had originated and expanded quickly, developing into a planet-encircling dust storm by July 11. Such dust storms have been observed on roughly one-third of the perihelion passages during the last decades, but never so early in the Martian year. On July 4, this very vigorous storm had covered roughly one hemisphere – the hemisphere that happened to be visible at the beginning of the Chandra observation. By the end of the observation, which lasted for one third of a Mars rotation, this hemisphere had mainly rotated away from our view. Thus, a comparison of the Chandra data from both regions would have revealed any influence of the dust storm on the X-ray flux.

There was, however, no change in the mean X-ray flux between the first and second half of the observation, where 150 and 157 photons were detected, respectively

(Dennerl, 2002). This implied that, if attodust particles were present in the upper Mars atmosphere, the dust storm did not lead to a local increase in their density, high enough to modify the observed X-ray flux significantly. No statement, however, could be made about the situation below ~80 km, as the solar X-rays do not reach these atmospheric layers. While the presence of some attodust in the upper atmosphere could not be ruled out by the Chandra observation, the fact that the X-ray spectrum of Mars was dominated by a single emission line showed that any contribution of such particles to the X-ray flux from Mars must be small compared to fluorescence, even in the process of a developing global dust storm.

While all the properties of the observed X-ray radiation from Mars were fully consistent with fluorescent scattering of solar X-rays in the upper atmosphere as the dominant source, there were, however, also indications for the presence of an additional component of a completely different origin. This will be discussed in the next section.

4. Charge Exchange Induced X-ray Emission

An exciting feature in Figure 11b is the gradual decrease of the X-ray surface brightness between 1 and ~3 Mars radii, indicating the presence of a soft X-ray halo around Mars. Although the excess of X-ray photons in this region was only $\sim 35 \pm 8$ relative to the background expected for this area, the spectral distribution of these photons was different from those of Mars and the sky background. This ruled out the possibility that the halo was caused by, e.g., the optics of the X-ray telescope. Furthermore, the presence of a component with the same spectral distribution as in the halo was indicated in the spectrum of the Mars disk. Within the very limited statistical quality, the spectrum of the halo resembled that of comets, in particular those which had been discovered in archival ROSAT data (Dennerl *et al.*, 1997). In analogy to the situation at comets, Dennerl (2002) interpreted these findings as evidence for the presence of an X-ray halo around Mars caused by charge exchange interactions between highly charged heavy ions in the solar wind and neutrals in the Martian exosphere (SWCX).

Among the protons, electrons and alpha particles, the solar wind also contains a small fraction, about 0.1%, of heavier particles in highly charged states, such as C^{6+}, N^{6+}, O^{6+}, Ne^{8+}, Si^{9+}, Fe^{11+}. They obtain this high degree of ionization in the hot solar corona, which has a temperature of several million degrees, before they leave the Sun at some hundred kilometers per second. On their trip through the solar system in the tenuous solar wind, these ions usually have no chance to capture the missing electrons. They remain in the highly ionized state until they hit sufficiently dense matter, e.g. a comet, where electrons are available in large numbers, mostly bound in H_2O molecules. When solar wind ions capture such electrons, they attain highly excited states and radiate a large fraction of the excitation energy in the extreme ultraviolet and X-ray range. The ionized molecule or atom may subsequently capture a free electron from the solar wind and return to its neutral state.

Triggered by the discovery of cometary X-ray emission (Lisse *et al.*, 1996; Dennerl *et al.*, 1997; Mumma *et al.*, 1997) and the subsequent finding that this emission is caused by SWCX (Cravens, 1997), the consequences of this process for the X-ray emission of Mars had already been investigated by several authors. Cravens (2000) predicted an X-ray luminosity of ~0.01 MW. Krasnopolsky (2000) estimated an X-ray emission of $\sim 4 \cdot 10^{22}$ ph s^{-1}. Adopting an average photon energy of 200 eV (e.g. Cravens, 1997), this corresponds to an X-ray luminosity of 1.3 MW. Holmström *et al.* (2001) computed a total X-ray luminosity of Mars due to charge exchange (within 10 Mars radii) of 1.5 MW at solar maximum, and 2.4 MW at solar minimum.

For the X-ray halo observed within 3 Mars radii, excluding Mars itself, the Chandra observation yielded a flux of $(0.9 \pm 0.4) \cdot 10^{-14}$ erg cm^{-2} s^{-1} in the energy range $E = 0.5$–1.2 keV (Dennerl, 2002). Assuming isotropic emission, this flux corresponds to a luminosity of 0.5 ± 0.2 MW. Taking all the uncertainties into account, this value agrees well with the predictions of Krasnopolsky (2000) and Holmström *et al.* (2001), in particular when the spectral shape is extrapolated to lower energies. Using the abundances of H, H_2, and hot oxygen in the Martian exosphere, Krasnopolsky and Gladstone (2005) compared the observed X-ray luminosity of the Martian X-ray halo with that expected for solar wind charge exchange, and obtained consistent results.

The estimates above indicate that the SWCX process is indeed able to produce the observed number of X-ray photons. In order to answer the question whether also their spatial distribution (Figure 11b) can be reproduced, simulations are necessary. Holmström *et al.* (2001) presented results of such simulations already prior to the Chandra observation of Mars. They used an empirical, axial symmetric model of the proton flow for computing the ion velocities and densities. Under the assumption that the magnetic field is frozen into the flow, they derived from the velocity model the global electric and magnetic fields, which enabled them to study the acceleration of the ions by the Lorentz force near Mars. This allowed them to calculate trajectories of heavy solar wind ions (e.g. Figure 13a), their charge exchange interactions with neutrals in the Martian exosphere, and the resulting X-ray emission.

The simulated SWCX images of Mars, reproduced in Figure 13b for different phase angles, show a dark Mars embedded in an extended, diffuse X-ray halo. This morphology is completely different from that of X-ray fluorescence (cf. Figures 8a–c, 9). The simulated SWCX images for a phase angle of 0° and 30° (top row in Figure 13b) show a good qualitative agreement with the observed X-ray brightness profiles (Figure 11b), even in the detail that the halo is brighter at the "dayside" than at the "nightside". The simulated SWCX images were computed by Holmström

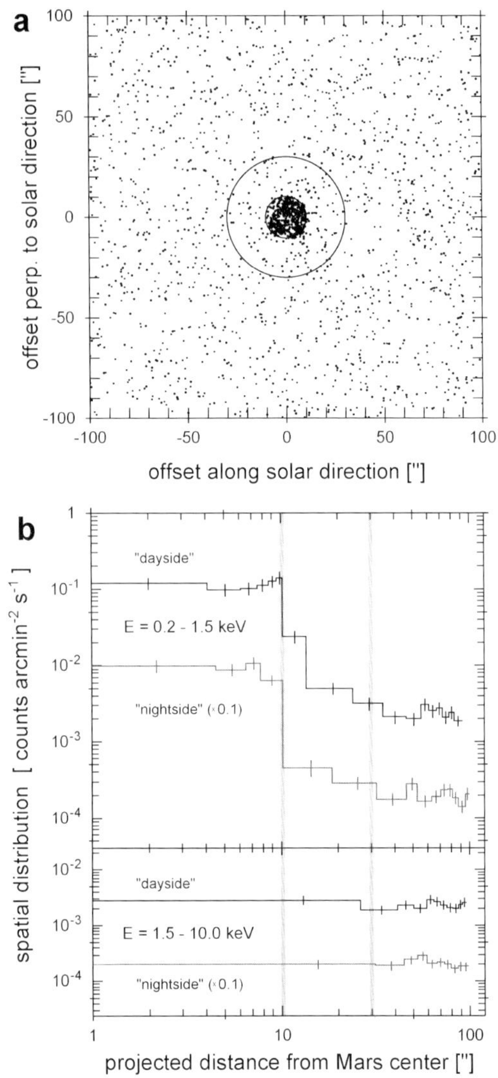

Figure 11. Radial distribution of photons around Mars in the Chandra observation. (a) All photons in the energy range 0.2–10.0 keV. The inner circle, with $r = 10.2''$, shows the geometric size of Mars, while the outer one, at $r = 30''$, illustrates the extent of the soft X-ray halo. (b) Spatial distribution of the photons in the soft ($E = 0.2$–1.5 keV) and hard ($E = 1.5$–10.0 keV) energy range, separately for the "dayside" (offset along projected solar direction > 0) and the "nightside" (offset < 0); note, however, that the phase angle was only 18.2°. For better clarity the nightside histograms were shifted by one decade downward. The thick vertical lines mark the radii $10.2''$ and $30''$ of the circles in (a) (from Dennerl, 2002).

et al. (2001) for average solar wind conditions at solar minimum. As these conditions and the phase angle were different during the Chandra observation, only qualitative agreement can be expected.

In order to make a direct comparison with the Chandra results, Gunell *et al.* (2004) computed the expected X-ray emission from Mars due to SWCX for the specific conditions of the Chandra observation. The solar wind parameters were estimated from data obtained by the WIND spacecraft two days before the X-ray observation. Since Mars was near opposition, plasma that was sampled by WIND near the Earth reached the Mars environment approximately two days later. The solar wind density was scaled with the square of the heliocentric distance ratio of Earth and Mars, and the direction of the magnetic field was assumed to follow a Parker spiral. The electric and magnetic fields around Mars were obtained by running a hybrid simulation of the interaction between the solar wind and Mars with 1.7×10^6 particles. Then, 10^9 trajectories of heavy solar wind ions in these electric and magnetic fields were calculated to get the corresponding SWCX induced X-ray emission, under the reasonable assumption of collisionally thin conditions. This calculation was done for the 15 ion species which were considered as the most important ones for SWCX induced X-ray emission.

The resulting X-ray image (Figure 12a) and radial surface brightness profile (Figure 12b) reproduced the overall shape of the observed X-ray photon distribution (Figure 11b) in the halo. Also the "dayside halo" was brighter than the "nightside halo", in agreement with the observation. However, a quantitative comparison revealed some differences: the calculated X-ray luminosity turned out to be higher than the observed one by a factor between one and three, and the simulated halo brightness dropped less steeply with radial distance than observed. Gunell *et al.* (2004) listed a number of uncertainties of the model that could cause deviations: the density of the neutral exosphere of Mars at the time of the observation, the solar wind parameters, the spatial resolution of the model, and uncertainties in the charge exchange cross sections.

The dependence of the simulation results on some of these parameters was investigated in a subsequent analysis by Gunell *et al.* (2005): the solar wind estimates derived from WIND data were compared with estimates resulting from MHD tomography based on interplanetary scintillation (IPS) data, and the influence of changes in the neutral exosphere was studied by varying the exobase temperature. This investigation was restricted to only two important ion species, O^{7+} and C^{6+}. It turned out that the IPS parameters produced a brighter and more compact X-ray image than the WIND parameters, with up to twice the surface brightness within 1.4 Mars radii, while the brightness distribution in the outer halo was similar for both parameter sets (Figure 14). Concerning the dependence on the exobase temperature, it was found that both the intensity and the extent of the X-ray emitting region increase with increasing exobase temperature (Figures 12c, 15). Another parameter influencing the properties of SWCX induced X-ray emission at Mars may be related to a specific property of the Martian crust: Holmström and Kallio

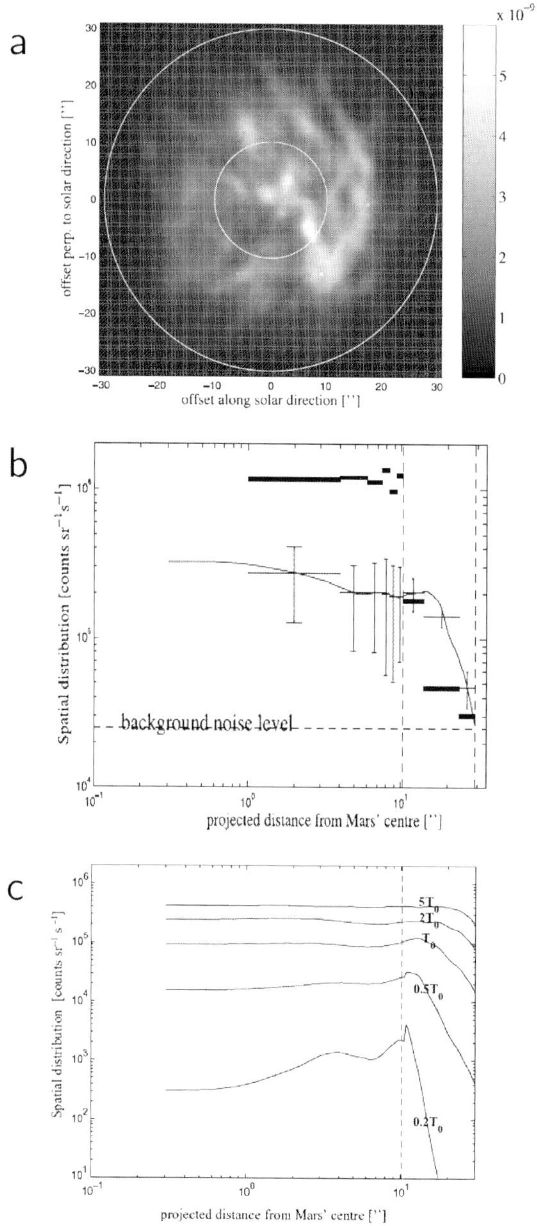

Figure 12. (a) Simulated X-ray image of Mars due to the SWCX process, computed for the conditions of the Chandra observation (Figure 11). (b) Corresponding radial distribution (thin solid curve) in comparison with the observed Chandra profile (thick horizontal lines). Note that the simulation does not take the flux contribution due to fluorescence into account (adapted from Gunell *et al.*, 2004). (c) Dependence of the radial distribution of simulated SWCX X-ray photons on the assumed exobase temperature (see Figure 15 for more information). From top to bottom, the nominal exobase temperatures have been scaled by a factor of 5; 2; 1; 0.5; and 0.2, respectively (from Gunell *et al.*, 2005).

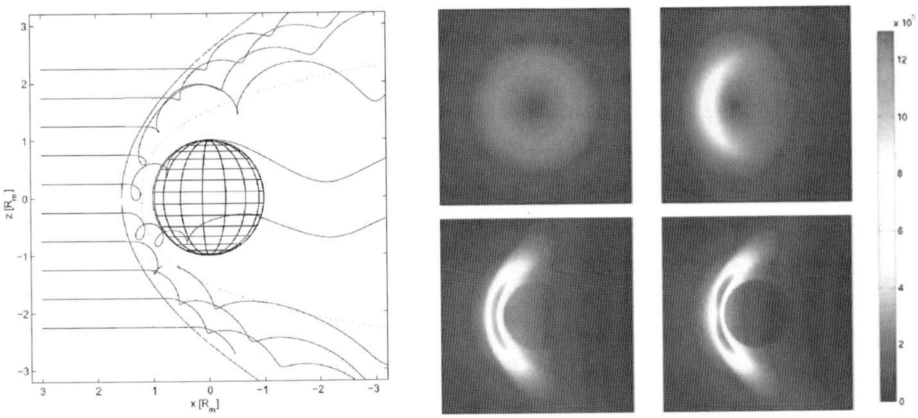

Figure 13. (a) Trajectories of solar wind O^{6+} ions around Mars for a magnetic field $B_{SW} = 4$ nT perpendicular to the image plane and for a velocity of 400 km s^{-1}. (b) Simulated images of the X-ray emission from Mars due to SWCX, for phase angles of 0° (top left), 30° (top right), 60° (bottom left) and 90° (bottom right), at solar minimum (from Holmström *et al.*, 2001).

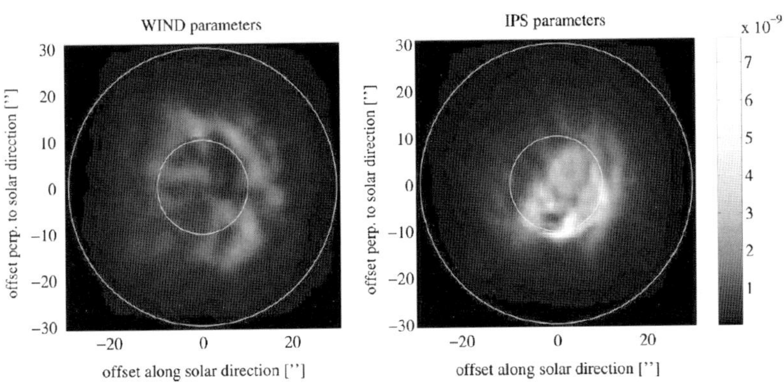

Figure 14. Dependence of the simulated SWCX X-ray images of Mars on the solar wind parameters, for a phase angle of 18.2°. Only the emission of O^{7+} and C^{6+} ions was modeled (from Gunell *et al.*, 2005).

(2004) noted that since crustal magnetizations are asymmetrically distributed, they will also introduce asymmetries in the solar wind flow around the planet and the production of X-rays.

Gunell *et al.* (2005) concluded that due to the sensitive dependence of the properties of the Martian X-ray halo on many parameters, the X-ray emission contains valuable information. However, despite its importance, the statistical quality of the available observational material was extremely poor: all the evidence about an X-ray halo around Mars was based on only $\sim 35 \pm 8$ excess photons and was thus near the sensitivity limit of the Chandra observation (Dennerl, 2002).

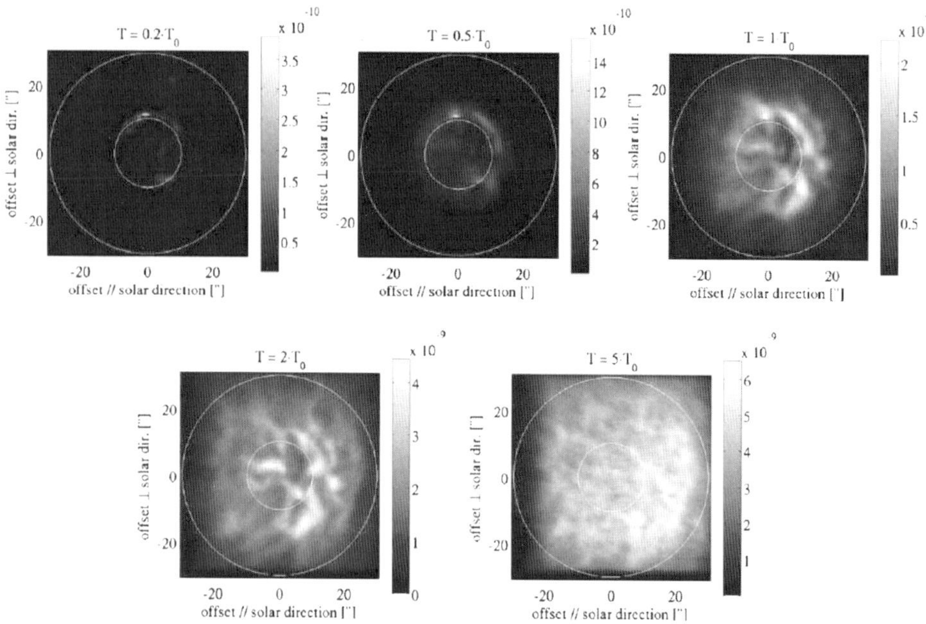

Figure 15. Dependence of the simulated SWCX X-ray images of Mars on the exobase temperature T_0, for a phase angle of 18.2° (cf. Figure 12c). The neutral exosphere was assumed to be spherically symmetric, consisting of atomic and molecular hydrogen and one hot and one thermal population of atomic oxygen, with the following densities ρ and temperatures T_0 at the exobase altitude of 170 km: $\rho(\mathrm{H}) = 3.1 \times 10^{10}\ \mathrm{m}^{-3}$, $T_0(\mathrm{H}) = 310$ K; $\rho(\mathrm{H}_2) = 4.3 \times 10^{11}\ \mathrm{m}^{-3}$, $T_0(\mathrm{H}_2) = 370$ K; $\rho(\mathrm{O_{th}}) = 3.3 \times 10^{14}\ \mathrm{m}^{-3}$, $T_0(\mathrm{O_{th}}) = 380$ K; $\rho(\mathrm{O_{hot}}) = 3.1 \times 10^{10}\ \mathrm{m}^{-3}$, $T_0(\mathrm{O_{hot}}) = 4600$ K. Only the emission of O^{7+} and C^{6+} ions was modeled (from Gunell *et al.*, 2005).

5. The First XMM-Newton Observation of Mars

This situation improved considerably with the first observation of Mars with the satellite XMM-Newton, on November 19–21, 2003 (Dennerl *et al.*, 2006a,b). XMM-Newton, launched on 10 December 1999, carries three aligned and tightly nested Wolter type I telescopes with a total effective area of more than 0.42 m^2 between 0.1 and 2.0 keV (Jansen *et al.*, 2001), providing the highest throughput currently flying. The X-rays are analysed with three instruments for direct imaging and spectroscopy (EPIC MOS 1, MOS 2, and PN: Turner *et al.*, 2001; Strüder *et al.*, 2001) and with two essentially identical Reflection Grating Spectrometers (RGS 1 and RGS 2), providing an energy resolution $E/\Delta E$ between 100 and and 600 in the energy range 0.33–2.1 keV (den Herder *et al.*, 2001). The main difference to Chandra is that all five X-ray instruments are operating simultaneously, providing higher sensitivity and higher spectral resolution at the expense of reduced spatial resolution and more stringent pointing restrictions.

Observations with XMM-Newton are restricted to targets at solar elongations between 70° and 110°. Thus, Mars cannot be observed around opposition, when it is closest to the Earth. Within the accessible range of solar elongations, values near 110° are most favourable, because Mars is then considerably closer to the Earth than at smaller elongations. XMM-Newton was pointed towards Mars from 2003 Nov 19, 23:47 to Nov 21, 05:05 UT. At that time, Mars had an apparent diameter of 12.2″. Its heliocentric distance was 1.43 AU, its geocentric distance 0.77 AU, and it was observed at a phase angle (Sun-Mars-Earth angle) of 41.2° and at a solar elongation of 108°.

The XMM-Newton observation confirmed the presence of the Martian X-ray halo and made for the first time a detailed analysis of its spectral, spatial, and temporal properties possible. It proved that the source of the X-ray emission is indeed charge exchange between highly charged solar wind ions and exospheric neutrals, providing the first definite detection of SWCX induced X-ray emission from the exosphere of another planet (Dennerl *et al.*, 2006a). Furthermore, it unambiguously showed that the X-ray radiation which we observe from the planetary disk is primarily due to scattered solar X-rays (Dennerl *et al.*, 2006b). These new findings are described in the next sections.

5.1. Temporal Properties of the X-ray Emission from Mars

The high sensitivity of the EPIC instruments onboard XMM-Newton made it possible to study the temporal properties of the X-ray emission from Mars with a time resolution of ~10 min. It turned out that the X-ray flux from Mars was quite variable, exhibiting several outbursts with a characteristic time scale of ~1 hour. The solar X-ray flux was also highly variable during this time, and the major outbursts show up clearly in the X-ray lightcurve observed from the Martian disk. This is direct proof that this flux is predominantly caused by scattered solar X-rays. At Jupiter (Bhardwaj *et al.*, 2005a) and Saturn, (Bhardwaj *et al.*, 2005b), a similar correlation has been seen between solar X-ray variability and the planetary X-ray flux.

X-rays were also observed from an extended region around Mars, covering several Mars radii. A lightcurve of the outer regions, between $r \sim 10''$ and $r \sim 50''$, led to an exciting discovery: this X-ray flux was also highly variable, but the variability was not correlated with that observed from the inner region. This is expected if the Martian exosphere responds to variations of the solar wind rather than the solar X-ray emission. Also the spectral properties during outbursts of the inner and outer regions were clearly different from each other, providing additional support to the idea that the X-rays from Mars are a superposition of two different components (Dennerl *et al.*, 2006a).

5.2. High Resolution X-ray Spectra of Mars

Although Mars is a very faint X-ray source, the sensitivity of the RGS instruments onboard XMM-Newton was sufficient for performing, for the first time ever, high

TABLE I
Emission lines in the XMM-Newton/RGS Mars spectra (from Dennerl *et al.*, 2006a)

Line ID				Line origin	
#	Abbr	Wavelength [Å]	Energy [eV]	Ion	Transition
1	Ne72	14.21	872.5	Ne^{7+}	$2p \rightarrow 1s$
2	O74	15.18	817.0	O^{7+}	$4p \rightarrow 1s$
3	O65	17.40	712.5	O^{6+}	$5p \rightarrow 1s$
4	O72	18.97	653.6	O^{7+}	$2p \rightarrow 1s$
5	N63	20.91	593.0	N^{6+}	$3p \rightarrow 1s$
6	O6r	21.60	574.0	O^{6+}	$2\,^1P_1 \rightarrow 1\,^1S_0$
7	O6i	21.81	568.5	O^{6+}	$2\,^3P_1 \rightarrow 1\,^1S_0$
8	O6f	22.11	560.9	O^{6+}	$2\,^3S_1 \rightarrow 1\,^1S_0$
9	CO2a	23.50	527.7	CO_2	$1\,\pi_g \rightarrow 1s$
10	CO2b	23.68	523.5	CO_2	$3\,\sigma_u \rightarrow 1s$
11	N62	24.78	500.3	N^{6+}	$2p \rightarrow 1s$
12	C55	26.36	470.4	C^{5+}	$5p \rightarrow 1s$
13	C54	26.99	459.4	C^{5+}	$4p \rightarrow 1s$
14	C53	28.47	435.6	C^{5+}	$3p \rightarrow 1s$
15	N22	31.51	393.5	N_2	$2p \rightarrow 1s$
16	C45	32.75	378.5	C^{4+}	$5p \rightarrow 1s$
17	C52	33.74	367.6	C^{5+}	$2p \rightarrow 1s$

resolution X-ray spectroscopy of the Martian atmosphere and exosphere (Dennerl *et al.*, 2006a). Figure 16a reveals that the emission of the Martian X-ray halo (at cross dispersion distances $15'' < |y| < 50''$) consists of many emission lines of similar flux. With the exception of lines #9 and #10, all these lines appear where emission is expected from the de-excitation of highly charged ions (Table I). The lines #9 and #10 are caused by fluorescence of CO_2. They appear in the halo spectrum only because of instrumental effects. However, these lines dominate the RGS spectra from regions close to Mars (at cross dispersion distances $|y| \leq 10''$ from its center; Figure 16b), where the N_2 fluorescence line (#15) is also indicated. These two high resolution spectra clearly show that the X-rays from Mars are composed of two completely different components.

In terms of spectral resolution, Figure 16a shows perhaps the best charge exchange spectrum ever obtained. The high spectral resolution is demonstrated in Figure 17a, which zooms to a small section covering only 3Å (70 eV). All the emission in this spectral region originates from inner shell electron transitions in oxygen, either six-fold ionized or neutral (embedded in CO_2). This spectrum was compiled for cross dispersion distances $|y| \leq 50''$ and includes also emission from Mars itself. It is dominated by CO_2 fluorescent emission,

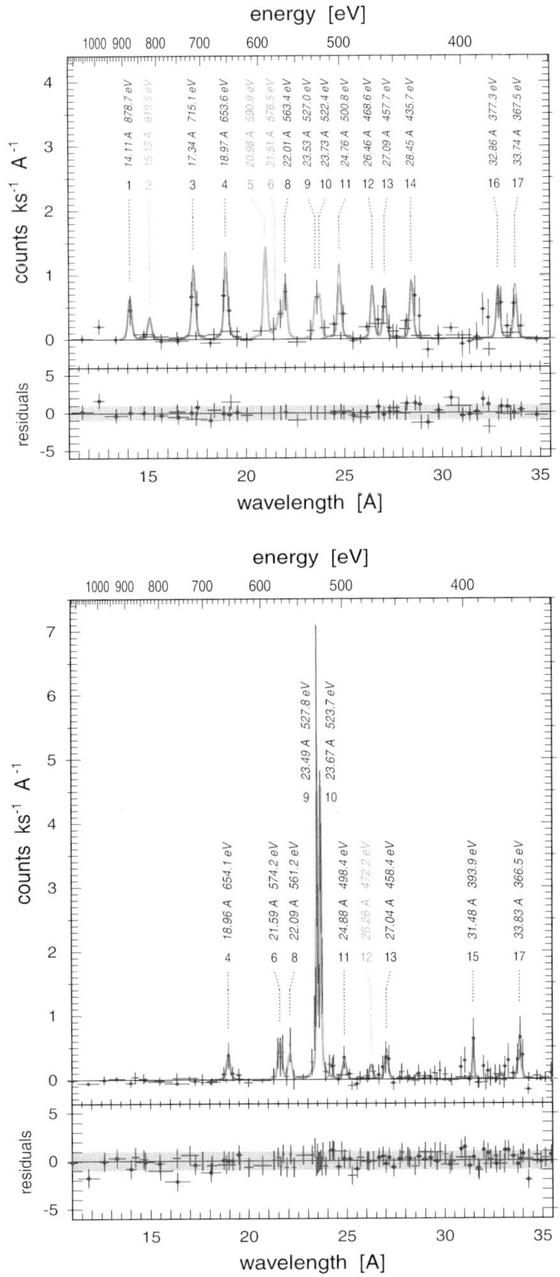

Figure 16. XMM-Newton/RGS spectra of Mars and its halo, accumulated at cross dispersion distances (a) $15'' < |y| < 50''$ and (b) $|y| \leq 10''$ from the center of Mars. Smooth curves show the corresponding folded model spectra, with a thick line for RGS 1 and a thin one for RGS 2. The positions resulting from the fit (in Å and eV) are written above each line, together with a running number, for easier reference (cf. Table I). A zoomed version of the central region around 23 Å/540 eV is shown in Figure 17a (from Dennerl *et al.*, 2006a).

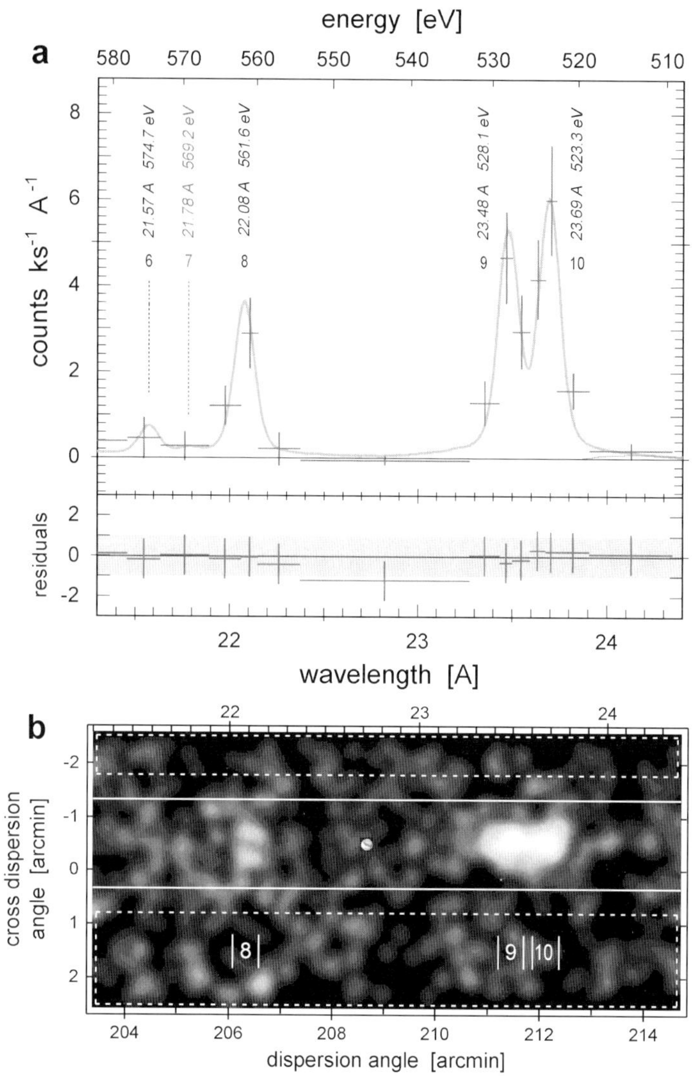

Figure 17. (a) XMM-Newton/RGS 1 spectrum of Mars and its halo at cross dispersion distances $|y| \leq 50''$ from the center of Mars, showing the region around the CO_2 doublet and the O^{6+} multiplet. (b) Dispersed image in the same wavelength/energy scale as in (a). This image was binned into $2'' \times 2''$ pixels and smoothed with a Gaussian function with $\sigma = 8'' \times 8''$. It contains the full FOV of RGS along the cross dispersion direction. The inner rectangle shows the extraction region for the $|y| \leq 50''$ spectra in (a); the background was taken from the two dashed bands above and below. An optical image (taken by the author with a 4" Newtonian telescope) is inserted at the center to illustrate the angular size of Mars during the observation; the observed X-ray images are blurred by the PSF of the telescope and the smoothing process. Note that due to the high dispersion of RGS, the X-ray image of the Martian halo (8) is completely detached from the fluorescent images (9, 10) (adapted from Dennerl *et al.*, 2006a).

which is clearly resolved into two lines (#9 and #10) of similar flux. This is the first astronomical measurement of fine structure in the X-ray fluorescence of CO_2.

The explanation for two peaks instead of one is that in the CO_2 atmosphere of Mars, the oxygen atom is embedded in a molecule, where additional energy states are available for the electrons. The line at 528 eV (#9) is caused by an electron transition from the $1\pi_g$ orbital (which is almost a pure $2p$ orbital around the oxygen atom, and thus similar to the isolated oxygen atom state) into the ground state, while the line at 523 eV (#10) is a superposition of transitions from three orbitals, $4\sigma_g$, $3\sigma_u$, and $1\pi_u$, into the ground state.

Of even higher diagnostic value in Figure 17a is the fine structure seen in the emission from O^{6+} ions (lines #6, #7, #8). These lines are the result of electron transitions between the $n = 2$ shell and the $n = 1$ ground state shell. As O^{6+} contains two electrons, there are two possible states of the ion, depending on the relative spin orientation of the electrons: singlet states (mainly $^1S_0$ and $^1P_1$) and triplet states (mainly $^3S_1$ and $^3P_{0,1,2}$). Because transitions from triplet states to the ground state require spin changes of the electrons, these are slow processes compared to transitions from singlet states.

The line #6 results from fast transitions from a singlet state ($^1P_1$) with a decay rate of $3.3 \cdot 10^{12}\ s^{-1}$, while the lines #7 and #8 are caused by slow transitions from triplet states. Particularly interesting is the line #8, because the corresponding transition starts from a metastable state ($^3S_1$) and has a decay rate of only $1.0 \cdot 10^3\ s^{-1}$. This state can easily be depopulated by collisions before the transition takes place. The fact that the line #8 is considerably brighter than the lines #6 and #7 excludes thermal or collisional excitation as the origin of the Martian X-ray halo emission. For hot plasmas, the flux ratio $G = (\#7 + \#8)/\#6$ of triplet to singlet transitions is usually less than one (e.g. Smith *et al.*, 2001).

However, if the emission lines result from electron capture by multi-charged ions colliding with neutral gas at low density, the situation is completely different. In this case, G is predicted to be in excess of three (Kharchenko *et al.*, 2003). The value of the G ratio for the O^{6+} emission induced by the interaction between the solar wind ions and heliospheric hydrogen gas has been evaluated as 6.7 (Pepino *et al.*, 2004), and for the cometary X-rays as 5.8 (Kharchenko, 2005). These values agree very well with that derived from the Mars RGS spectra: $G \sim 6$ for $|y| \leq 50''$. Thus, the high resolution X-ray spectrum in Figure 17a provides the direct proof that the X-ray emission of the Martian halo is indeed caused by the SWCX process.

5.3. X-Ray Images of Mars in Individual Emission Lines

As slitless spectrometers, the RGS produce in each spectral line an image of the observed object. Due to the high dispersion of the RGS and the small spatial extent of Mars, there is essentially no overlap between the individual images. This makes it possible to study the spatial structure of the X-ray emission in individual spectral lines. As fluorescence occurs in neutral atoms and molecules, while charge exchange involves highly charged ions, the energies/wavelengths of the corresponding emission lines are different. Thus, the contributions of fluorescence and charge exchange can be completely separated by this method.

Figure 17b shows the RGS image which corresponds to the zoomed spectrum in Figure 17a, at the same wavelength scale. To illustrate the spatial extent of Mars at the time of the observation, an optical image was inserted at the center. The spectral images of the CO_2 emission to the right (#9, #10) prove that this radiation originates close to the planet, as their brightness distributions peak at the position of Mars. Brightness profiles along the dispersion direction (Figure 18b) show that their extent is consistent with the size of Mars, if the instrumental blur is taken into account, and that there is no significant difference between both components.

The spectral image of the O^{6+}: $^3S_1 \rightarrow {}^1S_0$ transition (#8 in Figure 17b) is completely different from that of the CO_2 emission, exhibiting two distinct blobs along the cross dispersion direction (which is approximately the North-South direction on Mars), with practically no emission in between (Figure 18a). This means that the emission does not originate close to Mars or in an X-ray luminous extended shell around it, but at two well localized regions ~3000 km above both poles. For other emission lines, however, the morphology appears to be different. Figure 19

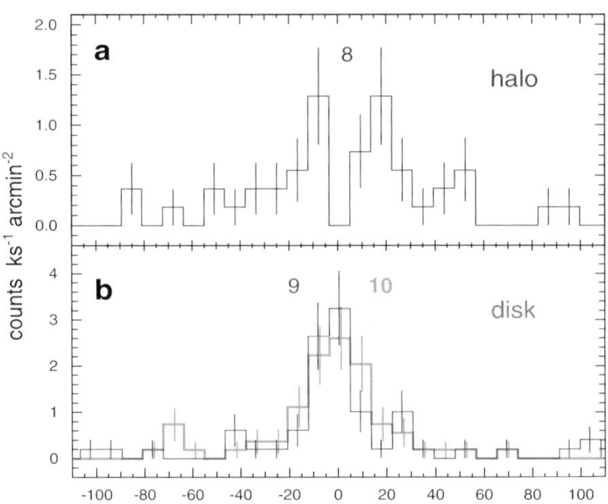

Figure 18. Brightness profiles (with 1σ error bars) along the cross dispersion direction of the Martian exosphere in the light of (a) the $2\,^3S_1 \rightarrow 1\,^1S_0$ transition of O^{6+} (#8) and (b) the CO_2 fluorescence doublet (#9, #10). The extraction geometry is illustrated in Figure 17b. While the CO_2 profiles are very similar, the O^{6+} profile has a quite different shape, characterized by a pronounced dip at the center (from Dennerl *et al.*, 2006a).

presents spectral images for the major emission lines, identified by the abbreviations listed in Table I. These images indicate differences in the spatial structures, not only between fluorescence and charge exchange, but also between different ions and ionization states.

The structure seen in the spectral image of O6f seems to be a specific property of emission from ionized oxygen, as the O72 image (Figure 19a) also shows two distinct blobs along the cross dispersion direction. Compared to O6f (Figure 19b), the O72 emission occurs at larger distances from Mars. There is also some evidence in Figure 19a that the peak of the O72 emission is shifted to the right with respect to Mars. Interpreted as redshift, this would indicate velocities of O^{8+} ions in excess of $\sim$400 km s^{-1} along the line of sight, as the dashed vertical lines in Figure 19a show. Alternatively, this shift may be interpreted as a spatial displacement.

Spectral images can also be obtained for the carbon emission lines C53 and C52 (Figure 19 d,e). Although the quality of these images is limited by statistical noise, they seem to indicate yet another morphology: there is again clear evidence for extended, unisotropic emission, but unlike the blobby O72 and O6f appearance, the C53 and C52 emissions exhibit a more band-like structure without a pronounced intensity dip at the position of Mars. There is also some evidence that at larger distances from Mars the emission is shifted towards the right. For comparison, the morphology of the fluorescent radiation (Figure 19 g–i) is clearly concentrated to the planet. While an interpretation of the fluorescence images is straightforward, an interpretation of the structures in the halo emission is not an easy task, since they depend on many parameters (e.g. Gunell *et al.*, 2005).

Figure 20 is a superposition of RGS images of Mars in seven emission lines between 19 Å and 34 Å (360–660 eV), with ionized oxygen coded in blue, ionized carbon coded in green, and fluorescence coded in yellow. The horizontal axis is approximately the dispersion direction, with wavelength increasing to the right. Along this direction, the angular and linear scale at lower right is valid only if the individual images are purely monochromatic and exhibit no wavelength/energy shift. Alternatively, the velocity scale at upper right illustrates the amount of horizontal displacement which would be caused by the Doppler shift. As the wavelength is not a linear function of dispersion in the RGS, the amount of displacement depends on the wavelength/energy. The two bars enclose the spectral range used, with the upper (shorter) bar referring to the O72 line at 18.97 Å/653.6 eV and the lower bar to the C52 line at 33.74 Å/367.6 eV. A displacement to the right (where wavelength increases) corresponds to a redshift.

Figure 20 reveals that the halo emission originates in an extended region which is elongated mainly along the cross dispersion direction. This alignment, however, is not precise: there are indications for some tilt of the upper and lower wings to the right. As this figure is composed of individual images obtained with a slitless spectrograph, there are generally two possibilities, spectrally or morphologically, for interpreting this tilt:

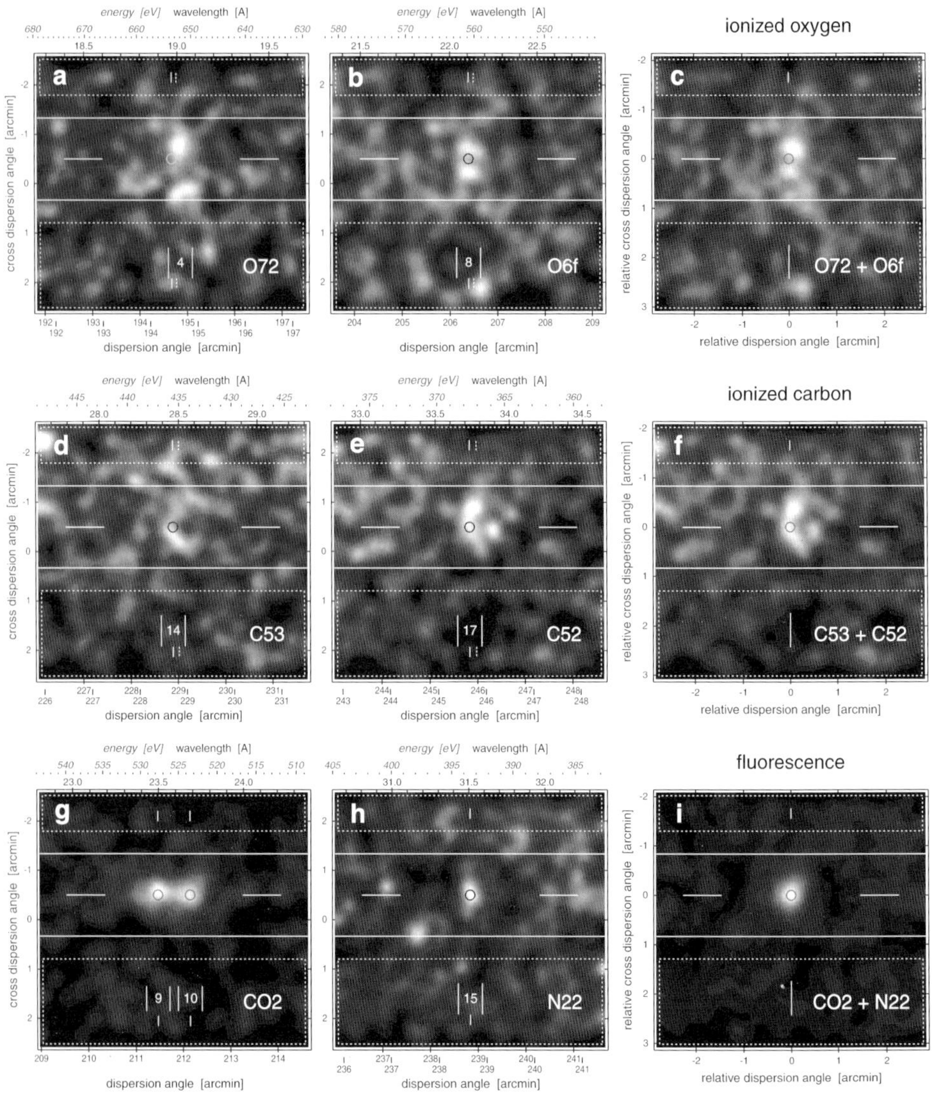

Figure 19. XMM-Newton/RGS images of Mars and its halo in the individual emission lines of ionized oxygen (top row), ionized carbon (middle row), and fluorescence of CO_2 and N_2 molecules (bottom row). The images were corrected for exposure variations, were binned into $2'' \times 2''$ pixels and smoothed with a Gaussian function with $\sigma = 8'' \times 8''$. All are displayed at the same angular scale; the dynamic scale, however, was individually adjusted. The images in the rightmost column are the sum of the images to their left. For the wavelengths/energies of the emission lines, the values in Table I were used. These are indicated by the short vertical lines and the circle in the middle, which illustrates the size and expected position of Mars at these emission lines. Dashed vertical lines in the frames a,b,d,e indicate the apparent shift along the dispersion direction for a redshift of 400 km s^{-1}. The projected direction to the Sun is towards the left (from Dennerl *et al.*, 2006a).

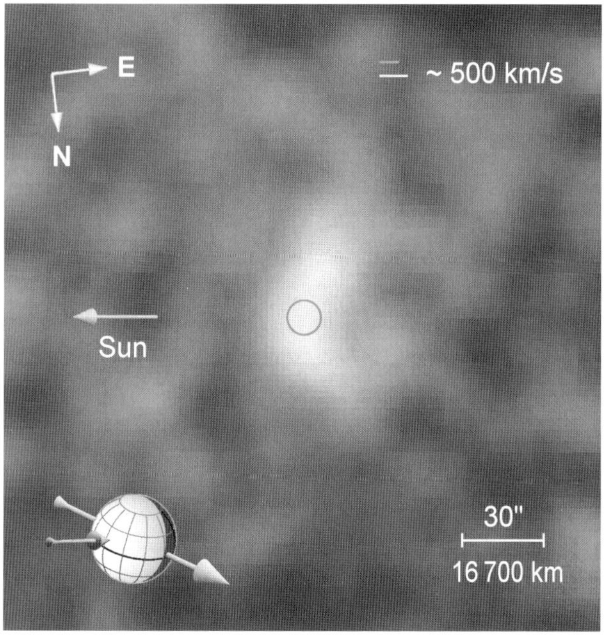

Figure 20. Superposition of the XMM-Newton/RGS images in Figure 19, each centered on the wavelength/energy of an individual emission line, with ionized oxygen coded in blue, ionized carbon coded in green, and fluorescence coded in yellow. The projected direction of the Sun is towards the left (horizontal arrow). The circle indicates the position and size of Mars; further details about the observing geometry are provided by the sphere at lower left: the grid shows areographic coordinates, with blue lines for the southern hemisphere (top) and red lines for the northern hemisphere (bottom). The bright part of the sphere is the sunlit side of Mars. A green arrow indicates its direction of motion, as seen from a stationary point at the position of the Earth. The yellow arrow illustrates the velocity of solar wind particles, emitted radially from the Sun with 400 km s^{-1} with respect to Mars (from Dennerl *et al.*, 2006a).

In the spectral interpretation, this would be evidence for a redshift. The redshift could be explained by the Doppler effect, as the excited solar wind ions are moving away from us. We should expect to observe a Doppler redshift of $\delta\lambda/\lambda \sim (v_i/c)\cos(\varphi)$ where $v_i \sim$ 400–800 km s^{-1} is the velocity of the solar wind ions and $\varphi = 41.2^\circ$ is the phase angle of Mars during the XMM-Newton observation. The Doppler shift of the 15–35 Å emission lines may reach the values of 0.015–0.035 Å for the slow solar wind and 0.03–0.07 Å for the fast solar wind. The fact that the observed redshifts decrease with decreasing distance from Mars (along the cross dispersion direction) would be evidence for the fact that the ion velocity decreases towards Mars. i.e., with increasing density of the Martian exosphere. This would be in agreement with model calculations of the solar wind interaction with Mars (e.g. Ma *et al.*, 2004). Part of the deceleration would be a direct consequence of momentum exchange as a by-product of the charge exchange interactions with atoms in the Martian exosphere.

In the morphological interpretation, we see an emission region which is most prominent above the poles, but somewhat tilted away from the Sun. This general appearance may be a consequence of the phase angle at which Mars was observed: if the X-ray emission originates preferentially at the sunward side of the Martian halo, then we should see a crescent-like structure when observing it at a phase angle of 41.2°. This was the result of numerical simulations by Holmström *et al.* (2001), which predicted a crescent-like structure (Figure 13b) that resembles (on a smaller scale) the observed emission in ionized carbon (Figure 19f). The fact that (i) the emission of ionized carbon extends far above the poles and that (ii) the emission of ionized oxygen is observed to occur almost exclusively above the poles could be understood as evidence for an asymmetric density structure in the Martian exosphere, which would be much denser above the poles and towards the night side than towards the Sun.

5.4. Luminosity of the Disk and the Halo

The observed fluxes in the individual emission lines can be converted into luminosities if the angular distribution of the emitted photons is known. For the halo, isotropic emission can be assumed, because the exosphere of Mars is optically thin to X-ray photons, and a halo luminosity $L_x(h) = 12.8 \pm 1.4$ MW is obtained in the spectral band 14–34 Å or 365–880 eV.

For the fluorescence radiation, the situation is different, because the atmosphere of Mars (and the planet itself) is optically thick to soft X-rays. In this case, the luminosity can be computed by spherically integrating the flux as a function of the phase angle (which can be obtained by modeling the individual scattering and absorption processes, as described in Dennerl *et al.*, 2002) over the phase angle (cf. Figure 9). This results in a disk luminosity $L_x(d) = 3.4 \pm 0.4$ MW for fluorescence of oxygen and nitrogen.

It is well possible that the disk spectrum contains also some contribution from elastic scattering of solar X-rays, in addition to fluorescence. This contribution, however, should be small, because no significant unambiguous evidence for elastic scattering is seen in the RGS spectrum. Therefore, the calculation of the disk luminosity was restricted to the oxygen and nitrogen fluorescence lines in the RGS bandpass.

This disk luminosity is significantly higher than the 1.4 ± 0.2 MW derived from the previous Chandra observation (Dennerl, 2002) for oxygen and nitrogen (the carbon fluorescence is just outside the RGS spectral band). An obvious explanation is found in Figure 3: although the XMM-Newton observation took place during the declining part of the solar cycle, this observation happened to fall into a period of extreme solar activity, where the mean solar 1–8 Å flux was almost one order of magnitude higher than during the Chandra observation.

The observed luminosity of the X-ray halo is even more than one order of magnitude higher than in the previous Chandra observation of Mars, which yielded 0.5 ± 0.2 MW in the energy range 0.5–1.2 keV (Dennerl, 2002). This lower value was in good agreement with theoretical predictions (Krasnopolsky, 2000; Holmström *et al.*, 2001; Krasnopolsky and Gladstone, 2005).

One reason for the much higher halo luminosity during the XMM-Newton observation is the fact that a considerable flux was observed at large distances from Mars: the Chandra flux was derived from within 3 Mars radii around its center, or $r = 30''$ at the time of the Chandra observation. Outside this radius, no significant excess of surface brightness relative to the Chandra background level was detectable. Applied to the XMM-Newton observation, this radius would correspond to only $18''$, due to the larger geocentric distance of Mars. Thus, the XMM-Newton 'halo region', which was defined here by cross dispersion distances $15'' < |y| \leq 50''$ (e.g. Figure 16a), would have been almost completely outside the Chandra halo region, and this region was found to contain most of the observed flux (cf. Figure 20): the X-ray luminosity in this region alone was 10.0 ± 1.2 MW.

Another reason for the high X-ray luminosity of the halo was the extreme solar activity in October–November 2003: over two solar rotations, ~80 fast coronal mass ejections (CMEs) were observed together with X-class flares, solar energetic particle events, and interplanetary shocks (Gopalswamy *et al.*, 2005). On November 18, a prolonged optical flare occurred, accompanied by two X-ray bursts at 07:52 and 08:31 UT and a CME with multiple components, starting with a comparatively weak ejection, but then followed by a much brighter, faster, and larger-scale partial-halo CME after ~08:40 UT and a third large CME after 09:26 UT (Chertok and Grechnev, 2005). It is very likely that this sequence of CMEs produced a highly disturbed solar wind environment at Mars during the XMM-Newton observation, which took place 1.63–2.85 days after the second CME. The XMM-Newton EPIC data (Dennerl *et al.*, 2006b) exhibit significant variability in the X-ray flux of the Martian halo, including several outbursts by a factor of ~4 with a duration of about one hour each. Unfortunately, Mars Express had not arrived yet at Mars in November 2003, so that no simultaneous in-situ plasma measurements exist.

6. Summary and Conclusions

X-rays from Mars consist of two different components: (i) solar X-rays scattered in the upper Martian atmosphere, and (ii) emission from highly charged heavy solar wind ions in excited states, resulting from charge exchange interactions with neutrals in the Martian exosphere. For both components, the pioneering observations with Chandra and XMM-Newton have shown how X-ray observations will provide novel methods for studying this planet: high resolution X-ray images of Mars in the fluorescence lines of C, N, O will make it possible to investigate the atmospheric layers above ~80 km, which are difficult to study otherwise, and their response to solar activity, while X-ray images of Mars in the lines of excited ions will enhance our knowledge about the Martian exosphere and its interaction with the solar wind.

It is remarkable that XMM-Newton has the capability to trace the exospheric X-ray emission, with high spectral resolution, out to ~8 Mars radii (~27 000 km), proceeding into exospheric regions far beyond those that have been observationally explored to date. This is particularly interesting because the X-ray emission results directly from charge exchange interactions between atmospheric constituents and solar wind ions, a process which is considered as an important nonthermal escape mechanism and which may be responsible for a significant loss of the Martian atmosphere. Although this escape process is mainly due to charge exchange with solar wind protons, which are ~1000 times more abundant than heavy ions and which do not produce X-rays, the X-ray observations by Chandra and XMM-Newton provide a useful tracer of this process. Despite this importance, our observational knowledge of the Martian exosphere is still poor. Thus, X-ray observations, providing a novel method for studying exospheric processes on a global scale, may lead to a better understanding of the present state of the Martian atmosphere and its evolution. They open up a completely new possibility of remote, global, imaging of planetary exospheres, and their spatial and temporal variability.

In addition to its importance to planetary studies, the possibility to obtain from X-ray observations of Mars not only charge exchange spectra with unprecedented spectral resolution, but to get at the same time also images of the morphological structures originating from specific electron transitions in individual ions, is likely to contribute to an improved understanding of the physics of charge exchange, which is of general importance to X-ray plasma diagnostics, both in the laboratory and in outer space.

Acknowledgements

The author would like to thank the anonymous referees for their help in improving the quality of the manuscript.

References

Bhardwaj, A., Branduardi-Raymont, G., Elsner, R. F., Gladstone, G. R., Ramsay, G., Rodriguez, P., *et al.*: 2005a, *Geophys. Res. Lett.* **32**, 3.

Bhardwaj, A., Elsner, R. F., Waite, J. H. J., Gladstone, G. R., Cravens, T. E., and Ford, P. G.: 2005b, *ApJ* **624**, L121.

Chertok, I. M., and Grechnev, V. V.: 2005, *Astronomy Reports* **49**, 155.

Cravens, T. E.: 1997, *Geophys. Res. Lett.* **24**, 105.

Cravens, T. E.: 2000, *Adv. Space Res.* **26**(10), 1443.

Cravens, T. E., and Maurellis, A. N.: 2001, *Geophys. Res. Lett.* **28**(15), 3043.

den Herder, J. W., Brinkman, A. C., Kahn, S. M., Branduardi-Raymont, G., Thomsen, K., Aarts, H., *et al.*: 2001, *A&A* **365**, L7.

Dennerl, K.: 2002, *A&A* **394**, 1119.

Dennerl, K., Burwitz, V., Englhauser, J., Lisse, C., and Wolk, S.: 2002, *A&A* **386**, 319.

Dennerl, K., Englhauser, J., and Trümper, J.: 1997, *Science* **277**, 1625.

Dennerl, K., Lisse, C. M., Bhardwaj, A., Burwitz, V., Englhauser, J., Gunell, H., *et al.*: 2006a, *A&A* **451**, 709.

Dennerl, K., Lisse, C. M., Bhardwaj, A., Burwitz, V., Englhauser, J., Gunell, H., *et al.*: 2006b, *A&A* in preparation.

Gopalswamy, N., Yashiro, S., Liu, Y., Michalek, G., Vourlidas, A., Kaiser, M. L., *et al.*: 2005, *J. Geophys. Res.* **110**, A09S15.

Grader, R. J., Hill, R. W., and Seward, F. D.: 1968, *J. Geophys. Res.* **73**, 7149 .

Gunell, H., Holmström, M., Kallio, E., Janhunen, P., and Dennerl, K.: 2004, *Geophys. Res. Lett.* **31**, 22801.

Gunell, H., Holmström, M., Kallio, E., Janhunen, P., and Dennerl, K.: 2005, *Adv. Space Res.* **36**, 2057.

Holmström, M., Barabash, S., and Kallio, E.: 2001, *Geophys. Res. Lett.* **28**(7), 1287.

Holmström, M., and Kallio, E.: 2004, *Adv. Space Res.* **33**, 187.

Jansen, F., Lumb, D., Altieri, B., Clavel, J., Ehle, M., Erd, C., *et al.*: 2001, *A&A* **365**, L1.

Justus, C. G., Duvall, A. L., and Johnson, D. L.: 2005, *Adv. Space Res.* **35.1**, 4.

Kharchenko, V.: 2005, in R. K. Smith (ed.), *X-ray Diagnostics of Astrophysical Plasmas: Theory, Experiment, and Observation*, Vol. 774, p. 271.

Kharchenko, V., Rigazio, M., Dalgarno, A., and Krasnopolsky, V. A.: 2003, *ApJ* **585**, L73.

Krasnopolsky, V.: 2000, *Icarus* **148**, 597.

Krasnopolsky, V. A., and Gladstone, G. R.: 2005, *Icarus* **176**, 395.

Krause, M. O.: 1979, *J. Phys. Chem. Ref. Data* **8**, 307.

Lisse, C. M., Dennerl, K., Englhauser, J., Harden, M., Marshall, F. E., Mumma, M. J., *et al.*: 1996, *Science* **274**, 205.

Ma, Y., Nagy, A. F., Sokolov, I. V., and Hansen, K. C.: 2004, *J. Geophys. Res.* **109**, A0721.

Metzger, A. E., Gilman, D. A., Luthey, J. L., Hurley, K. C., Schnopper, H. W., Seward, F. D., *et al.*: 1983, *J. Geophys. Res.* **88**, 7731.

Mewe, R., Gronenschild, E. H. B. M., and van den Oord, G. H. J.: 1985, *A&AS* **62**, 197.

Mumma, M. J., Krasnopolsky, V. A., and Abbott, M. J.: 1997, *ApJ* **491**, L125.

Nier, A. O., and McElroy, M. B.: 1977, *J. Geophys. Res.* **82**, 4341.

Owens, A., Parmar, A. N., Oosterbroek, T., Orr, A., Antonelli, L. A., Fiore, F., *et al.*: 1998, *ApJ* **493**, L47.

Pepino, R., Kharchenko, V., Dalgarno, A., and Lallement, R.: 2004, *ApJ* **617**, 1347.

Pfeffermann, E., Briel, U. G., Hippmann, H., Kettenring, G., Metzner, G., Predehl, P., *et al.*: 1986, *Proc. SPIE* **733**, 519.

Reilman, R. F., and Manson, S. T.: 1979, *ApJS* **74**, 815.

Sehnal, L.: 1990a, *Bull. Astron. Inst. Czechosl.* **41**, 115.

Sehnal, L.: 1990b, *Bull. Astron. Inst. Czechosl.* **41**, 108.

Smith, R. K., Brickhouse, N. S., Liedahl, D. A., and Raymond, J. C.: 2001, *ApJ* **556**, L91.

Strüder, L., Briel, U., Dennerl, K., Hartmann, R., Kendziorra, E., Meidinger, N., *et al.*: 2001, *A&A* **365**, L18.

Tobiska, W. K., Woods, T., Eparvier, F., Viereck, R., Floyd, L., Bouwer, D., *et al.*: 2000, *J. Atm. Solar Terr. Phys.* **62**, 1233.

Trümper, J.: 1983, *Adv. Space Res.* **2**(4), 241.

Turner, M. J. L., Abbey, A., Arnaud, M., Balasini, M., Barbera, N., Belsole, E., *et al.*,: 2001, *A&A* **365**, L27.

Wickramasinghe, N. C., and Hoyle, F.: 1996, *Ap&SS* **239**, 121.

Winckler, J. R., Peterson, L., Arnoldy, R., and Hoffman, R.: 1958, *Phys. Rev.* **110**, 1221.

Withers, P.: 2006, *Geophys. Res. Lett.* **33.2**, 10.1029/L02201.

ASYMMETRIES IN MARS' EXOSPHERE

Implications for X-ray and ENA Imaging

MATS HOLMSTRÖM
Swedish Institute of Space Physics, PO Box 812, SE-981 28 Kiruna, Sweden
(E-mail: matsh@irf.se)

(Received 6 April 2006; Accepted in final form 14 September 2006)

Abstract. Observations and simulations show that Mars' atmosphere has large seasonal variations. Total atmospheric density can have an order of magnitude latitudinal variation at exobase heights. By numerical simulations we show that these latitude variations in exobase parameters induce asymmetries in the hydrogen exosphere that propagate to large distances from the planet. We show that these asymmetries in the exosphere produce asymmetries in the fluxes of energetic neutral atoms (ENAs) and soft X-rays produced by charge exchange between the solar wind and exospheric hydrogen. This could be an explanation for asymmetries that have been observed in ENA and X-ray fluxes at Mars.

Keywords: Mars, energetic neutral atoms, X-rays, exospheres

1. Introduction

Traditionally, exospheric densities and velocity distributions are modelled by spherical symmetric analytical Chamberlain functions (Chamberlain and Hunten, 1987). Chamberlain theory assumes that gravity is the only force acting on the neutrals, that the exobase parameters (density and temperature) are uniform over a spherical exobase, and that no collisions occur above the exobase. Planetary exospheres are however not spherical symmetric due to non-uniform exobase parameters and due to effects such as photoionization, radiation pressure, charge exchange, recombination and planetary rotation. To account for these effects numerical simulations are needed. Using Monte Carlo test particle simulations it is possible to account for the above effects (if ion distributions are assumed).

Even though neutrals in the exospheres by definition do not collide often, collisions occur. Especially near the exobase the transition is gradual from collision dominated regions at lower heights (with Maxwellian velocity distributions) to essentially collisionless regions at greater heights. Using test particles one can model collisions with an assumed background atmospheric profile (Hodges Jr., 1994), but to account for collisions properly the test particle approach is not sufficient, and self consistent simulations are needed. One approach to model collisions is the direct simulation Monte Carlo (DSMC) method (Bird, 1976) for rarefied flows, that has been applied to exospheres by Krestyanikova and Shematovich (2005).

In this work we use the test particle approach to model the effects on the Martian exosphere from non-uniform exobase conditions, from photoionization, from

Space Science Reviews (2006) 126: 435–445
DOI: 10.1007/s11214-006-9036-7

radiation pressure, and from solar wind charge exchange. We launch test particles from the exobase and follow their trajectories. The forces on the particles are from gravity and radiation pressure. Along their trajectories the particles can be photoionized, and they can charge exchange with solar wind protons outside the bow shock. Exospheric column densities give us a qualitative estimate of how exospheric asymmetries effect solar wind charge exchange (SWCX) X-ray images. The energetic neutral atoms (ENAs) produced by charge exchange then gives us estimates of the ENA fluxes near Mars.

In this work we do not include the effects of collisions since it greatly increases the computational cost. Collisions and photochemical reactions are two channels, in addition to charge exchange, that produce an hot hydrogen population (Nagy *et al.*, 1990). By only including charge exchange outside the bow shock as a source of hot hydrogen in this work we therefore under estimate the extent of the hydrogen corona. Including all sources of hot hydrogen would result in more extended emissions of ENAs and X-rays, compared to the results presented here. An additional process that we do not consider is electron impact ionization, since it would require knowledge of electron fluxes and velocity distributions. However, this work should be seen as a first qualitative study of how asymmetries in exobase conditions at Mars effect the exosphere, and in turn the ENA and X-ray fluxes near Mars. To do a more accurate quantitative study is much more difficult. One then would need to specify the exact time, season and Mars–Sun distance; and have access to exobase conditions (at that time) from observations, global circulation models, and solar wind conditions, along with full knowledge of the ion fluxes near Mars.

1.1. ENA and X-ray Imaging

When the solar wind encounters a non-magnetized planet with an atmosphere, e.g., Mars or Venus, there will be a region of interaction, where solar wind ions collide with neutrals in the planet's exosphere. Two of the processes taking place are

- The production of ENAs by charge-exchange between a solar wind proton and an exospheric neutral (Holmström *et al.*, 2002), and
- The production of soft X-rays by SWCX between heavy, highly charged, ions in the solar wind and exospheric neutrals (Holmström *et al.*, 2001).

Images of ENAs and SWCX X-rays can provide global, instantaneous, information on ion-fluxes and neutral densities in the interaction region. It is however not easy to extract this information from the measured line of sight integrals that are convolutions of the ion-fluxes and the neutral densities. We need to introduce models that reduce the complexity of the problem. At Mars, the hydrogen exosphere is enlarged due to the planet's low gravity, and thus provide a large interaction region, extending outward several planet radii. Traditionally, most of the modeling of the outer parts of Mars' exosphere has been using analytical, spherical symmetric, Chamberlain profiles. Planetary exospheres are however not spherical symmetric to any good approximation, and asymmetries at Mars observed in ENAs by Mars Express and in X-rays by XMM-Newton could be due to asymmetries in the exosphere. The neutral particle imager, part of ASPERA-4 on-board Mars Express, has observed asymmetries in the ENA fluxes in the shadow of the planet (Brinkfeldt *et al.*, 2006). The decline of ENA fluxes when entering the shadow is different from the rise in flux when exiting the shadow. The XMM Newton X-ray telescope observed Mars in November 2003, and SWCX X-rays were positively identified from the hydrogen corona (Dennerl *et al.*, 2006). The morphology of the images are however different from what have been predicted by simulations (Holmström *et al.*, 2001). One of the differences is that the emissions are asymmetric with respect to the ecliptic plane. Here we investigate the asymmetries in exospheric densities at Mars due to various factors, and their impact on ENA and SWCX X-ray images.

We may note that although asymmetric exospheres have not been used often in modeling of solar wind-Mars interactions, they are well known in the engineering community since aerobraking and satellite drag is directly dependent on exospheric densities, and provides total density measurements (Justus *et al.*, 2002).

In Section 2 we describe in more detail the methods and parameters used in our simulations. In Section 3 we then present the results of our numerical experiments, and finally we present conclusions in Section 4.

2. Methods

Here we first describe the algorithms used to simulate Mars' hydrogen exosphere, and we then describe the detailed setup used in the numerical experiments.

2.1. The Simulation Algorithm

In what follows, the coordinate system used is Mars solar ecliptic coordinates, centered at the planet with the x-axis toward the Sun, the z-axis perpendicular to the planet's velocity, in the northern ecliptic hemisphere, and a y-axis that completes the right handed system. Based on this solar ecliptic coordinate system we define (longitude, latitude) coordinates, with the z-axis toward 90° latitude, the x-axis (sub solar point) at (0, 0), and the y-axis at (90, 0).

The simulation domain is bounded by two spherical shells centered at Mars. An inner boundary (the exobase) with a radius of 3580 km corresponding to a height of 200 km above the planet – assuming from now on a planet radius R_M of 3380 km – and an outer boundary with a radius of 10 R_M. At the start of the simulation the domain is empty of particles. Then meta-particles are launched from the inner boundary at a rate of 1000 meta-particles per second. Each meta-particles corresponds to N_m hydrogen atoms. The location on the inner boundary of each launched particle is randomly drawn with probability proportional to the local hydrogen exobase density. The velocity of each launched particle is randomly drawn from a probability distribution proportional to

$$(\mathbf{n} \cdot \mathbf{v})\, e^{-a|\mathbf{v}|^2},$$

where $\mathbf{n}$ is the local unit surface normal, $\mathbf{v}$ is the velocity of the particle, and $a = m/(2kT)$, m is the mass of a neutral, k is Boltzmann's constant, and T is the temperature (at the exobase position). Note that the distribution used is not a Maxwellian, but the distribution of the flux through a surface (the exobase) given a Maxwellian distribution at the location (Garcia, 2000).

After an hydrogen atom is launched from the inner boundary, we numerically integrate its trajectory with a time step of 5 s. To avoid energy dissipation, the time advance of the particles is done using a fourth order accurate symplectic integrator derived by Candy and Rozmus (1991).

Between time steps, the following events can occur for an exospheric atom

- Collision with an UV photon. Following Hodges Jr. (1994) this occurs as an absorption of the photon (Δv opposite the sun direction) followed by isotropic reradiation (Δv in a random direction). From Hodges Jr. (1994) we use a velocity change $\Delta v = 3.27$ m/s. The collision rate used is 10^{-3} s^{-1}, and the rate is zero if the particle is in the shadow behind the planet.
- Charge exchange with a solar wind proton. If the hydrogen atom is outside Mars' bow shock it can charge exchange with a solar wind proton, producing an ENA, at a rate of 8.4×10^{-8} s^{-1}. The ENA is randomly drawn from a Maxwellian velocity distribution with a bulk velocity of 450 km/s in the anti-sunward direction, and a temperature of 1.2×10^5 K. Thus, the original exospheric hydrogen atom is replaced by the ENA in the simulation. Following Slavin *et al.* (1991), we define the bow shock by the surface $(x, \rho)\, R_M$ such that

$$x = \frac{-x_0 + Le + x_0 e^2 - \sqrt{\rho^2 e^2 - \rho^2 L^2}}{e^2 - 1},$$

  where $L = 2.04\ R_M$, $e = 1.02$, and $x_0 = 0.55\ R_m$. Here $\rho = \sqrt{y^2 + z^2}$ is the distance to the x-axis (the Mars–Sun line). We can note that the charge exchange rate gives an average life time for an hydrogen atom of more than 100 days in the solar wind. The implication for our simulations is that few ENA meta-particles are produced. To handle this we increase the charge exchange rate by a factor of $f = 1000$ and when a charge exchange event occurs, the exospheric meta-particle with weight N_m is replaced by a meta-particle with weight $(1 - 1/f)N_m$ and an ENA with weight N_m/f.
- Photoionization by a solar photon occurs at a rate of 10^{-7} s when an exospheric hydrogen atom is outside the optical shadow behind the planet, and then the meta-particle is removed from the simulation.

All rates above are from Hodges Jr. (1994) for Earth, and average solar conditions, scaled by 0.43 to account for the smaller fluxes at Mars. For a given event rate, τ, after each time step, for each meta-particle, we draw a random time from an exponential distribution with mean τ, and the event occur if this time is smaller than the time step. Note that we only consider ENAs produced outside the bow shock, so the fluxes presented here is a lower bound. Additional ENAs are produced inside the bow shock, but including those would require a complete ion flow model. Anyhow, simulations (Holmström *et al.*, 2002) suggest that the ENA flux from the solar wind population is dominant in intensity. Also, we do not consider collisions between neutrals, as discussed in the introduction. We can note that omitting collisions means that the population of particles on satellite orbits will be small. The only generation mechanism for satellite particles will be radiation pressure.

2.2. The Simulation Setup

As stated in the introduction, the aim of this study is to make a qualitative study of the effects of non-uniform exobase conditions on the hydrogen exosphere, and the implications for ENA and SWCX X-ray fluxes. Thus, we choose to study a simplified model problem where we have artificially chosen a spatial distribution of exobase density and temperature, shown in Figure 1. We have constant density in three longitude bands, and three different temperature regions. This is an approximation of the conditions at southern summer solstice, and was chosen as follows. We use the density and temperature for solar minimum conditions from (Krasnopolsky, 2002, Figure 1) at a height of 200 km; 200 K and 4.2×10^5 cm^{-3} as a reference value. This is a day side average for a solar zenith angle of 60°. The corresponding density at 130 km (mostly CO_2) is 2.9 kg/km^3. Using the spatial variations from (Bougher *et al.*, 2000, Figure 5 and 10) we scale the reference

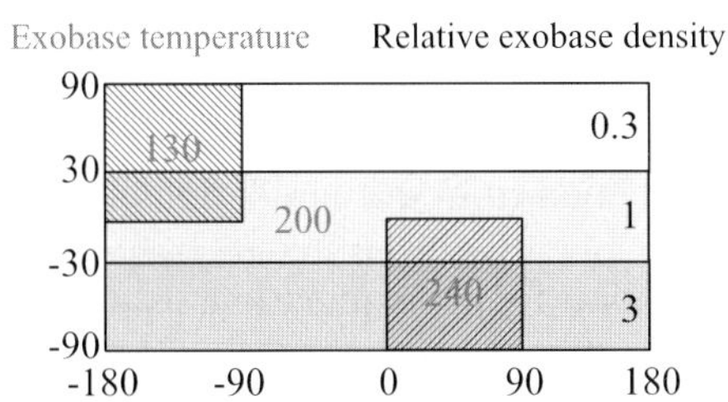

Figure 1. The non-uniform exobase density and temperature for Mars' hydrogen exosphere used in this study. The coordinate system is Mars solar ecliptic longitude and latitude, with (0, 0) corresponding to the sub solar point. The exobase temperature is shown in gray, with a value of 200 K, except for a region of increased temperature on the dayside, and a region of decreased temperature on the nightside. The exobase density is constant in three longitude bands, and shown in black is the density relative to a reference density of 4.2×10^5 cm^{-3}.

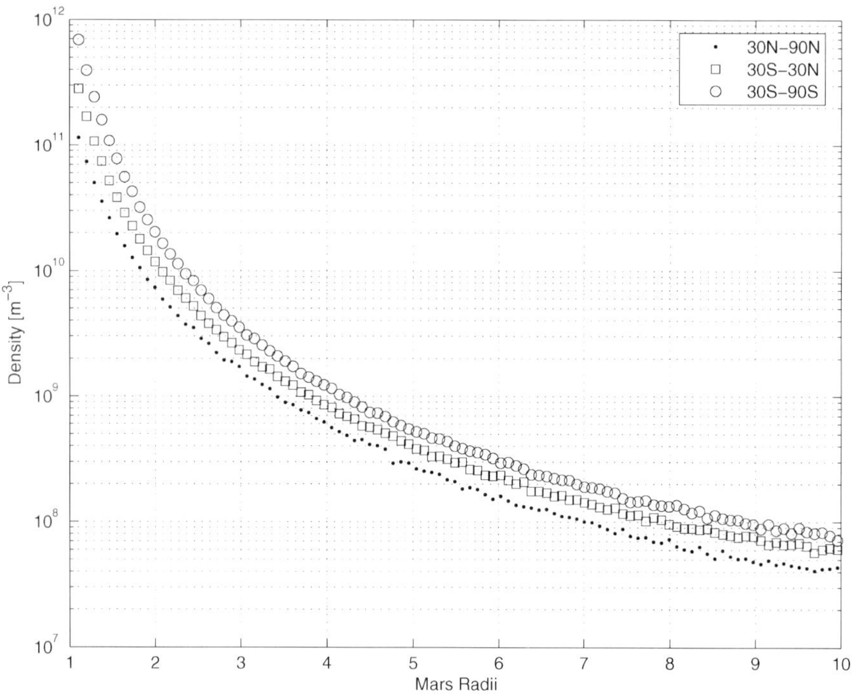

Figure 2. Average hydrogen number density [m^{-3}] as a function of planetocentric distance in R_M for non-uniform exobase conditions. The exosphere has been divided into three regions corresponding to the three latitude bands in Figure 1, and each of the profiles shown is an average over the corresponding region. Latitude 30 to 90 degrees (dots), -30 to 30 (squares), and -90 to -30 (circles).

values, and construct the exobase conditions shown in Figure 1. We will later denote this the non-uniform case, and the case when we use the reference values for all of the exobase will be the uniform case. The spatial variations in (Bougher *et al.*, 2000) are from a global circulation model of Mars' exosphere and is based on the observations available at that time. Later the model has been partially verified by observations (Lillis *et al.*, 2005). Note that these exobase parameters specify the upward velocity distribution of neutrals at the inner boundary (the exobase). The downward flux is then obtained from the simulation. Therefore, these parameters will differ from the values obtained from the converged simulation, e.g., we can see in Figure 2 that the number density at the inner boundary has the proportions 1, 3, and 7 in the different latitude bands.

3. Numerical Experiments

First we investigate the effects of non-uniform exobase conditions on the hydrogen exosphere. Then we study the implications for ENA and SWCX X-ray fluxes.

3.1. The Exosphere

Here we use the non-uniform exobase conditions shown in Figure 1. First of all we want to examine how far out from the planet the exosphere is non-uniform. Since the simulation particles are launched on ballistic trajectories, at any point there will be a mix of particles from different regions of the exobase. This will introduce a smoothing of the exobase boundary conditions, and it is not obvious how large this smoothing will be, i.e. how far from the planet the non-uniformity will persist. To investigate this we divide the exosphere into three regions corresponding to the three latitude bands in Figure 1, and plot profiles of the average hydrogen density for each of the regions. These profiles at a time of 10 h after the start of the simulation are shown in Figure 2. We see that the density variation at the exobase by a factor of 10 is reduced to approximately a factor of 3 at a planetocentric distance of 2 R_M, and a factor of 2 at 10 R_M. Thus, the exospheric densities get more uniform with distance to the planet, but large differences in density persist all through the simulation domain.

3.2. SWCX X-rays

To estimate the effects of the non-uniform exosphere on SWCX X-ray images, we note that the X-ray flux is a line-of-sight convolution of ion flux and hydrogen density. So in the unperturbed solar wind, outside the bow shock, the X-ray flux should be proportional to the hydrogen column density. In Figure 3 we show the hydrogen column density for the cases of uniform and non-uniform exobase conditions. These are then estimates of what SWCX X-ray images would look like, from Earth at Mars' opposition, at least away from the

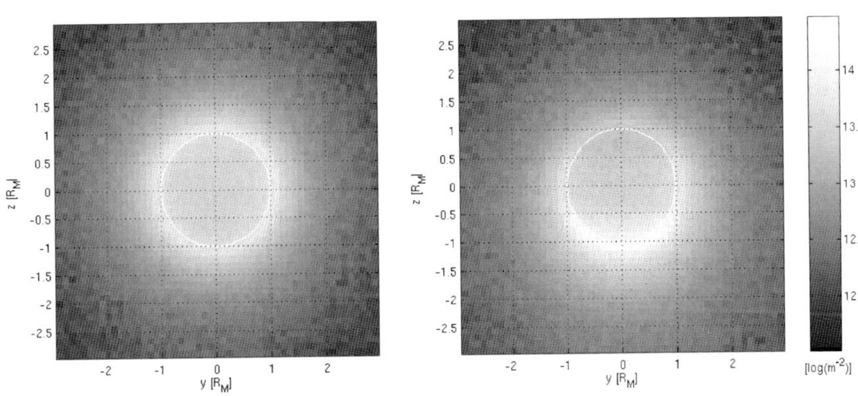

Figure 3. Logarithm of the hydrogen column density [m^{-2}] along the x-axis for uniform (left) and non-uniform (right) exobase conditions. The white circles shows the size of the exobase. The time is 10 h, the axes' units are R_M and the total number of meta-particles are 848320 and 893240. The maximum column density is 0.781 and 1.86 $\times$ 10^{14} m^{-2}.

planet (near the planet X-ray fluorescence dominate anyway). Note that the column density can vary by almost an order of magnitude, for constant planetocentric distances, even far away from the planet, in the non-uniform case. This would directly effect SWCX X-ray images and lead to asymmetries of the same magnitude.

3.3. ENA Fluxes

Here we investigate the fluxes of hydrogen ENAs that are created outside the bow shock by charge exchange between exospheric hydrogen and solar wind protons, with respect to any asymmetries induced by the asymmetric exosphere. One motivation for this investigation is that the neutral particle imager, part of ASPERA-4 on-board Mars Express, has seen asymmetries in the ENA fluxes in the tail behind the planet (Brinkfeldt *et al.*, 2006). In Figure 4 we compare the ENA fluxes through a plane at $x = -1.0$ for uniform and non-uniform exobase conditions. We can note the spherical symmetry for the case of uniform exobase conditions, apart from the statistical fluctuations associated with test particle Monte Carlo simulations. On the other hand, in the case of non-uniform exobase parameters the asymmetry of the ENA flux is clearly visible as enhanced, and extended, flux in the south corresponding to the higher densities in the southern hemisphere. There is also a suggestion of enhanced densities in the $+y$ hemisphere corresponding to the enhanced exobase temperature in that hemisphere. For a constant planetocentric distance in this plane we see that the ENA flux can vary by more than a factor of 2.

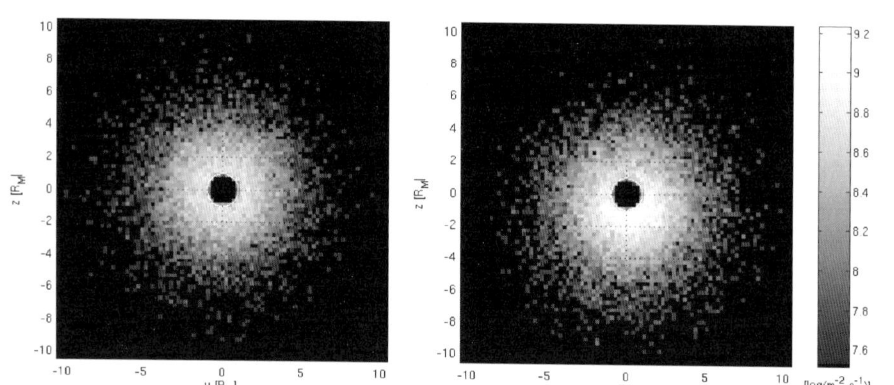

Figure 4. Logarithm of the ENA fluxes $[\mathrm{m}^{-2}\mathrm{s}^{-1}]$ through the yz-plane at $x = -1.0\ R_M$ for uniform exobase conditions (left), and for non-uniform conditions (right). The maximum flux is 0.912 and $1.79 \times 10^9\ \mathrm{m}^{-2}\ \mathrm{s}^{-1}$. The fluxes are computed by averages over all ENAs with $-1.05 < x < -0.95$, from time 0 to 10 h, the axes' units are R_M, and the total number of ENA meta-particles is 18796 and 22924. The white circle shows the size of the exobase.

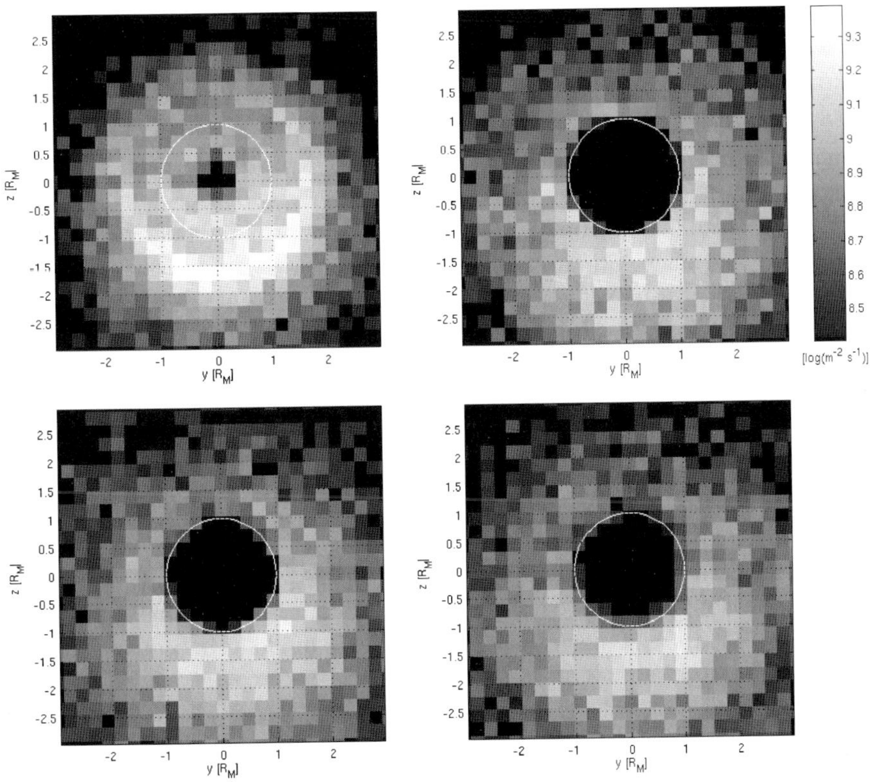

Figure 5. Logarithm of the ENA fluxes [$m^{-2}s^{-1}$] through the yz-planes at $x = 1\ R_M$ (upper left), $x = 0$ (upper right), $x = -1$ (lower left), and $x = -3$ (lower right) for non-uniform exobase conditions. All are averages over an x-width of 0.1 R_M. The maximum fluxes are 2.50, 1.63, 1.79, and 1.60×10^9 m^{-2} s^{-1}. The white circle shows the size of the exobase.

How does the ENA fluxes vary at different positions relative to the planet? In Figure 5 we plot the fluxes through the yz-planes at $x = 1.0, 0.0, -1.0$, and -3.0. The north-south asymmetry is visible in all plots, with the highest, most concentrated fluxes at $x = 1$. The area of large flux then spreads out slightly and has a bit lower intensity toward the tail. Note however that the flux at $x = -3$ is as large as the flux at $x = 0$. For all plots the maximum flux seems to be obtained at an approximate distance from the x-axis of 5000 km (about 1600 km outside the optical shadow). This is perhaps a bit surprising – that the maximum flux is not closer to the umbra, but is a consequence of the shape of the bow shock in combination with the exospheric profiles, as seen in the flux through $x = 1$ that is a crescent well outside the planet outline.

In all numerical experiments above, radiation pressure and photoionization was not included. It was found that including those events, as described in the previous section, did not change the results presented in any significant way.

4. Conclusions

Traditionally, modeling of the solar wind interaction with Mars' exosphere, and the production of SWCX X-rays and ENAs, has assumed a spherical symmetric exosphere. From observations and simulations we know however that the exosphere is not symmetric. From the results of our simple test particle model of Mars' exosphere we find that asymmetries in exobase density and temperature propagate to large heights (many Martian radii). Column densities can deviate by almost an order of magnitude from symmetry, implying similar asymmetries in SWCX X-ray images. We also find that the fluxes of ENAs that are produced in the solar wind can deviate by more than a factor of two from symmetry. These asymmetries could explain the asymmetries seen in X-ray images and in ENA observations, but further studies are needed to find out if that is the case. We also find that radiation pressure and photoionization are unimportant processes in comparison to asymmetries in exobase parameters. Finally, we can note that asymmetries in the exosphere could also possibly explain the low exospheric densities seen by the neutral particle detector on-board Mars Express, as reported in this issue by Galli *et al.* (2006), since that measurement was over the northern hemisphere during early northern spring (April 25, 2004), when exospheric densities should have been low due to the seasonal variations.

Acknowledgements

Parts of this work was accomplished while the author visited NASA's Goddard Space Flight Center during 2005, funded by the National Research Council (NRC). The software used in this work was in part developed by the DOE-supported ASC / Alliance Center for Astrophysical Thermonuclear Flashes at the University of Chicago.

References

Bird, G. A.: 1976, *Molecular Gas Dynamics*, Clarendon Press.

Bougher, S., Engel, S., Roble, R., and Foster, B.: 2000, *J. Geophys. Res.* **105**(E7), 17669.

Brinkfeldt, K., Gunell, H., Brandt, P., Barabash, S., Frahm, R., Winningham, J. *et al.*: 2006, *Icarus* **182**(2), 439.

Candy, J., and Rozmus, W. (1991), *J. Comput. Phys.* **92**, 230.

Chamberlain, J. W., and Hunten, D. M.: 1987, *Theory of Planetary Atmospheres*, Academic, San Diego, Calif.

Dennerl, K., Lisse, C., Bhardwaj, A., Englhauser, V. B. J., Gunell, H., Holmström, M. *et al.*: 2006, *Astronomy Astrophys.* **451**, 709.

Galli, A., Wurz, P., Lammer, H., Lichtenegger, H., Lundin, R., Barabash, S., *et al.*: 2006, *The Hydrogen Exospheric Density Profile Measured with ASPERA-3/NPD*, *Space Science Reviews*, this issue, doi: 10.1007/s11214-006-9089-7.

Garcia, A. L.: 2000, *Numerical Methods for Physics*, 2nd edn., Prentice Hall, Chapter 11.3.
Hodges Jr., R. R.: 1994, *J. Geophys. Res.* **99**(A12), 23229.
Holmström, M., Barabash, S., and Kallio, E.: 2001, *Geophys. Res. Lett.* **28**(7), 1287.
Holmström, M., Barabash, S., and Kallio, E.: 2002, *J. Geophys. Res.* **107**(10).
Justus, C., James, B., Bougher, S., Bridger, A., Haberle, R., Murphy, J., *et al.*: 2002, *Adv. Space Res.* **29**(2), 193.
Krasnopolsky, V. A.: 2002, *J. Geophys. Res.* **107**(E12), 5128, doi: 10.1029/2001JE001809.
Krestyanikova, M. A., and Shematovich, V. I.: 2005, *Solar Syst. Res.* **39**, 22, doi: 10.1007/s11208-005-0012-7.
Lillis, R. J., Engel, J. H., Mitchell, D. L., Brain, D. A., Lin, R. P., Bougher, S., *et al.*: 2005, *Geophys. Res. Lett.* **32**, doi: 10.1029/2005GL024337.
Nagy, A. F., Kim, J., and Cravens, T. E. 1990, *Ann. Geophys.* **8**(4), 251.
Slavin, J. A., Schwingenschuh, K., Riedler, W., and Eroshenko, E.: 1991, *J. Geophys. Res.* **96**, 11235.

THE HYDROGEN EXOSPHERIC DENSITY PROFILE MEASURED WITH ASPERA-3/NPD

A. GALLI[1,*], P. WURZ[1], H. LAMMER[2], H. I. M. LICHTENEGGER[2], R. LUNDIN[3], S. BARABASH[3], A. GRIGORIEV[3], M. HOLMSTRÖM[3] and H. GUNELL[3]

[1]*Physikalisches Institut, Sidlerstrasse 5, CH-3012, Bern, Switzerland*

[2]*Institut für Weltraumforschung, Österreichische Akademie der Wissenschaften, A-8042 Graz, Austria*

[3]*Swedish Institute of Space Physics, Box 812, SE-981 28 Kiruna, Sweden*

*(*Author for correspondence: E-mail: galli@space.unibe.ch)*

(Received 4 April 2006; Accepted in final form 25 October 2006)

Abstract. We have evaluated the Lyman-α limb emission from the exospheric hydrogen of Mars measured by the neutral particle detector of the ASPERA-3 instrument on Mars Express in 2004 at low solar activity (solar activity index = 42, $F_{10.7} = 100$). We derive estimates for the hydrogen exobase density, $n_H = 10^{10}$ m^{-3}, and for the apparent temperature, $T > 600$ K. We conclude that the limb emission measurement is dominated by a hydrogen component that is considerably hotter than the bulk temperature at the exobase. The derived values for the exosphere density and temperature are compared with similar measurements done by the Mariner space probes in the 1969. The values found with Mars Express and Mariner data are brought in a broader context of exosphere models including the possibility of having two hydrogen components in the Martian exosphere. The present observation of the Martian hydrogen exosphere is the first one at high altitudes during low solar activity, and shows that for low solar activity exospheric densities are not higher than for high solar activity.

Keywords: Martian exosphere, UV airglow measurements

1. Introduction

The physical properties of the Martian exosphere determine atmospheric loss rates, which are in turn necessary to understand the water inventory evolution of the planet (Lammer *et al.*, 2003). The thermal escape rate, for example, depends directly upon exospheric temperatures. The ASPERA-3 experiment on board the Mars Express spacecraft has been designed to study the interaction of the solar wind with the upper atmosphere (Barabash *et al.*, 2004) by measuring ions, electrons and energetic neutral atoms (ENAs). It was not designed to study the neutral exosphere directly, but the density and temperature of the hydrogen exosphere are the most important parameters that affect the production of ions and ENAs in the near-Mars space (Holmström *et al.*, 2002).

Unfortunately, there are still no *in situ* data available from the Martian hydrogen exosphere. All temperature and density values have been derived either from airglow measurements of the Lyman-α emission of neutral hydrogen or from aerobraking data (Lichtenegger *et al.*, this issue). The two Viking landers (Nier and McElroy, 1977) have performed mass spectrometer measurements below 200 km altitude,

Space Science Reviews (2006) 126: 447–467
DOI: 10.1007/s11214-006-9089-7

but it has been questioned if the measured temperatures of carbon oxides, nitrogen and argon apply to the exospheric hydrogen, too.

The first work to derive the temperature and the exobase density of the hydrogen corona was done by Anderson and Hord (1971), based on Mariner 6 and 7 Lyman-α airglow measurements at high solar activity (Barth *et al.*, 1971, 1972). Assuming one single temperature for the hydrogen exosphere they derived $T = 350$ K and $n_H = 3 \times 10^{10}$ m^{-3} at 250 km above the surface. Mariner 9 airglow measurements confirmed this temperature (Barth *et al.*, 1972). On the other hand, temperatures derived from mass spectrometer or aerobraking data (186 K and 145 K in the case of the Viking landers, 220 to 230 K for Mars Global Surveyor (Bougher *et al.*, 2000)) are considerably lower. Recent CO_2 dayglow measurements done with SPICAM (Leblanc *et al.*, 2006) for altitudes below 200 km also yield a temperature of only 200 K. In analogy to Venus, where the exospheric hydrogen density profile has been explained (Anderson, 1976; Bertaux *et al.*, 1978) by the presence of a cool bulk component and a tenuous hot population of atoms that have been heated up in photodissociation and dissociative recombination processes, the same approach is proposed by Lichtenegger *et al.* (this issue) for the Martian exosphere to resolve the discrepancy between Lyman-α airglow measurements and *in situ* measurements.

In this work we present a recent Lyman-α airglow measurement performed with the neutral particle detector on Mars Express in 2004 during low solar activity (solar activity index equal to 42). We interpret our measurements with a numerical exosphere model, and we compare the results to the Lyman-α airglow measurements from Mariner 6 and 7 in 1969.

We give a quick overview over the instrument (Section 2) and explain how we calibrated the Lyman-α sensitivity of the neutral particle detector (NPD) that was originally not intended as a UV detector (Section 3). After showing the observation condition for the Lyman-α airglow measurement (Section 4) we introduce the numerical exosphere model and the radiation transport equation that relates the modeled density profile to the measured UV emission (Section 5). We then interpret the UV limb emission: First we assume one single hydrogen component and search for the exosphere model whose temperature and exobase density fits best (Section 6.1). We compare these values for low solar activity to those found from Mariner measurements (Anderson and Hord, 1971) for high solar activity. Finally, we study what constraints the Mariner and Mars Express data put on an exosphere model with two hydrogen components (Section 6.2).

2. Instrumentation

The ASPERA-3 instrument on board the Mars Express spacecraft comprises four different sensors. The ion mass analyser and the electron spectrometer are used to measure local ion and electron densities, and the neutral particle detector (NPD) and the neutral particle imager (NPI) to detect energetic neutral atoms (ENAs)

(Barabash *et al.*, 2004). In the current report we concentrate on one single NPD data set.

The neutral particle detector consists of two identical sensors, NPD1 and NPD2, which are sensitive to ENAs in the energy range of 0.1 to 10 keV using the time-of-flight technique. Each sensor has one start and three stop surfaces, which provide an angular resolution of roughly 30° in azimuthal direction as shown in Figure 4. Although the NPD entrance system has been designed to suppress UV photon count rates the suppression is not complete, and an electronic signal can be triggered on the start and stop surface if the energy of a photon exceeds the ionization potential of the detector surface. The stop surface has an MgO coating with a bandgap of 7.8 eV (Deutscher *et al.*, 1999). Thus, UV radiation at wavelengths below $\lambda = 160$ nm is clearly visible in NPD data:

$$\lambda = hc/E_{\mathrm{crit}}, \quad E_{\mathrm{crit}} = 7.8\,\mathrm{eV} \tag{1}$$

The UV emission of the Martian atmosphere is known (Barth *et al.*, 1972, based on Mariner data) to have no contribution in the wavelength range below 160 nm except from the Lyman-α ($\lambda = 121.6$ nm) line that is caused by resonant scattering of solar UV light on neutral hydrogen atoms. The resonant scattering on helium atoms produces UV emission at even shorter wavelengths ($\lambda = 58.4$ nm), but the disk airglow intensity due to neutral helium has been found to reach only 40 to 70 Rayleigh (R) (Krasnopolsky and Gladstone, 1996, 2005) (1 R$\doteq 10^{10}/(4\pi)\,\mathrm{m}^{-2}\,\mathrm{sr}^{-1}\,\mathrm{s}^{-1}$), which is orders of magnitudes smaller than the hydrogen intensities. Therefore, we can use NPD data to map the neutral hydrogen densities when looking at the Mars exosphere.

3. Calibration of UV Sensitivity

The NPD sensor accumulates signals of incoming ENAs and UV photons in steps of one second by sampling all coincident count events. A start-stop signal pair is regarded as a correlated event, if the time gap between the two signals does not exceed 2048 ns. This time gap allows the detection of ENAs in the energy range between 0.1 and 10 keV. On the other hand, two different photons that by coincidence hit the start and the stop surface within 2048 ns mimic a correlated event as well. ENAs and UV background can be separated only by studying the entire TOF spectrum. This is demonstrated in Figure 1: After integrating the NPD count rates over several minutes a stream of ENAs produces a measurable peak in the TOF spectrum depending on mass and energy of the particles. UV photons, on the other hand, produce a flat noise level, which can appropriately be parameterized by a linear model. Since the NPD detector is only sensitive to hydrogen ENAs above 0.1 keV, corresponding to the first hundred TOF bins, the higher TOF bins of the spectrum can be used to estimate the background level due to UV photons.

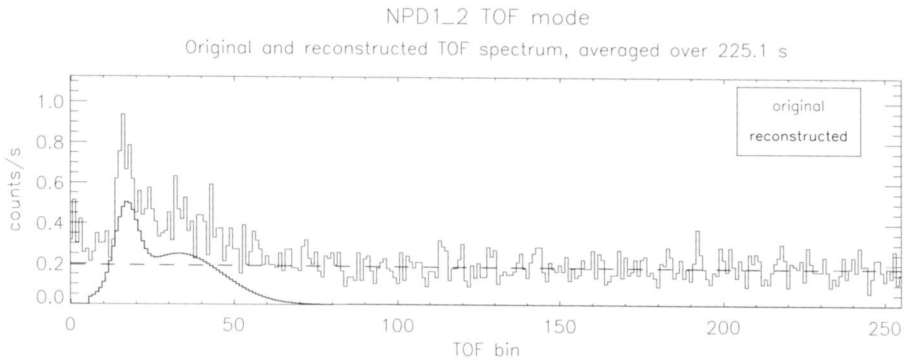

Figure 1. Time-of-flight spectrum measured by NPD1_2 from 12:29 until 12:33 UT on April 25, 2004. The peak between TOF bins 10 and 60 is due to hydrogen ENAs. The flat noise level of 0.2 counts per second is due to coincident photons. According to our calibration (Equation (3)) this level corresponds to a UV flux of 1.17 ± 0.22 kR.

To deduce the UV intensity from the registered background count rate (see Figure 1) the UV sensitivity of the instrument was calibrated using all NPD data from July 2003 to January 2004 during the cruise phase. For these measurements far away from any planetary atmosphere the background reflects only the known UV brightness of the sky. Figure 2 shows our reference UV map. It is a composite picture of UV measurements done by the SWAN experiment on SOHO (NASA, 2006) and gives an average UV intensity in Rayleigh for the year 2004 for all viewing directions. The map includes all non-planetary UV sources: the interstellar gas, single stars and the galactic background. The sensitivity range of the SWAN instrument includes wavelengths from 117 to 180 nm, but the vast majority of the photons belongs to the Lyman-α line of neutral hydrogen at 121.6 nm. The omnidirectional UV background of 300 to 600 R is due to resonant scattering of Lyman-α radiation on interstellar neutral hydrogen. The bright stars in the galactic plane emit UV radiation outside the Lyman-α line, but according to the diffuse sky brightness model of Leinert *et al.* (1998) the UV flux integrated from 140 to 180 nm does not exceed 60 R even in the galactic plane where the stellar UV brightness is highest. Thus, the UV map we have used for calibration (Figure 2) shows in good approximation the Lyman-α photon flux and the NPD instrument registers, at wavelengths below 160 nm, virtually only Lyman-α photons as well. In the following sections, 'UV' and 'Lyman-α' will be used as synonyms by default.

For the subsequent analysis of the Martian limb emission measurement we need to have a relation between the average UV flux inside the field-of-view of an NPD sector and the background count rate that is registered for this viewing direction. The measured background count rate in NPD must be proportional to the probability that two independent UV photons trigger a start and a stop pulse within a small

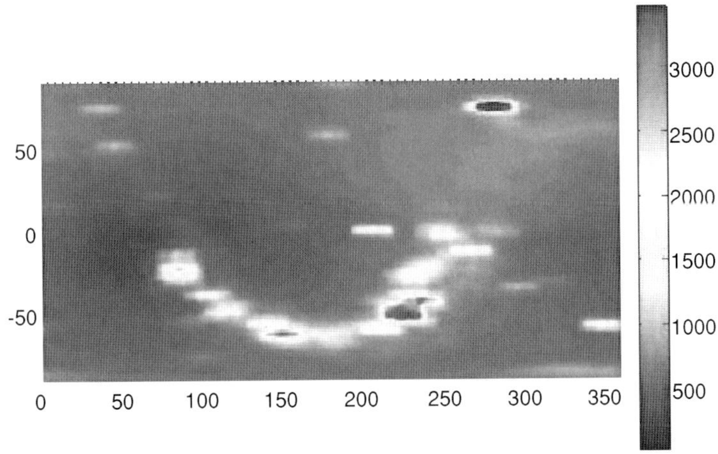

Figure 2. Full-sky image of the UV brightness in the wavelength range $\lambda = 117$ to 180 nm, adapted from measurements made with SWAN on SOHO (NASA, 2006). Coordinates are given in the ecliptic reference frame, the units of the color bar are given in Rayleigh. The dark background is due to the resonant scattering of Lyman-α radiation on interstellar hydrogen atoms, whereas the narrow U-bend shows the galactic plane with its bright UV stars.

time interval. More precisely, we expect the following relationship between the UV intensity I [kR] and the background count rate C [counts s^{-1}]:

$$C = c_1' I^2 + c_0 \text{ or } I = c_1 \sqrt{C - c_0}, \tag{2}$$

with c_0 the dark count rate.

To find the calibration constants c_0 and c_1 we have evaluated all available measurements from the cruise phase. Figure 3 shows the measurements and the least-squares fit curves for two of the six NPD directions, NPD1_1 (triangles) and NPD1_2 (circles). The error bars of the count rates reflect the statistical errors, the uncertainty of the UV intensity is due to possible short time variations of the solar UV irradiance and due to the size of the field-of-view ($4^\circ \times 30^\circ$) that may cover regions of variable UV brightness. This uncertainty is most pronounced for the six data points above 1 kR for which NPD1_1 was directed to the region of α Cen (the dark spot at $\lambda = 240^\circ$, $\beta = -45^\circ$ in Figure 2).

Within the individual error bars Equation (2) describes the UV sensitivity of the instrument. For NPD1_1 a least-squares fit yields for the dark count rate $c_0 = 0$ counts s^{-1} and $c_1 = 2.83 \pm 0.5$ kR s$^{1/2}$. The error bar takes into account the error of the count rates and of the UV intensities. In the following we will only use the calibration constants for NPD1_2 because the UV limb emission measurement was made with this sector. For NPD1_2 we find very similar calibration constants, $c_0 = 0.0037 \pm 0.003$ counts s^{-1} and $c_1 = 2.64 \pm 0.35$ kR s$^{1/2}$. Thus, I and C follow

$$I = (2.64 \pm 0.35)\sqrt{C - 0.0037 \pm 0.003}. \tag{3}$$

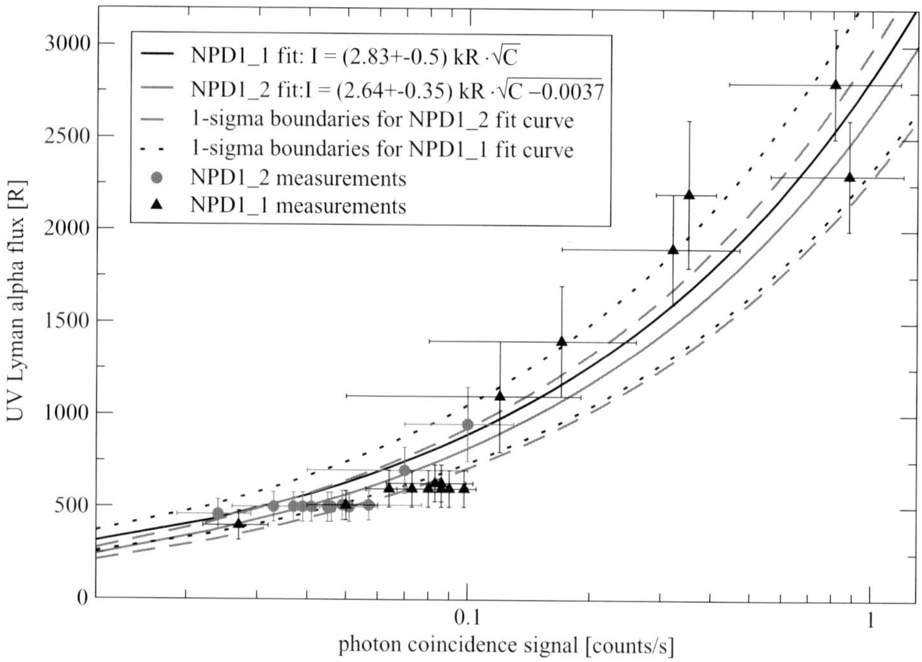

Figure 3. Relationship between UV intensity (Y-axis) and registered coincidence count rates (X-axis) for the NPD1_1 (triangles) and NPD1_2 (circles) sectors. The data have been obtained during the cruise phase from July 2003 to January 2004, when NPD measured the well known Lyman-α fluxes backscattered from interplanetary hydrogen and from stellar sources (values taken from Figure 2).

4. Limb Emission Measurement

For our study we have chosen the UV signal measured on April 25, 2004, from 12:00 until 13:10 UT, when the NPD field-of-view was directed to the North pole region of the Martian exosphere. The pointing direction of the NPD sensor is fixed with respect to the ecliptic reference frame while Mars with its exosphere slowly moves into the field-of-view of the sensor. This configuration ensures that the Lyman-α background due to the interplanetary hydrogen and the stellar radiation remains constant during the measurement. Figure 4 shows the observation condition for the begin and for the end of the measurement period in a spacecraft-centered view. The solar zenith angle remains constant at 145°.

Figure 5 illustrates the observation geometry: For any given time period the measured UV intensity is proportional to the UV limb emission averaged over the field-of-view and integrated along the line of sight through the Martian exosphere. The so-called tangential height, h_i, is the minimum distance between the line of sight and the Mars surface. For our observation it decreases within 75 minutes from 7250 to 1900 km, whereas the altitude of the spacecraft above the Mars surface,

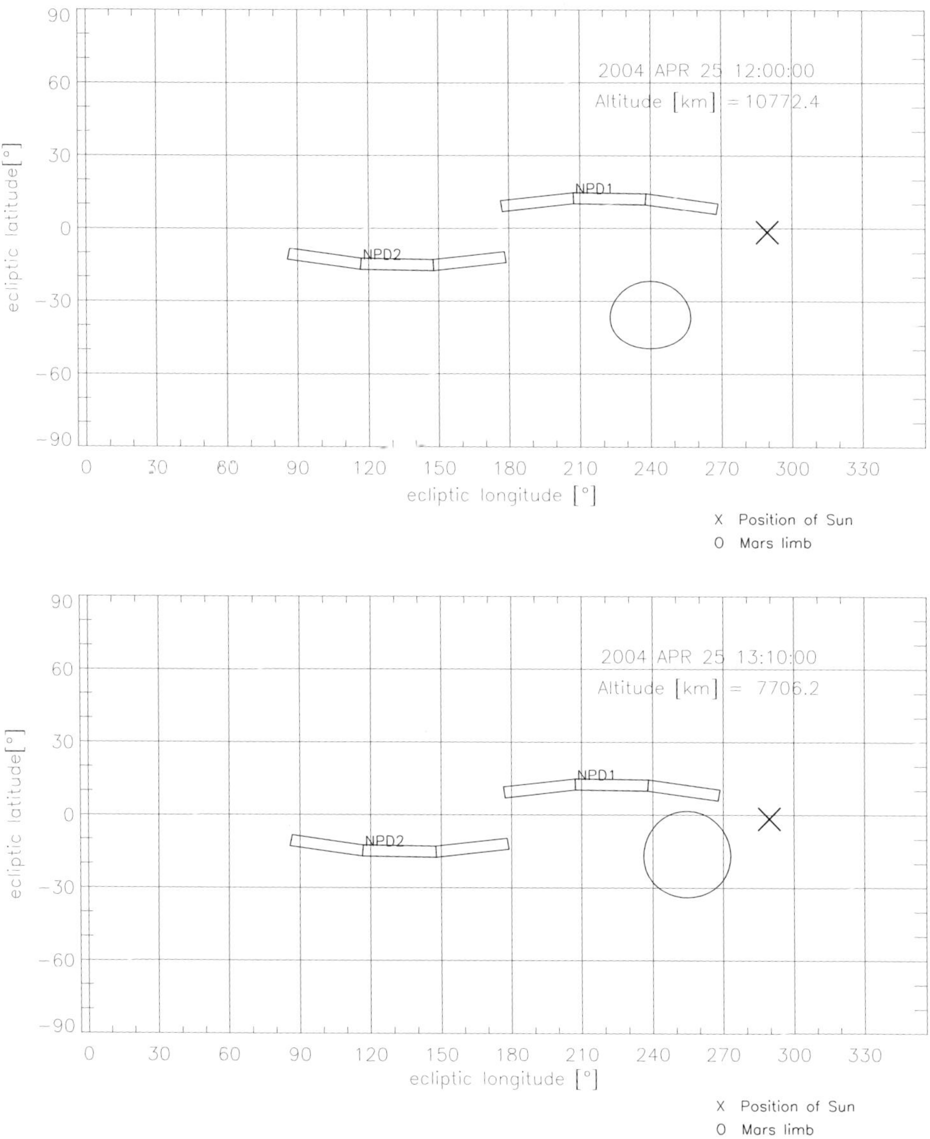

Figure 4. NPD observation directions on April 25, 2004, 12:00 UT (top) and at 13:10 UT (bottom), as seen from the spacecraft. The 2×3 elongated boxes are the fields-of-view of the two NPD sensors with their three angular channels ($4° \times 30°$ each). NPD1_2 is the sector closest to the Sun whose position is indicated with an X.

d_i, decreases from 10,800 to 7,700 km. Before 12:00 UT the instrument was not operating, and a few minutes after 13:15 the line of sight intersected the surface of Mars itself.

We have then divided the observation period into 15 consecutive intervals of 5 minutes duration and retrieved the height of the UV background for each

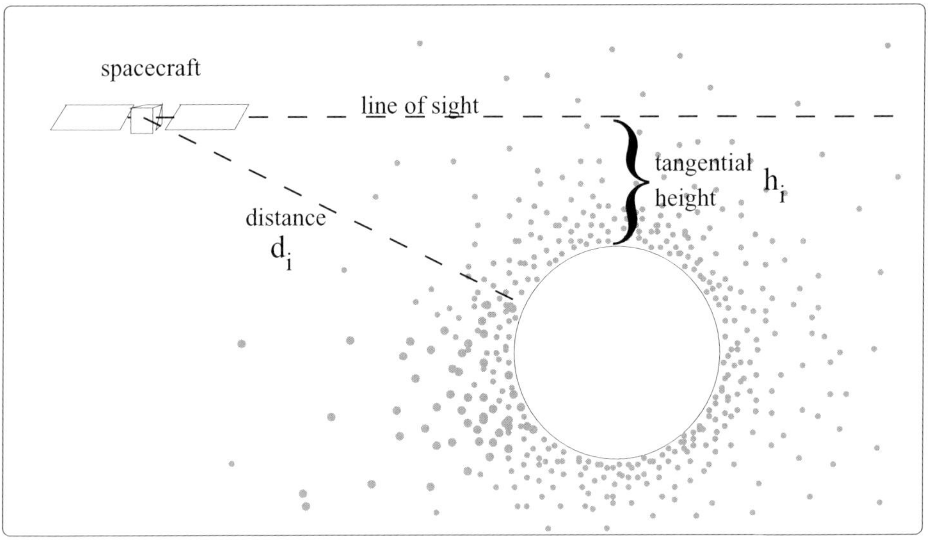

Figure 5. Observation geometry of the limb emission measurement.

measurement. One of the corresponding TOF spectra is shown in Figure 1, in which the dashed line indicates the height of the UV background. These count rates can be converted to Lyman-α intensities (in number of photons per second per steradian per square meter or in kR), since we have calibrated the UV sensitivity of the instrument.

With Equation (3) the 15 different measurements can be plotted as UV fluxes in kR (Figure 6). These fluxes are to be understood as UV emission of the neutral hydrogen integrated along the line of sight through the Martian exosphere. Because the hydrogen particle density is higher close to the planet the limb emission increases as the field-of-view covers deeper layers of the atmosphere. The non-planetary UV background (see Figure 2) has already been subtracted in Figure 6, it is a constant (Equation (10)) because the field-of-view retains its inertial pointing direction over the entire measurement.

5. UV Emission Model and Exosphere Model

If we are to deduce the temperature and the exobase density of the neutral hydrogen from the measured UV fluxes presented in Figure 6, we need two things: first we need a model that yields a hydrogen density profile $n_H(r)$ for a given temperature T and exobase density. Second, we need a radiation transport equation that relates the modeled density profile to a UV emission that can be compared to the measurements.

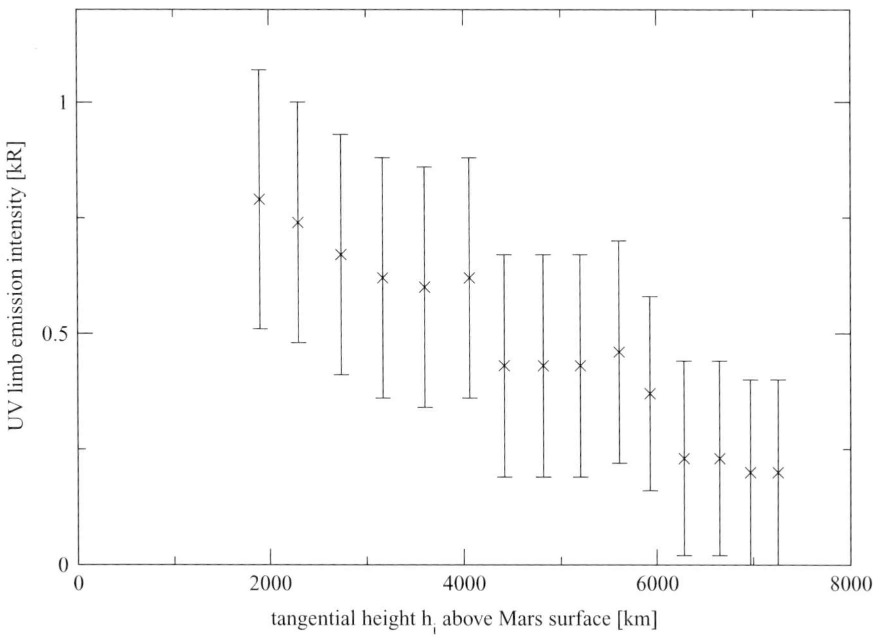

Figure 6. Measured limb emission of the exospheric hydrogen [kR] on April 25, 2004, from 12:00 until 13:15 UT. The error bars reflect the statistical errors, the error of the UV calibration (Equation (3)), and the uncertainty of the subtracted UV background (Equation (10)) from interstellar gas and stellar sources. The values for the height, h_i, are accurate to 300 km, the uncertainty being due to the finite aperture angle of the sensor.

For the exosphere model we assume a spherically symmetric distribution of neutral hydrogen with one single constant temperature. Several components of hydrogen are thought to co-exist in the exosphere (Lichtenegger *et al.*, this issue) but an exosphere with two components will simply be modeled as a linear combination of two different models, each of them with its own exobase density and temperature (see Section 6.2). We assume that the exobase lies at 220 km (Lichtenegger *et al.*, this issue) above the surface, and that the hydrogen atoms are fully thermalized at the exobase. The velocity distribution of particles in the exosphere is of course not Maxwellian. The exospheric density profile results from calculating the trajectories of many particles using a one-dimensional Monte Carlo code (Wurz and Lammer, 2003), for which the exobase density and the temperature are the two free parameters. The differences to a three-dimensional spherically symmetric code are not significant. Keep in mind, however, that a high spatial variability of temperatures and densities has to be expected. According to Holmström (this issue) the hydrogen exobase density may vary as much as an order of magnitude with geographic latitude and solar zenith angle, the lowest exospheric densities in spring 2004 are expected above the North pole.

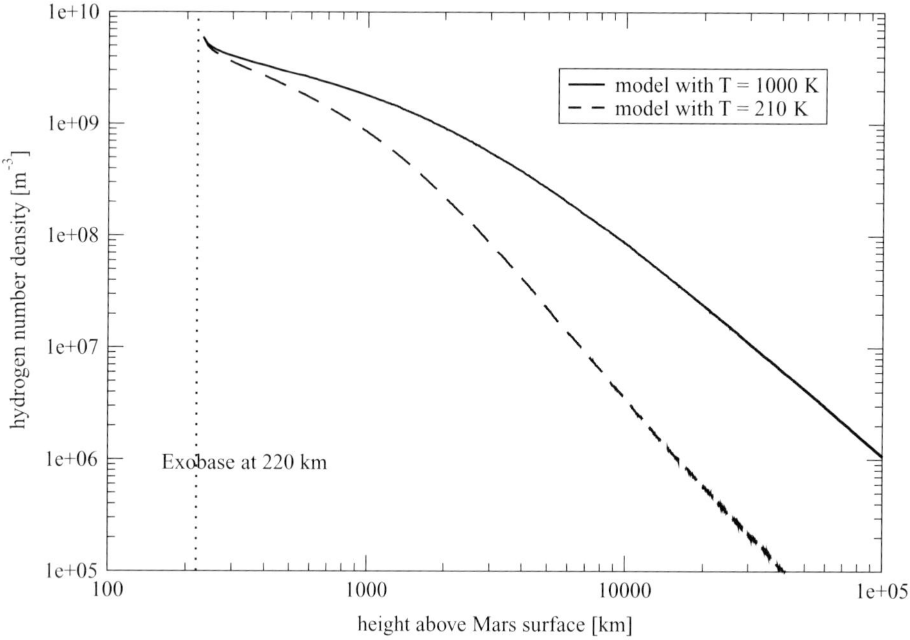

Figure 7. Modeled hydrogen density profiles for two different exospheric temperatures, $T = 210$ K (dashed line) and $T = 1000$ K (solid line). The exobase density is a free parameter and has been set to $n_H = 6 \times 10^9$ m^{-3}.

Two density profiles for different temperatures are presented in Figure 7. Empirically, the density profiles can be expressed as a set of continuous power law functions with differing exponents. For $T = 1000$ K, e.g., we find

$$
\begin{aligned}
&1.4 \times 10^{11}\ \mathrm{m}^{-3} r^{-2.92} && \text{for } 5.2 < r < 7, \\
&5.6 \times 10^{10}\ \mathrm{m}^{-3}\ r^{-2.47} && \text{for } 7 < r < 14.2, \\
&2.9 \times 10^{10}\ \mathrm{m}^{-3}\ r^{-2.22} && \text{for } 14.2 < r < 30, \\
&1.8 \times 10^{10}\ \mathrm{m}^{-3}\ r^{-2.07} && \text{for } 30 < r < 100,
\end{aligned}
\tag{4}
$$

with the radial distance $r = (h + R_M)/1000\,\mathrm{km}$.

To compare the modeled exosphere density profile with the UV limb emission measurements, we assume that the UV emission is entirely due to resonant scattering on hydrogen atoms. As discussed above, the Sun is outside the field-of-view and the line of sight is outside Mars eclipse. Moreover, the opacity is small, $\tau \ll 1$, and multiple scattering of UV photons can be disregarded. The opacity, or optical depth along a line of sight through the hydrogen exosphere with the local particle number density $n_H(r)$ calculates to:

$$\tau = \sigma_0 \int_{\mathbf{r}}^{\infty} n_H(r')dr'. \tag{5}$$

The cross-section for resonant scattering amounts to $\sigma_0 = 1.865 \times 10^{-17}$ m^2 (Meier, 1991). As we shall see in Section 6, $\tau < 0.2$ for all tangential heights in Figure 6.

In addition, we re-examine the two data sets from Mariner 6 and 7, presented by Anderson and Hord (1971), with our Monte Carlo model. For these data sets one has to be aware that the last few observations were obtained for lines of sight close to the exobase where the opacity is not negligible any more. For the present work we restrict the analysis to those regions where $\tau < 0.38$.

According to Thomas' (1963) and Bertaux' (1978) work on Lyman-α scattering in the geocorona the following linear relationship between particle density and intensity of UV limb emission is a good approximation under the assumption $\tau \ll 1$, even if the velocity distribution of the scattering particles is non-Maxwellian:

$$I(\mathbf{r}, \Omega) = \frac{f_{Ly-\alpha} \Delta\lambda_D \sqrt{\pi} \sigma_0}{4\pi} \int_{\mathbf{r}}^{\infty} n_{\mathrm{H}}(r') dr' \tag{6}$$

The emission rate factor in front of the integral consists of two natural constants ($\Delta\lambda_D = 0.001647$ nm being the Doppler width of the absorption process) times the central spectral solar Lyman-α irradiance $f_{Ly-\alpha}$ at 121.6 nm. In April 2004 at 1 AU distance the solar activity was low (solar activity index equal to 42) and the Lyman-α line irradiance was measured to be $F_{Ly-\alpha} = 4.22 \times 10^{11}$ cm^{-2} s^{-1} (NASA, 2004). According to the relation derived by Emerich *et al.* (2005) this corresponds to a central spectral irradiance of $f_{Ly-\alpha} = (3.65 \pm 0.10) \times 10^{12}$ cm^{-2} s^{-1} nm^{-1}. Finally, we have to scale the central spectral irradiance with the heliocentric distance of Mars in April 2004 to get the relevant number of Lyman-α photons at 121.6 nm for our UV limb emission measurement,

$$f_{Ly-\alpha} = \left(\frac{1\mathrm{AU}}{1.61\mathrm{AU}}\right)^2 3.65 \times 10^{12}\ \mathrm{s}^{-1}\,\mathrm{cm}^{-2}\ \mathrm{nm}^{-1} = 1.41 \times 10^{12}\ \mathrm{s}^{-1}\,\mathrm{cm}^{-2}\ \mathrm{nm}^{-1}. \tag{7}$$

Thus, we derive from Equation (6) the following relation between particle number densities and the predicted UV emission, I:

$$I(\mathbf{r}, \Omega) = b + k \int_{d}^{\infty} n_{\mathrm{H}}(r) dl, \tag{8}$$

with the emission rate factor k equal to

$$k = (6.1 \pm 0.2) \times 10^{-5}\ \mathrm{s}^{-1}\ \mathrm{sr}^{-1}, \tag{9}$$

where d is the distance of the spacecraft to the planet and l the axis along the NPD line-of-sight. The additive constant b denotes the UV background due to stars and Lyman-α backscattering from interplanetary hydrogen. For the pointing direction of NPD1_2 ($\lambda_{\rm ecl} = 255^\circ$, $\beta_{\rm ecl} = 10^\circ$) during the observation on April 25, 2004, b reaches according to Figure 2:

$$b = 0.74 \pm 0.1\,\text{kR}. \tag{10}$$

For a given density profile model with $n_{\rm H}(r)$ Equation (8) thus yields the UV emission we would expect in number of photons per seconds per steradian per square meter.

6. Results

To compare the airglow measurements to a modeled density profile, we must evaluate the path integral along the line of sight given in Equation (8). To simplify the notation let us first substitute for the distance of the spacecraft to the planet center $\tilde{d} \doteq d + R_M$, and for the tangential height $\tilde{h} \doteq h + R_M$, where R_M denotes the Mars radius. For the i-th measurement we define the column density as

$$S_{\rm model}(i) \doteq \int_{\tilde{d}_i}^{\infty} n_{\rm H}(r)dl. \tag{11}$$

Since we apply a spherically symmetric exosphere model, this path integral is symmetrical with respect to the radial distance to the center of the planet r, as long as the line of sight does not intersect with the planetary surface or with the eclipse. Thus, we convert the integration variable dl into an integration over dr. Since $\tilde{d}^2 = r^2 + l^2 \pm 2l\sqrt{r^2 - \tilde{h}^2}$

$$S_{\rm model}(i) \doteq \int_{\tilde{d}_i}^{\infty} n_{\rm H}(r)\frac{dl}{dr}dr = \tag{12}$$

$$= \int_{\tilde{d}_i}^{\tilde{h}i} n_{\rm H}(r)\frac{r}{\sqrt{r^2 - \tilde{h}_i{}^2}}dr + \int_{\infty}^{\tilde{h}_i} n_{\rm H}(r)\frac{r}{\sqrt{r^2 - \tilde{h}_i{}^2}}dr \tag{13}$$

evaluating the path integral piecewise from $r = \tilde{d}_i \ldots \tilde{h}_i$ and $r = \tilde{h}_i \ldots \infty$. We recognize that

$$\int_{\tilde{d}_i}^{\tilde{h}_i} n_{\rm H}(r)\frac{r}{\sqrt{r^2 - \tilde{h}_i^2}}dr = -\int_{\tilde{h}_i}^{\tilde{d}_i} n_{\rm H}(r)\frac{r}{\sqrt{r^2 - \tilde{h}_i^2}}dr, \tag{14}$$

and we find the following formula for the predicted column density along the line of sight for a given model exosphere:

$$S_{\rm model}(i) = 2\int_{\tilde{d}_i}^{\tilde{h}_i} n_{\rm H}(r)\frac{r}{\sqrt{r^2 - \tilde{h}_i^2}}dr + \int_{\infty}^{\tilde{d}_i} n_{\rm H}(r)\frac{r}{\sqrt{r^2 - \tilde{h}_i^2}}dr. \tag{15}$$

Because all heights $h_i < 10^4$ km, the second integral in Equation (15) can be approximated as:

$$\int_{10^5\,\mathrm{km}}^{\tilde{d}_i} n_{\mathrm{H}}(r)\frac{r}{\sqrt{r^2-\tilde{h}_i^2}}dr + \int_{\infty}^{10^5\,\mathrm{km}} n_{\mathrm{H}}(r)dr, \tag{16}$$

and since $n_{\mathrm{H}}(r) \sim 1/r^2$ for $r \gg 10^4$ km for any temperature, we can safely ignore the second term in expression 16, and we finally have:

$$S_{\mathrm{model}}(i) = 2\int_{\tilde{d}_i}^{\tilde{h}_i} n_{\mathrm{H}}(r)\frac{r}{\sqrt{r^2-\tilde{h}_i^2}}dr + \int_{10^5\,\mathrm{km}}^{\tilde{d}_i} n_{\mathrm{H}}(r)\frac{r}{\sqrt{r^2-\tilde{h}_i^2}}dr. \tag{17}$$

Equation (17) can be analytically integrated, since the model density profile $n_{\mathrm{H}}(r)$ can be expressed as a series of power law functions (see Equation (4)). We now search for the set of model parameters (temperature and exobase density) whose density profile $n_{\mathrm{H}}(r)$ minimizes the merit function

$$\chi^2 = \sum_{i=1}^{15}(S_{\mathrm{obs}}(i) - S_{\mathrm{model}}(i))^2/\sigma^2_{\mathrm{obs}}(i), \tag{18}$$

where $S_{\mathrm{model}}(i)$ is given by Equation (17), and S_{obs} with the corresponding standard deviation σ_{obs} is the integrated column density deduced from the i-th Lyman-α limb emission measurement according to Equation (8):

$$S_{\mathrm{obs}}(i) = \frac{I(i) - b}{k}. \tag{19}$$

To minimize the merit function in Equation (18) we first optimize for the exobase density for a given temperature, then the χ^2-deviations of the different model temperatures are compared to each other.

This is illustrated in Figure 8. Obviously, χ^2 for $T = 180$ K is much higher than for $T = 1000$ K, even with an optimized exobase density.

6.1. One Single Hydrogen Component

Several components of hydrogen with different temperatures are thought to co-exist in the Martian exosphere (see Section 1) but our measurement uncertainties urge us to test the one-component approach first. The derived values for the exospheric density and for the temperature will be dominated by the hotter hydrogen components because our observations are restricted to the upper exosphere (see the tangential heights in Figure 6). In Section 6.2 we will show the constraints the UV airglow measurements put on a two-component model. Let us first assume that there is only one component of neutral hydrogen with a temperature T and an exobase density $n_{\mathrm{H}}(r_{\mathrm{exo}})$ at 220 km above the surface. If we minimize the

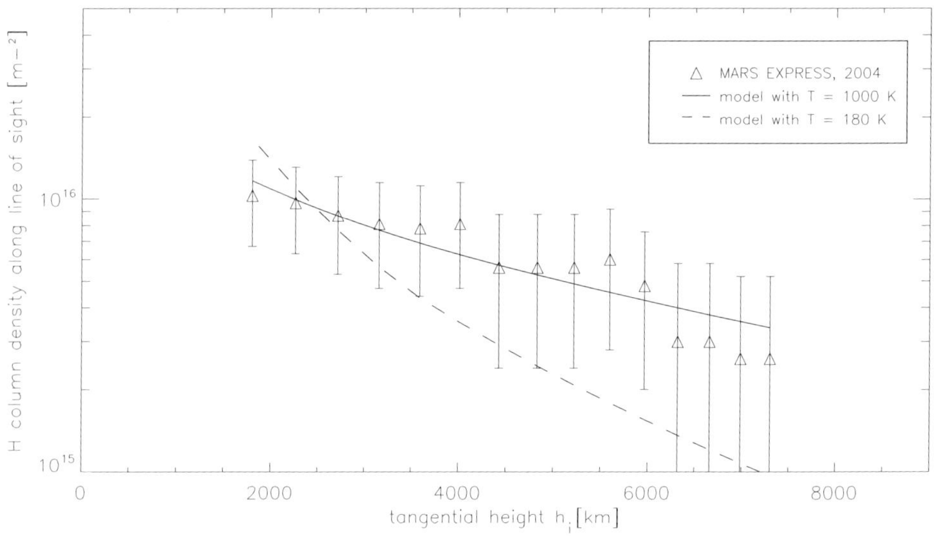

Figure 8. One-component models for a cool and for a hot hydrogen exosphere fitted to the NPD observations (triangles). The exobase densities have been chosen to minimize the χ^2-statistics: For $T = 180$ K we fit $n_H(r_{exo}) = 7.6 \times 10^{10}\,m^{-3}$ (dashed line), for $T = 1000$ K, on the other hand, we find only $n_H(r_{exo}) = 6.4 \times 10^9\,m^{-3}$ (solid line).

merit function (Equation (18)) we derive the following 1-σ boundaries for the exospheric hydrogen:

$$T = 1000^{+\infty}_{-400}\ \mathrm{K}, \tag{20}$$

$$n_H(r_{exo}) = 6.4^{+4.0}_{-0.6} \times 10^9\,\mathrm{m}^{-3}, \tag{21}$$

Note that the upper bound of the temperature estimate is ill constrained: An exospheric model with a deliberately high temperature provides almost as good a fit to the data as one with $T = 1000$ K. This is partly due to the large error bars of our measurement (see Figure 6); the other reason is that the modeled density profile does not change its slope substantially for temperatures above 1000 K. For $T \rightarrow \infty$, $n_H(r)$ in Equation (4) converges to $n_H(r) \sim r^{-2}$. Physically, of course, even the most energetic exospheric component cannot exceed a few 10^3 K (Lichtenegger *et al.*, this issue).

To re-evaluate the Mariner 6 and 7 measurements from 1969 published by Anderson and Hord (1971) we adopt their original values for the emission rate factor ($k = 8 \times 10^{-5}\,sr^{-1}\,s^{-1}$) and the UV background ($b = 300$ R) but we restrict ourselves to the data points above 3000 km where the optical depth of the exosphere can be neglected. We then compare the Mariner measurements to our exosphere model and we find, according to the same merit function (Equation (18)) and with the same statistical criteria:

$$T = 350^{+100}_{-50}\ \mathrm{K}, \tag{22}$$

$$n_H(r_{exo}) = 7.1^{+0.7}_{-1.6} \times 10^{10}\,\mathrm{m}^{-3}, \tag{23}$$

for Mariner 6. For Mariner 7, which approached Mars one week later, we find:

$$T = 425 \pm 50\ \mathrm{K} \tag{24}$$

$$n_H(r_{exo}) = 3.4^{+1.6}_{-0.5} \times 10^{10}\,\mathrm{m}^{-3}. \tag{25}$$

These results are plotted in Figure 9 and summarized in Table I. Anderson and Hord (1971) found, based on a Chamberlain exosphere (Chamberlain, 1963) model, $T = 350 \pm 100$ K and $n_H(r_{exo}) = (3.0 \pm 1.0) \times 10^{10}\,\mathrm{m}^{-3}$ for Mariner 6 and $n_H(r_{exo}) = (2.5 \pm 1.0) \times 10^{10}\,\mathrm{m}^{-3}$ for Mariner 7. Thus, the values derived with our exosphere model agree with the ones published by Anderson and Hord (1971) within a factor of 2. The error bars of the Mariner values are notably smaller because the Mariner measurements were obtained with a UV spectrometer that was designed for this task. Since our interpretation of the Mariner measurements is consistent with the one published 35 years ago by Anderson and Hord (1971) we conclude that the values derived from UV airglow measurements do not critically depend on the specific exosphere model.

TABLE I

Overview of fit results. The three upper lines refer to the single-component approach (see Equations (20) to (25)), whereas for the three lower rows we have assumed the co-existence of a cool and a hot hydrogen component. The temperature T_{cool} is not a fit value, it has been taken from Lichtenegger *et al.* (this issue).

Dataset	T_{hot} (K)	$n_{H,hot}$ (m^{-3})	T_{cool} (K)	Maximum $n_{H,cool}$ (m^{-3})
Mars Express	$1000^{+\infty}_{-400}$	$6.4^{+4.0}_{-0.6} \times 10^9$	–	–
Mariner 6	350^{+100}_{-50}	$7.1^{+0.7}_{-1.6} \times 10^{10}$	–	–
Mariner 7	425 ± 50	$3.4^{+1.6}_{-0.5} \times 10^{10}$	–	–
Mars Express	1000	6×10^9	180	$<1.2 \times 10^{10}$
Mariner 6	350	6.3×10^{10}	210	$<4.2 \times 10^{10}$
Mariner 7	425	3.1×10^{10}	210	$<3.4 \times 10^{10}$

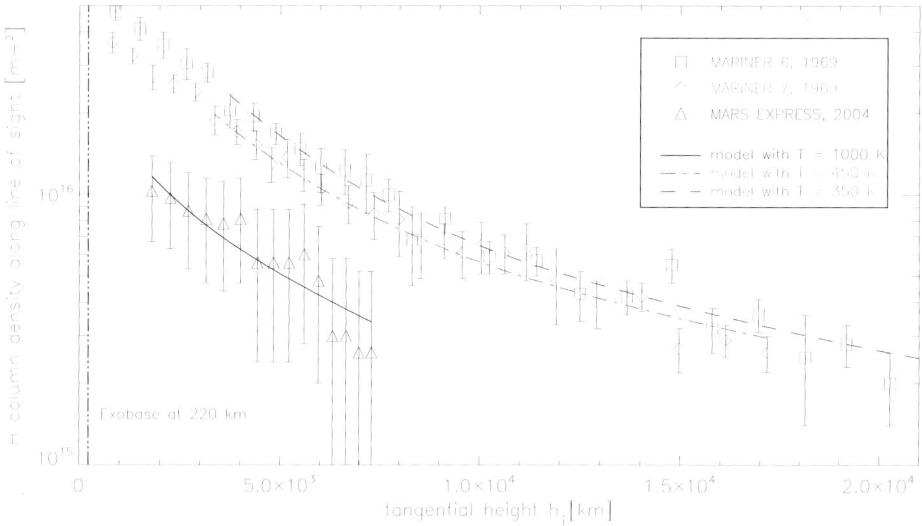

Figure 9. Measured column densities of neutral hydrogen (symbols) plotted against tangential height h_i. The solid lines represent models with values for the temperature and for the exobase density (denoted in Equation (20) to (25)) that minimize the χ^2-statistics. The most notable difference between the NPD measurements from 2004 (triangles) and the Mariner 6 (rectangles) and Mariner 7 (diamonds) data is the diminished exosphere density.

Generally, the upper exosphere in 2004 appears to be much thinner and hotter than in 1969. Note that we took into account the different solar irradiance, which was only 25% higher in 1969, and obviously falls short of explaining the 5 times higher hydrogen density during the Mariner mission compared to the Mars Express measurement. One reason why NPD measured lower UV fluxes might be the different observation geometry: The Mariner measurements were made from above the sun-illuminated Mars surface at solar zenith angles of 27° and 44°, whereas the NPD measurement was made from above the nightside at a solar zenith angle of 145° (see Figure 1). The exobase is expected to show a large spatial variability that results in an asymmetric exosphere. The model of Holmström (this issue) predicts a spatial variability of the exobase density of up to one order of magnitude. The North pole region where the NPD observation was made shows the lowest exospheric densities (Holmström, this issue). There are, however, other observations from spring 2004 where NPD measured the UV emission of the sun-illuminated Mars surface itself, which never exceeded 5 kR. From Mariner 9 measurements Barth *et al.* (1972) reported Lyman-α intensities from the Mars disk of up to 10 kR. This suggests that the Martian exosphere generally was thinner in 2004 than it was in 1969.

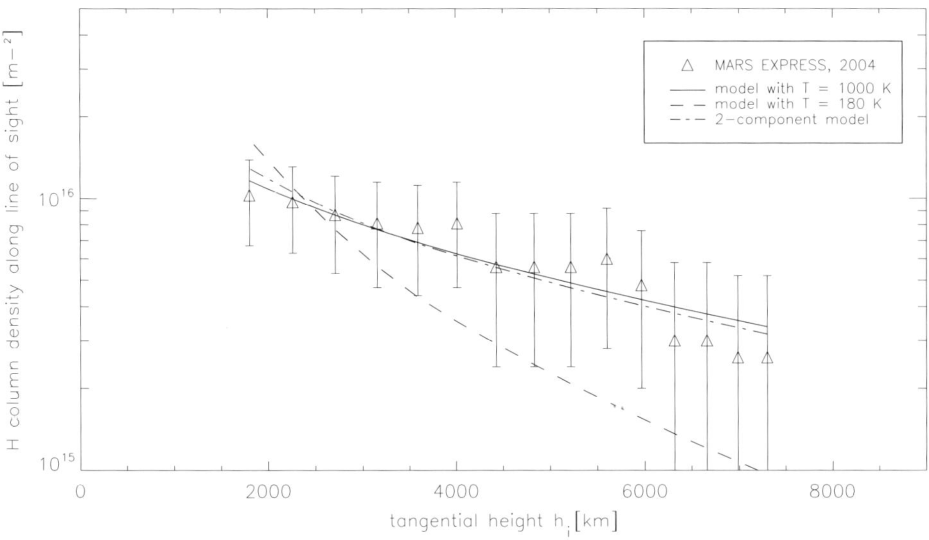

Figure 10. Example of a two-component model with very different temperatures that is still consistent with NPD data. The dashed-dotted line shows a density profile with $T_{cool} = 180$ K, $T_{hot} = 1000$ K, and with $n_{H,cool}(r_{exo}) = 1.2 \times 10^{10}\,m^{-3}$, $n_{H,hot}(r_{exo}) = 6 \times 10^9\,m^{-3}$.

6.2. Constraints on a Two-component Model

The temperatures derived in Section 6.1 for a one-component approach cannot be the bulk temperatures of the exosphere. The dominant CO_2 in the Martian atmosphere is a very efficient cooling agent (Lichtenegger *et al.*, 2002) and the aerobraking and mass spectrometry measurements at the exobase yield (see Lichtenegger *et al.* (this issue) for a summary) exospheric temperatures of only 150 to 230 K, for low and for high solar activity. The temperature of 350 K derived from Mariner measurements is considerably lower than the value found with the NPD data, but it is still significantly higher than the exospheric bulk temperature of ≤ 240 K that is derived from Mars Odyssey aerobraking data for a comparable solar activity (Lichtenegger *et al.*, this issue). To solve this paradoxon it is plausible to assume that several components of hydrogen co-exist in the Martian exosphere. If this approach is valid all UV airglow measurements have to be interpreted as a mixture of thermal (T close to the exobase bulk temperature of 150 to 230 K) atoms that are concentrated at the exobase and of hydrogen atoms with higher energies that have their origin in dissociative recombination and photodissociation processes. The Lyman-α airglow measurements are, of course, heavily influenced by the hottest component because the density of the cool component quickly drops with increasing altitudes. If we assume that our airglow measurements and those published by Anderson and Hord (1971) have given us correct estimates of the temperature of the hot component and that the temperature of the cool component is the one obtained from aerobraking and mass spectrometry data at the exobase, we can test a simple two-component approach: We estimate the maximum exobase density of a cool component, which still would be in agreement with the three different UV limb emission profiles. If the exobase density of the cool component is much higher than the hot component this influences notably the limb emission profile even at altitudes that fall within our airglow observation range. For this analysis we have set $T_{\rm cool} =$ 180 K for 2004 (solar minimum conditions) and $T_{\rm cool} = 210$ K for 1969 (solar maximum conditions) according to Lichtenegger *et al.* (this issue). $T_{\rm hot}$ is assumed to be the temperature fitted for the single component model (see Equations (20), (22), (24)). The results of this analysis are given in Table I in the three lower rows, Figure 10 illustrates the two-component approach for the NPD measuremet.

7. Conclusion

Although the NPD sensor was not designed as UV detector, the Lyman-α airglow measurements done by the Mariner 6 and 7 missions (Anderson and Hord, 1971; Barth *et al.*, 1971) have been reproduced successfully. We have fitted a model density profile of exospheric hydrogen to UV limb emission measurements done in 2004 during low solar activity, the height ranging from 1,900 to 7,250 km. In the case of a single hydrogen component our model allows for a well constrained value of the exobase density of $n_{\rm H}(r)_{\rm exo} = 6.4 \times 10^9$ m^{-3}. However, the optimum temperature of $T = 1000$ K is poorly constrained because of the statistical errors and because the modeled density profile shows no strong variation for temperatures much higher than 1000 K. In 2004 at low solar activity the hydrogen exosphere appeared to be thinner and hotter than in 1969 at high solar activity. The variability of the upper Mars exosphere is generally high. In 1969 the exobase density decreased by a factor of 2 (see Table I) within one single week, the solar zenith angles being similar. In 2004 the exosphere seems thinner by a factor of 5 than in 1969. The different observation direction or the spatial variability might explain this discrepancy; the minimum surface density for April 2004 is predicted (Holmström, this issue) for the region above the North pole where the NPD observation was made. Unfortunately, we do not have other limb emission measurements from NPD to track the spatial or temporal variability of the exosphere.

If there are several components of exospheric hydrogen with different scale heights the exosphere parameters we have inferred from the measurements are dominated by the hotter components. The Mariner data from 1969 seem to rule out the presence of a cool ($T = 210$ K) hydrogen component that is much denser than the observed hot one (see Table I); the temperature estimate does not increase if one restricts the evaluation step by step to the data points obtained at higher altitudes. Thus, the Mariner data on their own give no motivation to adopt a multi-component approach. Nonetheless, a two-component model with an average temperature of about 250 K and a total exobase density of 10^{11} m^{-3} hydrogen atoms may fit the

Mariner data as well as a one-component model. The NPD measurement in 2004 does not rule out the presence of a cool component with an exobase density higher than the observed hot component: A two-component exosphere model with a cool (180 K) and a hot (600 K) component with an average temperature of only 320 K would still be consistent with the NPD measurements. The hydrogen exobase temperatures that are derived from UV airglow measurements generally depend on the number of components one assumes, the derived temperature is bound to be dominated by the hotter components and should therefore not be used as an estimate for the exobase bulk temperature. Lichtenegger *et al.* (this issue) show that all temperature estimates for the Martian exosphere that were derived from Lyman-α airglow observations are significantly higher than the values derived from mass spectrometry and aerobraking measurements, which are sensitive around the exobase. The hydrogen exobase density is better constrained by our UV airglow measurement: Even if we allow for a cool hydrogen component our data rule out exobase densities above 2×10^{10} m^{-3} in 2004 above the North polar region.

Beside the Mariner 6 and 7 missions, there was only one Lyman-α airglow measurement at high altitudes before the arrival of Mars Express. In 1972, for medium solar activity (solar activity index of 70, $F_{10.7} = 120$), Dementyeva *et al.* (1972) found a hydrogen exobase density of only $n_{\rm H}(r)_{\rm exo} = 6 \times 10^9$ m^{-3}. They did, however, not calculate an optimal fit temperature; they assumed that $T = 350$ K, derived from Mariner 6 and 7 data, also applied to their observation (for an overview of temperatures derived for the martian exosphere over the last 35 years refer to Lichtenegger *et al.*, this issue, Figure 4a). Preliminary results from Lyman-α airglow data obtained with the SPICAM UV spectrograph on Mars Express (Chaufray *et al.*, 2006) yield exobase densities of a few to several 10^{10} m^{-3} above the dayside. For some observation configurations the emission profile requires two different hydrogen components with $T_{\rm hot} > 600$ and $T_{\rm cool} = 200$ K. The average exospheric temperature appears to be higher than 340 K (Chaufray *et al.*, 2006). The basic problem with comparing Lyman-α airglow data to each other may be that the exosphere is highly variable for different locations.

Although the accuracy of the NPD data is poorer than those measured by Mariner 6 and 7 it can be seen that the Mars exosphere above the North pole region was thinner and hotter for low solar activity than in 1969 at solar maximum. Preliminary results of Lyman-α airglow measurements done with SPICAM above the dayside yield exobase densities similar to the Mariner 6 and 7 data (Chaufray *et al.*, 2006). In any case we do not see (including the work of Dementyeva *et al.* (1972)) a trend to higher exobase densities for low solar activity, which was predicted by the exospheric models of Krasnopolsky and Gladstone (1996) and of Krasnopolsky (2002). If the Lyman-α data are interpreted as measurements of an energetic hydrogen population that is considerably hotter than the hydrogen bulk temperature, lower exospheric densities for low solar activity seem intuitive. In 2004 the solar activity was lower, the photodissociation rates were lower and thus the component of energetic hydrogen was less pronounced than in 1969. Krasnopolsky and

Gladstone (1996) use the Mariner measurements as input for the one-component hydrogen exosphere in their model and assume an extreme sensitivity to the solar cycle. Their model predicts for solar minimum conditions an exobase hydrogen density of $n_H = 10^{12}$ m^{-3} and $T = 200$ K. Models on atmospheric loss and on ENA production (see for instance (Barabash *et al.*, 2002; Holmström *et al.*, 2002; Lichtenegger *et al.*, 2002)) that use the values given by Krasnopolsky and Gladstone (1996) therefore have to be revised since their predictions critically depend upon the exosphere density and temperature. A more recent model of Krasnopolsky (2002) yields somewhat lower densities ($n_H = 2 \times 10^{11}$ m^{-3} for a solar activity index of 42), but these are still more than one order of magnitude away from our NPD observation.

The NPD measurement presented in this work is not the only experimental evidence that the Martian hydrogen exosphere is considerably thinner than previously modeled for low solar activity. We have already mentioned the Lyman-α airglow measurements done with SPICAM, but there are also non-photometric evidences: With NPD we also have mapped the outflow of hydrogen and oxygen ENAs from Mars (Galli *et al.*, this issue). We have found that the outflow is at least one order of magnitude below theoretical predictions that relied on the exosphere model of Krasnopolsky and Gladstone (1996). The detected H-ENA and O-ENA intensities are consistent with model predictions only if a thin neutral exosphere with hydrogen column densities of 10^{16} m^{-2} along the NPD line-of-sight is assumed. This is the value we also have found in this work based on the Lyman-α airglow (Figure 8).

Based on Mariner and NPD data we tentatively advise modelers to use a few 10^{10} m^{-3} as the spatially averaged exobase density for hydrogen, for low and for high solar activity. The expected spatial variability of the exosphere (Holmström, this issue) may be too notable to be neglected; many models on ENA and ion production have implied a constant exobase density and temperature over the entire planet.

The basic physical properties of the Martian exosphere obviously are not well known yet. There are clear disagreements between theoretical models and data and between different types of measurements. More measurements at the exobase are needed to understand the photochemistry and the heating mechanisms in the Martian atmosphere. The UV airglow measurements are only an indirect way of measuring the temperature and the density of the exosphere. The present situation could be comparable with the investigation of the Venusian exosphere some decades ago. The temperature of the hydrogen exosphere on Venus was also overestimated because the UV airglow measurements were dominated by hot hydrogen atoms far above the exobase. After the first direct mass spectrometer data the estimate of the exospheric temperature had to be reduced from 700 K down to 300 K (Lichtenegger *et al.*, this issue). It remains, however, to be shown why the hot hydrogen component has an exobase density that is comparable to the one of the cool component at Mars, whereas at Venus the exobase density of the cool component is 100 times higher (Bertaux *et al.*, 1978) than the hot one. A more accurate picture of the Mars exosphere will hopefully be gained one day from direct mass spectrometer data.

References

Anderson, D. E., and Hord, C. W.: 1971, *JGR* **76**(28), 6666.

Anderson, D. E.: 1976, *JGR* **81**(7), 1213.

Barabash, S., Holmström, M., Lukyanov, A., and Kallio, E.: 2002, *JGR* **107**(A10), 1280.

Barabash, S., *et al.*: 2004, ASPERA-3: analyser of space plasmas and energetic ions for Mars Express, 2004mesp.book, 121.

Barth, C. A., Hord, C. W., Pearce, J. B., Kelly, K. K., Anderson, G. P., and Stewart, A. I.: 1971, *JGR* **76**(10), 2213.

Barth, C. A., Stewart, A. I., and Hord, C. W.: 1972, *Icarus* **17**, 457.

Bertaux, J. L.: 1978, *PSS* **26**, 431.

Bertaux, J. L., Blamont, J., Marcelin, M., Kurt, V. G., Romanova, N. N., and Smirnov, A. S.: 1978, *PSS* **26**, 817.

Bougher, S. W., Engel, S., Roble, R. G., and Foster, B.: 2000, *JGR* **105**(E7), 17669.

Chaufray, J. Y., Quémerais, E., Bertaux, J. L., and Leblanc, F.: 2006, Sounding of the martian upper atmosphere with SPICAM on Mars Express, Europlanet conference 2006, abstract Nr. 00566.

Chamberlain, J. W.: 1963, *PSS* **11**, 901.

Dementyeva, N. N., Kurt, V. G., Smirnov, A. S., Titarchuk, L. G., and Chuvahin, S. D.: 1972, *Icarus* **17**, 475.

Deutscher, S. A., Borisov, A. G., and Sidis, V.: 1999, *Phys. Rev. A* **59**(6), 4446.

Emerich, C., Lemaire, P., Vial, J.-C., Curdt, W., Schühle, U., and Wilhelm, K.: 2005, *Icarus* **178**, 429.

Galli, A., Wurz, P., Barabash, S., Grigoriev, A., Gunell, H., Lundin, R., *et al.*: this issue, Space Sci. Rev., doi: 10.1007/s11214-006-9088-8.

Holmström, M., Barabash, S., and Kallio, E.: 2002, *JGR* **107**(A10), 1277.

Holmström, M.: this issue, Space Sci. Rev., doi: 10.1007/s11214-006-9036-7.

Krasnopolsky, V. A., and Gladstone, G. R.: 1996, *JGR* **101**(A7), 15765.

Krasnopolsky, V. A.: 2002, *JGR* **107**(E12), 5128.

Krasnopolsky, V. A., and Gladstone, G. R.: 2005, *Icarus* **176**, 395.

Lammer, H., Lichtenegger, H. I. M., Kolb, C., Ribas, I., Guinan, E. F., Abart, R., *et al.*: 2003, *Icarus* **165**, 9.

Leblanc, F., Chaufray, J. Y., Lilensten, J., Witasse, and Bertaux, J.-L.: 2006, *JGR* **111**(E9), 11.

Leinert, C., *et al.*: 1998, *A&A Sup. Ser.* **127**, 1.

Lichtenegger, H. I. M., Lammer, H., and Stumptner, W.: 2002, *JGR* **107**(A10), 1279.

Lichtenegger, H. I. M., Lammer, H., Kulikov, Yu. N., Kazeminejad, S., Molina-Cuberos, G. H., Rodrigo, R. *et al.*: this issue, Space Sci. Rev., doi: 10.1007/s11214-006-9082-1.

Lundin, R., Zakharov, A., Pellinen, R., Borg, H., Hultqvist, B., Pissarenko, N., *et al.*: 1989, *Nature* **341**, 609.

Meier, R. R.: 1991, *SSR* **58**, 1.

NASA: 2004, The Solar Radiation and Climate (SORCE) Experiment, http://lasp.colorado.edu/-sorce/sorce_data_access.

NASA: 2006, The Solar Wind Anisotropies (SWAN) experiment, http://sohowww.nascom.nasa.-gov/gallery/SWAN/index.html.

Nier, A. O., and McElroy, M. B.: 1977, *JGR* **82**(28), 4341.

Thomas, G. E.: 1963, *JGR* **68**(9), 2639.

Wurz, P., and Lammer, H.: 2003, *Icarus* **164**, 1.

EFFECTS OF LOW ENERGETIC NEUTRAL ATOMS ON MARTIAN AND VENUSIAN DAYSIDE EXOSPHERIC TEMPERATURE ESTIMATIONS

HERBERT I. M. LICHTENEGGER[1,*], HELMUT LAMMER[1], YURI N. KULIKOV[2], SHAHIN KAZEMINEJAD[1], GREGORIO H. MOLINA-CUBEROS[3], RAFAEL RODRIGO[4], BOBBY KAZEMINEJAD[1] and GOTTFRIED KIRCHENGAST[5]

[1]*Space Research Institute, Austrian Academy of Sciences, Schmiedlstr. 6, A-8042 Graz, Austria*

[2]*Polar Geophysical Institute, Russian Academy of Sciences, Khalturina Str. 15, 183010, Murmansk, Russian Federation*

[3]*Applied Electromagnetic Group, University of Murcia, Murcia, Spain*

[4]*Instituto de Astrofísica de Andalucía, CSIC, E-18080, Granada, Spain*

[5]*Institute for Geophysics, Astrophysics, and Meteorology, University of Graz, Universitätsplatz 5, A-8010 Graz, Austria*

(**Author for correspondence, E-mail: herbert.lichtenegger@oeaw.ac.at*)

(Received 28 March 2006; Accepted in final form 11 October 2006)

Abstract. The heating of the upper atmospheres and the formation of the ionospheres on Venus and Mars are mainly controlled by the solar X-ray and extreme ultraviolet (EUV) radiation ($\lambda = 0.1-102.7$ nm and can be characterized by the 10.7 cm solar radio flux). Previous estimations of the average Martian dayside exospheric temperature inferred from topside plasma scale heights, UV airglow and Lyman-α dayglow observations of up to $\sim$500 K imply a stronger dependence on solar activity than that found on Venus by the Pioneer Venus Orbiter (PVO) and Magellan spacecraft. However, this dependence appears to be inconsistent with exospheric temperatures ($<$250 K) inferred from aerobraking maneuvers of recent spacecraft like Mars Pathfinder, Mars Global Surveyor and Mars Odyssey during different solar activity periods and at different orbital locations of the planet. In a similar way, early Lyman-α dayglow and UV airglow observations by Venera 4, Mariner 5 and 10, and Venera 9–12 at Venus also suggested much higher exospheric temperatures of up to 1000 K as compared with the average dayside exospheric temperature of about 270 K inferred from neutral gas mass spectrometry data obtained by PVO. In order to compare Venus and Mars, we estimated the dayside exobase temperature of Venus by using electron density profiles obtained from the PVO radio science experiment during the solar cycle and found the Venusian temperature to vary between 250–300 K, being in reasonable agreement with the exospheric temperatures inferred from Magellan aerobraking data and PVO mass spectrometer measurements. The same method has been applied to Mars by studying the solar cycle variation of the ionospheric peak plasma density observed by Mars Global Surveyor during both solar minimum and maximum conditions, yielding a temperature range between 190–220 K. This result clearly indicates that the average Martian dayside temperature at the exobase does not exceed a value of about 240 K during high solar activity conditions and that the response of the upper atmosphere temperature on Mars to solar activity near the ionization maximum is essentially the same as on Venus. The reason for this discrepancy between exospheric temperature determinations from topside plasma scale heights and electron distributions near the ionospheric maximum seems to lie in the fact that thermal and photochemical equilibrium applies only at altitudes below 170 km, whereas topside scale heights are derived for much higher altitudes where they are modified by transport processes and where local thermodynamic equilibrium (LTE) conditions are violated. Moreover, from simulating the energy density distribution of photochemically

Space Science Reviews (2006) 126: 469–501
DOI: 10.1007/s11214-006-9082-1

produced moderately energetic H, C and O atoms, as well as CO molecules, we argue that exospheric temperatures inferred from Lyman-α dayglow and UV airglow observations result in too high values, because these particles, as well as energetic neutral atoms, transformed from solar wind protons into hydrogen atoms via charge exchange, may contribute to the observed planetary hot neutral gas coronae. Because the low exospheric temperatures inferred from neutral gas mass spectrometer and aerobraking data, as well as from CO_2^+ UV doublet emissions near 180–260 nm obtained from the Mars Express SPICAM UV spectrograph suggest rather low heating efficiencies, some hitherto unidentified additional IR-cooling mechanism in the thermospheres of both Venus and Mars is likely to exist.

Keywords: Venusian and Martian exosphere temperature, hot planetary coronae

1. Introduction

Due to the loss of the Japanese aeronomy Mars mission Nozomi there are still no accurate measurements by mass spectrometers allowing a direct exospheric temperature determination in the dayside upper atmosphere on Mars. However, a better understanding of the average exospheric temperature and its variation during minimum and maximum solar activity conditions is desirable since it will have important implications for the expected reactions of oxygen with the Martian surface soil (Lammer *et al.*, 2003) as well as the thermal (Jeans) loss of hydrogen.

The aim of this paper is to demonstrate that the Martian and Venusian exosphere contains thermal and non-thermal energy neutral particle populations which both have to be taken into account for the proper interpretation of the scale height temperature. In the remainder of the introduction a brief review of the various dayside exosphere temperature estimations based on many spacecraft observations at Mars and Venus is given. In Section 2 we derive the dayside exosphere temperature from electron density profiles obtained from radio occultation measurements aboard Mars Global Surveyor during high and moderate solar activity conditions. We then compare our results with the temperature values inferred from mass spectrometer data obtained by Viking 1 and 2, from aerobraking data of Mars Pathfinder (MPF), Mars Global Surveyor (MGS) and Mars Odyssey (MO), as well as from UV spectrograph data of Mars Express (MEX). As the upper atmospheres of Venus and Mars share similarities in composition and photochemical processes, we validate our methodology by applying it first to Venus and comparing the results with the exospheric temperatures based on the PVO mass density measurements.

Different photochemical processes which may produce hot hydrogen, oxygen and carbon populations at the exobase level are discussed in Section 3 and their effects on the apparent exobase temperature at Mars are investigated. Ionospheric profiles of OH^+ and CO_2H^+ ions are considered, which release low energetic hydrogen atoms through photochemical processes into the Martian atmosphere between 150–250 km altitude. Furthermore, we apply a Monte Carlo method to calculate the energy density distribution of photochemically produced hot H^*, C^*, O^* and CO^* at the Martian exobase for low and high solar activity and determine their daytime characteristic temperatures.

Finally, the problem of low heating efficiencies needed to model low thermospheric temperatures is briefly addressed in Section 4, followed by a summary of our results given in Section 5.

1.1. Mars

The first observations of the Lyman-α dayglow of Mars produced by resonance scattering of the 121.6 nm solar Lyman-α line by atomic hydrogen have been obtained by Mariner 6 and 7 in 1969 (Barth *et al.*, 1969, 1971; Anderson and Hord, 1971; Anderson, 1974). An extended atomic hydrogen corona was first detected around Mars when the closest point of the spectrometer optical axis to the planet was at a planetocentric distance of about 30000 km. On approaching Mars, the Lyman-α intensity increased and became a maximum when the field of view crossed the illuminated limb. Data from 200 to 24000 km altitude were analyzed to determine the structure of the Martian exosphere (Anderson and Hord, 1971). By applying a radiative transfer model together with considerations of orbit theory and solution of the collisionless Boltzmann equations, Anderson and Hord (1971) inferred an exospheric temperature of about 350 ± 100 K and a number density of atomic hydrogen of about $3.0 \pm 1 \times 10^4$ cm^{-3} at the exobase level of $\sim$250 km altitude. These values imply a thermal escape flux for hydrogen atoms of 1.8×10^8 cm^{-2} s^{-1} and an escape time of 10^4 s.

Further, Stewart (1972) analyzed the Mariner 6 and 7 UV spectrometer data of intense airglow emissions of CO_2^+, CO and O in the wavelength range between 190–430 nm and found that these species are excited predominantly due to the absorbtion of solar EUV photons by CO_2 and constitute a major energy loss mechanism for the thermosphere. Therefore, the airglow intensity is closely related to the altitude distribution of CO_2 and UV spectrometer data can thus yield reliable information about the Martian thermospheric temperature profile. Stewart (1972) concluded that the observed CO Cameron-band emission scale height of 19 ± 4.5 km suggests an average exospheric temperature of 315 ± 75 K, a value in close agreement with the one obtained by Anderson and Hord (1971).

However, recent studies by Fox and Hać (1999) and Fox and Bakalian (2001) indicate that the airglow emitting species analyzed by Stewart (1972) may have been involved in photochemical reactions like dissociative recombination, impact ionization or photodissociation, which yield energies in excess of the local thermal equilibrium condition (LTE) of the dayside atmosphere, so that hot carbon (Fox and Hać, 1999; Fox and Bakalian, 2001) and hot oxygen could populate the upper Martian exosphere as well (e.g., Ip, 1988; Nagy *et al.*, 1981; Nagy and Cravens, 1988; Lammer and Bauer, 1991; Kim *et al.*, 1998; Lammer *et al.*, 2000).

Other efforts to estimate the Martian dayside exospheric temperature were based on determination of the plasma scale height from ionospheric profiles obtained by means of radio occultation methods. McElroy (1967) argued that the ionosphere detected by Mariner 4 should not be interpreted as an F2 layer since this would require a cold exosphere (~100 K) and extremely low neutral densities not consistent with thermal calculations for the lower atmosphere. Rather, an E region interpretation was favored and a lower limit for the exospheric temperature of 400 K was concluded. Cloutier *et al.* (1969) used ionospheric profiles measured by Mariner 4 and inferred an exospheric temperature of about 490 K by assuming a dynamical model for the interaction of the solar wind with the Martian ionosphere and a radiative equilibrium model with a solar EUV heating efficiency of about 0.6 for a neutral atmosphere. On the other hand, assuming a lower heating efficiency between 0.25 and 0.45, Hogan and Stewart (1969) obtained temperatures of 267–322 K for conditions appropriate to the occultations of Mariner 4 on Mars.

Eddy transport of heat to lower altitudes as well as large eddy coefficients were proposed to explain lower temperatures. Later, Lindal *et al.* (1979) used radio occultation data of Mariner 4 and considered a lower exospheric temperature of about 260 K as reasonable.

Yet another method to determine the exospheric temperature relies on mass spectrometer observations. Temperatures inferred from mass spectrometer data obtained by the Viking Landers 1 and 2 during low solar activity period yielded variable values with an upper limit of about 225 K (e.g., Nier and McElroy, 1977; Hanson *et al.*, 1977; Fox and Dalgarno, 1979; Barth *et al.*, 1992). An even lower thermospheric peak temperature of 153 K for low solar activity was obtained during the nighttime descent of Mars Pathfinder (MPF) in 1997 (Schofield *et al.*, 1997; Magalhaes *et al.*, 1999). Moreover, scale heights derived from aerobraking data of Mars Global Surveyor above 150 km altitude implied exospheric temperatures of about 220 K during moderate solar activity conditions (e.g., Keating *et al.*, 1998b; Bougher *et al.*, 2000).

In order to explain the apparently large variations of the Martian dayside exospheric temperature of 150–350 K from low to moderate and high solar activity periods, Bougher *et al.* (2000) considered seasonal-solar cycle extremes for the exobase temperature calculations and estimated the temperature to vary from 200 K near aphelion during solar minimum to 370 K near perihelion during solar maximum. However, the results of various studies by Bauer and Hantsch (1989), Krasnopolsky and Gladstone (1996), Breus *et al.* (2004) and Lichtenegger *et al.* (2004) indicate that such extreme variations of the average dayside exospheric temperature are in disagreement with the observed solar cycle dependence of the ionospheric peak plasma density as measured by the Viking Lander 1, the Viking Orbiters 1 and 2, Mariner 4, 6, 7, the Mars 2, 3, 4, 6 and Mariner 9 spacecraft.

It should be noted that all neutral gas temperatures inferred from topside plasma scale heights, Lyman-α dayglow and UV airglow observations on Mars imply a much stronger dependence on solar activity (expressed by the $F_{10.7}$ index[1]) than

[1]The radio emission from the sun at a wavelength of 10.7 cm (2800 MHz) is measured in units of 10^{-22} W m^{-2} Hz^{-1}; this unit will be omitted in the rest of the paper.

was found for Venus, although both exospheres contain similar constituents (e.g., Bauer *et al.*, 1979; Bauer and Taylor, 1981). On the other hand, estimations of the Martian exosphere temperature based on the data obtained by mass spectrometers of the landing probes or on aerobraking data yield much lower values and also a weaker dependence on solar activity than is expected from previous observations.

1.2. VENUS

First estimations of the exospheric temperature on Venus were based on Lyman-α dayglow observations of Mariner 5 and resulted in extremely high exospheric temperatures of up to 700 K for the outer Lyman-α component (e.g., Barth, 1968; Barth *et al.*, 1968; McElroy, 1968; Stewart, 1968; Wallace, 1969). Using both Mariner 5 and Venera 4 data, the upper atmospheric temperature was calculated by Stewart (1968) through integration of the thermal conduction equation with inclusion of radiative losses by CO_2, CO and O that resulted in a value of 700 ± 100 K. The uncertainty of this result reflected the uncertainties in the CO_2 vibrational deactivation coefficient due to collisions and in the assumed heating efficiency.

It was noticed by Barth (1968) and Barth *et al.* (1968) that the Mariner 5 measurements could be fitted with the barometric-exospheric equation only if a two component model atmosphere was used, where the Lyman-α radiation was assumed to be produced both by resonant scattering from atomic hydrogen and by photodissociation of molecular hydrogen. While atomic hydrogen was identified to dominate at large distances, molecular hydrogen was favored to prevail closer to the planet and a single temperature of 650 K was attributed to both constituents. However, the conjecture of H_2 being the second species besides atomic H was considered to be untenable by Donahue (1969) and McElroy and Hunten (1969), since the large production rate of atomic hydrogen due to photodissociation implied by this model could not be reconciled with the small observed atomic hydrogen densities. Rather, these authors proposed deuterium at 650–700 K as the second component, suggesting the D/H ratio to be about 10 at the base of the exosphere. In addition, Wallace (1969) also obtained a reasonable fit to the Mariner 5 data on the basis of a two-component hydrogen-deuterium model with a subsolar exospheric temperature of ~640 K and deuterium as the dominant species below ~3000 km altitude. Moreover, this model required a significant day/night asymmetry in the densities at the exobase level, the deuterium and atomic hydrogen concentrations on the dayside being about 10 and 2 times larger, repectively. However, subsequent rocket observations of the Venus ^2H and ^1H lines did not support the deuterium model (Wallace *et al.*, 1971).

Later, Broadfoot *et al.* (1974) analyzed airglow observations in the wavelength interval between 20–170 nm obtained by means of an objective grating spectrometer aboard Mariner 10. The data revealed the presence of significant concentrations of H, He, C and O atoms in the upper atmosphere of Venus. Based on the analysis of the hydrogen data, an exospheric temperature of about 400 K was deduced. According to Sze and McElroy (1975), an exospheric source of hot hydrogen atoms would have offered the most likely explanation for the extended Lyman-α emission measured by Mariner 5. The atoms could have been produced predominantly by reaction (6) of Section 3.1 and the apparently high temperature of the hydrogen corona could have reflected the temperature of ionospheric O^+ ions.

By investigating the hydrogen emissions observed by Venera 9 and 10, Bertaux *et al.* (1978) estimated a Venus exospheric temperature of about 500 ± 100 K and concluded that the measured sharp increase of the linewidth above 300 km could be interpreted as the signature of an additional hot component created by charge exchange collisions between the neutral atoms and the solar wind. From Venera 11 and 12 observations of the EUV emission, Bertaux *et al.* (1981) obtained two hydrogen populations with temperatures of 400 and 700 K. Later, the dayside exospheric temperature at the polar region was guessed to be 300 ± 25 K(Venera 11) and 275 ± 25 K (Venera 12) together with a non-thermal component of 1000 K (Bertaux *et al.* 1982). This temperature of the cooler component was distinctly smaller than previous temperature estimations and comparable with those inferred from neutral gas measurements of the Pioneer Venus Orbiter (Nieman *et al.*, 1979a,b,1980; von Zahn *et al.*, 1980; Hedin, 1983; Hedin *et al.* 1983). Kumar and Hunten (1974) presented a fit to the Mariner 5 electron density data which implied a cool temperature of around 350 K and argued for the existence of an additional non-thermal atomic hydrogen component produced by several ion-neutral reactions of H_2. Using a complex radiation transfer model for the interpretation of the Mariner 5 Lyman-α measurements, Anderson (1976) suggested that Venus exosphere is highly asymmetric and may consist of two different hydrogen components with day- and nightside temperatures of 275 ± 50 K/1020 ± 100 K and $150 \pm 50/1500 \pm 200$ K, respectively. The density of these components at the exobase was estimated as $2 \pm 1 \times 10^5/1.3 \times 10^3$ cm^{-3} and $2 \pm 1 \times 10^5/1 \times 10^3$ cm^{-3}, respectively.

Finally, also the existence of a hot oxygen corona was confirmed by the 130.4 nm observation of the UV spectrometer on board of PVO (Nagy *et al.*, 1981), suggesting that the temperatures inferred from airglow data can also be influenced by photochemically produced atoms like hot O (e.g., Nagy *et al.*, 1981, 1988) or C (Fox and Bakalian, 2001) which populate the upper Venusian atmosphere.

A comparison of the earlier estimated high Venusian exospheric temperatures of up to 700 K from Lyman-α dayglow and UV airglow observations with the temperatures of less than 300 K inferred from PVO and Magellan data (see Bougher *et al.*, 1999 and references therein) thus suggests that Venus dayside upper atmosphere is populated by a cool hydrogen background as well as by hot atoms of moderate energy (e.g. H^*, O^*, C^*). Moreover, energetic neutral atoms (ENAs) like H^{**} and O^{**}, which originate from charge exchange processes between solar wind protons and neutral atmospheric species are also likely to be present and to play a role in the upper atmosphere.

Since similar photochemical processes are expected in the upper atmospheres of Venus and Mars (e.g., Nagy *et al.*, 1981; Nagy and Cravens, 1988), photochemically produced low energetic as well as energetic atoms will also populate the Martian exosphere (Barabash *et al.*, 2002; Lichtenegger *et al.*, 2002; Holmström *et al.*, 2002; Lichtenegger *et al.*, 2004); in particular, species like O^*, C^*, H^*, H^{**} and O^{**} should be found (McElroy, 1972; Sze and McElroy, 1975; Wallis, 1978; Nagy *et al.*, 1981; Ip, 1988; Nagy and Cravens, 1988; Lammer and Bauer, 1991; Kim *et al.*, 1998; Hodges, 2000; Lammer *et al.*, 2000; Fox and Bakalian, 2001). One can thus anticipate that the average high daytime exospheric temperature values on Mars of about 315 ± 75 K inferred from Mariner 6 and 7 UV airglow data by Stewart (1972) were influenced by hot C^*, CO^* and O^* populations. In a similar way, the effect of a hot hydrogen population, as well as of ENAs, on the data interpretation of the early Lyman-α dayglow observations on Mars can be expected.

2. Exospheric Temperatures Inferred from Ionospheric Peak Plasma Densities and Aerobraking Data

2.1. Neutral Gas Temperatures Inferred from Ionospheric Peak Plasma Densities

For the estimation of the temperature at the Martian exobase at about 220 km, we determine the neutral gas temperature at the peak of the ionosphere, which is found at about 140–150 km altitude. It should be kept in mind that at this altitude the ionosphere is in thermal and photochemical equilibrium and ion transport processes are negligible (e.g., Hanson and Mantas, 1988; Bauer and Lammer, 2004). One can therefore infer the neutral gas temperature close to the exobase from fits of the electron density profiles, which are obtained from radio-occultations during various solar $F_{10.7}$ fluxes, to Chapman profiles. These latter profiles are given by

$$N(h,\chi) = N_m(\chi)\exp\left\{\frac{1}{2}\left[1 - \frac{h-h_m(\chi)}{H} - \exp\left(-\frac{h-h_m(\chi)}{H}\right)\right]\right\}, \qquad (1)$$

where $N(h,\chi)$ is the electron density as function of altitude h and solar zenith angle χ. Here, $N_m(\chi)$ is the electron density at the ionospheric peak at altitude $h_m(\chi)$, both depending on χ. For known values of $N_m(\chi)$ and $h_m(\chi)$, the scale height H can be adjusted to give a best fit to Equation (1). The neutral gas temperature T_n is embedded in the parameter $H = kT_n/(mg)$, where g is the altitude dependent acceleration of gravity, m the effective mass of the main atmospheric constituents and k the Boltzmann constant.

The Chapman layer theory is nominally valid within the altitude range of about one scale height H above and below the electron density peak. Taking the average neutral atmosphere scale height value $H \sim 10$ km and the measured peak altitudes for Mars (Keating *et al.*, 1998b), we get a height range of about 20 to 30 km around the measured peak altitudes of the electron density where we can expect the Chapman approximation to be true (see also Bauer and Hantsch, 1989). Due to these limitations, outside this range substantial deviations of the idealized Chapman layer from the measured electron density profiles may be expected. However, these deviations do not affect the determination of a correct H value near the electron density peak. Also, at altitudes below ~175 km, where the ratio of the ion (N) to neutral (n) number density is small, the ion temperature T_i will be approximately equal to the neutral gas temperature T_n (i.e. $T_\mathrm{i} \approx T_\mathrm{n}$) due to efficient collisional energy transfer between ions and neutrals. At altitudes above ~175 km where N/n becomes large, T_i will be larger than T_n since Coulomb collisions between electrons and ions is the principal energy transfer process.

A verification of the method to determine the neutral gas temperature by fitting a Chapman layer to the electron density profiles can be achieved by estimating Venus' exospheric temperature. At Venus, neutral density measurements by PVO were possible only when the orbital periapsis was below about 300 km, i.e. in 1978–1980 at solar maximum conditions ($F_{10.7} \approx$ 180–200) and again in fall 1992 at solar medium conditions ($F_{10.7} \approx 120$). The average dayside exospheric temperature on Venus during moderate solar activity is found to be about 270 K, while the temperature during solar minimum ($F_{10.7} \approx 80$), as derived from Magellan aerobraking data, is about 240–250 K (Kasprzak *et al.*, 1997; Keating *et al.*, 1998a; Bougher *et al.*, 1999). A similar temperature is obtained by a Chapman fit to the electron density for low solar activity as illustrated in Figure 1a, where the data are taken from the model of Fox and Sung (2001).

On December 9, 1978, the structure and composition of the upper atmosphere between 130–650 km altitude were measured for a solar zenith angle of approximately 60 degrees by the bus neutral mass spectrometer (BNMS) on board the Pioneer Venus multiprobe (von Zahn *et al.*, 1980). The solar activity at the time of the multiprobe entry was high ($F_{10.7} = 189.6$) and an almost constant neutral gas temperature of ~ 275 K was derived from the BNMS He number densities measurements on the dayside of Venus in an altitude range between 160–500 km (von Zahn *et al.*, 1980). Similar exospheric daytime temperatures of about 285–300 K were inferred from the orbiter neutral mass spectrometer (ONMS) instrument (Nieman *et al.*, 1979a,b; 1980; Hedin, 1983; Hedin *et al.*, 1983; Fox and Sung, 2001) and for the first three PVO diurnal cycles from the orbiter atmospheric drag (OAD) data (Keating *et al.*, 1980) and for the re-entry (Kasprzak *et al.*, 1997). The Venus exospheric temperature over the solar cycle was also determined by the UVS instrument (Kasprzak *et al.*, 1997). Again these temperatures closely coincide with those obtained by fitting the electron density by a Chapman function, as seen in Figure 1b. The data in this figure are taken according to the Venus International Reference Atmosphere (VIRA) and correspond to average dayside electron densities for a solar zenith angle between 50 and 70 degrees and high solar activity (Bauer *et al.*, 1985). These results indicate that the temperature is not likely to attain the expected high values inferred from the early UV airglow and

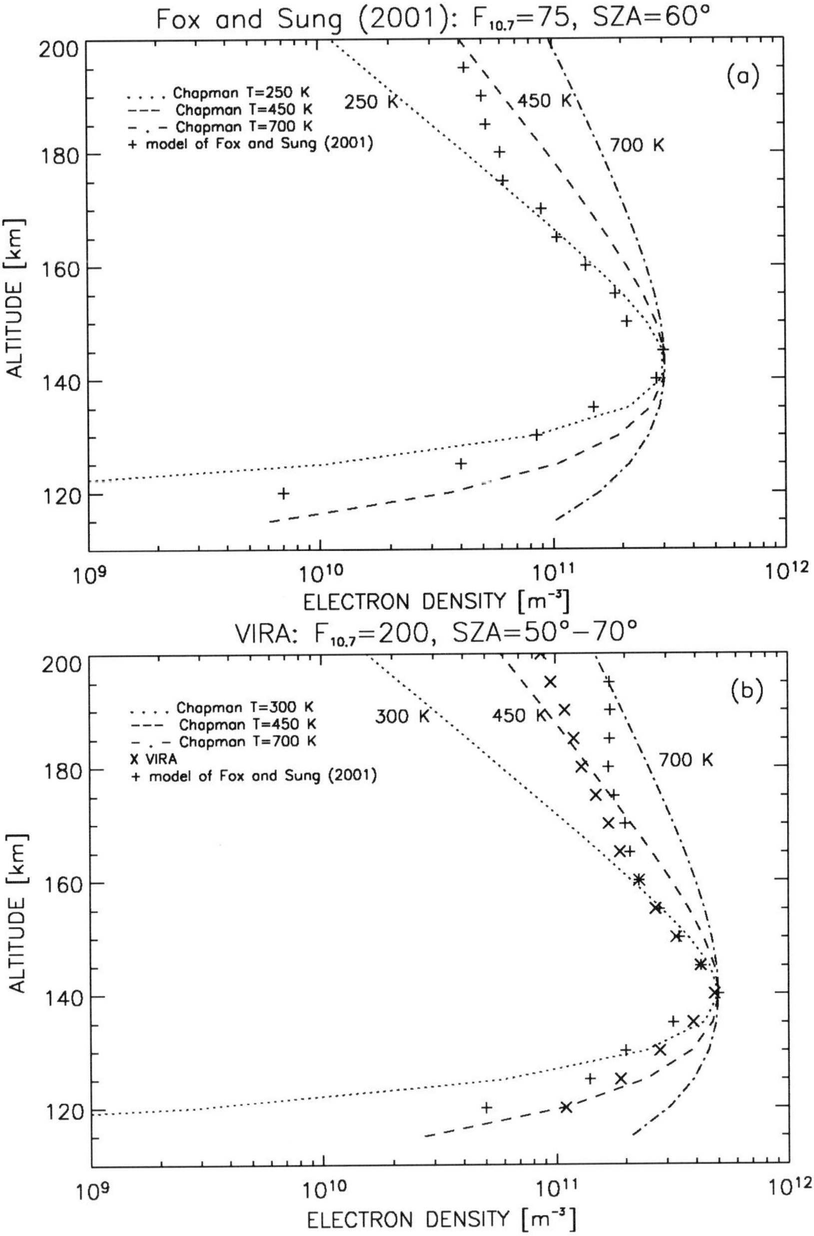

Figure 1. Electron density profiles based on PVO data at Venus for low (a) and high (b) solar activity. Crosses correspond to the results of the model of Fox and Sung (2001) (a) and to the VIRA model (b) and the lines represent various temperature fits.

Lyman-α dayglow observations of about 400 K (Broadfoot *et al.*, 1974; Bertaux *et al.*, 1981) and 700 K (Stewart, 1968; Wallace, 1969), respectively. A comparison of the observations performed during solar maximum and medium conditions reveals that the upper atmosphere on the dayside and also the nightside of Venus varies only slightly with solar activity ($\Delta T \approx 55$ K), reflecting relatively small dayside changes unlike in the Earth's upper atmosphere (Kasprzak *et al.*, 1997).

2.2. Martian Neutral Gas Temperature Inferred from Aerobraking Data and Entry Probes

Viking 1 and 2 landed on the Martian surface on July 20 and September 3, 1976 respectively (Soffen, 1977). The structure of the Martian daytime atmosphere between 200–100 km altitude was derived from in situ measurements of the neutral mass spectrometers (Nier and McElroy, 1977) and in the altitude range of 120–26 km from the measurements of the three-axis accelerometers (Seiff and Kirk, 1977). The results provide exospheric temperatures of 185 K and 145 K, respectively, during low solar activity periods and these are the only values for Mars which were obtained from mass spectrometer measurements.

Another approach to infer the neutral gas temperature from the atmospheric structure is aerobraking. An aerobraking spacecraft passes through the atmosphere near the periapsis of its orbit and experiences an aerodynamic drag force, which decreases the energy and thus the semimajor axis of the spacecraft's orbit without the need for significant fuel consumption. Another advantage of this maneuver is that the onboard accelerometer measures the aerodynamic forces on the spacecraft, which can be processed to generate an atmospheric density profile along each flight path through the atmosphere. Assuming hydrostatic equilibrium and using the ideal gas law, one can derive in addition a pressure and temperature profile. Aerobraking therefore permits in situ atmospheric studies usually not possible from a typical spacecraft orbit.

The Mars Global Surveyor spacecraft aerobraking data comprise two phases, which cover two distinct Mars seasons: phase 1 consists of 7 months approaching perihelion from southern spring to early summer (Ls = 180–300),[2] and phase 2 covers 4.5 months near aphelion from northern spring to early summer (Ls = 30–95) (Bougher and Keating, 1999; Keating *et al.*, 1998b). The two aerobraking data sets yield modelled exospheric temperatures during medium solar activity of about 220 and 230 K, respectively (Bougher *et al.*, 2000).

The entry, descent and landing of the Mars Pathfinder spacecraft during low solar activity on July 4, 1997, provided the first opportunity to make *in situ*

[2] Mars seasons are denoted by the "Ls" parameter corresponding to Mars aerocentric longitude. This is an angular measurement for the advance of its seasons (Ls = 0, 180 equinox; Ls = 90, northern summer solstice; Ls = 270, southern summer solstice)

measurements of the Martian atmospheric structure since the 1976 Viking mission. The MPF atmospheric structure instrument (ASI) provided a detailed snapshot of the density, pressure, and temperature of the Martian atmosphere on the nightside for a wide range of altitudes extending from the thermosphere down to the lower atmosphere along one trajectory and at one time (Spencer *et al.*, 1998; Magalhães *et al.*, 1999; Withers *et al.*, 2003). The reconstructed temperature profile shows a thermospheric increase between 125–134 km with a peak value of 153 K, which is 30 K colder than the corresponding Viking 1 value.

Mars Odyssey was launched on April 7, 2001 and arrived at Mars during high solar activity on October 24, 2001 (Saunders *et al.*, 2004). After orbit insertion, MO performed a series of orbit changes to drop the low point of its orbit into the upper fringes of the Martian atmosphere at an altitude of about 110 kilometers. MO used the aerobraking technique over a period of three months to transit from an elliptical orbit into a 400 km nearly circular orbit intended for mapping. Aerobraking ended in January 2002, and its science mapping mission began in February 2002. Preliminary results of the reconstructed atmospheric density and temperature from MO accelerometer measurements are provided at the atmospheric node of the Planetary Data System (PDS).[3] The application of the MO data to thermospheric modelling yields an average global exospheric temperature at high solar activity of about ≤ 240 K (Bougher, 2005; private communication), which is much lower than the early estimations of about 350 K inferred from Lyman-α dayglow emissions of neutral atomic hydrogen (Barth *et al.*, 1969; Barth *et al.*, 1971; Anderson and Hord, 1971; Anderson, 1974).

Table I summarizes the Martian exospheric temperatures inferred from various methods. It is obvious that aerobraking measurements combined with thermospheric model simulations (AERO/MOD) and temperatures inferred from neutral mass spectrometer (NMS) measurements and models (MOD) for entry, descent and landing probes (EDL/NMS and EDL/MOD) yield lower exospheric temperatures than those obtained from plasma scale heigths, Lyman-α dayglow observations and UVS airglow observations.

The Mars Exploration Rovers (MER) "Spirit" and "Opportunity" carried two 3-axis accelerometers and two 3-axis gyroscopes to trigger events, such as parachute deployment, during atmospheric entry in early 2004. These instruments were designed and operated for engineering and operational purposes; they were not science instruments. However, their measurements can be used to derive vertical profiles of atmospheric density, pressure and temperature, but due to the higher entry mass compared to MPF the atmospheric structure reconstruction could only be achieved below altitudes of about 100 km (Withers and Smith, 2005), which does not allow so far any accurate prediction of exospheric temperatures.

[3]http://atmos.nmsu.edu/PDS/review/odya$_0$001/catalog/.

TABLE I

Martian dayside exospheric temperatures inferred from plasma scale heights H_{pl}, Lyman-α dayglow observations, UVS airglow data, entry, descent and landing probes combined with neutral gas mass spectrometer (EDL/NMS and EDL/MOD), and aerobraking measurements combined with model simulations (AERO/MOD). The heliocentric distance (HD) of Mars at the time of observation is given in astronomical units (AU); SZA is the solar zenith angle.

Mission	Date	HD [AU]	$F_{10.7}$	SZA[deg]	Altidute[km]	T_{exo} [K]	Method	Literature
Mariner 4	June 14, 1965	1.55	77	67	–	490	H_{pl}	Cloutier *et al.* (1969)
			57	–	150–200	265	H_{pl}	Lindal *et al.* (1979)
Mariner 6	July 31, 1969	1.42	167	0–44	200–24000	350 ± 100	Lyman-α	Anderson and Hord (1971)
				0–44	–	315 ± 75	UVS airglow	Stewart (1972)
Mariner 7	Aug. 5, 1969	1.42	188	0–44	200–24000	350 ± 100	Lyman-α	Anderson and Hord (1971)
				0–44	–	315 ± 75	UVS airglow	Stewart (1972)
Mariner 9	Nov. – Dec. 1971	1.44	103	47–56	–	335	H_{pl}	Kliore *et al.* (1973)
Mars 2, 3	Winter 1971	1.45	120	–	–	310	H_{pl}	Bauer and Hantsch (1989)
Mariner 9 extended I	May–June 1972	1.63	100	>70	–	330	H_{pl}	Kliore *et al.* (1973)
Mars 4, 6	Spring 1974	1.55	80	–	–	265	H_{pl}	Bauer and Hantsch (1989)
Viking L1	July 20, 1976	1.65	69	44	120–200	< 200	EDL/NMS	Nier and McElroy (1977)
Viking L2	Sept. 3, 1976	1.61	76	44	115–200	< 200	EDL/NMS	Nier and McElroy (1977)
Mars Pathfinder	July 4, 1997	1.56	70	135	–	153	EDL/MOD	Bougher *et al.* (2000)
Mars Global Surveyor	Jan. 16, 1998	1.38	93	73	<170	220	AERO/MOD	Bougher *et al.* (2000)
	Oct. 27, 1998	1.65	127	57		230	AERO/MOD	Bougher *et al.* (2000)
Mars Odyssey	Oct. 2001	1.4	175	–	–	240	AERO/MOD	Bougher (2005) private com.
Mars Express	Apr. 25, 2004	1.61	100	145	>1900	>600	Lyman-α	Galli *et al.* (this issue)
	Oct. 2004 –Mar. 2005	1.47–1.65	89–133	28–82	150–190	201±10[a]	UVS dayglow	Leblanc *et al.* (2006)
						252±13[b]	UVS dayglow	Leblanc *et al.* (2006)

[a]from CO_2^+ ($B^2\Sigma^+ - X^2\Pi$) doublet emission.

[b]from CO Cameron band system emission.

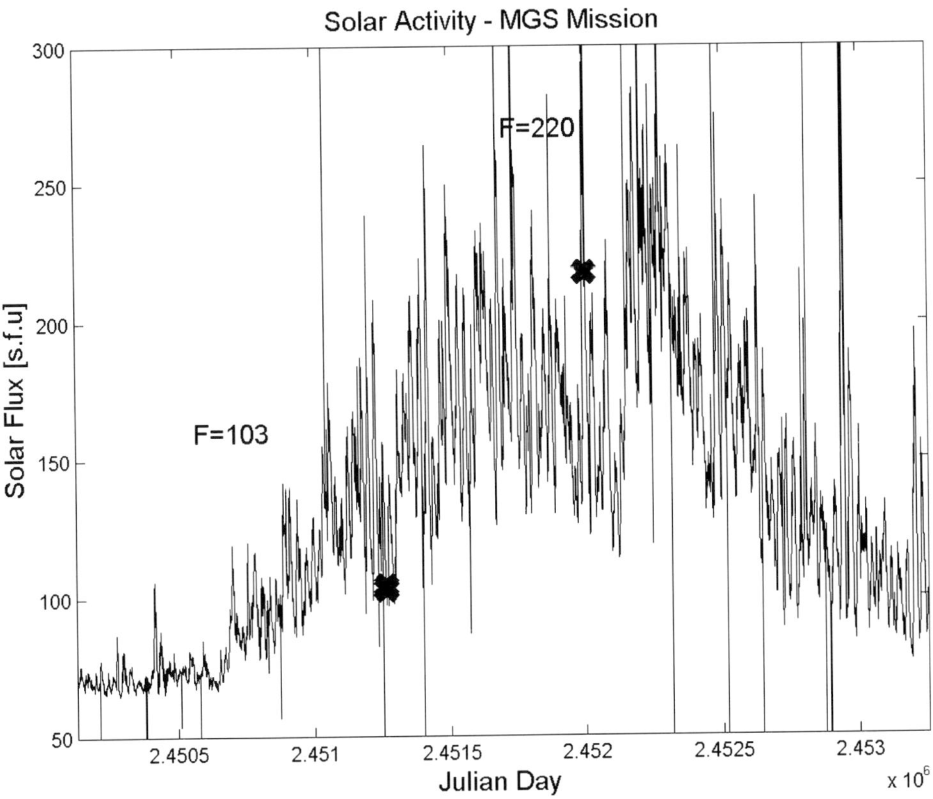

Figure 2. Solar activity ($F_{10.7}$) observed by the Solar and Heliospheric Observatory (SOHO) during the time period of the MGS mission. The $F_{10.7} = 103$ and $F_{10.7} = 220$ values are indicated by crosses.

2.3. Comparison of Obtained Exospheric Temperatures

The solar activity index ($F_{10.7}$) observed by the Solar and Heliospheric Observatory (SOHO)[4] during the time period of the MGS mission is shown in Figure 2. Since this mission lasted over the whole solar maximum, electron density profiles could be obtained for $F_{10.7}$ flux values between 100 and 220 as indicated in Figure 2.

Figures 3a and 3b display vertical electron density profiles obtained by the MGS radio science experiment[5] during medium ($F_{10.7} = 103$, SZA $= 77.8°$) and high solar ($F_{10.7} = 220$, SZA $= 71.9°$) activity, respectively. The observations were made with the radio occultation technique, which involves sounding of the Martian atmosphere with microwave radiation transmitted by MGS. The measured values are represented by black crosses while the solid lines show our fits for a temperature of 190 K (Figure 3a) and 220 K (Figure 3b) based on Equation (1); the dotted

[4]http://sohowww.nascom.nasa.gov/.
[5]http://nova.stanford.edu/projects/mod/public.html.

lines correspond to a temperature of 350 K. The best fits for the medium solar activity in Figure 3a correspond to 190–200 K, which is in good agreement with exospheric temperatures inferred from models using MGS aerobraking data during medium solar activity (Bougher *et al.*, 2000). Even at very high solar activity periods ($F_{10.7} = 220$), the neutral gas temperature at about 150 km does not exceed ~220 K (Figure 3b) and temperatures of 350 K or more appear very unlikely. Although the average exospheric temperature of the dayside atmosphere can be slightly higher than the derived temperature at an altitude near the ionization peak, it is hard to believe that it can surpass a value of 250 K at the exobase altitude of about 220 km. The reason for this is that above 150 km a rise of the equilibrium thermospheric temperature due to solar heating is strongly restrained by the high downward thermal heat conduction in the region above the ionospheric peak up to the exobase, as model simulations clearly demonstrate (see, for example, Bougher

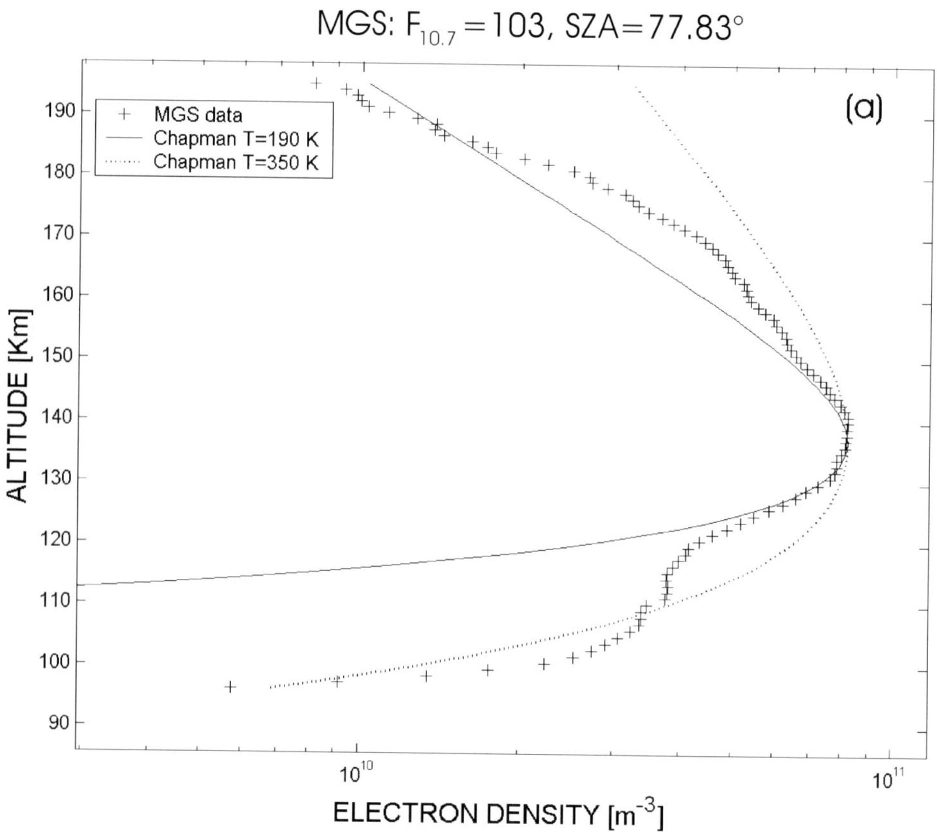

Figure 3. Electron density profiles obtained by the MGS radio science experiment during (a) medium ($F_{10.7} = 103$) and (b) high ($F_{10.7} = 220$) solar activity. Crosses show MGS data and the solid and dotted lines represent fits for different temperature values. (*Continued on next page*)

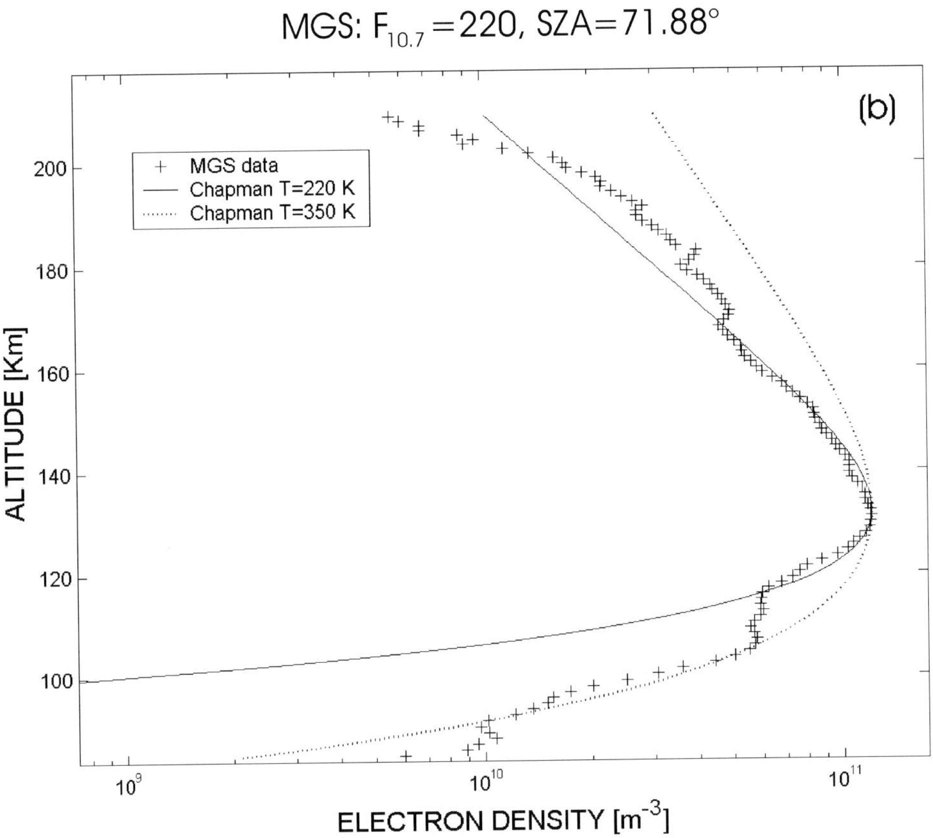

Figure 3. (*Continued*)

et al., 2000) – in their Figure 12 the temperature rise above 150 km is not more than about 15 K for solar minimum conditions. Moreover, our results for high solar activity closely coincide with temperatures inferred from MO aerobraking data.

Further, it should be emphasized that dayside neutral gas temperatures on Mars derived from peak ionospheric plasma densities during low solar activity have values less than 190 K and are also in good agreement with the in situ measurements by the Viking 1 and 2 landers (Lichtenegger *et al.*, 2004; Kazeminejad, 2005).

An exospheric temperature of about 350 K would imply a temperature increase between the ionospheric peak and the exobase of ∼130 K. However, as mentioned above, this appears unrealistic since a strong thermal heat conduction above the ionospheric peak inhibits a considerable temperature rise at the exobase due to solar heating and thus reduces it to values less than ΔT ∼100 K. Our results are also consistent with those of Nier and McElroy (1977), who used parametric fits for the neutral mass spectrometer measured CO_2, N_2, CO, O_2, and NO densities on both Viking landers and found that the thermal structure of the upper atmosphere was variable with an average temperature below 200 K. Furthermore, Hanson *et al.*

(1977) studied the ion and electron data from the retarting potential analyzers (RPAs) during the Viking 1 and 2 entries and concluded that Viking 1 measured the ion temperature of approximately 150 K near the Martian F_1 ion peak at about 130 km altitude. While the ion temperature increased to a value of about 210 K near 175 km, above this altitude, departures from the local thermal equilibrium between the neutral gas and the ions occurred and the ion temperature increased rapidly to more than 1000 K.

As illustrated in Figure 4a, excluding the estimations based on Lyman-α, airglow and plasma scale height data, the Martian exospheric temperature variations during low and high solar activity are in the range of about 170–240 K. Our results agree with previous investigations by Bauer (1999), Bauer and Hantsch (1989) and Lichtengger *et al.* (2004) which have shown that the Martian neutral gas temperature at the ionospheric peak varies between 180–220 K over the solar cycle ($F_{10.7} \approx 50$–200). In addition, the temperature inferred from Chapman fits to the peak electron densities both at low and high solar activity is in close agreement with those obtained via aerobraking and mass spectrometer data. Thus, we can conclude that the response of the upper atmospheric temperature near the ionization maximum on Mars to solar activity is essentially the same as on Venus ($\Delta T \approx 60$ K).

It therefore seems obvious to suggest that the main hydrogen population together with other atmospheric species found at the Martian exobase has an average "cold" exospheric temperature of about 200 K, which is distinctly less than the 350 K inferred from Lyman-α and CO Cameron bands dayglow observations. A recent analysis of the SPICAM UV spectrograph data obtained by Mars Express also indicates a systematic lower scale height associated with the CO_2^+ ($B^2\Sigma^+ - X^2\Pi$) doublet emission than the scale height associated with the CO Cameron band system (Leblanc *et al.*, 2006).

An overview of dayside exospheric temperature estimations of previous Venus missions, as shown in Figure 4b, also clearly demonstrates that UV airglow and Lyman-α observations yielded much higher values compared to those obtained at the exobase altitude from neutral mass spectrometer data, aerobraking measurements, and our Chapman fits to electron density profiles.

In the following Section we identify possible sources of hot components which may have influenced the Lyman-α and CO Cameron bands observations and also model their energy density distributions and number density profiles in the exosphere.

3. Hot Particle Populations in the Martian Exosphere

3.1. Low and High Energy Hydrogen Atoms

Recalling the striking difference between the early interpretations of the high daytime exospheric temperatures of up to 700 K of the Venus Lyman-α dayglow and UVS airglow observations and the much lower exospheric temperatures of about

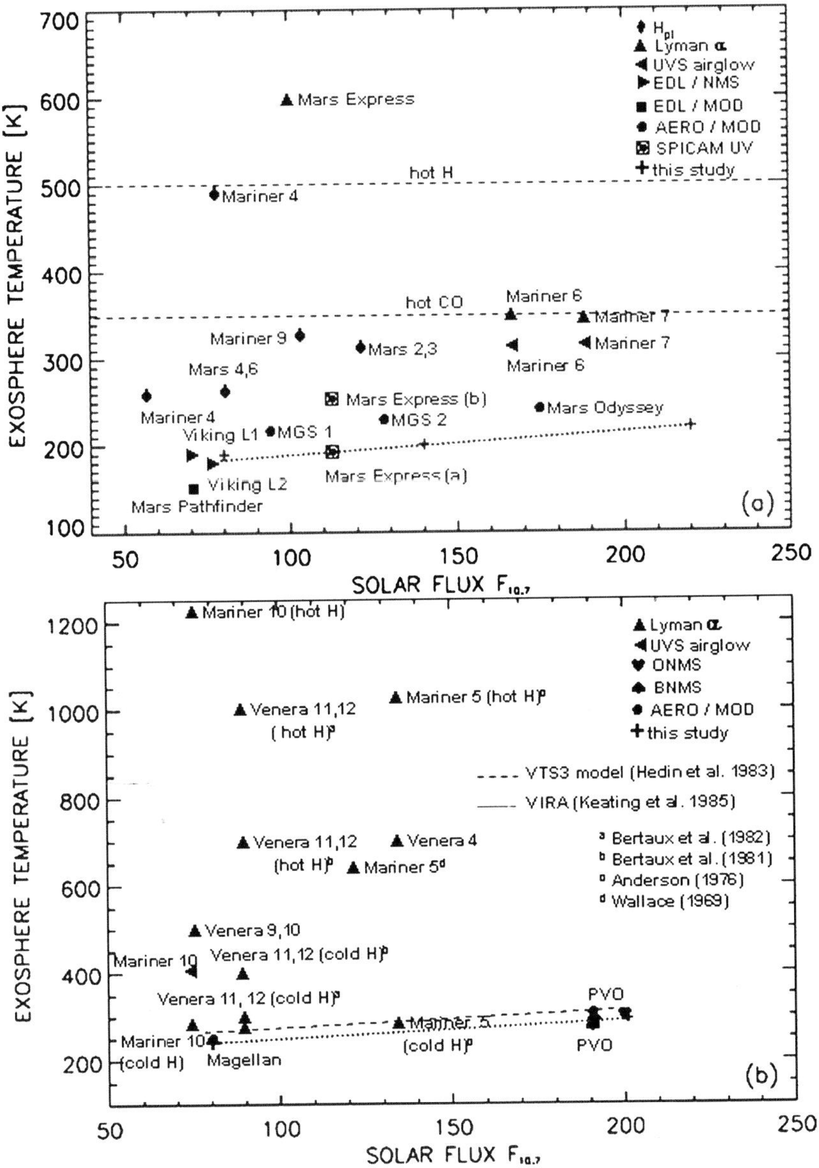

Figure 4. Exospheric daytime temperature estimations based on the data of various missions during the solar cycle for Mars (a) and Venus (b) (see also Tables I and II). The two dashed lines in (a) indicate the temperature for H* and CO* according to our Monte Carlo model.

240–300 K inferred from the PVO neutral gas density measurements (BNMS, ONMS, and OAD) and the Magellan aerobraking maneuvers (see Table II), it is very likely that the seemingly high exospheric temperatures on Mars (Table I) might be of similar origin. By inspection of Table II it can be seen that among all

TABLE II

Venusian dayside exospheric temperatures inferred from Lyman-α dayglow observations, UVS airglow data, neutral gas mass spectrometer (ONMS and BNMS), and aerobraking measurements combined with model simulations (AERO/MOD).

Mission	Date	$F_{10.7}$	SZA[deg]	Altidute[km]	T_{exo} [K]	Method	Literature
Venera 4	Oct. 18, 1967	135	–	–	700 ± 100	Lyman-α	Stewart (1968)
Mariner 5	Oct. 19, 1967	120	–	<18000	640	Lyman-α	Wallace (1969)
				<3000	275 ± 50; cold	Lyman-α	Anderson (1976)
				>3000	1020 ± 100; hot	Lyman-α	Anderson (1976)
Mariner 10	Feb. 5, 1974	75	–	<25000	400	UVS airglow	Broadfoot *et al.* (1974)
	Feb. 5, 1974		–	–	275 ± 50; cold	Lyman-α	Takacs *et al.* (1980)
	Feb. 5, 1974		–	–	1250 ± 100; hot	Lyman-α	Takacs *et al.* (1980)
Venera 9, 10	Oct.–Nov. 1975	75–80	–	<1500	500 ± 100; cold	Lyman-α	Bertaux *et al.* (1978)
Venera 11, 12	Dec. 1978	89	70–100	–	400; cold	Lyman-α	Bertaux *et al.* (1981)
					700; hot	Lyman-α	Bertaux *et al.* (1981)
					275–300 ± 25; cold	Lyman-α	Bertaux *et al.* (1982)
				>4000	1000; hot	Lyman-α	Bertaux *et al.* (1982)
Pioneer Venus Orbiter	Dec. 1978–Aug. 1980	180–200	–	150–250	285	ONMS	Nieman *et al.* (1980)
	Dec. 9, 1978	190	60	130–650	275	BNMS	von Zahn *et al.* (1980)
	Dec. 1978–Aug. 1979	180–200	–	140–190	300	AERO/MOD	Keating *et al.* (1980)
	Dec. 1978–Aug. 1980	200	–	170–250	300	ONMS	Hedin *et al.* (1983)
Magellan	Oct. 1994	80	–	<200	240–250	AERO/MOD	Keating *et al.* (1998a)

exospheric temperature estimates based on airglow observations of Venus, the lower temperature of the two component hydrogen model of Anderson (1976) agrees best with the corresponding results inferred later from the PVO (von Zahn *et al.*, 1980; Hedin *et al.*, 1983; Nieman *et al.*, 1979a, b, 1980; Keating *et al.*, 1980; Kasprzak *et al.*, 1997) and Magellan aerobraking data (Keating *et al.*, 1998a; Bougher *et al.*, 1999).

Since the upper atmospheres of Venus and Mars near the ionospheric maximum are similar in composition, temperature, and pressure, it is likely that the same ionospheric reactions occur on both planets. In the following we, therefore, assume that hydrogen atoms in the vicinity of Mars are made up of three different components: (a) a thermal population with an average global exospheric temperature of $\sim$200 K which is the most abundant at the exobase level of about 220 km altitude; (b) low energetic neutral atoms with energies between 0.1 and 8 eV produced by photochemical reactions (2)–(8) specified below; (c) energetic neutral atoms (ENAs) originating from the solar wind due to charge exchange with planetary hydrogen atoms, having energies comparable to the typical solar wind energy in the magnetosheath, since they essentially take over the energy of the parent protons (Barabash *et al.*, 2002; Holmström *et al.*, 2002; Lichtenegger *et al.*, 2002).

The main reactions involved in the production of low energetic and hot H^* atoms are given by (McElroy *et al.*, 1982)

$$CO_2^+ + H_2 \rightarrow CO_2H^+ + H^* \tag{2}$$

$$O^+ + H_2 \rightarrow HO^+ + H^* \tag{3}$$

$$CO_2H^+ + e \rightarrow CO_2 + H^* \tag{4}$$

$$OH^+ + e \rightarrow O + H^* \tag{5}$$

$$H^+ + O \rightleftarrows O^+ + H^* \tag{6}$$

$$H^+ + H \rightleftarrows H^* + H^+ \tag{7}$$

$$O^* + H \rightleftarrows O^* + H^*. \tag{8}$$

In the following we calculate the energy distribution of various low energetic neutral hydrogen components at the exobase in order to estimate their effect on the Martian exospheric temperature. For this purpose the neutral density and temperature profiles were taken from the models of Rodrigo *et al.* (1990) and Nair *et al.* (1994), whereas the densities for the major ions CO_2^+, CO^+, and O^+ as well as the electron concentration for solar minimum and maximum conditions were obtained from Fox *et al.* (1996). These values serve as input for a chemical scheme illustrated in Figure 5a, including vertical transport, that models the production of the remaining CO_2H^+ and OH^+ ions involved in the reactions (4)–(5) upon assuming that the chemistry does not affect the concentration of the input species.

The mathematical transport equations used in our model are those developed by Stubbe (1973) to study the ionic constituents in the terrestrial atmosphere. Later on Rodrigo *et al.* (1986, 1990) and Molina-Cuberos *et al.* (2003) applied it to the study of atmospheric composition on Earth and Mars.

The concentration of each neutral and ion species is calculated from the continuity and momentum equations by assuming steady state conditions and can be expressed as

$$P_i = n_i l_i + \frac{\partial}{\partial z} n_i v_i, \tag{9}$$

$$v_i = -D_i \left(\frac{1}{n_i} \frac{\partial n_i}{\partial z} + \frac{1}{H_i} \right) - K_i \left(\frac{1}{n_i} \frac{\partial n_i}{\partial z} + \frac{1}{H} \right), \tag{10}$$

where the subscript i denotes the ith constituent, P_i the production rate, n_i the concentration, l_i the specific loss, z the altitude and v_i is the mean vertical velocity. Further, D_i and K_i are the molecular and eddy diffusion coefficients and H_i and H are the individual and atmospheric scale heights, respectively.

Vertical transport is mainly produced by turbulent and molecular diffusion, the former being more effective at lower levels. The molecular diffusion coefficients D_i were calculated following a diffusion theory (e.g., Chapman and Cowling, 1970),

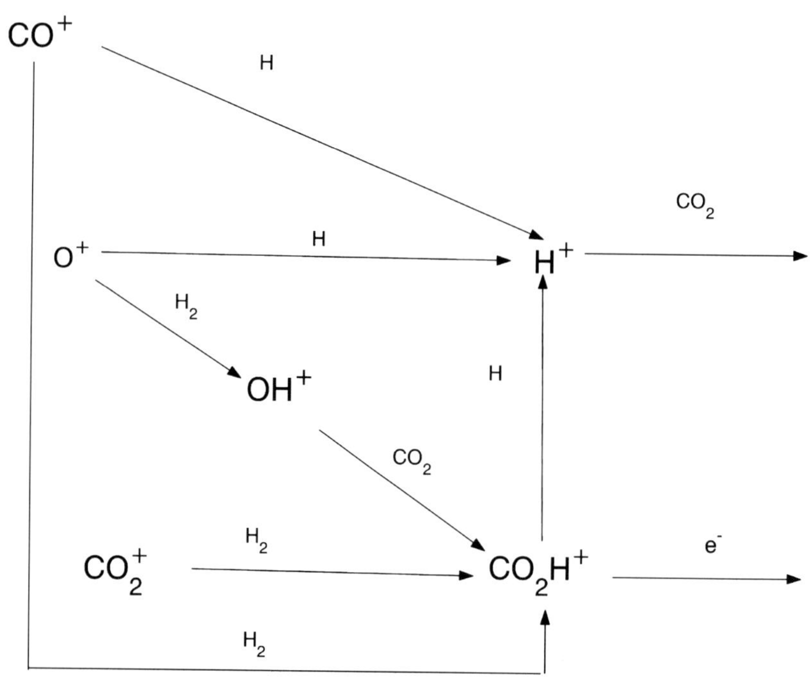

Figure 5. (a) Chemical scheme used to calculate H^+, OH^+ and CO_2H^+, (b) Density profiles for OH^+ and CO_2H^+ ions. The solid and dotted lines correspond to solar maximum and minimum conditions, respectively.

(*Continued on next page*)

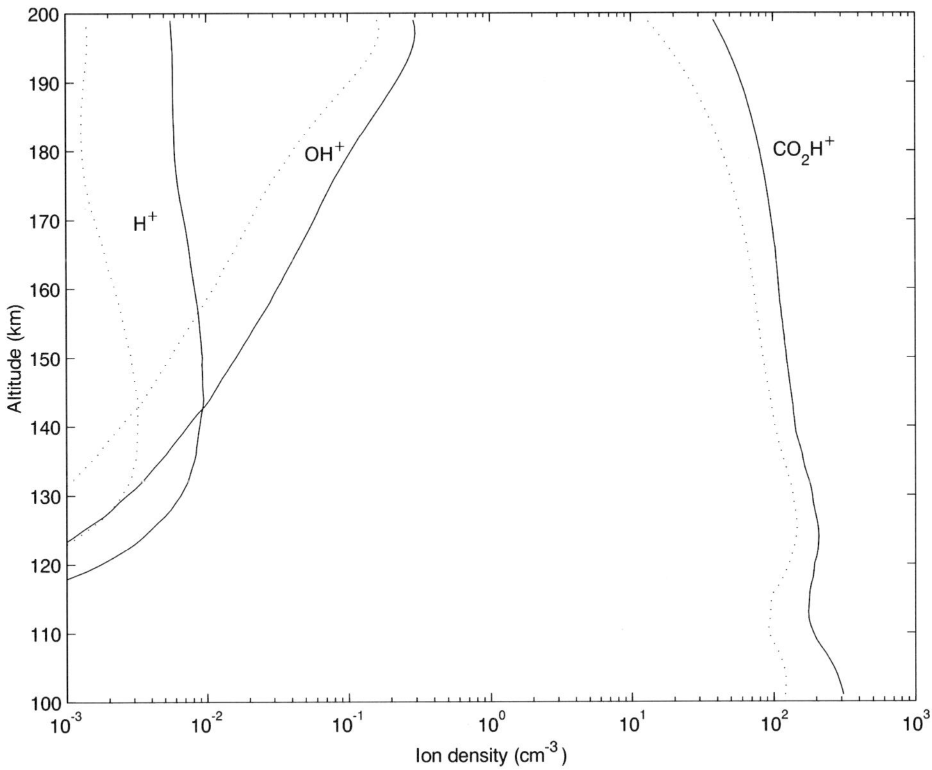

Figure 5. (b) (*Continued*)

while the turbulent diffusion is parameterized by means of the eddy diffusion coefficient K calculated by Rodrigo *et al.* (1990). At the lower boundary, chemical equilibrium was assumed for all the ions.

The ion densities of $OH^+$and CO_2H^+ according to our calculations for both solar minimum and maximum (dotted and solid lines, respectively) are displayed in Figure 5b. The low number densities of CO_2H^+ and OH^+ between 100–200 km altitude are in agreement with earlier model simulations of Izakov *et al.* (1981) and Pavlov (1985).

For the calculation of the energy density distribution of hot H^* atoms at the Martian exobase the Monte Carlo model of Lammer *et al.* (2000) is used. Collision probabilities, particle direction and energy loss after each collision between the background gas and a new H^* atom created from reactions specified in (2)–(8) are simulated by generating random numbers. The newly born hot hydrogen atoms are assumed to become eventually thermalized by a series of elastic hard sphere collisions with the background gases such as CO_2 or O, while inelastic collisions are negligibly small at these low energies. It should be noted that the absolute values of the number densities are not divided by a factor of 2 – up to energies smaller than the escape energy – in order to account for the downward component of the hot O^* atoms.

At the exobase level z_e, the distribution of the hot H^* flux $F(E, z)$ as a function of the kinetic energy E is converted into the corresponding energy density distribution function $f(E, z_e)$ through the equation

$$f(E, z_e) = \frac{F(E, z_e)}{v(E)}, \tag{11}$$

where $v(E)$ is the velocity of the H^* atoms corresponding to their energy. The velocity space for the hot H^* atoms was divided into cells with an energy spacing of 0.05 eV. The characteristic temperatures of the hot particle populations are obtained by fitting Equation (11) to Maxwellian distribution functions.

In the Monte Carlo model, the neutral gas densities and the CO_2^+ and O^+ ion profiles for low and high solar activity are taken from Figures 1 and 3 of Fox and Bakalian (2001), whereas the H_2 number density is obtained both from Far Ultraviolet Spectroscopic Explorer (FUSE) observations and from Figure 2 of Krasnopolsky and Feldman (2001). Figures 6a and 6b show the energy distribution function for hot H^* atoms produced from reactions (2) and (3) at an altitude of 240 kilometer. The dotted and dashed lines correspond to low and high solar activity, respectively. The lower efficiency in the production of hot H^* atoms at high solar activity is due to the fact that the H_2 number density at Mars is expected to decrease with increasing solar activity (Krasnopolsky and Feldman, 2001). Moreover, since CO_2^+ is more abundant in the Martian ionosphere than O^+ up to the exobase, reaction (2) is also more efficient in the production of hot H^* atoms than (3). Finally, Figure 6c displays the energy distribution resulting from reaction (4). Due to the small CO_2H^+ density, its efficiency is already rather low at the exobase level and is thus negligible compared to the rates produced by (2)–(3). The same is true for the remaining reactions (5)–(8).

Figure 7 illustrates the corresponding density of hot hydrogen at Mars according to our calculations, where the profiles of Krasnopolsky and Gladstone (1996) are included for comparison. The major population is characterized by a temperature of $\sim$500 K and is rather insensitive to solar activity. This is due to the fact that while the CO_2^+ density increases with solar activity, the H_2 density decreases with it (according to Krasnopolsky's model) and vice versa. Therefore, the production rate of H^* is nearly constant, since an increase (decrease) of CO_2^+ is approximately compensated by a decrease (increase) of H_2.

The density of the hot component is lower than the densities obtained by Krasnopolsky and Gladstone (1996) for the $T = 200$ (solar minimum) and $T = 350$ (solar maximum) values by 1 through 3 orders of magnitude at the exobase. However, at higher altitudes, the hot component comes closer to the "cool" solar maximum population. Further, it should be noted that the hydrogen densities of Krasnopolsky and Gladstone (1996) may be somewhat overestimated (Galli *et al.*, this issue). By analyzing the Lyman-α limb emission of exospheric hydrogen on

Mars observed by the neutral particle detector of the ASPERA-3 instrument on board of Mars Express at low solar activity, Galli *et al.* (this issue) estimated a temperature of > 600 K for the hot hydrogen component above 1900 km.

Another energetic neutral component is expected to consist of solar wind protons converted into neutral hydrogen atoms due to charge transfer to atmospheric constituents. These ENAs are characterized by high temperatures taken over from the solar wind and appear to modify the temperature mainly in a localized region

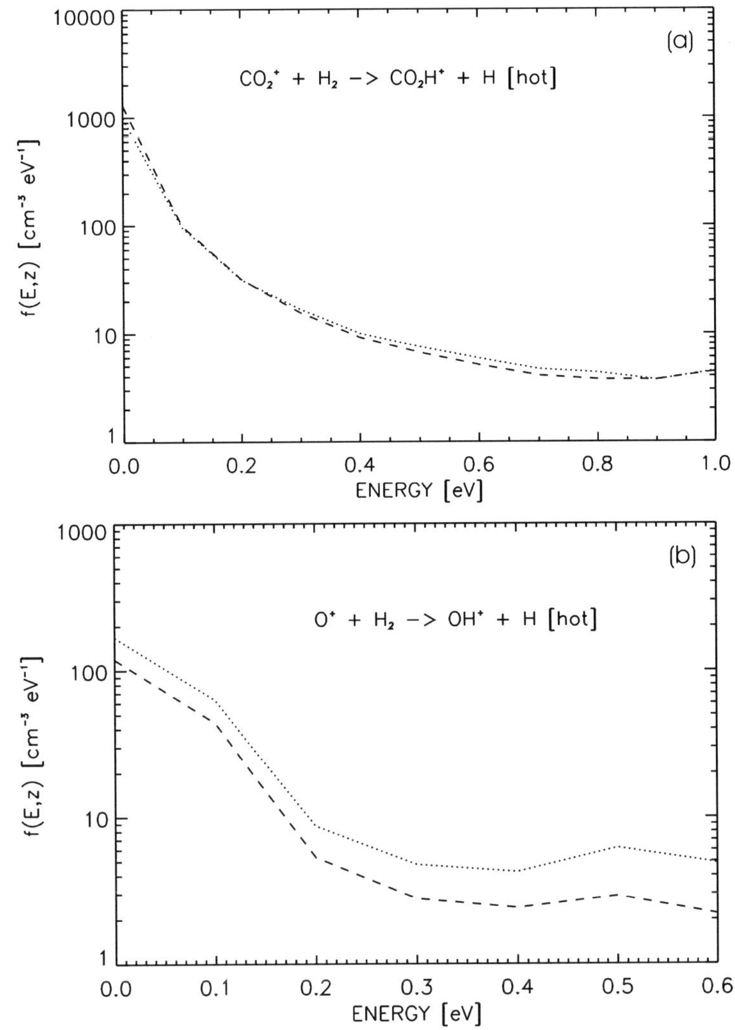

Figure 6. Energy distribution of hot H* produced by the reactions (2)–(4). The dashed and dotted lines correspond to solar maximum and minimum conditions, respectively.

(*Continued on next page*)

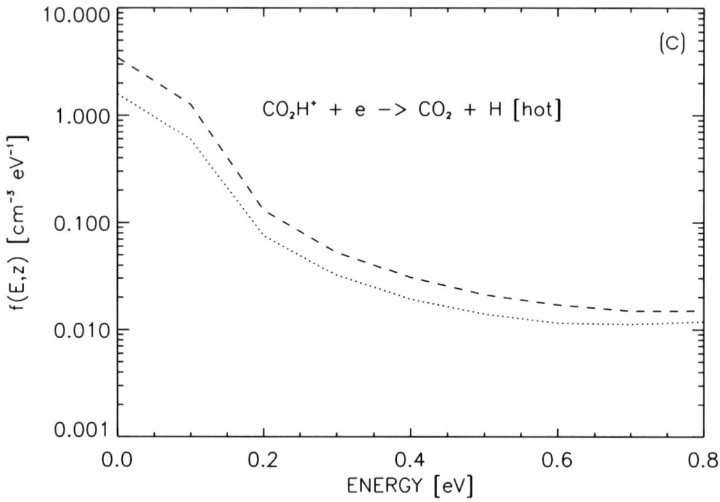

Figure 6. (*Continued*)

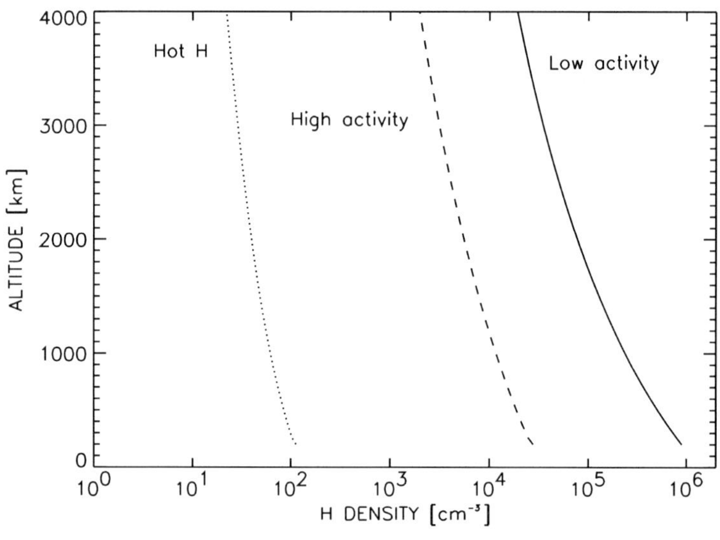

Figure 7. Density profile of the hot hydrogen component with a characteristic temperature of ~500 K. For comparison, the hydrogen densities for low and high solar activity (solid and dashed line, respectively) of Krasnopolsky and Gladstone (1996) are also shown.

around the bow shock downstream of the terminator (Lichtenegger *et al.*, 2004). Planetary ENAs originating from the "cool" neutral hydrogen atmosphere via ionization, subsequent acceleration and final neutralization through charge exchange can also acquire ion temperatures, but may be negligible due to their very low densities.

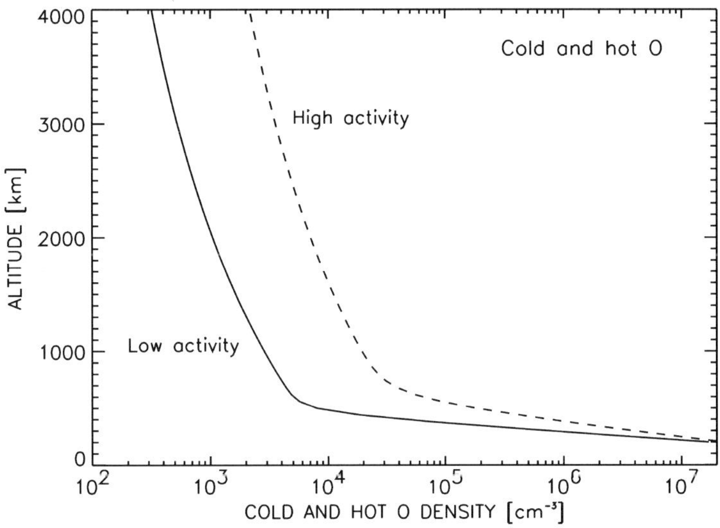

Figure 8. Density profiles of the sum of cold and hot oxygen for low and high solar activity (solid and dashed line, respectively). The two hot components correspond to temperatures of $\sim$750 and $\sim$ 5000 K, respectively.

3.2. Low Energetic Oxygen Atoms

The most important mechanism for the production of hot atomic oxygen in the Martian exosphere is dissociative recombination of ionospheric O_2^+ ions (e.g., Nagy and Cravens,1988; Fox and Hać, 1997; Kim *et al.*, 1998, Lammer *et al.*, 2000)

$$O_2^+ + e \rightarrow O^* + O^*. \tag{12}$$

Figure 8 shows the corresponding density altitude profiles for cold and hot O^* atoms based on reaction (12) for low (solid line) and high (dashed line) solar activity. The major contributions to this hot population is due to two components with $\sim$ 750 and $\sim$ 5000 K, respectively. The colder component is significant at lower altitudes (represented by the parts of the curves with a smaller inclination or smaller scale height), while the hotter component corresponds to those parts of the curves with a higher inclination or a larger scale height. The relative importance of the cold and hot components changes at the "kink" of the curves.

3.3. Low Energetic Carbon Atoms and Carbon Monoxide

McElroy (1972), Wallis (1989), Fox and Hać (1999) and Fox and Bakalian (2001) identified a number of photochemical sources that can produce carbon atoms in the energy range between 0.2–2.9 eV, including dissociative recombinations of CO^+

$$CO^+ + e \rightarrow C^* + O^*. \tag{13}$$

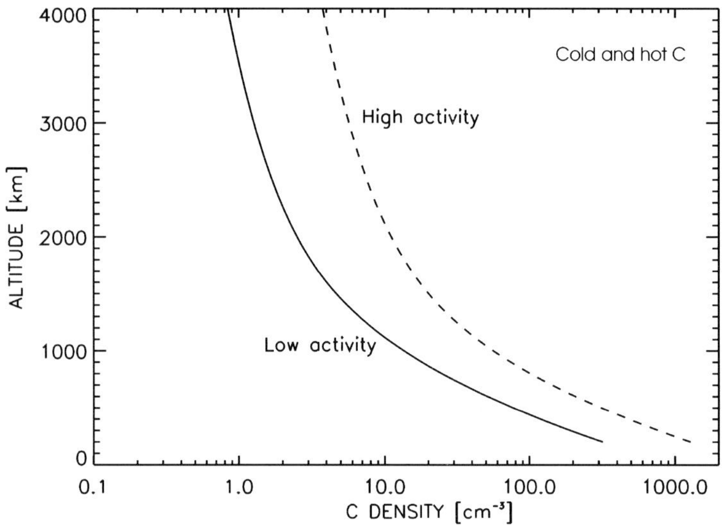

Figure 9. Density profile of cold and hot carbon for low and high solar activity (solid and dashed line, respectively). Two hot components of ~900 and ~ 4000 K are implied.

Photoelectron impact dissociation, photodissociation, photodissociative ionization and photoelectron impact dissociative ionization of CO are also considered:

$$CO + e^* \rightarrow C^* + O + e, \tag{14}$$

$$CO + h\nu \rightarrow C + O, \tag{15}$$

$$CO + h\nu \rightarrow C + O^+ + e, \tag{16}$$

$$CO + e^* \rightarrow C + O^+ + 2e. \tag{17}$$

Based on the neutral gas density and the CO_2^+ ion profiles for low and high solar activity from Figures 1 and 3 of Fox and Bakalian (2001), the temperature dependent dissociative recombination coefficient for CO^+ molecular ions ($2.75 \times 10^{-7}(300/T_e)^{0.55}$ cm^{-3}s^{-1}) measured by Rosén *et al.* (1998), as well as a photodissociation coefficient of 1.9×10^{-7}s^{-1} and 5.2×10^{-7}s^{-1} for low and high solar flux (Fox and Black, 1989), our Monte Carlo simulation yields the energy density distribution of hot C^* atoms at the exobase.

The corresponding densities of cold and hot carbon atoms for low (solid line) and high (dashed line) solar activity are illustrated in Figure 9 and imply two hot components of ~900 and ~4000 K. We note that these densities closely match those of Nagy *et al.* (2001).

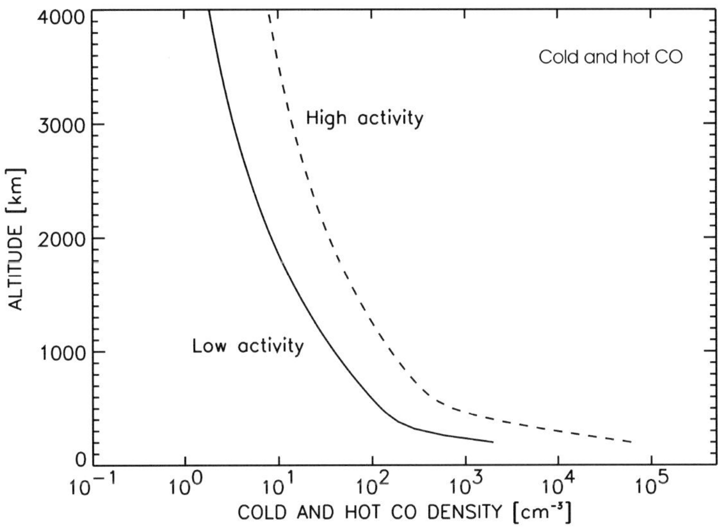

Figure 10. Density profile of cold and hot CO for low and high solar activity (solid and dashed line, respectively). The hot component has a temperature of ~ 350 K.

A source of energetic CO^* atoms is given by the reactions

$$CO_2^+ + e \rightarrow CO^* + O^* \tag{18}$$

$$CO_2^+ + h\nu \rightarrow CO^* + O^*. \tag{19}$$

The computed densities for solar minimum and maximum (solid and dashed line, respectively) corresponding to these reactions are shown in Figure 10; the cold background gas has a temperature of ~ 180 and ~ 230 K, respectively, for low and high solar activity and the hot component conforms to a temperature of ~ 350 K. This temperature of the hot CO^* component calculated from the energy density distribution is in good agreement with the exospheric temperature of about 325 K derived by Stewart *et al.* (1972) from the CO (a^3P) Cameron band intensities measured by Mariner 9. Recent analysis of the Martian dayglow data observed by the SPICAM UV spectrograph aboard Mars Express indicates that exospheric temperatures inferred from the CO Cameron band emissions close to the Martian exobase altitude at about 170–190 km result in values which are also high. Leblanc *et al.* (2006) estimated the mean temperature value of about 252 ± 13 K from the CO Cameron band system emission. By calculating the temperature in the upper thermosphere from the CO_2^+ ($B^2\Sigma^+ - X^2\Pi$) doublet emission they obtained a lower neutral gas temperature of about 200 ± 10 K. They found no evident solar zenith angle dependence of these temperatures. The CO Cameron band emission might be produced in a significant proportion by dissociative recombination of CO_2^+ above the ionospheric peak, where the CO_2^+ ($B^2\Sigma^+ - X^2\Pi$) doublet UV emission contributes less than 10 % (Leblanc *et al.*, 2006). Therefore, the temperature obtained from scale heights of the CO_2^+ ($B^2\Sigma^+ - X^2\Pi$) doublet UV emission is most likely closer to the neutral atmosphere temperature of the main atmospheric gas than the temperature obtained from the CO Cameron band emission. This value of ~200 K is in reasonable accord with temperatures inferred from aerobreaking data and those obtained by Chapman fits to ionospheric peaks.

4. Heating Efficiency in a CO_2-Rich Thermosphere

For studies of the thermospheric energy budget, the knowledge of the solar ultraviolet heating efficiency is of fundamental concern. It is conventional to define this parameter as the fraction of the solar EUV energy absorbed at a given altitude which appears locally as heat.

Based on Viking and PVO data, Fox and Dalgarno (1979, 1981) have examined the absorption of solar radiation in the thermospheres of Mars and Venus. They calculated the EUV heating efficiencies in the thermosphere for different assumptions about the fraction of excess energy converted to vibrational excitation of atmospheric molecules in elementary molecular processes. For Mars, Fox and Dalgarno (1979) found the heating efficiencies in the range between 8–32% below 120 km and between 16–27% above 125 km. Later Fox (1988), using detailed theoretical and observational considerations of atmospheric ion-molecular internal energy kinetics, reconsidered the earlier heating efficiencies and found them for Venus to be between 16–22% below 125 km altitude and between 22–25% above 130 km.

First models of the neutral temperature in the Venus thermosphere after the estimation of the EUV heating efficiency by Fox and Dalgarno (1979, 1981) were published by Hollenbach *et al.* (1985) and Dickinson and Bougher (1986). The major aim of both models was to reproduce the low daytime thermospheric temperatures observed by the Pioneer Venus Orbiter, where exospheric temperatures in the range 275–300 K have been reported at high solar activity (see Section 2.1). Hollenbach *et al.* (1985) derived heating efficiences of 10–15% from reproducing the observed neutral temperatures by including strong eddy cooling. Dickinson and Bougher (1986) estimated an altitude-independent heating efficiency in the Venus thermosphere of 9.5% with no eddy cooling assumed.

For their later Mars thermospheric simulations Bougher *et al.* (1999, 2000) assumed a 22% heating efficiency and obtained seasonal-solar cycle maximum exobase temperatures of up to 380 K which are not compatible with the equilibrium CO_2 exobase temperatures of less than 250 K for solar maximum estimated in the present study. Moreover, besides the EUV solar heating, also near-IR solar heating is present in the lower CO_2-rich thermosphere. According to Taylor *et al.* (1983) this near-IR heating in the Venus upper atmosphere is a dominant heating process below ~130 km.

It should be also noted that Bougher *et al.* (2000) adopted in their simulations a high CO_2-O coefficient of 3×10^{-12} cm^3/s for the collisional excitation of the 15 – μm CO_2 fundamental band emission, which is about 3 times larger than the value of $\sim 10^{-12}$ cm^3/s needed to model the energy balance in the Earth's thermosphere (Gordiets *et al.*; 1982, Bougher and Roble, 1991). Therefore, these discrepancies strongly point out to the possible presence of some hitherto unidentified additional IR-cooling mechanism in the thermospheres of both Venus and Mars. Among the possible candidates, a strong non-LTE CO_2 isotopic and hot bands cooling observed recently by means of the Mars Express PFS instrument (Formisano *et al.*, 2006) may be expected.

5. Summary and Conclusion

Beginning with the first observations of Mariner 4, the estimations of the exospheric temperatures on Venus and Mars have been summarized. It is emphasized that all calculations based on airglow data resulted in Venus temperatures distinctly higher than those based on the PVO and Magellan neutral gas mass spectrometer, aerodynamic drag and aerobraking data, suggesting that the airglow temperatures were overestimated due to the influence of one or more hot components. Moreover, Chapman fits to the ionospheric plasma peak data obtained by PVO also indicate temperatures much lower than those inferred from Lyman-α observations.

In a similar way, the upper atmosphere of Mars is expected to consist of a cold component, as well as of extended hot components, the latter giving rise to a temperature in excess of that of the cold component when based on airglow data. All estimations resting upon neutral gas mass spectrometers, aerodynamic drag, and aerobraking measurements yield significantly lower exospheric temperatures than Lyman-α and UVS airglow observations (Tables 1 and 2 and Figure 4). These lower values are in agreement with temperatures of the present study inferred from ionospheric peak plasma densities and imply a similar dependence of the exospheric temperature on solar activity for both Venus and Mars. Moreover, our calculations of the photochemically produced hot CO energy distribution at the Martian exobase resulted in a temperature of $\sim$350 K, which agrees with the the estimated temperatures between 300 and 400 K inferred from the CO(a^3P) Cameron bands space observations. We thus favor the interpretation that the Martian dayside upper atmosphere temperature does not exceed 250 K even at solar maximum and that the airglow data yield a combined temperature of the cold and hot components since in the exosphere due to low density there is no thermodynamic equilibrium between them.

We also conjecture that unrealistically low heating efficiencies required to reproduce low thermospheric temperatures on Venus and Mars may result from a common deficiency in all recent thermospheric heat balance models of infrared radiation cooling. This deficiency can be either due to inadequate modelling of the observed enhanced non-LTE radiation of a CO_2 molecule in its numerous hot and isotopic rotational-vibrational bands, or may result from the presence of trace amounts of some so far unidentified molecules in the thermospheres of both planets being able to contribute effectively to thermospheric cooling at low temperatures.

Acknowledgements

We thank F. Bakalian for valuable discussions about photodissociation and dissociative recombination of CO_2. H. I. M. Lichtenegger, H. Lammer and Yu. N. Kulikov, thank the "Österreichischer Austauschdienst", which supported this work by the projects I.12/04. These authors acknowledge also the support by the Austrian Academy of Sciences, "Verwaltungsstelle für Auslandsbeziehungen" and by the Russian Academy of Sciences. Yu. N.Kulikov also thanks the Russian Foundation for the Basic Research, which partially supported this study as a joint Russian-Austrian project No. 03-05-20003 "Solar-planetary relations and space weather".

The authors are also indebted to the two anonymous referees for their constructive comments which helped to improve the paper.

References

Anderson, D. E., Jr., and Hord, C. W.: 1971, *J. Geophys. Res.* **76**, 6666.

Anderson, D. E, Jr.: 1974, *J. Geophys. Res.* **79**, 1513.

Anderson, D. E., Jr.: 1976, *J. Geophys. Res.* **81**, 1213.

Barabash, S., Holmström, M., Lukyanov, A., and Kallio, E.: 2002, *J. Geophys. Res.* **107**, 1280, DOI10.1029/2001JA000326.

Barth, C. A.: 1968, *J. Atmos. Sci.* **25**, 564.

Barth, C. A., Wallace, L., and Pearce, J. B.: 1968, *J. Geophys. Res.* **73**, 2541.

Barth, C. A., Fastie, W. G., Hord, C. W., Pearce, J. B., Kelly, K. K., Stewart, A. I., *et al.*: 1969, *Science* **165**, 1004.

Barth, C. A., Hord, C. W., Pearce, J. B., Kelly, K. K., Anderson, G. P., and Stewart, A. I.: 1971, *J. Geophys. Res.* **76**, 2213.

Barth, C. A., Stewart, A. I. F., Bougher, S. W., Hunten, D. M., Bauer, S. J., and Nagy, A. F.: 1992, in: Mars, Univ. Arizona Press, pp. 1054.

Bauer, S. J., Donahue, T. M., Hartle, R. E., and Taylor, H. A.: 1979, *Science* **205**, 109.

Bauer, S. J., and Taylor, H. A.: 1981, *Geophys. Res. Lett.* **8**, 840.

Bauer, S. J., Brace, L. M., Taylor Jr., H. A., Breus, T., Kliore, A. J., *et al.*: 1985, in: The Venus international reference atmosphere, Kliore, A. J., Moroz, V. I., and Keating, G. M. (eds)., Adv. Space Res, Vol. 5, pp. 233.

Bauer, S. J.: 1999, *Anz. Österreichischen Akademie der Wissenschaften, Klasse*, Vol. 136, pp. 19.

Bauer, S. J., and Hantsch, M. H.: 1989, *Geophys. Res. Lett.* **16**, 373.

Bauer, S. J., and Lammer, H.: 2004, *Planetary Aeronomy: Atmosphere Environments in Planetary Systems*, Springer Verlag, Heidelberg, New York.

Bertaux, J. L., Blamont, J., Marcelin, M., Kurt, V. G., Romanova, N. N., and Smirnov, A. S.: 1978, *Planet. Space Sci.* **26**, 817.

Bertaux, J. L., Blamont, J., Lepine, V. M., Kurt, V. G., Romanova, N. N., and Smirnov, A. S.: 1981, *Planet. Space Sci.* **29**, 149.

Bertaux, J. L., Lepine, V. M., Kurt, V. G., and Smirnov, A. S.: 1982, *Icarus* **52**, 221.

Bougher, S. W., and Roble, R. G.: 1991, *J. Geophys. Res.* **96**, 11045.

Bougher, S. W., and Keating, G. M.: 1999, The Fifth International Conference on Mars, July 19–24, 1999, Pasadena, California, abstract no. 6010.

Bougher, S. W., Engel, S., Roble, R. G., and Foster, B.: 1999, *J. Geophys. Res.* **104**, 16591.

Bougher, S. W. Engel, S., Roble, R. G., and Foster, B.: 2000, *J. Geophys. Res.* **105**, 17669.

Breus, T. K., Krymskii, A. M., Crider, D. H., Ness, N. F., Hinson, D., and Barashyan, K. K.: 2004, *J. Geophys. Res.* **109**, A09310, doi:10.1029/2004JA010431.

Broadfoot, A. L., Kumar, S., Belton, M. J. S., and McElroy, M. B.: 1974, *Science* **183**, 1315. Chapman, S., and T.

Cloutier, P. A., McElroy, M. B., and Michel, F. C.: 1969, *J. Geophys. Res.* **74**, 6215.

Chapman, S., and Cowling, T. G.: 1970, *The Mathematical Theory of Non-Uniform Gases*, 3rd edition, Cambridge University Press, New York.

Dickinson, R. E., and Bougher, S. W.: 1986, *J. Geophys. Res.* **91**, 70.

Donahue, T. M.: 1969, *J. Geophys. Res.* **74**, 1128.

Formisano, V., Maturilli, A., Giuranna, M., D'Aversa, E., and Lopez-Valverde, M. A.: 2006, *Icarus* **182**, 51.

Fox, J. L., and Dalgarno, A.: 1979, *J. Geophys. Res.* **84**, 7315.

Fox, J. L., and Dalgarno, A.: 1981, *J. Geophys. Res.* **86**, 629.

Fox, J. L.: 1988, *Planet. Space Sci.* **36**, 37.

Fox, J. L., and Black, J. H.: 1989, *Geophys. Res. Lett.* **16**, 291.

Fox, J. L., Zhou, P., and Bougher, S. W.: 1996, *Adv. Space Res.*, 17(11), 203.

Fox, J. L., and Hać, A.: 1997, *J. Geophys. Res.* **102**, 24005.

Fox, J. L., and Hać, A.: 1999, *J. Geophys. Res.* **104**, 24729.

Fox, J. L., and Bakalian, F. M.: 2001, *J. Geophys. Res.* **106**, 28785.

Fox, J. L., and Sung, K. Y.: 2001, *J. Geophys. Res.* **106**, 21305.

Galli, A., Wurz, P., Lammer, H., Lichtenegger, H. I. M., Lundin, R., and Barabash, S., *et al.*: *Space Sci. Rev.*, this issue, doi: 10.1007/s11214-006-9089-7.

Gordiets, B. F., Kulikov, Yu. N., Markov, M. N., and Marov, M. Ya.: 1982, *J. Geophys. Res.* **87**, 4504.

Hanson, W. B., Santani, S., and Zuccaro, D. R.: 1977, *J. Geophys. Res.* **82**, 4351.

Hanson, W. B., and Mantas, G. P.: 1988, *J. Geophys. Res.* **93**, 7538.

Hedin, A. E.: 1983, *J. Geophys. Res.* **88**, 10170.

Hedin, A. E., Nieman, H. B., Kasprzak, W. T., and Seiff, A.: 1983, *J. Geophys. Res.* **88**, 73.

Hodges, R., and R., Jr.: 2000, *J. Geophys. Res.* **105**, 6971.

Hogan, J. S., and Stewart, R. W.: 1969, *J. Atmos. Sci.* **26**, 332.

Hollenbach, D. J., Prasad, S. S., and Witten, R. C.: 1985, *Icarus* **64**, 205.

Holmström, M., Barabash, S., and Kallio, E.: 2002, *J. Geophys. Res.* **107**, 1277, DOI10.1029/2001JA000325.

Ip, W.-H.: 1988, *Icarus* **76**, 135.

Izakov, M. N., Krasitskii, O. P., and Pavlov, A. V.: 1981. *Kosmicheskie Issledovaniia*, **19**, 733.

Kasprzak, W. T., Keating, G. M., Hsu, N .C., Stewart, I. F., Coldwell, W. B., and Bougher, S. W.: 1997, in: *Venus II*, Bougher, S. W., Hunten, D. M., and Phillips, R. J. (eds.), Univ. Arizona Press, pp. 225.

Kazeminejad, S.: 2005, Exospheric temperature estimation and atmospheric loss: A comparative study of Mars and Venus, Ms. Thesis, University of Graz, Austria.

Keating, G. M., Nicholson, J. Y., and Lake, L. R.: 1980, *J. Geophys. Res.* **85**, 7941.

Keating, G. M., Bertaux, J. L., Bougher, S. W., Dickinson, R. E., Cravens, T. E., and Hedin, A. E.: 1985, *Adv. Space Res.* **5**, 117.

Keating, G. M., Tolson, R. H., Schellenberg, T. J., Hsu, N. C., and Bougher, S. W.: 1998a. Study of Venus upper atmosphere using Magellan drag measurements, Second Ann. Progress Rep. NAG5-6081, NASA, Washington DC.

Keating, G. M., Bougher, S. W., Zurek, R. W., Tolson, R. H., Cancro, G. J., and Noll, S. N., *et al.*: 1998b. *Science* **279**, 1672.

Kim, J., Nagy, A. F., Fox, J. L., and Craven T.: 1998, *J. Geophys. Res.* **103**, 29339.

Kliore, A. J., Fjeldbo, G., Seidel, B. L., Sykes, M. J., and Woiceshyn, P. M.: 1973, *J. Geophys. Res.* **78**, 4331.
Krasnopolsky, V. A., and Gladstone, G. R.: 1996, *J. Geophys. Res.* **101**, 15765.
Krasnopolsky, V. A., and Feldman, P. D.: 2001, *Science* **294**, 1914.
Kumar, S., and Hunten, D. M.: 1974, *J. Geophys. Res.* **79**, 2529.
Lammer, H., and Bauer, S. J.: 1991, *J. Geophys. Res.* **96**, 1819.
Lammer, H., Stumptner, W., and Bauer, S. J.: 2000, *Planet. Space Sci.* **48**, 1473.
Lammer, H., Lichtenegger, H. I. M., Kolb, C., Ribas, I., Guinan, E. F., and Bauer, S. J.: 2003, *Icarus* **165**, 9.
Leblanc, F., Chaufray, J. Y., Lilensten, J., Witasse, O., and Bertaux, J.-L.: 2006, *J. Geophys. Res* **111**, E09S11, doi:10.1029/2005JE002664.
Lichtenegger, H. I. M., Lammer, H., and Stumptner, W.: 2002, *J. Geophys. Res.* **107**, 1279, doi:10.1029/2001JA000322.
Lichtenegger H. I. M, Lammer, H., Vogl, D. F., and Bauer, S. J.: 2004, *Adv. in Space Res.* **33**, 140.
Lindal, G. F., Hotz, H. B., Sweetnam, D. N., Shippony, Z., Brenkle, J. P., and Spear, R. T.: 1979, *J. Geophys. Res.* **84**, 8443.
Magalhães, J. A., Schofield, J. T., and Seiff, A.: 1999, *J. Geophys. Res.* **104**, 8943.
McElroy, M. B.: 1967, *Astrophys. J.* **73**, 1125.
McElroy, M. B.: 1968, *J. Geophys. Res.* **73**, 1513.
McElroy, M. B., and Hunten, D. M.: 1969, *J. Geophys. Res.* **74**, 1720.
McElroy, M. B.: 1972, *Science* **175**, 443.
McElroy, M. B., Prather, M. J., and Rodriguez, J. M.: 1982, *Science* **215**, 1614.
Molina-Cuberos, G. J., Witasse, O., Lebreton, J.-P., Rodrigo, R., and López-Moreno, J. J.: 2003, *Planet. Space Sci.* **51**, 239.
Nagy, A. F., Cravens, T. E., Yee, J.-H., and Stewart, A. I. P.: 1981, *Geophys. Res. Lett.* **8**, 629.
Nagy, A. F., and Cravens, T. E.: 1988, *Geophys. Res. Lett.* **15**, 433.
Nagy, A. F., Liemohn, M. W., Fox, J. L., and Kim, J.: 2001, *J. Geophys. Res.* **106**, 21565.
Nair, H., Allen, M., Anbar, A. D., Yung, Y. L., and Clancy, R. T., 1994, *Icarus* **111**, 124.
Nieman, H. B., Hartle, R. E., Kasprzak, W. T., Spencer, N. W., Hunten, D. M., and Carignan, G. R.: 1979a, *Science* **203**, 770.
Nieman, H. B., Hartle, R. E., Hedin, A. E., Kasprzak, W. T., Spencer, N. W., and Hunten, D. M.: *et al.*, 1979b, *Science* **205**, 54.
Niemann, H. B., Kasprzak, W. T., Hedin, A. E., Hunten, D. M., and Spencer, N. W.: 1980, *J. Geophys. Res.* **85**, 7817.
Nier, A. O., and McElroy, M .B.: 1977, *J. Geophys. Res.* **82**, 4341.
Pavlov, A. V.: 1985, *Kosmicheskie Issledovaniya* **23**, 276.
Rodrigo, R., López-Moreno, J. J., López-Puertas, M., Moreno, F., and Molina, A.: 1986, *Planet. Space Sci.* **34**, 723.
Rodrigo, R., Garcí-Álvarez, E., López-González, J. J., and López-Moreno, J. J.: 1990, *J. Geophys. Res.* **95**, 14795.
Rosén, S., Peverall, R., Larsson, M., Le Padellec, A., Semaniak, J., and Larson, Å, *et al.*: 1998, *Phys. Rev. A* **57**, 4462.
Saunders, R. S., Arvidson, R. E., Badhwar, G. D., Boynton, W. V., Christensen, P. R., and Cucinotta, F. A., *et al.*: 2004, *Space Sci. Rev.* **110**, 1.
Schofield, J. T., Barnes, J. R., Crisp, D., Haberle, R. M., Larsen, S., and Magalhães, J. A., *et al.*: 1997, *Science* **278**, 1752.
Seiff, A., and Kirk, D. B.: 1977. *J. Geophys. Res.* **82**, 4364.
Soffen, G. A.: 1977, *J. Geophys. Res.* **82**, 3959.
Spencer, D. A., Blanchard, R. C., Thurmann, S. W., Braun, R. D., Peng, C.-Y., and Kallemeyn, P. H.: 1998, Mars Pathfinder Atmospheric Entry Reconstruction. NASA Technical Report, AAS 98.

Stewart, R. W.: 1968, *J. Atmosph. Sci.* **25**, 578.
Stewart, A. I.: 1972, *J. Geophys. Res.* **77**, 54.
Stubbe, P.: 1973, Sci. Rep., 418, Ionosphere REs. Lab., Pa. State Univ., University Park.
Sze, N. D., and McElroy, M. B.: 1975, *Planet. Space Sci.* **23**, 763.
Takacs, P. Z., Broadfoot, A. L., Smith, G. R., and Kumar, S.: 1980, *Plant. Space Sci.* **28**, 687.
Taylor, F. W., Hunten, D. M., and Ksanfomaliti, L. V.: 1983, in: Hunten, D. M., Colin, L., Donahue, T. M., and Moroz, V. I. (eds.). University of Arizona Press, Tucson, Arizona, pp. 650.
von Zahn, U., Fricke, K. H., Hunten, D. M., Krankowsky, D., Mauersberger, K., and Nier, A. O.: 1980, *J. Geophys. Res.* **85**, 7829.
Wallace, L.: 1969, *J. Geophys. Res.* **74**, 115.
Wallace, L., Stuart, F. E., Nagel, R. H., and Larson, M. D.: 1971, *Astrophys. J.* **168**, L29.
Wallis, M. K.: 1978, *Planet. Space Sci.* **26**, 949.
Wallis, M. K.: 1989, *Sci. Lett.* **93**, 321.
Withers, P., Towner, M. C., Hathi, B., and Zarnecki, J. C.: 2003, *Planet. Space Sci.* **51**, 541.
Withers, P., and Smith, M. D.: 2005, Submitted to Icarus.

Erratum

EFFECTS OF LOW ENERGETIC NEUTRAL ATOMS ON MARTIAN AND VENUSIAN DAYSIDE EXOSPHERIC TEMPERATURE ESTIMATIONS

HERBERT I. M. LICHTENEGGER, HELMUT LAMMER, YURI N. KULIKOV, SHAHIN KAZEMINEJAD, GREGORIO H. MOLINA-CUBEROS, RAFAEL RODRIGO, BOBBY KAZEMINEJAD and GOTTFRIED KIRCHENGAST

DOI: 10.1007/s11214-006-9082-1

The original version of this article unfortunately contained a mistake. The presentation of Figure 8 was incorrect. The correct Figure 8 is displayed below.

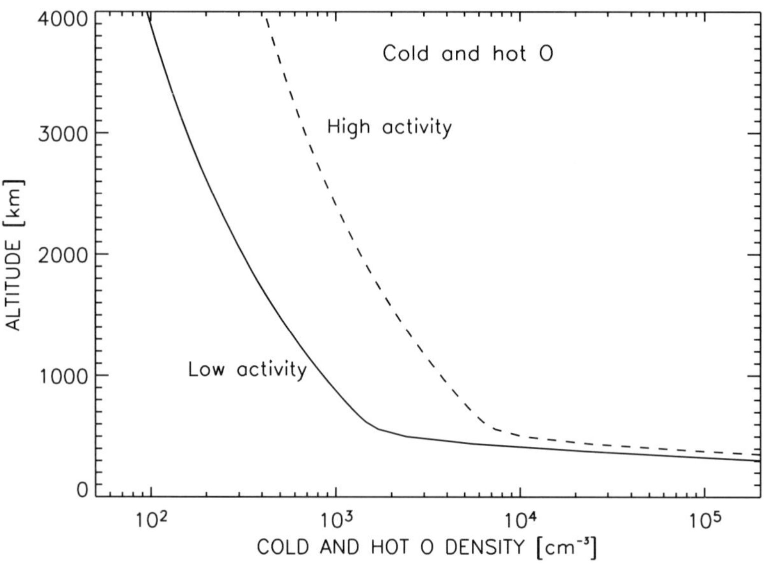

The online version of the original article can be found at http://dx.doi.org/10.1007/s11214-006-9082-1

Space Science Reviews (2007) 126: 503
DOI: 10.1007/s11214-007-9158-6